Rare Earth and Actinide Complexes

Special Issue Editors

Stephen Mansell
Steve Liddle

Special Issue Editors
Stephen Mansell
Heriot-Watt University
UK

Steve Liddle
The University of Manchester
UK

Editorial Office
MDPI AG
St. Alban-Anlage 66
Basel, Switzerland

This edition is a reprint of the Special Issue published online in the open access journal *Inorganics* (ISSN 2304-6740) from 2015–2015 (available at: http://www.mdpi.com/journal/inorganics/special_issues/Rare-Earth-Actinide-Complexes).

For citation purposes, cite each article independently as indicated on the article page online and as indicated below:

Author 1; Author 2; Author 3 etc. Article title. *Journalname*. **Year**. Article number/page range.

ISBN 978-3-03842-328-7 (Pbk)
ISBN 978-3-03842-329-4 (PDF)

Table of Contents

About the Guest Editors

Stephen Mansell's research interests focus on bringing together aspects of main group, transition metal and f-block chemistry through the design of unconventional ligands for these metals. His main goal is to explore the catalytic properties of these unusual compounds in order to better understand how to design new catalysts. Since September 2013, he has been an Assistant Professor at Heriot-Watt University in Edinburgh. He obtained his MChem degree from Imperial College London in 2005 and his PhD from The University of Bristol in 2009. After post-doctoral work in boron chemistry and small molecule activation mediated by organometallic complexes of uranium, he worked briefly as part of a large interdisciplinary team designing new functional materials at The University of Bath.

Steve Liddle is Professor and Head of Inorganic Chemistry and co-Director of the Centre for Radiochemistry Research at The University of Manchester. He was elected a Fellow of the Royal Society of Chemistry in 2011 and as Vice President to the Executive Committee of the European Rare Earth and Actinide Society (2012–now). He was Chairman of COST Action CM1006, a 22-country research group network for f-block chemistry, and he has been awarded a number of prizes, including most recently the RSC Corday–Morgan prize. He has been awarded ERC Starter and Consolidator Grants and he currently holds an EPSRC Established Career Fellowship. He has published over 150 research articles, reviews, and book chapters, the majority of which are concerned with the chemistry of f-elements and in particular that of uranium.

inorganics   MDPI

Editorial
Rare Earth and Actinide Complexes

Stephen M. Mansell [1,*] and Stephen T. Liddle [2,*]

1   Institute of Chemical Sciences, School of Engineering and Physical Sciences, Heriot-Watt University, Edinburgh, EH14 4AS, UK
2   School of Chemistry, The University of Manchester, Oxford Road, Manchester, M13 9PL, UK
*   Correspondences: s.mansell@hw.ac.uk (S.M.M.); steve.liddle@manchester.ac.uk (S.T.L.);
    Tel.: +44-131-451-4299 (S.M.M.); +44-161-275-4612 (S.T.L.)

Academic Editor: Duncan H. Gregory
Received: 4 October 2016; Accepted: 12 October 2016; Published: 14 October 2016

The rare earth metals (scandium, yttrium, lanthanum and the subsequent 4f elements) and actinides (actinium and the 5f elements) are vital components of our technology-dominated society. Examples include the fluorescent-red europium ions used in euro banknotes to deter counterfeiting [1], the radioactive americium used in smoke detectors [2] that save countless lives every year as well as neodymium used in the strongest permanent magnets [3]. However, the rare earth and actinide elements remain poorly recognised by non-scientists, and even by many undergraduates in chemistry.

The similar radii of the respective +3 cations (Figure 1) belies their individually unique spectral [4] and magnetic [5] properties that contribute to their fascinating chemistry. In this Special Issue, devoted to molecular rare earth and actinide complexes, work from Natrajan and co-workers [6] has explored how fluorinated ligands improve the luminescence of 4f complexes, while Baker and co-workers [7] investigated the optical properties, as well as structure, of a new class of uranyl selenocyanate. Pointillart and co-workers' article [8] bridges the areas of lanthanide optical and magnetic properties—literally—by using bridging tetrathiafulvalene derivatives. The growing field of Single Molecule Magnetism originates in the d-block, but recent interest in the f-elements has been growing. Powell and co-workers [9] explore the use of dimeric dysprosium (which has a highly anisotropic f-electron distribution) compounds with a "hula-hoop" geometry, defined by the ligand that sits in an equatorial plane around both Dy atoms. The main current medical use for the lanthanides, as Magnetic Resonance Imaging (MRI) contrast agents, also relies on the unique electronic properties of the lanthanides. The article by Parac-Vogt and co-workers [10] demonstrates the combination of gadolinium for MRI imaging (thanks to its seven unpaired electrons) connected to a luminescent BODIPY fragment in order to explore combined MRI and optical imaging, addressing the drawbacks of both techniques through their complementary properties.

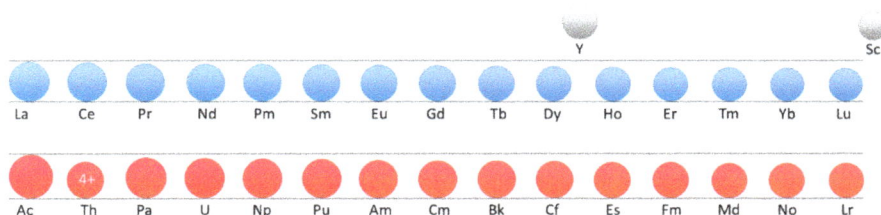

Figure 1. The ionic radii of the 6-coordinate M^{3+} cations of the rare earth and actinide metals (except for Th which is Th^{4+}) [11–13].

Another growing application of molecular lanthanide complexes is in catalysis, and in this issue, Kostakis and co-workers [14] report the use of lanthanide coordination polymers as catalysts in a domino reaction. As with the d-block, organometallic lanthanide chemistry has proven to be of vital use in the development of homogeneous catalysis. This issue reflects this growing interest with papers demonstrating the synthesis of organometallic lanthanide complexes using imide (Anwander and co-workers) [15], amidinate (Edelman and co-workers) [16], reduced bipyridine (Mills and co-workers) [17] and metallocene (Ce^{4+} complexes by Gordon and co-workers [18], the reactivity of Sm^{2+} by Maron and co-workers [19] and U^{3+}/U^{4+} bromides by Kiplinger and co-workers [20]) ligand frameworks. A review from Eisen and co-workers [21] is devoted to actinide catalysis and an article from Visseaux and co-workers [22] details the extension of organometallic Nd catalysis into the solid state demonstrating the numerous current applications of these interesting species. The review by Turner [23] highlights N_2 and P_4 activation chemistry of the f-block, an area of great future catalytic potential.

We hope you will enjoy the breadth of chemistry offered in this open access Special Issue that highlights the many differences between complexes of the rare earths and actinides. However, the similarity of ionic radii is inescapable for the +3 oxidation state, which gives rise to one of the most challenging remaining problems for f-block chemists and the potential renaissance of nuclear power (as well as tackling historical problems). The separation of highly radioactive and frustratingly long-lived heavier actinides from shorter-lived radioactive isotopes of the lanthanides would greatly aid planning for the long-term storage of nuclear waste. It is promising that the current resurgence of interest in the fundamental chemistry of the f-block can feed into the goal of discriminating between the actinides and lanthanides based on differences in bonding and reactivity thereby boosting efforts in the area of nuclear waste separation and storage. In fact, the article by Beekmeyer and Kerridge compares covalency in $[CeCl_6]^{n-}$ and $[UCl_6]^{n-}$ as a means of shedding light on this very problem [24]. We look forward to many more academic and practical advances in all of the above fields of research in the near future.

References

1. Rare Earths: Neither Rare, Nor Earths. Available online: http://www.bbc.co.uk/news/magazine-26687605 (accessed on 30 September 2016).
2. Smoke Detectors and Americium. Available online: http://www.world-nuclear.org/information-library/non-power-nuclear-applications/radioisotopes-research/smoke-detectors-and-americium.aspx (accessed on 30 September 2016).
3. Brown, D.N. Fabrication, processing technologies, and new advances for RE-Fe-B magnets. *IEEE Trans. Magn.* **2016**, *52*, 1–9. [CrossRef]
4. Bünzli, J.-C.G.; Piguet, C. Taking advantage of luminescent lanthanide ions. *Chem. Soc. Rev.* **2005**, *34*, 1048–1077. [CrossRef] [PubMed]
5. Sessoli, R.; Powell, A.K. Strategies towards single molecule magnets based on lanthanide ions. *Coord. Chem. Rev.* **2009**, *253*, 2328–2341. [CrossRef]
6. Swinburne, A.; Langford Paden, M.; Chan, T.; Randall, S.; Ortu, F.; Kenwright, A.; Natrajan, L. Optical properties of heavily fluorinated lanthanide tris β-diketonate phosphine oxide adducts. *Inorganics* **2016**, *4*, 27. [CrossRef]
7. Nuzzo, S.; Browne, M.; Twamley, B.; Lyons, M.; Baker, R. A structural and spectroscopic study of the first uranyl selenocyanate, $[Et_4N]_3[UO_2(NCSe)_5]$. *Inorganics* **2016**, *4*, 4. [CrossRef]
8. Pointillart, F.; Speed, S.; Lefeuvre, B.; Riobé, F.; Golhen, S.; Le Guennic, B.; Cador, O.; Maury, O.; Ouahab, L. Magnetic and photo-physical properties of lanthanide dinuclear complexes involving the 4,5-bis(2-pyridyl-*N*-oxidemethylthio)-4',5'-dicarboxylic acid-tetrathiafulvalene-, dimethyl ester ligand. *Inorganics* **2015**, *3*, 554. [CrossRef]
9. Peng, Y.; Mereacre, V.; Anson, C.; Powell, A. Tuning of hula-hoop coordination geometry in a Dy dimer. *Inorganics* **2016**, *4*, 2. [CrossRef]
10. Ceulemans, M.; Nuyts, K.; De Borggraeve, W.; Parac-Vogt, T. Gadolinium(III)-DOTA complex functionalized with BODIPY as a potential bimodal contrast agent for MRI and optical imaging. *Inorganics* **2015**, *3*, 516. [CrossRef]

11. Shannon Effective Ionic Radii. Available online: http://v.web.umkc.edu/vanhornj/shannonradii.htm (accessed on 10 October 2016).
12. Bilewicz, A. The ionic radius of No^{3+}. *J. Nucl. Radiochem. Sci.* **2002**, *3*, 147–149. [CrossRef]
13. Bilewicz, A. Ionic radii of heavy actinide(III) cations. *Radiochim. Acta* **2004**, *92*, 69–72. [CrossRef]
14. Loukopoulos, E.; Griffiths, K.; Akien, G.; Kourkoumelis, N.; Abdul-Sada, A.; Kostakis, G. Dinuclear lanthanide (III) coordination polymers in a domino reaction. *Inorganics* **2015**, *3*, 448. [CrossRef]
15. Schädle, D.; Maichle-Mössmer, C.; Törnroos, K.; Anwander, R. Holmium(III) supermesityl-imide complexes bearing methylaluminato/gallato ligands. *Inorganics* **2015**, *3*, 500. [CrossRef]
16. Sroor, F.; Hrib, C.; Edelmann, F. New lanthanide alkynylamidinates and diiminophosphinates. *Inorganics* **2015**, *3*, 429. [CrossRef]
17. Ortu, F.; Zhu, H.; Boulon, M.-E.; Mills, D. Synthesis and reactivity of a cerium(III) scorpionate complex containing a redox non-innocent 2,2'-bipyridine ligand. *Inorganics* **2015**, *3*, 534. [CrossRef]
18. Sutton, A.; Clark, D.; Scott, B.; Gordon, J. Synthesis and characterization of cerium(IV) metallocenes. *Inorganics* **2015**, *3*, 589. [CrossRef]
19. Kefalidis, C.; Maron, L. On the dehydrocoupling of alkenylacetylenes mediated by various samarocene complexes: A charming story of metal cooperativity revealing a novel dual metal σ-bond metathesis type of mechanism (DM I σ-BM). *Inorganics* **2015**, *3*, 573. [CrossRef]
20. Lichtscheidl, A.; Pagano, J.; Scott, B.; Nelson, A.; Kiplinger, J. Expanding the chemistry of actinide metallocene bromides. Synthesis, properties and molecular structures of the tetravalent and trivalent uranium bromide complexes: $(C_5Me_4R)_2UBr_2$, $(C_5Me_4R)_2U(O-2,6-{}^iPr_2C_6H_3)(Br)$, and $[K(THF)][(C_5Me_4R)_2UBr_2]$ (R = Me, Et). *Inorganics* **2016**, *4*, 1. [CrossRef]
21. Karmel, I.; Batrice, R.; Eisen, M. Catalytic organic transformations mediated by actinide complexes. *Inorganics* **2015**, *3*, 392. [CrossRef]
22. Russell, S.; Loiseau, T.; Volkringer, C.; Visseaux, M. Luminescent lanthanide metal organic frameworks for *cis*-selective isoprene polymerization catalysis. *Inorganics* **2015**, *3*, 467. [CrossRef]
23. Turner, Z. Molecular pnictogen activation by rare earth and actinide complexes. *Inorganics* **2015**, *3*, 597. [CrossRef]
24. Beekmeyer, R.; Kerridge, A. Assessing covalency in cerium and uranium hexachlorides: A correlated wavefunction and density functional theory study. *Inorganics* **2015**, *3*, 482. [CrossRef]

inorganics

MDPI

Review

Catalytic Organic Transformations Mediated by Actinide Complexes

Isabell S. R. Karmel, Rami J. Batrice and Moris S. Eisen *

Schulich Faculty of Chemistry, Technion—Israel Institute of Technology, Technion City, Haifa 32000, Israel; karmel@campus.technion.ac.il (I.S.R.K.); batricer@gmail.com (R.J.B.)
* Author to whom correspondence should be addressed; chmoris@tx.technion.ac.il; Tel./Fax: +972-4-829-2680.

Academic Editors: Stephen Mansell and Steve Liddle
Received: 16 September 2015; Accepted: 9 October 2015; Published: 30 October 2015

Abstract: This review article presents the development of organoactinides and actinide coordination complexes as catalysts for homogeneous organic transformations. This chapter introduces the basic principles of actinide catalysis and deals with the historic development of actinide complexes in catalytic processes. The application of organoactinides in homogeneous catalysis is exemplified in the hydroelementation reactions, such as the hydroamination, hydrosilylation, hydroalkoxylation and hydrothiolation of alkynes. Additionally, the use of actinide coordination complexes for the catalytic polymerization of α-olefins and the ring opening polymerization of cyclic esters is presented. The last part of this review article highlights novel catalytic transformations mediated by actinide compounds and gives an outlook to the further potential of this field.

Keywords: organoactinides; actinide coordination complexes; homogeneous catalysis; hydroelementations; polymerization of olefins; ROP; activation of heterocumulenes

1. Introduction

The beginning of modern organoactinide chemistry is often attributed to the synthesis of uranocene, [(η^8-C$_8$H$_8$)$_2$U] in 1968, as the analogous compound to ferrocene and other transition metal metallocenes [1,2]. Since then, the organometallic and coordination chemistry of the early actinide elements thorium and uranium has reached a high level of sophistication, including the synthesis and application of these compounds in stoichiometric and catalytic organic transformations [3–14]. Owing to the large ionic radii of the actinides and the presence of 5f orbitals, actinide coordination complexes often exhibit different catalytic activities, as well as complementary reactivities to their early and late transition metal analogues. This in turn leads to different chemo- and regio-selectivities, thus expanding the scope of accessible products obtained in catalytic organic processes [3]. The extraordinary catalytic reactivity of organoactinides was first exemplified in the homogeneous polymerization of butadiene, mediated by the organometallic uranium complex [(η^3-allyl)$_3$UCl] in combination with the co-catalyst [(C$_2$H$_5$)AlCl] in 1974. This process gave poly-1,4-butadiene with a higher *cis* content and therefore with superior mechanical properties, than the polymers obtained with transition metal catalysts [15–19]. In general, the reactivity of actinide coordination complexes is influenced by steric and electronic factors. The steric hindrance in actinide complexes has been described by the "packing saturation model", which attributes the stability of the respective coordination complex to the sum of the ligands cone angles. Hence, coordinative "over saturated" complexes display lower stabilities [20–22]. In addition, de Matos introduced the "steric coordination number" for actinide complexes, which implies a pure ionic bonding model, and is based on the ligands cone angles [23]. Furthermore, the reactivity of actinide compounds can be further elucidated by using thermochemical studies, which allow for a prediction of new reaction pathways by taking into account the metal ligand bond disruption enthalpies [24–30]. It has been also shown that due to the high energy orbital impediment to undergo oxidative addition and reductive elimination reactions, catalytic processes mediated by neutral organoactinide complexes proceed via an ordered four-membered transition state (Scheme 1) [31]. In this review, we will briefly present a survey of homogeneous catalytic transformations mediated by organoactinides and actinide coordination complexes.

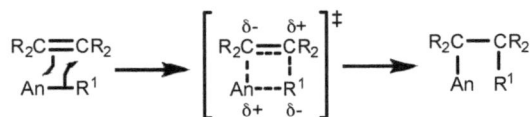

Scheme 1. Operative mechanism in actinide mediated catalytic transformations (R^1: Alkyl, Benzyl, NR_2; R_2: Alkyl, Aryl, $SiMe_3$).

2. Application of Actinide Complexes in Catalytic Transformations

2.1. Hydroelementation Reactions

2.1.1. Hydroalkoxylation/Cyclization of Alkynyl Alcohols

Hydroalkoxylation reactions present an atom economic route for the preparation of new C–O bond by the addition of an O–H bond across a C–C double, or triple bond [32,33]. Recently, Marks *et al.*, reported the thorium mediated hydroalkoxylation/cyclization of alkynyl alcohol using primary, and secondary alcohols, as well as terminal and internal alkynes as substrates [34]. Kinetic studies revealed a zero order dependence of the reaction rate on substrate and a first order dependence on catalyst. Moreover, the reaction rates with primary alcohols and terminal alkynes was faster, as compared to the rate when using sterically more encumbered substrates, corroborating that the transition state of the reaction is dominated by steric interactions. The reaction was studied with three different thorium catalysts (Figure 1), among which the constrained geometry catalyst (CGC)Th(NMe$_2$)$_2$ **2** exhibited the highest catalytic activity, which was attributed to the more open coordination sphere of complex **2**. All substrate and catalyst combinations reported displayed a high selectivity for the exclusive formation of the *exo*-methylene product (*Markovnikov* selectivity), while internal alkynes afforded alkene products with an E-orientation.

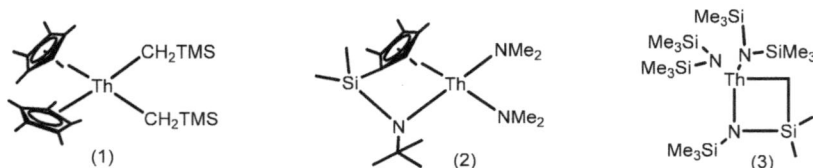

Figure 1. Thorium complexes **1–3** used as pre-catalysts in the hydroalkoxylation/cyclization of alkynyl alcohols.

The mechanism of the reaction (Scheme 2) includes an activation step, which affords the catalytically active Th–OR species **A** after hydrolysis of the benzyl or amide pro-ligands with an incoming alcohol. A subsequent insertion of the alkyne into the Th–OR bond via a four membered transition state leads to the formation of intermediate **C**. The insertion of the alkyne is the thermodynamically most demanding step of the catalytic cycle, hence it presents the rate limiting step of this mechanism. The Th–C bond of intermediate **C** is cleaved rapidly by an incoming substrate, furnishing the product under regeneration of the active catalyst **A** [34].

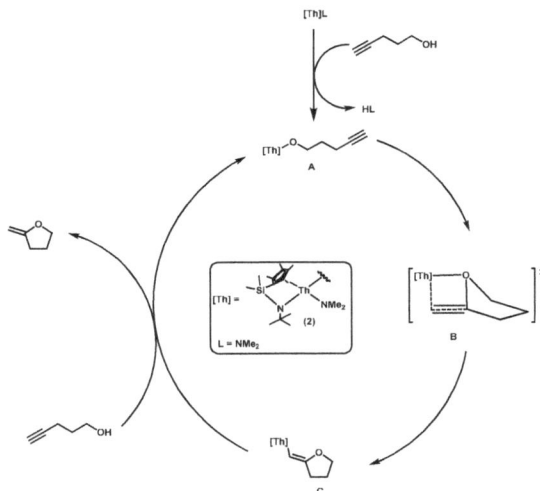

Scheme 2. Proposed mechanism for the thorium mediated hydroalkoxylation/cyclization of alkynyl alcohols. Adapted with permission from Wobser, S.D.; Marks, T.J. *Organometallics* **2013**, *32*, 2517–2528 [34]. Copyright 2013 American Chemical Society.

2.1.2. Hydrothiolation of Terminal Alkynes

The hydrothiolation of terminal alkynes gives an easy access to vinyl sulfides through the addition of S–H bonds across the C≡C triple bond of an alkyne. While this reaction has been studied with a large variety of early and late transition metal catalysts, achieving selectivity for the formation of the Markovnikov addition product still remains a challenge [35–37]. Additionally, most transition metal catalysts have a limited substrate scope, proceeding with low conversions when using aliphatic thiols [38–44]. Hence, the hydrothiolation of terminal alkynes was studied using the cyclopentadienyl based actinide catalysts **1** and **4** (Figure 2) [45–47].

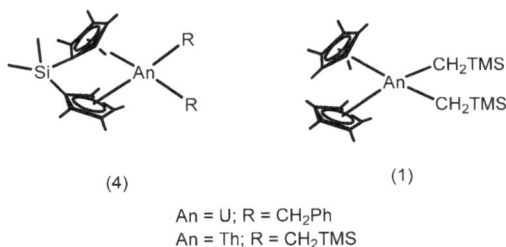

An = U; R = CH₂Ph
An = Th; R = CH₂TMS

Figure 2. Actinide catalysts **4** and **1** used in the catalytic hydrothiolation of terminal alkynes [45–47].

Bond enthalpy calculations have shown that the insertion of alkynes into An–SR bond is strongly exothermic, allowing for an actinide mediated catalytic hydrothiolation [6,45–47]. The most noticeable aspect of the actinide catalyzed hydrothiolation using the *ansa*-bridged thorium complex **4** as pre-catalyst is the wide substrate scope, which includes aliphatic, benzylic and aromatic thiols, as well as the high selectivity for the formation of the Markovnikov addition product. The Markovnikov selectivity can be mainly attributed to steric interactions, as well as to bond polarity mismatches in the highly ordered four-membered transition state of the reaction (Figure 3). The terminal R-substituent of the alkyne is oriented away from the sterically demanding *ansa*-cyclopentadienyl ligands in the transition state **A**, which leads to the formation of the Markovnikov vinyl sulfide. Furthermore, the larger the electron density on the α-carbon atom to the metal center at the transition state **A**, leads to a better stabilization of the highly electrophilic actinide center [45–47].

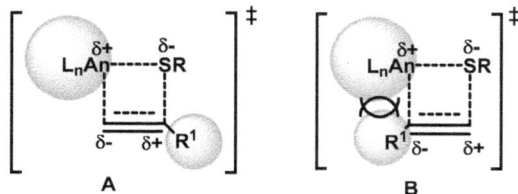

Figure 3. Markovnikov (**A**) and anti-Markovnikov (**B**) transition states for the actinide mediated hydrothiolation of alkynes. Adapted with permission from Weiss, C.J.; Marks, T.J. *Dalton Trans.* **2010**, *39*, 6576–6588 [47]. Copyright 2010 Royal Society of Chemistry.

The general mechanism for the hydrothiolation of terminal alkynes mediated by the thorium catalyst **4** is presented in Scheme 3. After an activation step, which leads to the formation of the catalytically active thorium species **A**, the alkyne inserts into the Th–SR bond of **A** to give intermediate **C**, which is the rate limiting step of the catalytic cycle. Subsequent protonolysis with an incoming thiol furnishes the vinyl sulfide product under regeneration of the active catalyst **A**.

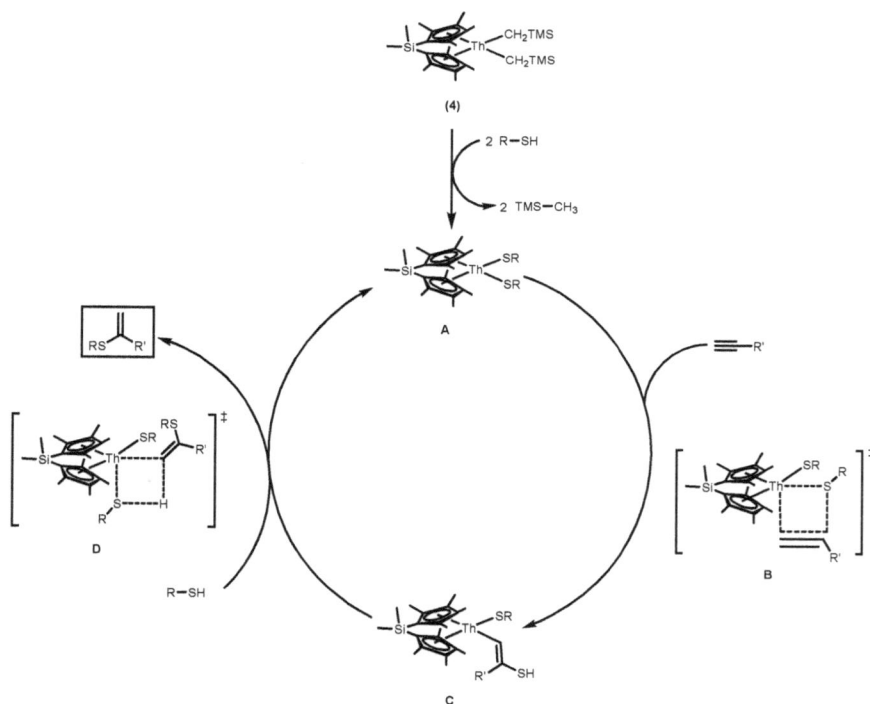

Scheme 3. Proposed mechanism for the thorium mediated hydrothiolation. Adapted with permission from Weiss, C.J.; Wobser, S.D.; Marks, T.J. *J. Am. Chem. Soc.* **2009**, *131*, 2062–2063 [45]. Copyright 2009 American Chemical Society.

2.1.3. Inter- and Intra-Molecular Hydroamination

The intermolecular hydroamination of terminal alkynes with primary aliphatic amines has been explored by Eisen *et al.*, using the $Cp*_2AnMe_2$ (An = Th, U; $Cp* = C_5Me_5$) complexes **5–6** as catalysts [48,49]. The chemo- and regio-selectivity of the products obtained showed a strong dependence on the nature of the amine and the actinide center, but not on the substituents of the respective alkyne. While the use of $Cp*_2UMe_2$ (**5**) lead to the formation of imine products with a *syn*-regiochemistry, the analogous thorium complex $Cp*_2ThMe_2$ (**6**) furnished the respective imine products, as well as dimeric and trimeric alkyne oligomers (Scheme 4). Kinetic studies showed a first order dependence of the reaction on the respective actinide pre-catalysts $Cp*_2AnMe_2$, a zero order dependence on alkyne concentration, and a reversed first order dependence on amine [48,49].

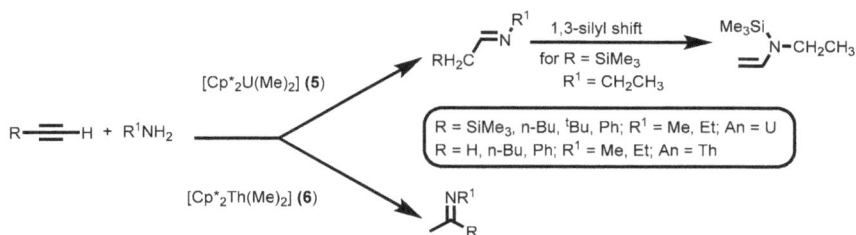

Scheme 4. Chemo- and regio-selectivity of the intermolecular hydroamination mediated by the organoactinide complexes $Cp*_2AnMe_2$ (An = U, Th). Adapted with permission from Haskel, A.; Straub, T.; Eisen, M.S. *Organometallics* **1996**, *15*, 3773–3775 [48]. Copyright 1996 American Chemical Society.

The mechanism of the intermolecular hydroamination (Scheme 5) has been shown to go over the formation of an actinide imido species of the type $Cp*_2An(=NR)$ (An = Th, U) (**D**), which have been isolated and structurally characterized [49]. The first step of the catalytic cycle is the protonolysis reaction with an incoming amine, furnishing the bis(amido)amine intermediate **A**, as well as two equivalents of methane gas. Intermediate **A** can react via two competitive pathways. In the first case, complex **B** undergoes a σ-bond metathesis with the terminal alkynes, via the bis(acetylide) complex **C**, resulting in the formation of the respective alkyne oligomerization products. The second pathway, which has been shown to be the rate determining step of the catalytic cycle, includes the elimination of one equivalent of amine to generate the actinide imido complex **D**. Intermediate **D** undergoes a rapid bond metathesis with an incoming alkyne (*anti*-insertion) forming the actinide metallacycle **E**, followed by a rapid protonolysis with an incoming amine to furnish the actinide amido intermediate **F**. Complex **F** can now either react with another equivalent of amine, generating an enamine product (**I**) that will rapidly isomerize to the respective imine **H**, under regeneration of the catalytically active complex **B**. In addition, complex **F** can undergo a 1,3-hydride shift furnishing intermediate **G** and a subsequent reaction with an incoming amine to give the imine product **H** under regeneration of the catalytically active species **B** (Scheme 5) [48,49].

In addition to the intermolecular hydroamination of terminal acetylenes, Marks *et al.*, have investigated the intramolecular hydroamination/cyclization of aminoalkenes aminoalkynes, and aminoallenes with the constrained geometry actinide catalysts $(CGC)An(NR_2)_2$ (An = Th (**2**); U (**7**)) with various amido ligands [50,51]. The constrained geometry catalysts **2** and **7** display a more open coordination sphere as compared to the respective pentamethyl cyclopentadienyl compounds $Cp*_2AnR_2$ (An = U (**5**); Th (**6**)) leading to an increased catalytic activity in the hydroamination/cyclization. Furthermore, Marks and coworkers have shown that the nature of the actinide center influences the rate of the intermolecular hydroamination/cyclization, displaying higher rates for the respective thorium complexes than for their uranium analogues. This was attributed to the more open coordination sphere in $(CGC)Th(NR_2)_2$ as compared to the isostructural

(CGC)U(NR$_2$)$_2$, due to the larger ionic of thorium as compared to uranium. Investigations with a large variety of R substituents on the amido ligands in (CGC)An(NR$_2$)$_2$ (An = Th or U) catalysts have shown, that the nature of the R substituent does not influence the reaction rate, which was attributed to the rapid protonolysis of the respective amido ligands with an incoming substrate (Scheme 6). Mechanistic studies have led to proposed mechanism shown in Scheme 6, which displays similar steps to the mechanism of the intermolecular hydroamination/cyclization mediated by organolanthanide complexes. The first step of the mechanism comprises a rapid protonolysis of both amido ligands NR$_2$ with two equivalents of incoming substrate, generating the active species **A**. Subsequent insertion of the C=C double, or C≡C triple bond into the An–NHR^1 bonds of complex **A** over a four membered transition state (**B**), furnishes intermediate **C**. The intermolecular insertion of the unsaturated C=C, or C≡C bonds into the An–NHR^1 bond of **A** is the rate determining step of the reaction mechanism, followed by a rapid protonolysis with an incoming substrate, which leads to the formation of product **D** under regeneration of the catalytically active species **A**. In comparison to the analogous mechanism using lanthanide (III) complexes as catalysts, the use of CGC(An)(NR$_2$)$_2$ catalysts, allows for dual simultaneous substrate cyclization, due to the availability of two amido ligands NR$_2$, whereas lanthanocenes exhibit only one NR$_2$ ligand, which can be replaced by only one equivalent of substrate [50,51].

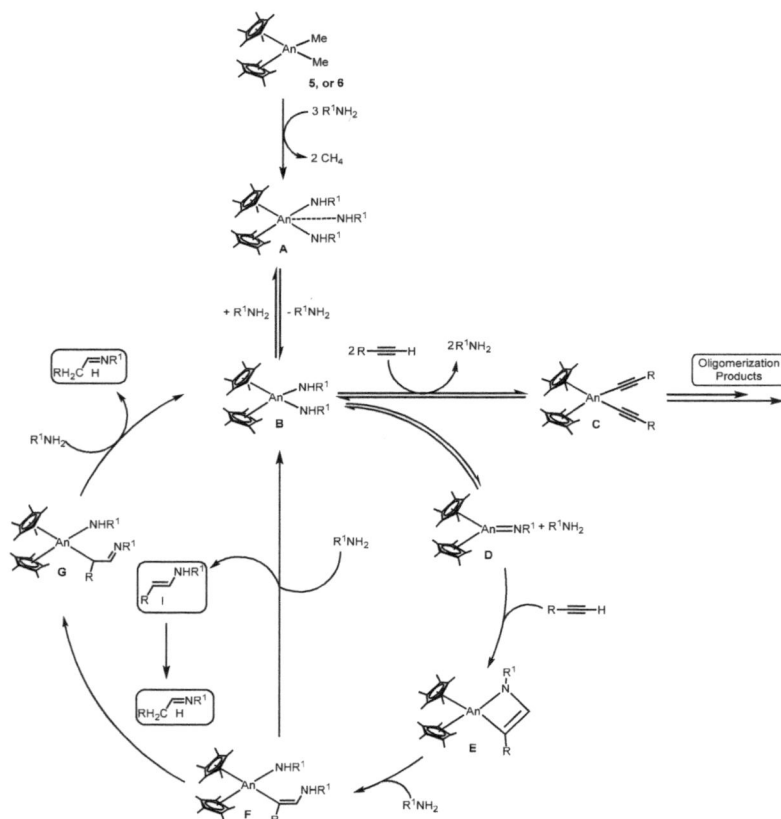

Scheme 5. Proposed mechanism for the intermolecular hydroamination of terminal alkynes mediated by the organoactinide complexes **4** and **5**. Adapted with permission from Haskel, A.; Straub, T.; Eisen, M.S. *Organometallics* **1996**, *15*, 3773–3775 [48]. Copyright 1996 American Chemical Society.

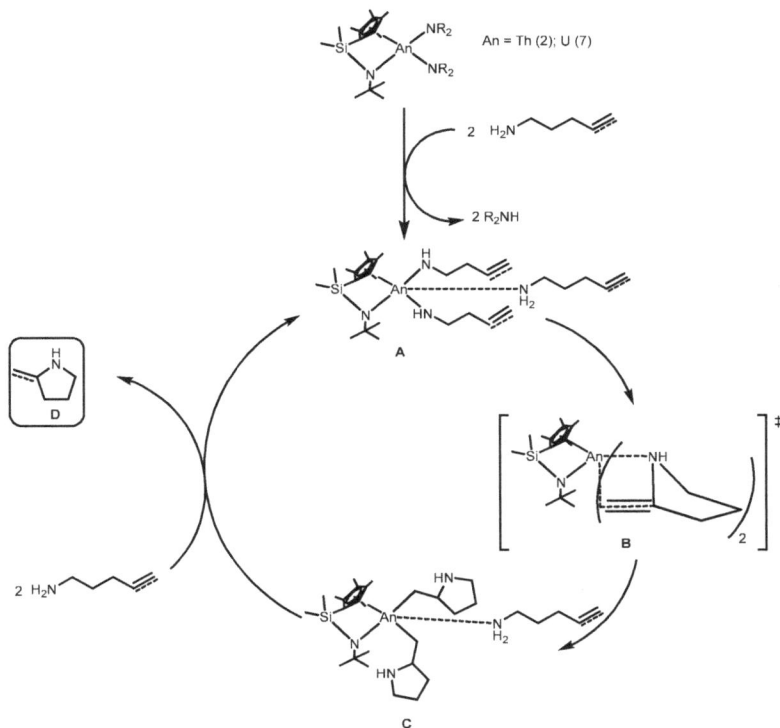

Scheme 6. Intermolecular hydroamination mediated by the constrained geometry actinide complexes **2** and **7**. Stubbert, B.D.; Stern, C.L.; Marks, T.J. *Organometallics* **2003**, *22*, 4836–4838 [50]. Copyright 2003 American Chemical Society.

In order to compare the active mechanism in the inter- and intra-molecular hydroamination reaction, Diaconescu *et al.*, used the ferrocene-diamide supported uranium (IV) complex (fcc)U(CH$_2$Ph)$_2$ **8** (Figure 4) in the intramolecular hydroamination of acetylenes with anilines, as well as in the intermolecular hydroamination/cyclization of aminoalkenes and aminoalkynes [52]. While the intramolecular has been shown to proceed via the imido-mechanism proposed by Eisen *et al.*, (Scheme 5) [48,53,54], an unambiguous result for the intermolecular hydroamination/cyclization with aminoalkenes and aminoalkynes could not be obtained. Furthermore, Leznoff and co-workers have recently applied diamido-ether supported actinide complexes **9** and **10** (Figure 4) for the hydroamination/cyclization of aminoalkenes under mild conditions, displaying high catalytic activities [55].

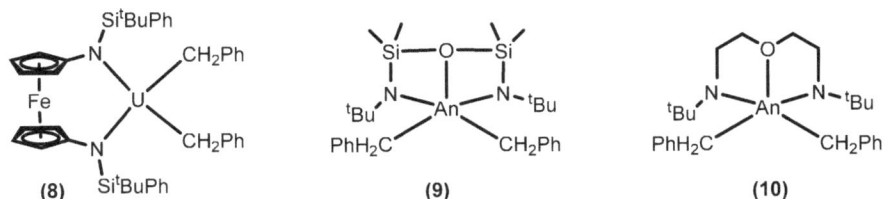

Figure 4. Actinide complexes **8–10** used as catalysts for hydroamination reactions [52,55].

2.1.4. Hydrosilylation of Terminal Alkynes

The hydrosilylation of terminal alkynes can proceed to give three different isomeric products (Scheme 7), which comprise the *cis*, and *trans* vinylsilanes, and the geminal vinylsilane, which are obtained due to the 1,2 *syn* and *anti* and the 2,1-insertion modes, respectively. The product distribution obtained has been shown to depend strongly on the nature of the metal catalyst, the substrates and the reaction conditions [56–59].

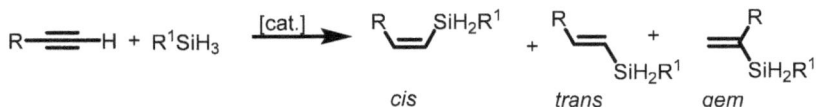

Scheme 7. Possible products in the catalytic hydrosilylation of terminal alkynes. Adapted with permission from Dash, A.K.; Wang, J.Q.; Eisen, M.S. *Organometallics* **1999**, *18*, 4724–4741 [60]. Copyright 1999 American Chemical Society.

Several actinide complexes have been explored as catalysts in the hydrosilylation of terminal acetylenes and olefins with $PhSiH_3$, displaying different chemo- and regio-selectivities in dependence of the actinide center, the ancillary ligands, the substrates used and the reaction conditions. The organoactinide compounds $Cp^*_2AnMe_2$ (An = U (**5**); Th (**6**)) were shown to mediate the hydrosilylation of terminal alkynes with $PhSiH_3$, leading to a different product distribution in dependence of the reaction temperature [60]. When the reaction was carried out at room temperature, the *trans* vinylsilane **D** (Scheme 8) was obtained as the major product, as well as silylalkynes and the respective alkene as byproducts. At higher temperatures (50–80 °C), the product distribution depends on the respective actinide center in $Cp^*_2AnMe_2$. While the use of $Cp^*_2ThMe_2$ (**6**) as catalyst displayed the same regio- and chemo-selectivity like at room temperature, the isostructural uranium complex $Cp^*_2UMe_2$ (**5**) led to the formation of the *cis* vinylsilane, as well as the double hydrosilylated product in addition to the products observed at room temperature. These observations suggest a competing reaction pathway at higher temperatures that yields the *cis* vinylsilane, as well as the alkene and silylalkyne. The proposed mechanism for the catalytic hydrosilylation of terminal acetylenes mediated by $Cp^*_2ThMe_2$ (**6**) is presented in Scheme 8. The first step of the catalytic cycle comprised the reaction of $Cp^*_2ThMe_2$ (**6**) with two equivalents of acetylene, generating the bis(acetylide) compound **A**, which upon reaction with $PhSiH_3$ leads to the formation of the silylalkyne and thorium hydride **B**. After reinsertion of the silylalkyne into **B**, the thorium hydride complex **B** is equilibrium with intermediate **F**. In addition the thorium hydride complex **B** undergoes a rapid insertion of an alkyne into the Th–H bond, forming the thorium alkenyl-acetylide complex **C**. A subsequent reaction of **C** with $PhSiH_3$, which is the rate determining step of the catalytic cycle, regenerates the catalytically active species **A** with formation of the *trans* vinylsilane **D**. However, the thorium alkenyl-acetylide complex **C** can also react with an additional equivalent of alkyne, generating the alkene product **E** and the thorium bis(acetylide) intermediate **A**. The double hydrosilylated product **H**, which is observed at higher temperatures, is obtained by phenysilane insertion into intermediate **F**, regenerating thorium-hydride intermediate **B**. Additionally, intermediate **F** can also react with an equivalent of alkyne at higher temperatures, furnishing the *cis* vinylsilane **G** and intermediate **A** (Scheme 8) [60].

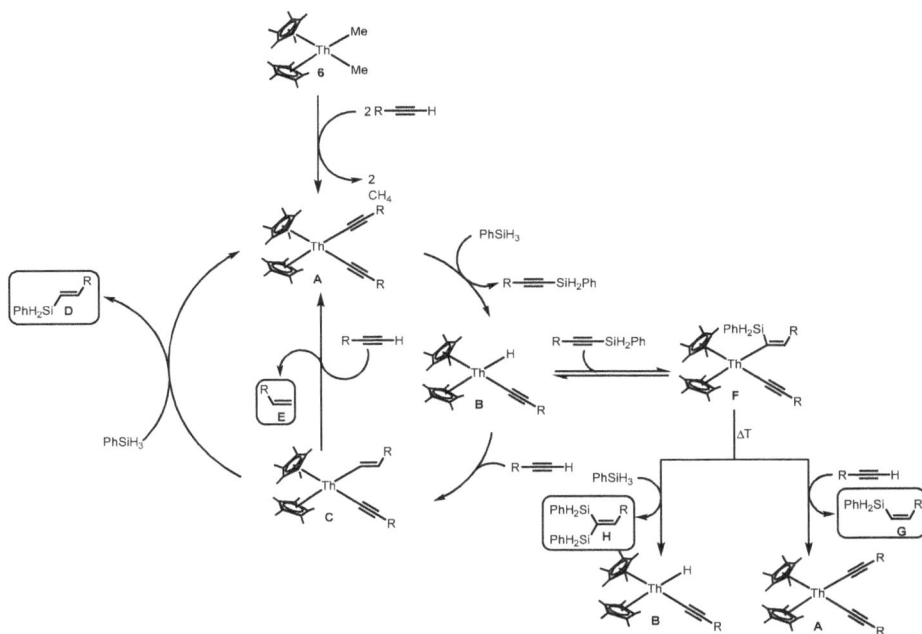

Scheme 8. Proposed mechanism for the hydrosilylation of terminal alkynes, mediated by Cp*$_2$ThMe$_2$ (**6**). Adapted with permission from Dash, A.K.; Wang, J.Q.; Eisen, M.S. *Organometallics* **1999**, *18*, 4724–4741 [60]. Copyright 1999 American Chemical Society.

An improvement in the reaction rate, as well as in the chemo- and regio-selectivity of the reaction was achieved by using the *ansa*-bridged thorium complex **4** as catalyst for the hydrosilylation of terminal alkynes and alkenes with PhSiH$_3$ [61,62]. The increased reaction rate was attributed to the more open coordination sphere in the *ansa*-bridged complex **4**, as compared the Cp*$_2$ThMe$_2$ (**6**). The increased chemo- and regio-selectivity, which is exhibited in the formation of the *trans* vinylsilane, or the 1-silylalkane as major products in the hydrosilylation of terminal alkynes and alkenes, respectively, was attributed to the more hindered equatorial plane in **4**. The disposition of the methyl groups in the ligand backbone forces the incoming substrates to react in a specific regiochemistry, allowing for a fine-tuning of the organoactinide reactivity by modifying the ancillary ligand backbone. While the hydrosilylation of terminal alkynes proceeds with excellent chemo- and regio-selectivity, generating the respective *trans* vinylsilane as the only product, the hydrosilylation of alkenes shows a high regioselectivity toward the formation of 1-silylalkanes. However only a moderate chemoselectivity is achieved in the later case since the respective alkane is obtained in nearly equimolar amounts. The mechanism for the hydrosilylation of alkenes is shown in Scheme 9. The first step of the catalytic cycle comprises the formation of the catalytically active thorium-hydride complex **A** by reaction of the pre-catalyst **6** with PhSiH$_3$ under formation of PhSiH$_2$Me. Intermediate **A** reacts with an incoming alkene to furnish the thorium-alkyl compound **B**. The thorium-alkyl species **B** can now react via three different pathways: the first option includes a reaction with another equivalent of alkene to give the π-alkene complex **G**. In addition, complex **B** can undergo a metathesis reaction with a Si–H moiety, to give the 1-silylalkane product **F** and complex **A**, or a protonolysis reaction with a Si–H moiety to give intermediate **D** and the respective alkane **C**. Since these two steps have similar energies of activation, both can take place, furnishing an equimolar amount of 1-silylalkane **F** and alkane **D** (Scheme 9) [61,62].

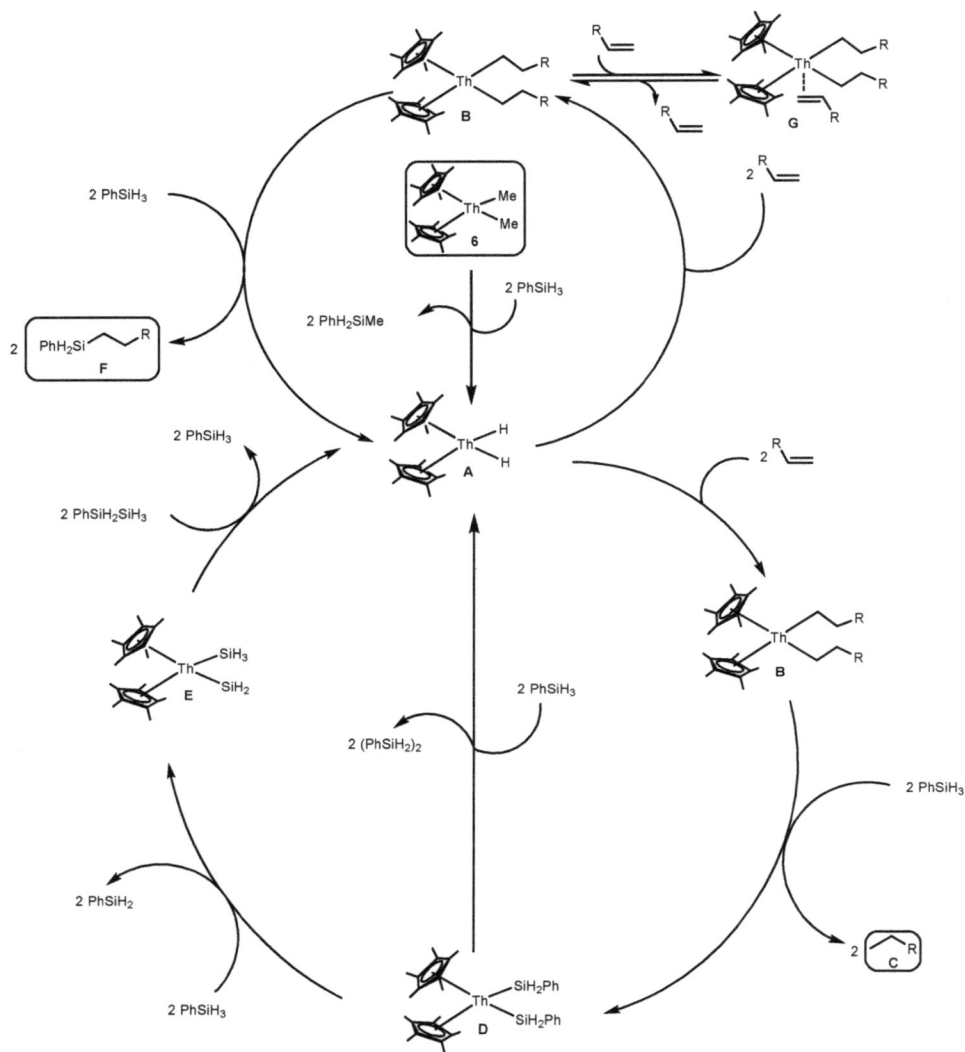

Scheme 9. Proposed mechanism for the thorium mediated hydrosilylation of alkenes Adapted with permission from Dash, A.K.; Gourevich, I.; Wang, J.Q.; Wang, J.; Kapon, M.; Eisen, M.S. *Organometallics* **2001**, *20*, 5084–5104, Copyright 2001 American Chemical Society [61].

In addition, the cationic uranium (IV) complex [(Et$_2$N)$_3$U] [BPh$_4$] (**11**) was explored as a pre-catalyst in the hydrosilylation of terminal acetylenes, displaying a similar chemo- and regio-selective to the organoactinide system Cp*$_2$UMe$_2$ (**5**) [63].

2.2. Coupling Reactions

2.2.1. Coupling of Terminal Acetylenes

The coupling of terminal alkynes to give oligomeric products through actinide mediated C–H activation processes has been extensively investigated by Eisen and co-workers, using the organoactinides Cp*$_2$AnMe$_2$ (**5** and **6**) [53,54,64], as well as the cationic uranium amide [(Et$_2$N)$_3$U] [BPh$_4$] (**11**) [65–67] as pre-catalysts. While the chemo- and regio-selectivity of the oligomerization process depends on the steric encumbrance of the R substituent on the alkyne RCCH, leading to a different regioselectivity for bulky and non-bulky alkynes, the uranium complex **5** and the isostructural thorium compound **6** displays a similar reactivity [53,64]. When the sterically encumbered alkyne *t*BuCCH is used as substrate, the head-to-tail dimer is obtained. However, the use of TMSCCH leads to the formation head-to-tail geminal dimer, and to the head-to-tail-to-head trimer. Similarly, when sterically not encumbered alkynes, such as PhCCH, and *i*PrCCH are applied, a myriad of products is obtained. The high regioselectivity obtained when using *t*BuCCH can be explained by the insertion of the alkyne into the An–C bond (Figure 5), in which the bulky *tert*-butyl group points away from the sterically encumbering pentamethyl cyclopentadienyl ligands.

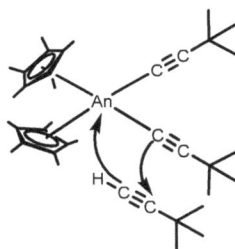

Figure 5. Regioselective insertion of *t*BuCCH into An–C bond. Adapted with permission from Haskel, A.; Straub, T.; Dash, A.K.; Eisen, M.S. *J. Am. Chem. Soc.* **1999**, *121*, 3014–3024 [53]. Copyright 1999 American Chemical Society.

In general, the extent of oligomerization, which describes ratio between the formation of dimers (**E**), or trimers (**D**) *versus* higher oligomers, depends on the difference of the energies of activation $\Delta\Delta G^{\ddagger}$. The difference in the energies of activation $\Delta\Delta G^{\ddagger}$ in turn depends on the size of the metal center, as well as on the steric encumbrance of the R substituents of the alkynes. The mechanism for the actinide mediated oligomerization of alkynes is depicted in Scheme 10. The first step of the mechanism is the formation of the catalytically active actinide bis(acetylide) complex **A** by a protonolysis of Cp*$_2$AnMe$_2$ with two equivalents of alkyne, forming two equivalents of methane. Subsequently, an incoming alkyne inserts into the An–C bond of complex **A** in a 1,2-head-to-tail fashion, the actinide bis(alkenyl) intermediate **B**. Complex **B** can now either undergo a σ-bond metathesis with an incoming alkyne, forming the geminal dimer **E** under regeneration of complex **A**, or an additional 2,1-tail-to-head insertion of an incoming alkyne, furnishing the bis(dienyl) actinide species **C**. A subsequent σ-bond metathesis with an incoming alkyne generates the trimeric product **D** under regeneration of the catalytically active actinide bis(acetylide) complex **A** (Scheme 10) [53,64].

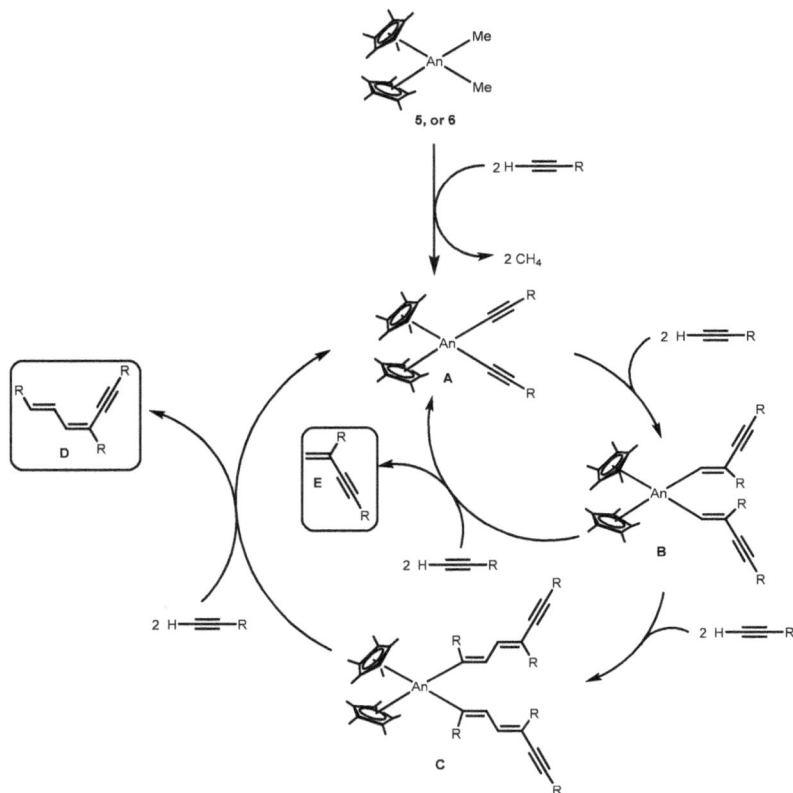

Scheme 10. Proposed mechanism for the actinide mediated oligomerization of terminal alkynes. Adapted with permission from Haskel, A.; Straub, T.; Dash, A.K.; Eisen, M.S. *J. Am. Chem. Soc.* **1999**, *121*, 3014–3024 [53]. Copyright 1999 American Chemical Society.

In order to obtain exclusively short oligomers Eisen *et al.*, developed a methodology, which comprises the addition of external amines to the reaction mixture, leading to a controlled formation of short oligomeric products [54]. The added primary and secondary amines act as chain transfer reagents without being incorporated into the product, and therefore also don't need to be eliminated from the obtained oligomer. The effect of the amine on the chemo- and regio-selectivity of the reaction depends on the nature of the actinide center, and the nature of the alkyne. While the use of $Cp^*_2UMe_2$ (5) as pre-catalyst lead to the formation of the respective hydroamination product upon addition of an external amine, the use of $Cp^*_2ThMe_2$ (6) as the pre-catalyst, yields the desired short oligomers. When aromatic alkynes and non-bulky primary amines are used in combination with the organothorium complex 6, only the *trans* dimer is obtained as product. Per contra, aliphatic alkynes with bulky amines and the pre-catalyst 6 furnished only the geminal dimer, and trimer. The thorium mediated oligomerization of alkynes in addition of an external amine source is presented in Scheme 11. In the first step of the catalytic cycle the catalytically active thorium bis(amido) complex **B** is formed by a protonolysis reaction between $Cp^*_2ThMe_2$ (6) and two equivalents of amine. The thorium bis(amido) complex **B**, which is in equilibrium with the thorium bis(amido) amine compound **A**, subsequently reacts with one equivalent of alkyne, generating intermediate **C**. The thorium complex **C** can now undergo a rapid insertion of an incoming alkyne into the Th–σ–carbyl bond, furnishing the thorium-alkynyl-amido intermediate **D**. Complex **D** can now react either in a protonolysis reaction with an incoming amine, forming dimer **E** under regeneration of the catalytically active species **B**, or insert a further equivalent of alkyne into the Th–C bond of **D**, followed by σ-bond protonolysis by an incoming amine to furnish the respective trimer and complex **B** (Scheme 11) [54].

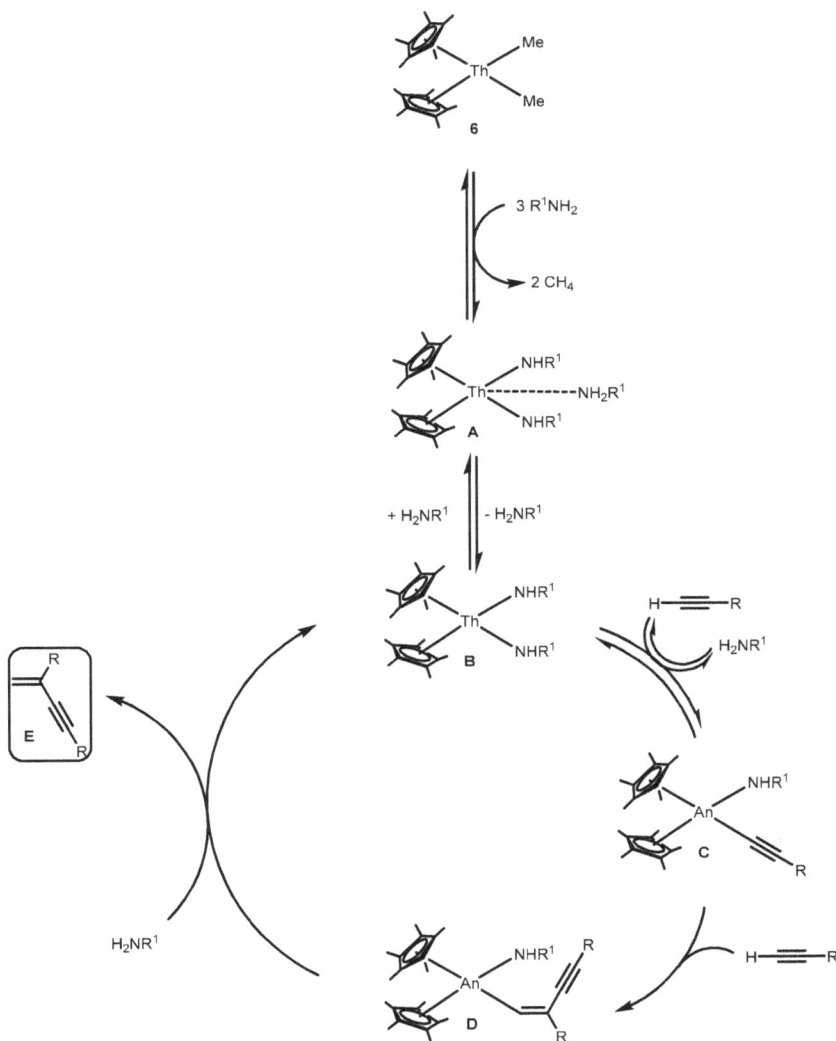

Scheme 11. Proposed mechanism for the thorium mediated dimerization of alkynes using external primary amines. R = Alkyl, Aryl, or SiMe₃. The regioselectivity of the alkyne insertion into the Th–C bond is explained in Figure 5 (*vide supra*). Adapted with permission from Haskel, A.; Wang, J.Q.; Straub, T.; Neyroud, T.G.; Eisen, M.S. *J. Am. Chem. Soc.* **1999**, *121*, 3025–3034 [54]. Copyright 1999 American Chemical Society.

In addition to the organoactinides **5** and **6**, the cationic uranium complex [(Et₂N)₃U] [BPh₄] (**11**) was successfully applied in the oligomerization and cross oligomerization of terminal alkynes, leading to the spectroscopic characterization of the first π-alkyne uranium complex as the key intermediate in the catalytic cycle [65–67]. Recently, Meyer *et al.*, investigated the reactivity of a uranium (III) coordination complex toward various terminal acetylenes, which led to the formation of binuclear uranium (IV) species that could be isolated and structurally characterized [68].

An interesting reactivity in the σ-bond metathesis of silylated alkynes was obtained using the oxo-bridged uranium complex [(η^5-(C$_5$Me$_4$)$_2$SiMe$_2$)U(^{n}Bu)(μ-O)]$_2$ (**12**) as pre-catalyst [69]. The product distribution using terminal, or internal acetylenes and catalyst **12** is presented in Scheme 12, and the mechanism of the reaction in Scheme 13. At first, the catalytically active actinide acetylide complex **A** is formed by a protonolysis of complex **12** with TMSCCH, generating two equivalents of butane. The uranium acetylide **A** reacts with an incoming molecule of TMSCCH over the four-membered transition state **B**. This leads to the cleavage of the Si–C bond and the formation of the uranium acetylide species **D** under elimination of the internal alkyne product **C**. Subsequently, complex **D** undergoes a protonolysis reaction with an additional equivalent of TMSCCH, furnishing acetylene (**E**) and regenerating the active catalyst **A** (Scheme 12) [69].

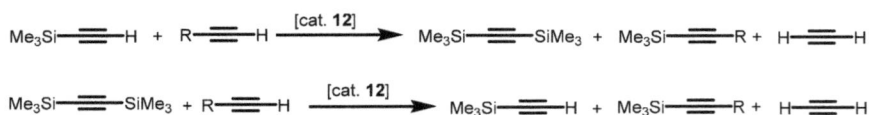

Me$_3$Si—≡—H + R—≡—H $\xrightarrow{\text{[cat. 12]}}$ Me$_3$Si—≡—SiMe$_3$ + Me$_3$Si—≡—R + H—≡—H

Me$_3$Si—≡—SiMe$_3$ + R—≡—H $\xrightarrow{\text{[cat. 12]}}$ Me$_3$Si—≡—H + Me$_3$Si—≡—R + H—≡—H

Scheme 12. Product distribution for the cross metathesis of silylated alkynes, mediated by complex **12**. Adapted with permission from Wang, J.; Gurevich, Y.; Botoshansky, M.; Eisen, M.S. *J. Am. Chem. Soc.* **2006**, *128*, 9350–9351 [69]. Copyright 2006 American Chemical Society.

2.2.2. Coupling of Terminal Alkynes and Isonitriles

The coupling of terminal alkynes with isonitriles to furnish α,β-unsaturated aldimines was studied with the organoactinides Cp*$_2$AnMe$_2$ (An = U, Th) **5** and **6**, as well as cationic uranium complex [(Et$_2$N)$_3$U] [BPh$_4$] (**11**). The product distribution depends on the catalyst, and the alkyne to isonitrile ratio [70]. While the cationic uranium complex [(Et$_2$N)$_3$U] [BPh$_4$] (**11**) furnishes selectively the *trans* mono coupling product **D** between one equivalent of alkyne and isonitrile, the organouranium catalyst Cp*$_2$UMe$_2$ (**5**) generates the double insertion product of two equivalents of isonitrile and one equivalent of alkyne besides the mono coupling product **D** (Scheme 14). When the isostructural Cp*$_2$ThMe$_2$ (**6**) is used as pre-catalyst, product **D** is obtained, as well as the double insertion product between two equivalents of alkyne and one equivalent of isonitrile. The different product distributions of Cp*$_2$AnMe$_2$ and [(Et$_2$N)$_3$U] [BPh$_4$] indicate different operative mechanisms for neutral and cationic actinide complexes in this catalytic transformation. The proposed mechanism using Cp*$_2$AnMe$_2$ (An = Th, U) as pre-catalyst is depicted in Scheme 14. The catalytically active actinide bis(acetylide) species **A** is formed by a protonolysis reaction of Cp*$_2$AnMe$_2$ with two equivalents of terminal alkyne, generating two equivalents of methane. A subsequent 1,1-insertion of isonitrile into the An–C bond of **A**, which is the rate determining step of the catalytic cycle, generates the iminoacyl intermediate **B**. Due to the steric hindrance of the pentamethylcyclopentadienyl ligands, the R group of the isonitrile displays a *syn* regiochemistry to the alkyne in the insertion process. Protonolysis of compound **B** with an incoming alkyne leads to the formation of the *cis* mono insertion product **C**, which isomerizes to yield the more stable *trans* product **D**, under regeneration of the active catalyst **A**. Under alkyne starvation conditions, however, the last protonolysis step is slow, hence the iminoacyl complex **B** can undergo an additional 1,1-insertion with a second isonitrile equivalent, furnishing intermediate **E**. A subsequent protonolysis with an additional equivalent of alkyne yields the double insertion product **F**, which isomerizes to the more stable E,E-isomer, under regeneration of the active catalyst **A**. When an excess of alkyne is used the catalytically active species **A** can insert the triple bond of product **D** into the Th–C bond, generating intermediate **G**, followed by a rapid protonolysis step to furnish product **H** under regeneration of the catalytically active thorium bis(acetylide) species (Scheme 14) [70].

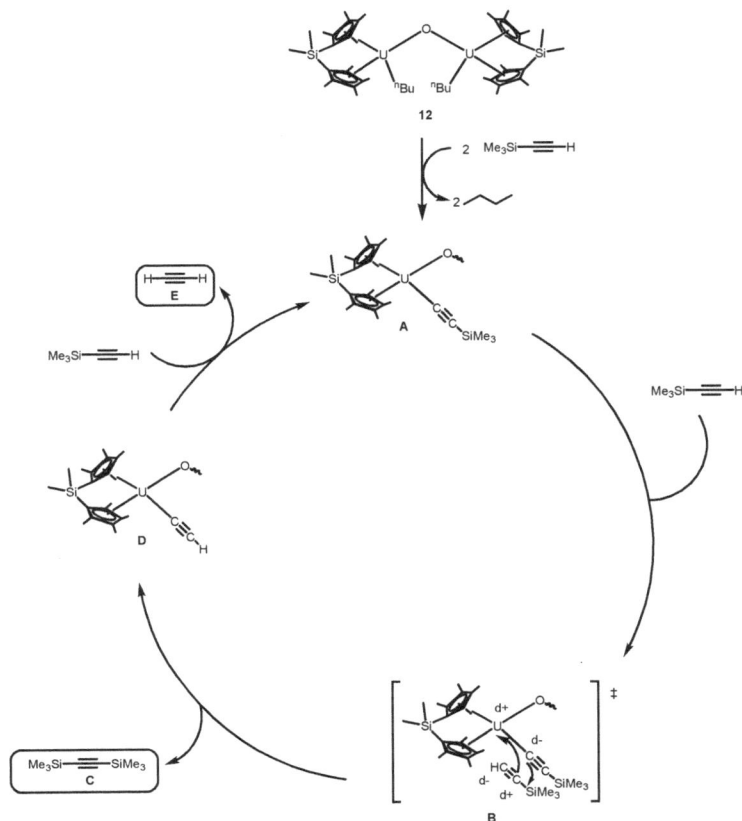

Scheme 13. Proposed mechanism for the cross metathesis of silylated alkynes mediated by complex **12**. A symmetric half of complex **12** has been omitted for clarity. Adapted with permission from Wang, J.; Gurevich, Y.; Botoshansky, M.; Eisen, M.S. *J. Am. Chem. Soc.* **2006**, *128*, 9350–9351 [69]. Copyright 2006 American Chemical Society.

In contrary, the use of $[(Et_2N)_3U]$ $[BPh_4]$ (**11**) as pre-catalyst, leads to the selective formation of the mono insertion product **D**, as well as traces of alkyne oligomerization products [70]. Additionally, the product distribution does not depend on the steric encumbrance of the alkyne substituents, suggesting a coordinative more open active catalyst, in which the ligand does not interfere with the approach of the respective substrate. The mechanism for the catalytic coupling terminal acetylenes with isonitriles mediated by complex **11** is presented in Scheme 15. The first step is a rapid equilibrium, leading to the formation of the uranium-acetylide complex **A** by reversible elimination of Et_2NH from $[(Et_2N)_3U]$ $[BPh_4]$ (**11**). A subsequent 1,1-insertion of an incoming isonitrile into the U–C bond of **A**, which is the rate determining step of the catalytic cycle, furnishes intermediate **B**. Intermediate **B** can undergo a rapid protonolysis either with an incoming alkyne, or with Et_2NH. Due to the stronger basicity of Et_2NH as compared to terminal acetylenes, a protonolysis of complex **B** is faster with an additional equivalent of amine, leading to the formation of a stronger M–N bond. Hence, the protonolysis step with Et_2NH yields the mono insertion product **D** under regeneration of the pre-catalyst $[(Et_2N)_3U]$ $[BPh_4]$ (**11**). In addition, the rapid protonolysis by Et_2NH prevents a further insertion of isonitrile into intermediate **B**, avoiding the formation of double insertion products [70].

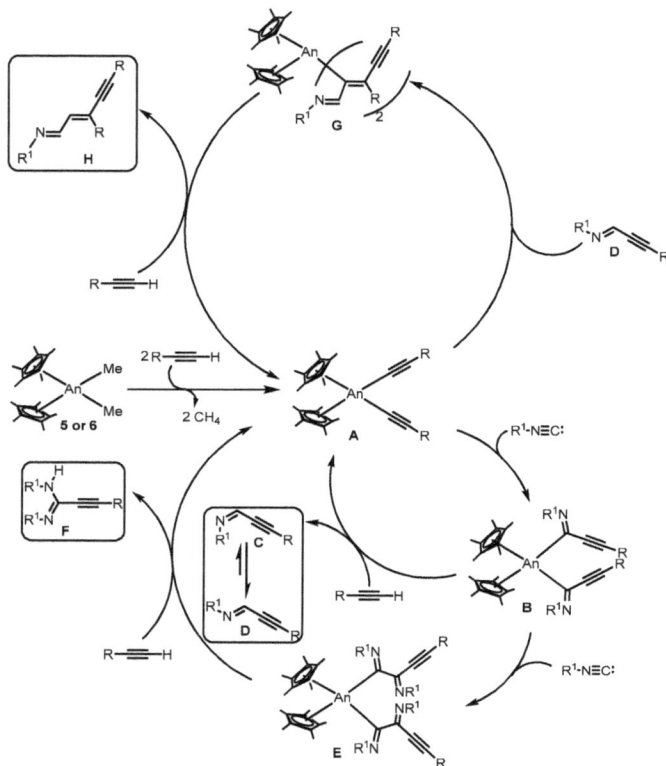

Scheme 14. Proposed mechanisms for the coupling of terminal acetylenes with isonitriles mediated by the organoactinide complexes Cp*$_2$AnMe$_2$ **5** and **6**. Adapted with permission from Barnea, E.; Andrea, T.; Berthet, J.-C.; Ephritikhine, M.; Eisen, M.S. *Organometallics* **2008**, *27*, 3103–3112 [70]. Copyright 2004 American Chemical Society.

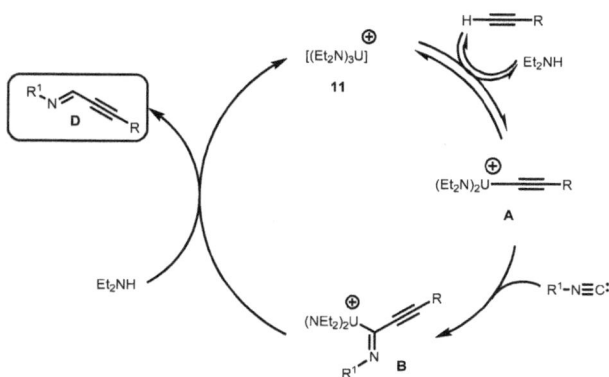

Scheme 15. Proposed mechanism for the coupling of isonitriles and terminal acetylenes mediated by the cationic uranium complex [(Et$_2$N)$_3$U] [BPh$_4$] (**11**). Adapted with permission from Barnea, E.; Andrea, T.; Berthet, J.-C.; Ephritikhine, M.; Eisen, M.S. *Organometallics* **2008**, *27*, 3103–3112 [70]. Copyright 2004 American Chemical Society.

Inorganics **2015**, *3*, 1–33

2.2.3. Coupling of Aldehydes: Tishchenko Reaction

Despite the wide application organoactinides have found in various catalytic processes (*vide supra*), transformations involving oxygen containing molecules have been usually been excluded. This can be mainly attributed to the high oxophilicity of the early actinide elements thorium and uranium, which results in a decreased catalytic activity due to the formation of thermodynamically stable An=O (An = actinides) species, which are catalytically inactive [71]. Marks *et al.*, showed, that the replacement of one of the methyl ligands in Cp*$_2$ThMe$_2$ by an alkoxo ligand (OR) leads to decrease in the activity by a factor of 4000 in the hydrogenolysis reaction [71]. Eisen *et al.*, broke the myth of catalytically inactive actinide oxo species by developing a methodology based on a nearly thermoneutral reaction, which comprises the reaction of an An–O species with an oxygen containing molecule. Hence, a new An–O bond with similar bond energy is formed, and the reaction is governed by entropy. This methodology has been applied in the catalytic Tishchenko reaction, which comprises the dimerization of two equivalents of aldehyde to furnish the respective ester [72,73]. Thus, the Tishchenko reaction presents an atom economic route for the preparation of esters. Eisen and co-workers have used the thorium catalysts Cp*$_2$ThMe$_2$ (**6**), Th(NMeEt)$_4$ (**13**), and the *ansa*-bridged complex Me$_2$Si(C$_5$Me$_4$)Th(n-Bu)$_2$ (**4**) as pre-catalyst for the dimerization of a variety of aromatic aldehydes. The catalytic activity was shown to increase in the order Cp*$_2$ThMe$_2$ (**6**) < Th(NMeEt)$_4$ (**13**) < Me$_2$Si(C$_5$Me$_4$)Th(n-Bu)$_2$ (**4**), corroborating an increased catalytic activity with a more open coordination sphere of the respective actinide complex [73]. Additionally, the nature of the substituents on the aromatic ring of the respective aldehyde have been shown to influence the reaction rate, displaying higher rates for electron withdrawing substituents on the aromatic ring. While the Tishchenko reaction with various aromatic aldehydes yielded the respective esters in moderate to high yields, the use of two different aldehydes to give the asymmetrically substituted ester did not prove successful. The mechanism for the thorium mediated dimerization of aldehydes is presented in Scheme 16. In the first step two equivalents of aldehyde insert into Th–C bond of complex **6** forming intermediate **A**, followed by an insertion of two further equivalents of aldehyde into the Th–O bond of **A**, furnishing the thorium-alkoxo species **B**. A hydride transfer over a six membered transition state (**C**), leads to the formation of two equivalents of the α-substituted ester **D**, and the catalytically active thorium alkoxo complex **E**. A subsequent insertion of two equivalents of aldehydes into the Th–O bond of **E** (only one active site of the catalyst is shown for clarity), furnishes the thorium alkoxo intermediate **F**, which can undergo a hydride transfer with an incoming aldehyde, furnishing the ester **H** under regeneration of the active catalyst **E** [72,73].

In addition, Eisen *et al.*, have applied the mixed bis(pentamethyl(cyclopentadienyl)) thorium mono(imidazolin-2-iminato) complex Cp*$_2$Th(Im^{Dipp}N)(Me) (**14**) (Figure 6) (Dipp = di-*iso*-propylphenyl) as pre-catalyst in the dimerization of aromatic, heteroaromatic, cyclic and branched aliphatic aldehydes [74]. Owing to the ability of the imidazolin-2-iminato ligand to donate further electron density to the thorium center, the oxophilicity of complex **14** is slightly decreased, leading to an enhanced catalytic activity with oxygen containing molecules. While Cp*$_2$ThMe$_2$ (**6**) exhibits two catalytically active methyl groups, Cp*$_2$Th(Im^{Dipp}N)(Me) (**14**) has only one labile ligand, allowing for an evaluation of the catalytic activity per active site, which was higher for the mixed ligand thorium complex Cp*$_2$Th(Im^{Dipp}N)(Me) (**14**) than for the structurally similar Cp*$_2$ThMe$_2$ (**6**) [74].

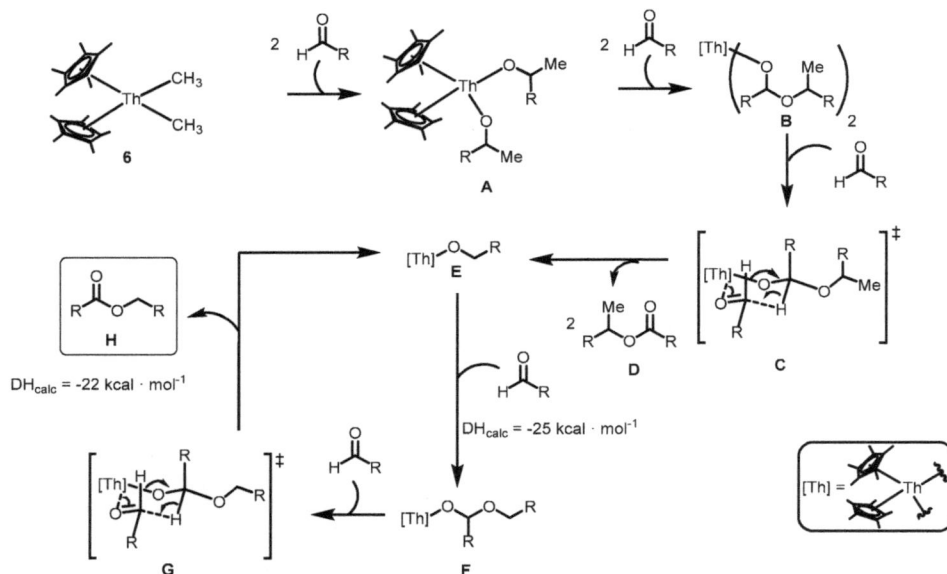

Scheme 16. Proposed mechanism for the catalytic dimerization of aldehydes mediated by Cp*$_2$ThMe$_2$ (**6**). Adapted with permission from Andrea, T.; Barnea, B.; Eisen, M.S. *J. Am. Chem. Soc.* **2008**, *130*, 2454–2455 [72]. Copyright 2008 American Chemical Society.

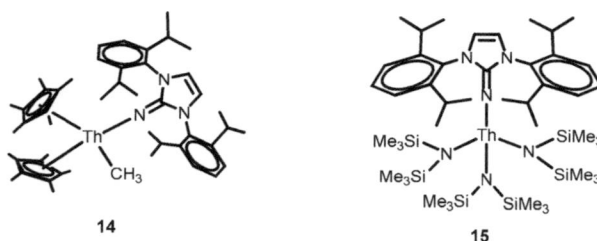

Figure 6. Imidazolin-2-iminato thorium complexes **14** and **15** applied as pre-catalysts for the Tishchenko reaction [74,75].

The Tishchenko reaction was rendered selective toward the formation of asymmetrically substituted esters (Scheme 17) by applying the mono(imidazolin-2-iminato) thorium complex (Im^{DiPP}N)Th(N(SiMe$_3$)$_2$)$_3$ (**15**) and an excess of aromatic aldehyde (RCHO/ArCHO 50/200) [75]. While the mechanism using complex **15**, follows the cycle described in Scheme 16, the selectivity toward the formation of the asymmetrically substituted ester could be ascribed to the large difference between the reactions rates for the dimerization of aliphatic aldehydes, as compared to their aromatic counterparts. Hence, the thorium alkoxo species **A** (Scheme 17) will react preferentially with an aliphatic aldehyde, forming a thorium alkoxo intermediate (**B**), which will transfer a hydride to an incoming aromatic aldehyde over the six membered transition state **C**, since aromatic aldehydes are better hydride acceptors than their aliphatic counterparts. Thus, the asymmetrically substitutes ester **D** is obtained, under regeneration of the catalytically active thorium alkoxo species **A** (Scheme 17) [75].

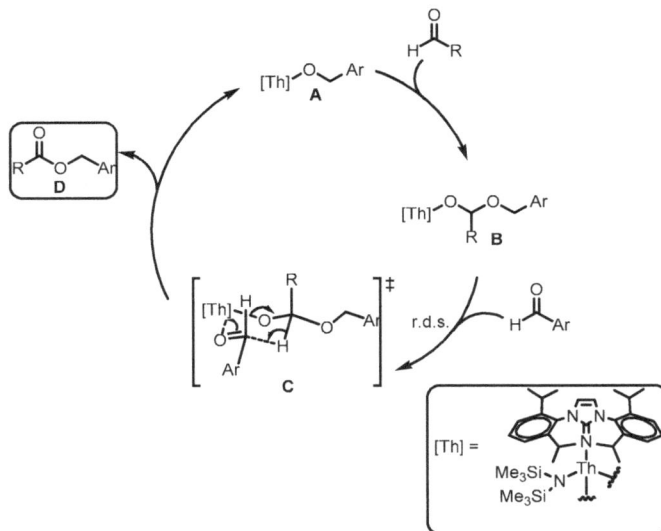

Scheme 17. Mechanism for the crossed Tishchenko reaction mediated by $(Im^{DiPP}N)Th(N(SiMe_3)_2)_3$ (**15**). Adapted with permission from Karmel, I.S.R.; Fridman, N.; Tamm, M.; Eisen, M.S. *J. Am. Chem. Soc.* **2014**, *136*, 17180–17192 [75]. Copyright 2014 American Chemical Society.

2.2.4. Dehydrocoupling of Amines with Silanes

Dehydrocoupling reactions of amines with silanes to form silazanes have gained the interest of the scientific community, since silazanes are important staring materials for the production of silicon nitride materials [76]. Eisen *et al.*, have investigated the use of the cationic uranium amide $[(Et_2N)_3U]$ $[BPh_4]$ (**11**) as pre-catalyst for the catalytic coupling of primary and secondary amines with $PhSiH_3$, furnishing aminosilanes with the general formula $PhSiH_{3-n}(NHR)_n$ (n = 1–3) [77]. The yield of the reaction and product distribution depends on the experimental conditions, as well as on the amine, showing higher catalytic activities for the coupling of $PhSiH_3$ with primary amines than with secondary amines. In addition, the steric hindrance of the R substituent of the respective amine exhibited an influence on the product obtained. While small substituents, such as *n*-propyl furnished a mixture of $PhSiH(NH^nPr)_2$ and $PhSi(NH^nPr)_3$, the sterically demanding *tert*-butyl amine, afforded exclusively $PhSiH_2NH^tBu$ as product. A plausible mechanism for this reaction is presented in Scheme 18. In the first step, the catalytically active species **A** is formed by a transamination reaction of $[(Et_2N)_3U]$ $[BPh_4]$ (**11**) with an excess of incoming amine RNH_2. Subsequent reaction of **A** with $PhSiH_3$ furnishes the uranium hydride complex **C** and one equivalent of $PhSiH_2NHR$ (**B**). The uranium hydride intermediate **C** reacts with one equivalent of incoming amine to give one equivalent of H_2, under regeneration of the active catalyst **A** (Scheme 18). Different polyaminosilanes of the formula $PhSiH_{3-n}(NHR)_n$ are obtained, when complex **A** reacts with $PhSiH_{4-n}(NHR)_{n-1}$, instead of $PhSiH_3$ (Scheme 18) [77].

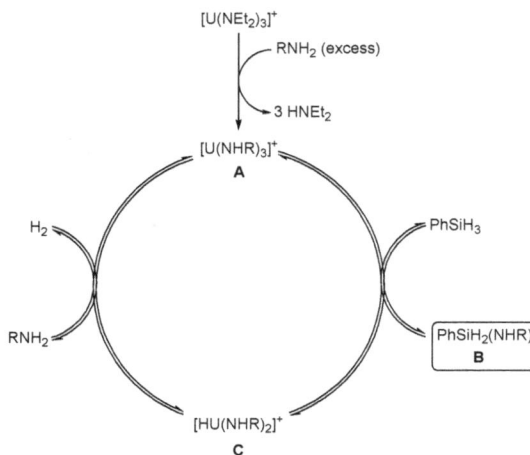

Scheme 18. Mechanism for the dehydrocoupling of amines and silanes, mediated by complex **11**.

2.3. Polymerization Reactions

2.3.1. Polymerization of α-Olefins

The polymerization of α-olefins has been initially studied with the cationic actinide complexes of the type [Cp*$_2$ThMe] [BPh$_4$], and [Cp*$_2$ThMe] [B(C$_6$F$_5$)$_4$], in which a dependence of the catalytic activity on the nature of the counter-anion was observed [78,79]. Hence, a large variety of counter-anions with different steric and electronic properties has been synthesized, in order to find out the optimal polymerization conditions using organoactinide complexes. Nevertheless, the catalytic activity of cyclopentadienyl based organoactinide complexes in the polymerization of α-olefins remained several orders of magnitude lower, as compared to isostructural group IV metal complexes [80]. An improvement was recently achieved with the use diamido ether uranium (IV) dialkyl complexes **8**, **16**, and **17** (Figure 7), for which catalytic activities up to 560 g·mol^{-1}·h^{-1}·atm^{-1} were obtained in the polymerization of ethylene under ambient conditions [81]. Surprisingly, the addition of co-catalyst, such as MMAO (modified methylalumoxane), Et$_2$AlCl, or B(C$_6$F$_5$)$_3$ led to either a reduced catalytic activity, or to a deactivation of the catalyst.

In addition, the actinide bis(amidinate) complexes **18** and **19** have been explored as pre-catalysts for the polymerization of ethylene with different co-catalysts, such MAO (methylalumoxane), B(C$_6$F$_5$)$_3$ with small amounts of MAO, and TIBA (triisobutyl aluminium), displaying the highest catalytic for the production of high density polyethylene with TIBA as co-catalyst [82]. A series of mechanistic studies, and radical trapping experiments led to the proposed mechanism shown in Scheme 19. In the first step, the alkylation of pre-catalyst **18** with TIBA furnishes the bis(alkyl) compound **A**. A subsequent reduction of **A** to the uranium (III) intermediate **B**, leads to the formation of an isobutyl radical, which was confirmed by trapping experiments with C$_{60}$. Intermediate **B** is oxidized back to the U(IV) complex **C** by ethylene. Complex **C** will undergo further radical insertions of ethylene to give compound **D**. A subsequent β-H-elimination of the oligomeric chain radical, furnishes the metal hydride species **E**. The formation of the radical chain was confirmed by C$_{60}$ trapping experiments. The reaction of complex **F** with an additional equivalent of TIBA generates the catalytically active cationic complex **G** [82].

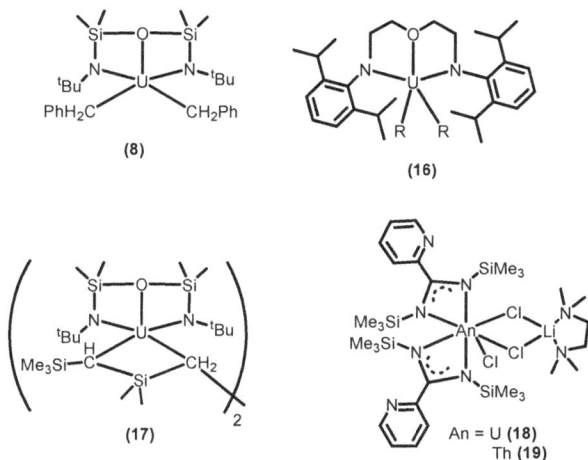

Figure 7. Diamino ether uranium (IV) complexes and bis(amidinate) actinide complexes used as pre-catalysts in the polymerization of ethylene [81,82].

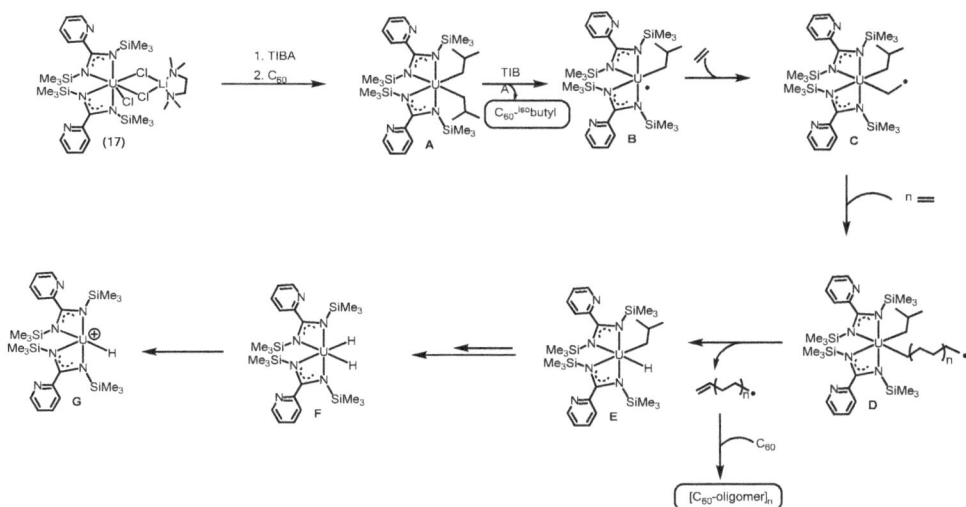

Scheme 19. Proposed mechanism for the polymerization of ethylene mediated by the bis(amidinate) uranium complex **18**. Adapted with permission from Domeshek, E.; Batrice, R.J.; Aharonovich, S.; Tumanskii, B.; Botoshansky, M.; Eisen, M.S. *Dalton Trans.* **2013**, *42*, 9069–9078 [82]. Copyright 2013 Royal Society of Chemistry.

2.3.2. Ring Opening Polymerization of Cyclic Esters

Biodegradable polymers such as poly(ε-caprolactone) and polylactide have received growing attention in the course of the past two decades, which can be attributed to the availability of the respective monomers, as well as to the various fields of applications, such environmentally friendly packaging materials [83], microelectronics [84] and adhesives [85]. While main group, transition metal, and lanthanide complexes have been thoroughly studied as pre-catalysts in the ROP (Ring Opening Polymerization) of cyclic esters [86], actinide compounds have only been investigated in the course of the last decade. Ephritikhine *et al.*, carried out a comparative study using a series of tris(amidinate) lanthanide (III) complexes and the isostructural uranium complex for the ROP of ε-caprolactone [87]. While the lanthanide complexes displayed a very high catalytic activity under mild reaction conditions, the uranium analogue was almost completely inactive, which was attributed to the higher oxophilicity of the uranium center [87]. The ROP of ε-caprolactone and L-lactide was rendered catalytic by Eisen *et al.*, using the actinide metallocenes Cp*$_2$AnMe$_2$ (An = U (**5**), Th (**6**)), as well as the cationic uranium complex [(Et$_2$N)$_3$U] [BPh$_4$] (**11**) [88]. The highest catalytic activity was displayed by the cationic complex **11**, due to the more open coordination sphere as compared to **5** and **6**. In addition, Cp*$_2$ThMe$_2$ (**6**) was found to be more active than Cp*$_2$UMe$_2$ (**5**), which was attributed to the higher oxophilicity of uranium as compared to thorium. The polymers obtained displayed high molecular weights and narrow molecular weight distributions, indicative of a living polymerization process. The coordination-insertion mechanism for the ROP of ε-caprolactone mediated by the organoactinides Cp*$_2$AnMe$_2$ is presented in Scheme 20. The first step of the mechanism includes the abstraction of the α-hydrogen of the monomer, furnishing the actinide enolate **B** and two equivalents of methane. Subsequent insertion of further ε-caprolactone monomers into the An–O bond of **B** gives the open chain intermediate **C**, which upon addition of methanol gives the respective polymer **D** [88]. Similarly, the mixed ligand thorium metallocene Cp*$_2$Th(Im^{DiPP}N)(Me) (**14**) also operates via a coordination insertion mechanism, with one active methyl group per catalyst molecule [74].

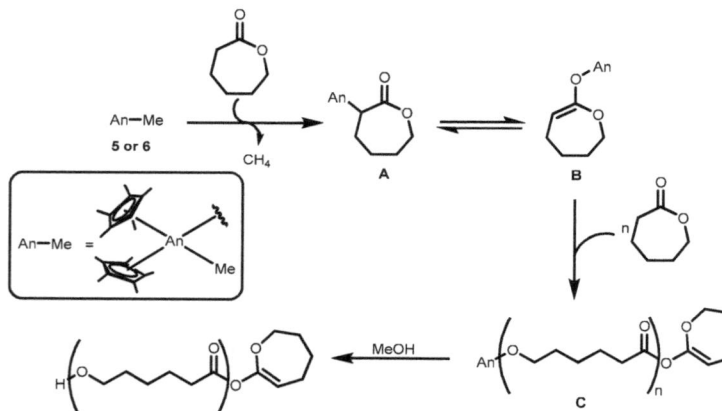

Scheme 20. Proposed mechanism for the ROP (Ring Opening Polymerization) of ε-caprolactone, mediated by the organoactinide complexes Cp*$_2$AnMe$_2$ (**5** and **6**). Adapted with permission from Barnea, E.; Moradove, D.; Berthet, J.-C.; Ephritikhine, M.; Eisen, M.S. *Organometallics* **2006**, *25*, 320–322 [88]. Copyright 2006 American Chemical Society.

In addition, several bis(amidinate) actinide compounds have been investigated as pre-catalysts in the ROP of ε-caprolactone. The thorium bis(pyridylamidinate) **19** has been found to promote the cyclooligomerization of ε-caprolactone, in which complex **19** reacts via the amidine moiety (Scheme 21) [89]. The oligomerization reaction mediated by complex **19** furnished two different oligomer fractions: the first fraction contains cyclic pentamers, whereas the second fraction is composed of cyclic undecamers to tridecamers, indicating that two active sites are operative, while the respective fractions are obtained by a similar mechanism. The mechanism for the cyclooligomerization of ε-caprolactone mediated by complex **19** is depicted in Scheme 21. The first step of the mechanism includes the coordination of ε-caprolactone to the thorium center, followed by a displacement of μ-chloro ligand by a molecule of ε-caprolactone, furnishing intermediate **A**. A subsequent nucleophilic attack of an amidinate ligand on the carbonyl moiety of the coordinate ε-caprolactone over a six-membered transition state (**B** and **C**) yields the thorium compound **D**. Intermediate **D** is in an equilibrium with complex **E**, which is further stabilized by an additional coordination of the pyridyl moiety to the thorium center. Insertion of further ε-caprolactone yield complex **F** and additional monomer insertions into the Th–O bond of **F**, gives complex **I** after re-coordination of the cyclic oligomer **G**. Intermediate **I** can react with ε-caprolactone, regenerating the catalytically active species **A**. However, if complex **F** reacts via the second amidinate ligand, then both amidinate moieties are independently active, furnishing the thorium intermediate **H**. Complex **H** can either eliminate the cyclic oligomer **G** under regeneration of complex **F**, or eliminate both cyclic oligomers from both active cites, generating complex **I** [89].

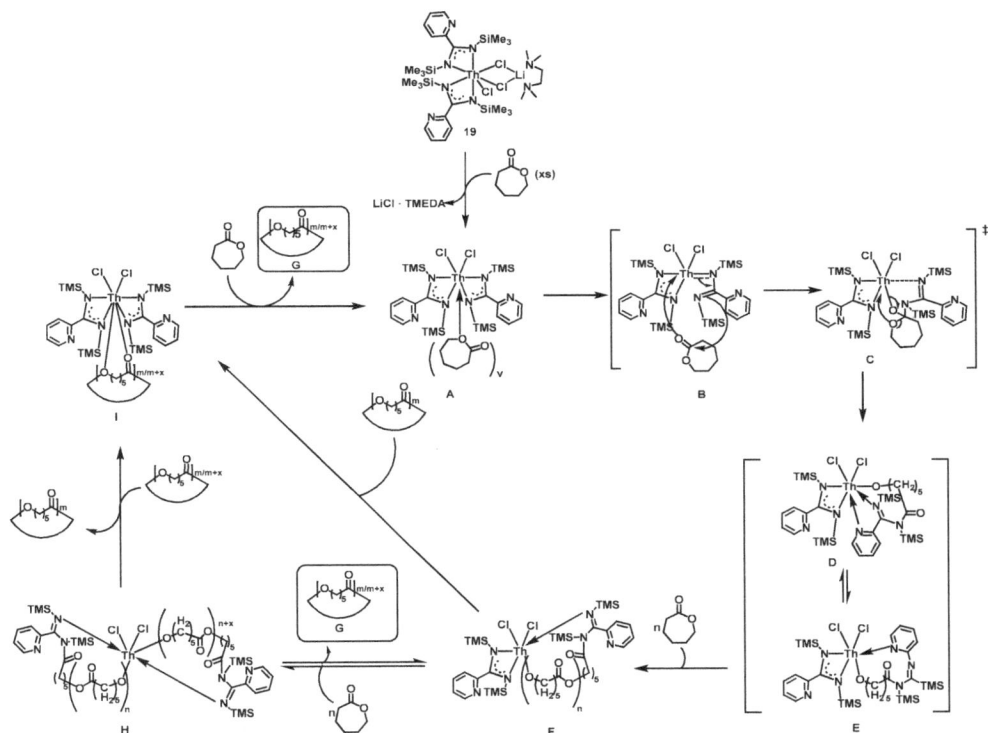

Scheme 21. Mechanism for the thorium mediated cyclooligomerization of ε-caprolactone. Adapted with permission from Rabinovich, E.; Aharonovich, S.; Botoshansky, M.; Eisen, M.S. *Dalton Trans.* **2010**, *39*, 6667–6676 [89]. Copyright 2010 Royal Society of Chemistry.

The actinide bis(amidinate) systems (Figure 8) with functionalized side arms on the amidinate ligand have shown to promote the ROP of ε-caprolactone, furnishing linear polymers with narrow molecular weight distributions [90,91]. A systematic study of a series of bis(amidinate) thorium and uranium complex (**21–25**) with a dimethylamine side arms has shown an increase in the catalytic activity in the order **23** > **24** > **25** > **21** > **22** [91]. Hence, the thorium complexes **23–25** are more active than their uranium analogues **21–22**, and complexes with a phenyl at the *ipso*-position (**21, 23–24**) are more active than the analogous compounds with a pyridyl moiety at the *ipso*-position (**22** and **25**). Furthermore, complexes **21** and **23** with a shorter carbon linker chain (*n* = 2) display a higher catalytic activity than their analogues **23, 24–25** with a longer linker chain (*n* = 3), suggesting an increase of the catalytic activity with increasing electron density of the metal center.

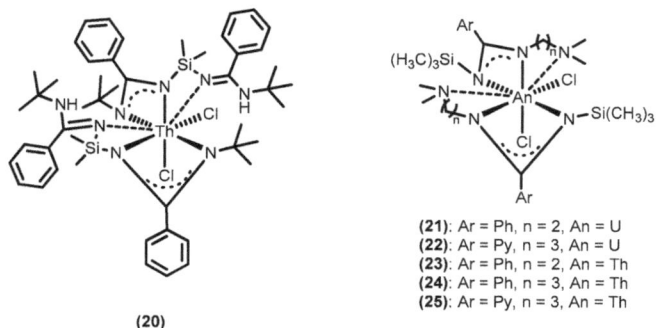

(20)

(21): Ar = Ph, n = 2, An = U
(22): Ar = Py, n = 3, An = U
(23): Ar = Ph, n = 2, An = Th
(24): Ar = Ph, n = 3, An = Th
(25): Ar = Py, n = 3, An = Th

Figure 8. Actinide bis(amidinate) complexes **20–25** applied in the ROP of ε-caprolactone [90,91].

The mechanism for the ROP of ε-caprolactone mediated by the actinide bis(amidinate) complexes **20–25** is presented in Scheme 22. The respective actinide complex acts as a Lewis acid; hence none of the ligands is displaced by an ε-caprolactone monomer. Instead the monomer is activated by the Lewis acidic actinide center, generating the actinide alkoxo-caprolate intermediate **A**. A subsequent nucleophilic attack of an incoming monomer on complex **A**, gives the open chain intermediate **B**, which is the rate determining step of the reaction. Further insertion of ε-caprolactone units into the An–O bond of **B**, furnishes compound **C**, which after the elimination of the polymer (**D**) regenerates the active catalyst **A** [90,91]. Also actinide inclusion complexes have been recently shown to operate via the mechanism depicted in Scheme 22 [92].

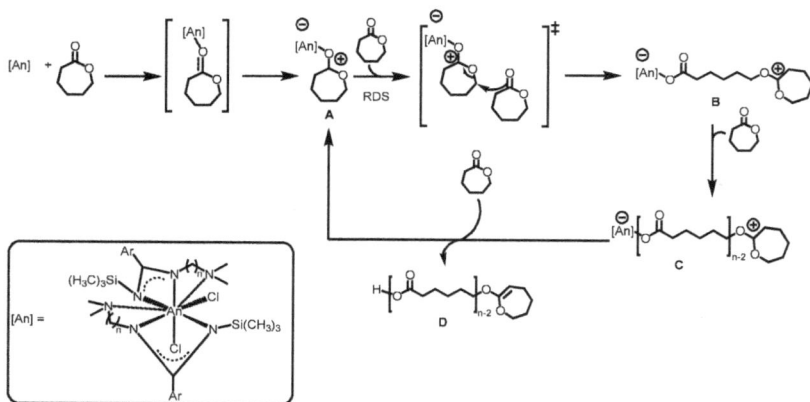

Scheme 22. Mechanism for the Lewis acid catalyzed ROP of ε-caprolactone. Adapted with permission from Karmel, I.S.R.; Elkin, T.; Fridman, N.; Eisen, M.S. *Dalton Trans.* **2014**, *43*, 11376–11387 [90]. Copyright 2014 Royal Society of Chemistry.

Recently, Baker *et al.*, have studied the use of the uranyl complex **26** as pre-catalyst for the cyclooligomerization of various lactones, and the proposed mechanism was further studied using computational methods [93]. While complex **26** mediated the ROP oligomerization of ε-caprolactone and δ-valerolactone to give cyclic oligomers, no catalytic activity was achieved using lactones with smaller ring sizes, such as β-butyrolactone, or γ-butyrolactone. The lack of activity with β-butyrolactone, and γ-butyrolactone was attributed to the endothermic enthalpy of displacement of a coordinated THF molecule by an incoming monomer. The bimetallic mechanism for the cyclooligomerization of ε-caprolactone mediated by **26** is depicted in Scheme 23. The first step of the mechanism includes the displacement of a coordinated THF molecule by an incoming monomer furnishing intermediate **A**. This step has been calculated to be almost thermoneutral for δ-valerolactone, and ε-caprolactone. The subsequent insertion of the monomer into the U–O bond furnishes the tetrahedral intermediate **B**, which after an intramolecular initiation gives the open chain intermediate **C**. The propagation of the polymerization reaction takes place by the addition of further monomer units, as well as a second catalyst molecule. Thus, two equivalents of catalyst **26** are necessary for the intermolecular propagation, rendering the second part of the mechanism bimetallic (Scheme 23).

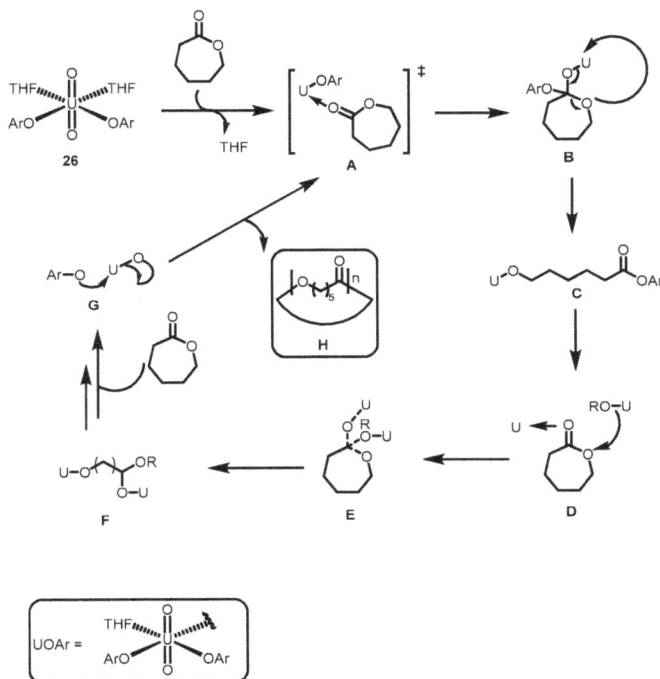

Scheme 23. Mechanism for the cyclooligomerization of ε-caprolactone mediated by the uranyl complex **26**. Adapted with permission from Walshe, A.; Fang, J.; Maron, L.; Baker, R.J. *Inorg. Chem.* **2013**, *52*, 9077–9086 [93]. Copyright 2013 American Chemical Society.

The ROP of *rac*-lactide has been studied with the thorium metallocene complexes **27–29** (Figure 9), furnishing atactic polymers with narrow molecular weight distributions under mild reaction conditions [94,95]. Also the actinide diamido ether compounds have been applied as pre-catalysts for the ROP of *rac*-lactide and L-lactide [96]. The catalytic activity showed a dependence on the actinide center, as well as on the ligand backbone.

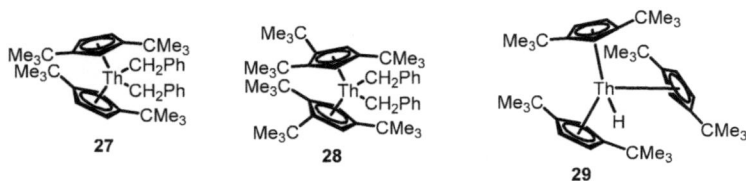

Figure 9. Thorium metallocenes used as pre-catalysts for the ROP of rac-lactide [94,95].

2.4. Further Actinide Mediated Catalytic Transformations

2.4.1. Reduction of Azides and Hydrazines

The catalytic reduction of organic azides to amines requires the transfer of two electrons, which has been achieved by using the high valent U(VI) bis(amido) complex **30** [97]. The catalytic cycle (Scheme 24) consists in the reduction of the U(VI) complex **30** with molecular hydrogen, to furnish the uranium bis(amido) intermediate **A**. Subsequent oxidation of **A** by an incoming organic azide, regenerates the active catalyst **30** under concomitant formation of the primary amine product **B**. Additionally, Cp*$_2$U(=NAd)$_2$ (**30**) has been applied in the catalytic reduction of *N*,*N'*-diphenyl hydrazine, generating aniline and azobenzene. Interestingly, this reaction proceeds without the addition of hydrogen, indicating that *N*,*N'*-diphenyl hydrazine acts as reducing and oxidizing agent [97].

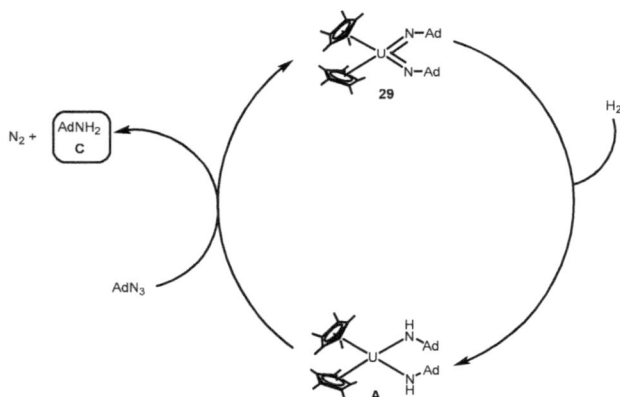

Scheme 24. Catalytic reduction of azides mediated by complex **30**. Ad: Adamantyl. Adapted with permission from Peters, R.G.; Warner, B.P.; Burns, C.J. *J. Am. Chem. Soc.* **1999**, *121*, 5585–5586 [97]. Copyright 1999 American Chemical Society.

2.4.2. ROP of Epoxides

The ring opening polymerization of propylene oxide and cyclohexene oxide using the uranyl compounds [UO$_2$(OAr)$_2$(THF)$_2$] (**26**), [UO$_2$Cl$_2$(THF)$_3$] (**31**), and [UO$_2$Cl$_2$(THF)$_2$]$_2$ (**32**) as pre-catalysts has recently been investigated by Baker *et al.* [98,99]. When using the alkoxo complex **26**, the reaction is expected to be thermodynamically favored, since only U–O bonds are broken, while new U–O bonds with similar bond energies are formed, leading to entropically controlled polymerization. The polymers obtained using catalysts **26** and **31**–**32** display narrow molecular weight distributions; however the polymerization process doesn't proceed via a living polymerization mechanism. The proposed bimetallic polymerization mechanism is presented in Scheme 25. The first step of the mechanism is the displacement of a coordinated THF molecule by a molecule of propylene oxide, furnishing intermediate **B**. A subsequent nucleophilic attack on the coordinated propylene oxide monomer represents the rate determining step of the reaction mechanism, giving rise to the open chain intermediate **C**. Further intermolecular nucleophilic attacks on the coordinated monomer via a chain shuttling mechanism generate intermediate **D**. The polymer **E** is obtained after the addition of methanol (Scheme 25) [98,99].

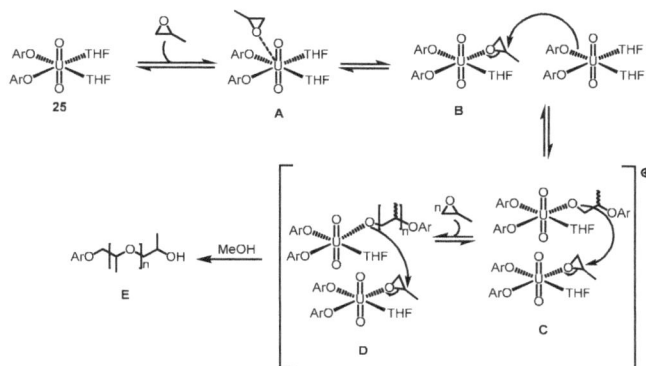

Scheme 25. Proposed mechanism for the ROP of epoxides. Adapted with permission from Baker, R.J.; Walshe, A. *Chem. Commun.* **2012**, *48*, 985–987 [98]. Copyright 2012 Royal Society of Chemistry.

2.4.3. Insertion of Nucleophilic E–H (E = N, P, S) Bonds into Heterocumulenes

The insertion of E–H bonds (E = N, P, S) across heterocumulene systems, such as carbodiimides, isocyanates and isothiocyanates represents and atom-economic route for the generation of a variety of organic synthons, such as guanidines and phosphaguanidines. Therefore this process has been widely studied over the past decade using lanthanide catalysts [100–106]. However, only recently Eisen *et al.*, have reported the application of a thorium catalyst for this type of atom-economic catalytic transformation [107]. The mono(imidazolin-2-iminato) thorium (IV) complex [(Im^{DiPP}N)Th(N(SiMe$_3$)$_2$)$_3$] (**15**) was successfully applied as pre-catalyst for addition of aromatic and aliphatic amines, phosphines and thiols to carbodiimides, isocyanates and isothiocyanates, furnishing the respective addition products in moderate to high yields under mild reaction conditions. Additionally, the thorium complex **15** displayed an unusual high tolerance toward functional groups, and heteroatoms, allowing for a wide scope of products to be accessed. The mechanism for the actinide mediated insertion of phosphines into carbodiimides is presented in Scheme 26. The first step of the reaction mechanism comprises the insertion of two equivalents of carbodiimide into the Th–N$_{amido}$ bonds, furnishing the thorium bis(guanidine) intermediate **A**. Subsequent protonolysis by two equivalents of phosphine leads to the formation of the catalytically active thorium species **C** under generation of two equivalents of guanidine **B**. The insertion of two equivalents of carbodiimide into the Th–P bond intermediate **C** gives rise to the thorium bis(phosphaguanidine) compound **D**. The rate determining step of the reaction mechanism is the protonolysis reaction of **D** with two equivalents of phosphine, which furnishes the phosphaguanidine product **E** under regeneration of the active catalyst **C** (Scheme 26) [107].

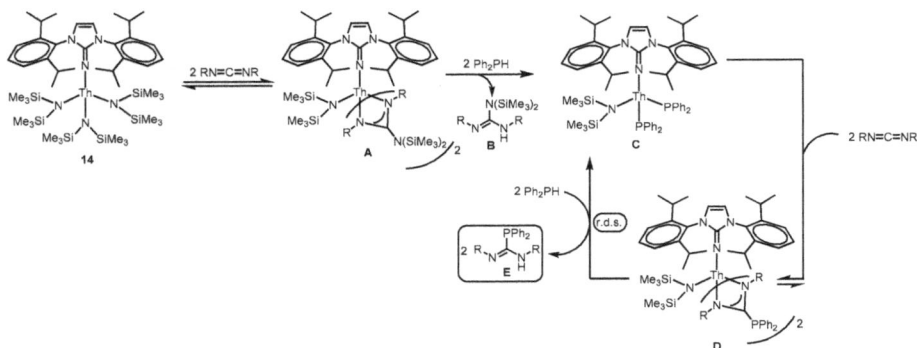

Scheme 26. Proposed mechanism for the actinide mediated insertion of phosphines into carbodiimides [107].

3. Conclusions

The use of organoactinide complexes in catalytic organic transformations has undergone a long journey, since the use of [(η^3-allyl)$_3$UCl] in the homogeneous polymerization of butadiene. The first milestone in organoactinide catalysis was set by the application of bis(pentamethylcyclopentadienyl) actinide complexes for various catalytic organic transformations, giving rise to new reactivities and selectivities. Moreover, the use of actinide catalysts often yields a complementary product distribution as compared to their respective transition metal analogues, allowing for different product classes to be obtained. More recently, the use of novel post-metallocene actinide catalysts has opened a new era for the application of organoactinides and actinide coordination complexes as catalysts in homogeneous organic transformations, allowing for systematic structure–reactivity studies, as well as studies concerning the involvement of f-electrons in the reactivity of the early actinides. Owing to knowledge acquired throughout the past two decades regarding the preparation and reactivity of organoactinide, new catalytic processes involving actinide catalysts can be designed. Furthermore, the unique reactivity patterns, and product distributions observed when using actinide complexes as catalysts, enable the preparation of organic synthons, which are not accessible by traditional transition metal, or organocatalysis. Despite the progress made in the field of actinide catalysis during the last two decades, many new reactivity pathways are yet to be explored using actinide coordination complexes as catalysts.

Acknowledgments: This work was supported by the Israel Science Foundation Administered by the Israel Academy of Science and Humanities under Contract No. 78/14; and by the PAZY Foundation Fund (2015) administered by the Israel Atomic Energy Comission.

Author Contributions: This manuscript was prepared through an equal contribution of the authors Isabell S. R. Karmel, Rami J. Batrice and Moris S. Eisen.

Conflicts of Interest: The authors declare no conflict of interest.

References

1. Streitwieser, A.; Mueller-Westerhoff, U. Bis(cyclooctatetraenyl)uranium (uranocene). A new class of sandwich complexes that utilize atomic *f* orbitals. *J. Am. Chem. Soc.* **1968**, *90*, 7364–7364. [CrossRef]
2. Zalkin, A.; Raymond, K.N. Structure of di-π-cyclooctatetraeneuranium (uranocene). *J. Am. Chem. Soc.* **1969**, *91*, 5667–5668. [CrossRef]
3. Barnea, E.; Eisen, M. Organoactinides in catalysis. *Coord. Chem. Rev.* **2006**, *250*, 855–899. [CrossRef]
4. Andrea, T.; Eisen, M.S. Recent advances in organothorium and organouranium catalysis. *Chem. Soc. Rev.* **2008**, *37*, 550–567. [CrossRef] [PubMed]

5. Fox, A.R.; Bart, S.C.; Meyer, K.; Cummins, C.C. Towards uranium catalysts. *Nature* **2008**, *455*, 341–349. [CrossRef] [PubMed]

6. Sharma, M.; Eisen, M.S. Metallocene organoactinide complexes. In *Structure and Bonding*; Springer Science + Business Media: Berlin, Germany, 2008; pp. 1–85.

7. Eisen, M.S. Catalytic C–N, C–O, and C–S bond formation promoted by organoactinide complexes. In *C–X Bond Formation*; Springer Science + Business Media: Berlin, Germany, 2010; pp. 157–184.

8. Batrice, R.; Karmel, I.; Eisen, M.; Fuerstner, A.; Hall, D.; Marek, I.; Oestreich, M.; Stoltz, B.; Schaumann, E. Product class 13: Organometallic complexes of the actinides. *Sci. Synth. Knowl. Updat.* **2013**, *4*, 99–211.

9. Ephritikhine, M. Recent advances in organoactinide chemistry as exemplified by cyclopentadienyl compounds. *Organometallics* **2013**, *32*, 2464–2488. [CrossRef]

10. Jones, M.B.; Gaunt, A.J. Recent developments in synthesis and structural chemistry of nonaqueous actinide complexes. *Chem. Rev.* **2013**, *113*, 1137–1198. [CrossRef] [PubMed]

11. Edelmann, F.T. Lanthanides and actinides: Annual survey of their organometallic chemistry covering the year 2012. *Coord. Chem. Rev.* **2014**, *261*, 73–155. [CrossRef]

12. Hayes, C.E.; Leznoff, D.B. Actinide coordination and organometallic complexes with multidentate polyamido ligands. *Coord. Chem. Rev.* **2014**, *266–267*, 155–170. [CrossRef]

13. Kaltsoyannis, N.; Kerridge, A. Chemical bonding of lanthanides and actinides. In *Fundamental Aspects of Chemical Bonding*; Wiley-Blackwell: Oxford, UK, 2014; pp. 337–356.

14. Meihaus, K.R.; Long, J.R. Actinide-based single-molecule magnets. *Dalton Trans.* **2015**, *44*, 2517–2528. [CrossRef] [PubMed]

15. Lugli, G.; Mazzei, A.; Poggio, S. High 1,4-*cis*-polybutadiene by uranium catalysts, 1. Tris(π-allyl)uranium halide catalysts. *Makromol. Chem.* **1974**, *175*, 2021–2027. [CrossRef]

16. De Chirico, A.; Lanzani, P.C.; Raggi, E.; Bruzzone, M. High 1,4-*cis*-polybutadiene by uranium catalysts, 2. Bulk and solution crystallization of polymers. *Makromol. Chem.* **1974**, *175*, 2029–2038. [CrossRef]

17. Guiliani, G.P.; Sorta, E.; Bruzzone, M. High 1,4-*cis*-polybutadiene by uranium catalyst. III. Strain induced crystallization and processability. *Angew. Makromol. Chem.* **1976**, *50*, 87–99. [CrossRef]

18. Gargani, L.; Giuliani, G.P.; Mistrali, F.; Bruzzone, M. High 1,4-*cis*-polybutadiene by uranium catalyst. IV. Stretch induced crystallization and ultimate properties. *Angew. Makromol. Chem.* **1976**, *50*, 101–113. [CrossRef]

19. Marks, T.J.; Ernst, R.D. 21—Scandium, yttrium and the lanthanides and actinides. In *Comprehensive Organometallic Chemistry*; Abel, E.W., Stone, E.G.A., Wilkinson, G., Eds.; Pergamon: Oxford, UK, 1982; pp. 173–270.

20. Li, X.-F.; Xu, Y.-T.; Feng, X.-Z.; Sun, P.-N. Steric packing and molecular geometry. I. Simulation on tetrahedral structures of weak covalent bonding. *Inorg. Chim. Acta* **1986**, *116*, 75–83. [CrossRef]

21. Li, X.-F.; Feng, X.-Z.; Xu, Y.-T.; Wang, H.-T.; Shi, J.; Liu, L.; Sun, P.-N. Cone packing model—A geometrical approach to coordination and organometallic chemistry of lanthanides and actinides. *Inorg. Chim. Acta* **1986**, *116*, 85–93. [CrossRef]

22. Li, X.-F.; Guo, A.-L. The nature of seat-ligand fitting in coordination space. V. Steric hindrances and reaction mechanisms—A further discussion on the structure and chemistry of compounds containing three π-bonded cyclopentadienyl groups. *Inorg. Chim. Acta* **1987**, *134*, 143–153. [CrossRef]

23. Marçalo, J.; de Matos, A.P. A new definition of coordination number and its use in lanthanide and actinide coordination and organometallic chemistry. *Polyhedron* **1989**, *8*, 2431–2437. [CrossRef]

24. Marks, T.J.; Gagne, M.R.; Nolan, S.P.; Schock, L.E.; Seyam, A.M.; Stern, D. What can metal-ligand bonding energetics teach us about stoichiometric and catalytic organometallic chemistry? *Pure Appl. Chem.* **1989**, *61*, 1665–1672. [CrossRef]

25. Leal, J.P.; Marques, N.; Pires de Matos, A.; Calhorda, M.J.; Galvao, A.M.; Simoes, J.A.M. Uranium-ligand bond dissociation enthalpies in uranium(IV) polypyrazolylborate complexes. *Organometallics* **1992**, *11*, 1632–1637. [CrossRef]

26. King, W.A.; Marks, T.J.; Anderson, D.M.; Duncalf, D.J.; Cloke, F.G.N. Organo-f-element bonding energetics. Large magnitudes of metal arene bond enthalpies in zero-valent lanthanide sandwich complexes. *J. Am. Chem. Soc.* **1992**, *114*, 9221–9223. [CrossRef]

27. Jemine, X.; Goffart, J.; Berthet, J.-C.; Ephritikhine, M. Absolute uranium? Ligand bond-disruption enthalpies of [U(C_5H_4R)$_3$X] complexes (X = I or H, R = Bu^t or SiMe$_3$). *Dalton Trans.* **1992**, 2439–2440. [CrossRef]

28. Jemine, X.; Goffart, J.; Ephritikhine, M.; Fuger, J. Organo-f-element thermochemistry. Thorium-ligand bond disruption enthalpies in {(CH$_3$)$_3$SiC$_9$H$_6$}$_3$ThX (X = H or D) and in {(CH$_3$)$_3$SiC$_5$H$_4$}$_3$ThH complexes. *J. Organomet. Chem.* **1993**, *448*, 95–98. [CrossRef]

29. Leal, J.P.; Simoes, J.A.M. Uranium-ligand bond-dissociation enthalpies of uranium(IV) poly(pyrazolyl)borate complexes. *Dalton Trans.* **1994**, 2687–2691. [CrossRef]

30. King, W.A.; Marks, T.J. Metal-silicon bonding energetics in organo-group 4 and organo-f-element complexes. Implications for bonding and reactivity. *Inorg. Chim. Acta* **1995**, *229*, 343–354. [CrossRef]

31. Marks, T.J.; Day, V.W. *Fundamental and Technological Aspects of Organo-F-Element Chemistry*; Springer Science + Business Media: Berlin, Germany, 1985.

32. Alonso, F.; Beletskaya, I.P.; Yus, M. Transition-metal-catalyzed addition of heteroatom–hydrogen bonds to alkynes. *Chem. Rev.* **2004**, *104*, 3079–3160. [CrossRef] [PubMed]

33. Hashmi, A.S.K. Gold-catalyzed organic reactions. *Chem. Rev.* **2007**, *107*, 3180–3211. [CrossRef] [PubMed]

34. Wobser, S.D.; Marks, T.J. Organothorium-catalyzed hydroalkoxylation/cyclization of alkynyl alcohols. Scope, mechanism, and ancillary ligand effects. *Organometallics* **2013**, *32*, 2517–2528. [CrossRef]

35. Griesbaum, K. Problems and possibilities of the free-radical addition of thiols to unsaturated compounds. *Angew. Chem. Int. Ed. Eng.* **1970**, *9*, 273–287. [CrossRef]

36. Benati, L.; Montevecchi, P.C.; Spagnolo, P. Free-radical reactions of benzenethiol and diphenyl disulphide with alkynes. Chemical reactivity of intermediate 2-(phenylthio)vinyl radicals. *J. Chem. Soc. Perkin Trans.* **1991**, *1*, 2103–2109. [CrossRef]

37. Kondoh, A.; Takami, K.; Yorimitsu, H.; Oshima, K. Stereoselective hydrothiolation of alkynes catalyzed by cesium base: Facile access to (Z)-1-alkenyl sulfides. *J. Org. Chem.* **2005**, *70*, 6468–6473. [CrossRef] [PubMed]

38. Kondo, T.; Mitsudo, T.-A. Metal-catalyzed carbon–sulfur bond formation. *Chem. Rev.* **2000**, *100*, 3205–3220. [CrossRef] [PubMed]

39. Cao, C.; Fraser, L.R.; Love, J.A. Rhodium-catalyzed alkyne hydrothiolation with aromatic and aliphatic thiols. *J. Am. Chem. Soc.* **2005**, *127*, 17614–17615. [CrossRef] [PubMed]

40. Malyshev, D.A.; Scott, N.M.; Marion, N.; Stevens, E.D.; Ananikov, V.P.; Beletskaya, I.P.; Nolan, S.P. Homogeneous nickel catalysts for the selective transfer of a single arylthio group in the catalytic hydrothiolation of alkynes. *Organometallics* **2006**, *25*, 4462–4470. [CrossRef]

41. Delp, S.A.; Munro-Leighton, C.; Goj, L.A.; Ramírez, M.A.; Gunnoe, T.B.; Petersen, J.L.; Boyle, P.D. Addition of S–H bonds across electron-deficient olefins catalyzed by well-defined copper(I) thiolate complexes. *Inorg. Chem.* **2007**, *46*, 2365–2367. [CrossRef] [PubMed]

42. Shoai, S.; Bichler, P.; Kang, B.; Buckley, H.; Love, J.A. Catalytic alkyne hydrothiolation with alkanethiols using wilkinson's catalyst. *Organometallics* **2007**, *26*, 5778–5781. [CrossRef]

43. Kondoh, A.; Yorimitsu, H.; Oshima, K. Palladium-catalyzed anti-hydrothiolation of 1-alkynylphosphines. *Org. Lett.* **2007**, *9*, 1383–1385. [CrossRef] [PubMed]

44. Fraser, L.R.; Bird, J.; Wu, Q.; Cao, C.; Patrick, B.O.; Love, J.A. Synthesis, structure, and hydrothiolation activity of rhodium pyrazolylborate complexes. *Organometallics* **2007**, *26*, 5602–5611. [CrossRef]

45. Weiss, C.J.; Wobser, S.D.; Marks, T.J. Organoactinide-mediated hydrothiolation of terminal alkynes with aliphatic, aromatic, and benzylic thiols. *J. Am. Chem. Soc.* **2009**, *131*, 2062–2063. [CrossRef] [PubMed]

46. Weiss, C.J.; Wobser, S.D.; Marks, T.J. Lanthanide- and actinide-mediated terminal alkyne hydrothiolation for the catalytic synthesis of markovnikov vinyl sulfides. *Organometallics* **2010**, *29*, 6308–6320. [CrossRef]

47. Weiss, C.J.; Marks, T.J. Organo-f-element catalysts for efficient and highly selective hydroalkoxylation and hydrothiolation. *Dalton Trans.* **2010**, *39*, 6576–6588. [CrossRef] [PubMed]

48. Haskel, A.; Straub, T.; Eisen, M.S. Organoactinide-catalyzed intermolecular hydroamination of terminal alkynes. *Organometallics* **1996**, *15*, 3773–3775. [CrossRef]

49. Straub, T.; Haskel, A.; Neyroud, T.G.; Kapon, M.; Botoshansky, M.; Eisen, M.S. Intermolecular hydroamination of terminal alkynes catalyzed by organoactinide complexes. Scope and mechanistic studies. *Organometallics* **2001**, *20*, 5017–5035. [CrossRef]

50. Stubbert, B.D.; Stern, C.L.; Marks, T.J. Synthesis and catalytic characteristics of novel constrained-geometry organoactinide catalysts. The first example of actinide-mediated intramolecular hydroamination. *Organometallics* **2003**, *22*, 4836–4838. [CrossRef]

51. Stubbert, B.D.; Marks, T.J. Mechanistic investigation of intramolecular aminoalkene and aminoalkyne hydroamination/cyclization catalyzed by highly electrophilic, tetravalent constrained geometry 4d and 5f complexes. Evidence for an M–N σ-bonded insertive pathway. *J. Am. Chem. Soc.* **2007**, *129*, 6149–6167. [CrossRef] [PubMed]

52. Broderick, E.M.; Gutzwiller, N.P.; Diaconescu, P.L. Inter- and intramolecular hydroamination with a uranium dialkyl precursor. *Organometallics* **2010**, *29*, 3242–3251. [CrossRef]

53. Haskel, A.; Straub, T.; Dash, A.K.; Eisen, M.S. Oligomerization and cross-oligomerization of terminal alkynes catalyzed by organoactinide complexes. *J. Am. Chem. Soc.* **1999**, *121*, 3014–3024. [CrossRef]

54. Haskel, A.; Wang, J.Q.; Straub, T.; Neyroud, T.G.; Eisen, M.S. Controlling the catalytic oligomerization of terminal alkynes promoted by organoactinides: A strategy to short oligomers. *J. Am. Chem. Soc.* **1999**, *121*, 3025–3034. [CrossRef]

55. Hayes, C.E.; Platel, R.H.; Schafer, L.L.; Leznoff, D.B. Diamido-ether actinide complexes as catalysts for the intramolecular hydroamination of aminoalkenes. *Organometallics* **2012**, *31*, 6732–6740. [CrossRef]

56. Fengyu, B.; Kanno, K.-I.; Takahashi, T. Early transition metal catalyzed hydrosilation reaction. *Trends Org. Chem.* **2008**, *12*, 1–17.

57. Rooke, D.A.; Menard, Z.A.; Ferreira, E.M. An analysis of the influences dictating regioselectivity in platinum-catalyzed hydrosilylations of internal alkynes. *Tetrahedron* **2014**, *70*, 4232–4244. [CrossRef]

58. Iglesias, M.; Fernández-Alvarez, F.J.; Oro, L.A. Outer-sphere ionic hydrosilylation catalysis. *ChemCatChem* **2014**, *6*, 2486–2489. [CrossRef]

59. Greenhalgh, M.D.; Jones, A.S.; Thomas, S.P. Iron-catalysed hydrofunctionalisation of alkenes and alkynes. *ChemCatChem* **2014**, *7*, 190–222. [CrossRef]

60. Dash, A.K.; Wang, J.Q.; Eisen, M.S. Catalytic hydrosilylation of terminal alkynes promoted by organoactinides. *Organometallics* **1999**, *18*, 4724–4741. [CrossRef]

61. Dash, A.K.; Gourevich, I.; Wang, J.Q.; Wang, J.; Kapon, M.; Eisen, M.S. The catalytic effect in opening an organoactinide metal coordination sphere: Regioselective dimerization of terminal alkynes and hydrosilylation of alkynes and alkenes with PhSiH$_3$ promoted by Me$_2$SiCp"$_2$Th^{n}Bu$_2$. *Organometallics* **2001**, *20*, 5084–5104. [CrossRef]

62. Dash, A.K.; Gurevizt, Y.; Wang, J.Q.; Wang, J.; Kapon, M.; Eisen, M.S. Organoactinides—Novel catalysts for demanding chemical transformations. *J. Alloys Compd.* **2002**, *344*, 65–69. [CrossRef]

63. Dash, A.K.; Wang, J.X.; Berthet, J.C.; Ephritikhine, M.; Eisen, M.S. Diverse catalytic activity of the cationic actinide complex [(Et$_2$N)$_3$U][BPh$_4$] in the dimerization and hydrosilylation of terminal alkynes. Characterization of the first f-element alkyne π-complex [(Et$_2$N)$_2$U(C≡C^{t}Bu)(η^2-HC≡C^{t}Bu)][BPh$_4$]. *J. Organomet. Chem.* **2000**, *604*, 83–98. [CrossRef]

64. Straub, T.; Haskel, A.; Eisen, M.S. Organoactinide-catalyzed oligomerization of terminal acetylenes. *J. Am. Chem. Soc.* **1995**, *117*, 6364–6365. [CrossRef]

65. Wang, J.Q.; Dash, A.K.; Berthet, J.C.; Ephritikhine, M.; Eisen, M.S. Selective dimerization of terminal alkynes promoted by the cationic actinide compound [(Et$_2$N)$_3$U][BPh$_4$]. Formation of the alkyne π-complex [(Et$_2$N)$_2$U(C≡C^{t}Bu)(η^2-HC≡C^{t}Bu)][BPh$_4$]. *Organometallics* **1999**, *18*, 2407–2409. [CrossRef]

66. Wang, J.; Kapon, M.; Berthet, J.C.; Ephritikhine, M.; Eisen, M.S. Cross dimerization of terminal alkynes catalyzed by [(Et$_2$N)$_3$U][BPh$_4$]. *Inorg. Chim. Acta* **2002**, *334*, 183–192. [CrossRef]

67. Wang, J.; Dash, A.K.; Kapon, M.; Berthet, J.-C.; Ephritikhine, M.; Eisen, M.S. Oligomerization and hydroamination of terminal alkynes promoted by the cationic organoactinide compound [(Et$_2$N)$_3$U][BPh$_4$]. *Chem. Eur. J.* **2002**, *8*, 5384–5396. [CrossRef]

68. Kosog, B.; Kefalidis, C.E.; Heinemann, F.W.; Maron, L.; Meyer, K. Uranium(III)-mediated C–C-coupling of terminal alkynes: Formation of dinuclear uranium(IV) vinyl complexes. *J. Am. Chem. Soc.* **2012**, *134*, 12792–12797. [CrossRef] [PubMed]

69. Wang, J.; Gurevich, Y.; Botoshansky, M.; Eisen, M.S. Unique σ-bond metathesis of silylalkynes promoted by an *ansa*-dimethylsilyl and oxo-bridged uranium metallocene. *J. Am. Chem. Soc.* **2006**, *128*, 9350–9351. [CrossRef] [PubMed]

70. Barnea, E.; Andrea, T.; Berthet, J.-C.; Ephritikhine, M.; Eisen, M.S. Coupling of terminal alkynes and isonitriles by organoactinide complexes: Scope and mechanistic insights. *Organometallics* **2008**, *27*, 3103–3112. [CrossRef]

71. Lin, Z.; Marks, T.J. Metal, bond energy, and ancillary ligand effects on actinide–carbon σ-bond hydrogenolysis. A kinetic and mechanistic study. *J. Am. Chem. Soc.* **1987**, *109*, 7979–7985. [CrossRef]

72. Andrea, T.; Barnea, E.; Eisen, M.S. Organoactinides promote the tishchenko reaction: The myth of inactive actinide-alkoxo complexes. *J. Am. Chem. Soc.* **2008**, *130*, 2454–2455. [CrossRef] [PubMed]

73. Sharma, M.; Andrea, T.; Brookes, N.J.; Yates, B.F.; Eisen, M.S. Organoactinides promote the dimerization of aldehydes: Scope, kinetics, thermodynamics, and calculation studies. *J. Am. Chem. Soc.* **2011**, *133*, 1341–1356. [CrossRef] [PubMed]

74. Karmel, I.S.R.; Fridman, N.; Tamm, M.; Eisen, M.S. Mixed imidazolin-2-iminato–Cp* thorium(IV) complexes: Synthesis and reactivity toward oxygen-containing substrates. *Organometallics* **2015**, *34*, 2933–2942. [CrossRef]

75. Karmel, I.S.R.; Fridman, N.; Tamm, M.; Eisen, M.S. Mono(imidazolin-2-iminato) actinide complexes: Synthesis and application in the catalytic dimerization of aldehydes. *J. Am. Chem. Soc.* **2014**, *136*, 17180–17192. [CrossRef] [PubMed]

76. Marciniec, B.; Chojnowski, J. *Progress in Organosilicon Chemistry*; Gordon and Breach Science Publishers: Basel, Switzerland, 1995.

77. Wang, J.X.; Dash, A.K.; Berthet, J.C.; Ephritikhine, M.; Eisen, M.S. Dehydrocoupling reactions of amines with silanes catalyzed by $[(Et_2N)_3U][BPh_4]$. *J. Organomet. Chem.* **2000**, *610*, 49–57. [CrossRef]

78. Yang, X.; Stern, C.; Marks, T.J. Models for organometallic molecule-support complexes. Very large counterion modulation of cationic actinide alkyl reactivity. *Organometallics* **1991**, *10*, 840–842. [CrossRef]

79. Jia, L.; Yang, X.; Stern, C.; Marks, T.J. Cationic d0/f0 metallocene catalysts. Properties of binucleaoordinating counteranions derived therefrom. *Organometallics* **1994**, *13*, 3755–3757. [CrossRef]

80. Jia, L.; Yang, X.; Stern, C.L.; Marks, T.J. Cationic metallocene polymerization catalysts based on tetrakis(pentafluorophenyl)borate and its derivatives. Probing the limits of anion "noncoordination" via a synthetic, solution dynamic, structural, and catalytic olefin polymerization study. *Organometallics* **1997**, *16*, 842–857. [CrossRef]

81. Hayes, C.E.; Leznoff, D.B. Diamido-ether uranium(IV) alkyl complexes as single-component ethylene polymerization catalysts. *Organometallics* **2010**, *29*, 767–774. [CrossRef]

82. Domeshek, E.; Batrice, R.J.; Aharonovich, S.; Tumanskii, B.; Botoshansky, M.; Eisen, M.S. Organoactinides in the polymerization of ethylene: Is TIBA a better cocatalyst than MAO? *Dalton Trans.* **2013**, *42*, 9069–9078. [CrossRef] [PubMed]

83. Ikada, Y.; Tsuji, H. Biodegradable polyesters for medical and ecological applications. *Macromol. Rapid Commun.* **2000**, *21*, 117–132. [CrossRef]

84. Hedrick, J.L.; Magbitang, T.; Connor, E.F.; Glauser, T.; Volksen, W.; Hawker, C.J.; Lee, V.Y.; Miller, R.D. Application of complex macromolecular architectures for advanced microelectronic materials. *Chem. Eur. J.* **2002**, *8*, 3308–3319. [CrossRef]

85. Joshi, P.; Madras, G. Degradation of polycaprolactone in supercritical fluids. *Polym. Degrad. Stab.* **2008**, *93*, 1901–1908. [CrossRef]

86. Arbaoui, A.; Redshaw, C. Metal catalysts for ε-caprolactone polymerisation. *Polym. Chem.* **2010**, *1*, 801–826. [CrossRef]

87. Villiers, C.; Thuery, P.; Ephritikhine, M. A comparison of analogous 4f- and 5f-element compounds: Syntheses, X-ray crystal structures and catalytic activity of the homoleptic amidinate complexes $[M\{MeC(NCy)_2\}_3]$ (M = La, Nd or U). *Eur. J. Inorg. Chem.* **2004**, *2004*, 4624–4632. [CrossRef]

88. Barnea, E.; Moradove, D.; Berthet, J.-C.; Ephritikhine, M.; Eisen, M.S. Surprising activity of organoactinide complexes in the polymerization of cyclic mono- and diesters. *Organometallics* **2006**, *25*, 320–322. [CrossRef]

89. Rabinovich, E.; Aharonovich, S.; Botoshansky, M.; Eisen, M.S. Thorium 2-pyridylamidinates: Synthesis, structure and catalytic activity towards the cyclo-oligomerization of ε-caprolactone. *Dalton Trans.* **2010**, *39*, 6667–6676. [CrossRef] [PubMed]

90. Karmel, I.S.R.; Elkin, T.; Fridman, N.; Eisen, M.S. Dimethylsilyl bis(amidinate)actinide complexes: Synthesis and reactivity towards oxygen containing substrates. *Dalton Trans.* **2014**, *43*, 11376–11387. [CrossRef] [PubMed]

91. Karmel, I.S.R.; Fridman, N.; Eisen, M.S. Actinide amidinate complexes with a dimethylamine side arm: Synthesis, structural characterization, and reactivity. *Organometallics* **2015**, *34*, 636–643. [CrossRef]

92. Das, R.K.; Barnea, E.; Andrea, T.; Kapon, M.; Fridman, N.; Botoshansky, M.; Eisen, M.S. Group 4 lanthanide and actinide organometallic inclusion complexes. *Organometallics* **2015**, *34*, 742–752. [CrossRef]

93. Walshe, A.; Fang, J.; Maron, L.; Baker, R.J. New mechanism for the ring-opening polymerization of lactones? Uranyl aryloxide-induced intermolecular catalysis. *Inorg. Chem.* **2013**, *52*, 9077–9086. [CrossRef] [PubMed]

94. Ren, W.; Zhao, N.; Chen, L.; Song, H.; Zi, G. Synthesis, structure, and catalytic activity of an organothorium hydride complex. *Inorg. Chem. Commun.* **2011**, *14*, 1838–1841. [CrossRef]

95. Ren, W.; Zhao, N.; Chen, L.; Zi, G. Synthesis, structure, and catalytic activity of benzyl thorium metallocenes. *Inorg. Chem. Commun.* **2013**, *30*, 26–28. [CrossRef]

96. Hayes, C.E.; Sarazin, Y.; Katz, M.J.; Carpentier, J.-F.; Leznoff, D.B. Diamido-ether actinide complexes as initiators for lactide ring-opening polymerization. *Organometallics* **2013**, *32*, 1183–1192. [CrossRef]

97. Peters, R.G.; Warner, B.P.; Burns, C.J. The catalytic reduction of azides and hydrazines using high-valent organouranium complexes. *J. Am. Chem. Soc.* **1999**, *121*, 5585–5586. [CrossRef]

98. Baker, R.J.; Walshe, A. New reactivity of the uranyl ion: Ring opening polymerisation of epoxides. *Chem. Commun.* **2012**, *48*, 985–987. [CrossRef] [PubMed]

99. Fang, J.; Walshe, A.; Maron, L.; Baker, R.J. Ring-opening polymerization of epoxides catalyzed by uranyl complexes: An experimental and theoretical study of the reaction mechanism. *Inorg. Chem.* **2012**, *51*, 9132–9140. [CrossRef] [PubMed]

100. Behrle, A.C.; Schmidt, J.A.R. Insertion reactions and catalytic hydrophosphination of heterocumulenes using α-metalated *N,N*-dimethylbenzylamine rare-earth-metal complexes. *Organometallics* **2013**, *32*, 1141–1149. [CrossRef]

101. Tu, J.; Li, W.; Xue, M.; Zhang, Y.; Shen, Q. Bridged bis(amidinate) lanthanide aryloxides: Syntheses, structures, and catalytic activity for addition of amines to carbodiimides. *Dalton Trans.* **2013**, *42*, 5890–5901. [CrossRef] [PubMed]

102. Li, Z.; Xue, M.; Yao, H.; Sun, H.; Zhang, Y.; Shen, Q. Enol-functionalized *N*-heterocyclic carbene lanthanide amide complexes: Synthesis, molecular structures and catalytic activity for addition of amines to carbodiimides. *J. Organomet. Chem.* **2012**, *713*, 27–34. [CrossRef]

103. Cao, Y.; Du, Z.; Li, W.; Li, J.; Zhang, Y.; Xu, F.; Shen, Q. Activation of carbodiimide and transformation with amine to guanidinate group by Ln(OAr)$_3$(THF)$_2$ (Ln: Lanthanide and yttrium) and Ln(OAr)$_3$(THF)$_2$ as a novel precatalyst for addition of amines to carbodiimides: Influence of aryloxide group. *Inorg. Chem.* **2011**, *50*, 3729–3737. [CrossRef] [PubMed]

104. Zhang, X.; Wang, C.; Qian, C.; Han, F.; Xu, F.; Shen, Q. Heterobimetallic dianionic guanidinate complexes of lanthanide and lithium: Highly efficient precatalysts for catalytic addition of amines to carbodiimides to synthesize guanidines. *Tetrahedron* **2011**, *67*, 8790–8799. [CrossRef]

105. Du, Z.; Zhou, H.; Yao, H.; Zhang, Y.; Yao, Y.; Shen, Q. The first bridged lanthanide carbene complex formed through reduction of carbodiimide by diamine-bis(phenolate) ytterbium(II) complex and its reactivity to phenylisocyanate. *Chem. Commun.* **2011**, *47*, 3595–3597. [CrossRef] [PubMed]

106. Yi, W.; Zhang, J.; Hong, L.; Chen, Z.; Zhou, X. Insertion of isocyanate and isothiocyanate into the Ln–P σ-bond of organolanthanide phosphides. *Organometallics* **2011**, *30*, 5809–5814. [CrossRef]

107. Karmel, I.S.R.; Tamm, M.; Eisen, M.S. Actinide-mediated catalytic addition of E–H bonds (E = N, P, S) to carbodiimides, isocyanates, and isothiocyanates. *Angew. Chem. Int. Ed.* **2015**, *54*, 12422–12425. [CrossRef] [PubMed]

inorganics | MDPI

Review

Molecular Pnictogen Activation by Rare Earth and Actinide Complexes

Zoë R. Turner

Chemistry Research Laboratory, Department of Chemistry, University of Oxford, Mansfield Road, Oxford OX1 3TA, UK; zoe.turner@chem.ox.ac.uk; Tel.: +44-1865-285-157

Academic Editors: Stephen Mansell and Steve Liddle
Received: 1 October 2015; Accepted: 9 December 2015; Published: 21 December 2015

Abstract: This review covers the activation of molecular pnictogens (group 15 elements) by homogeneous rare earth and actinide complexes. All examples of molecular pnictogen activation (dinitrogen, white phosphorus, yellow arsenic) by both rare earths and actinides, to date (2015), are discussed, focusing on synthetic methodology and the structure and bonding of the resulting complexes.

Keywords: group 3; rare earth; lanthanide; actinide; dinitrogen; white phosphorus; yellow arsenic; pnictogen; small molecule activation

1. Introduction

Rare earth (scandium, yttrium and the lanthanides) and actinide complexes remain underexplored with respect to the transition metals and main group elements but often demonstrate both unique reactivity and molecular properties. Understanding of the bonding and electronic structure of these complexes has particular significance for separation of metals in nuclear waste streams [1].

Activation of molecular pnictogens (group 15 elements) is an area of growing importance; atmospheric dinitrogen (N_2) and white phosphorus (P_4) are principal sources of N- and P-containing compounds (e.g., polymers, pharmaceuticals, agrochemicals, explosives, and specialty chemicals) but are both very challenging to selectively activate. Metal-arsenic, -antimony and -bismuth complexes remain rare [2–4], while the study of metal-pnictogen complexes, including these heavier pnictogen homologues, is also of fundamental importance with respect to Ln and An-pnictogen bonding and electronic structure.

Fixation of N_2, the six electron reduction to two molecules of more reactive ammonia, is necessary for further formation of N-element bonds. In nature, nitrogenase enzymes containing metalloproteins (Fe, Mo or V) fix N_2 through proton-coupled electron transfer under ambient conditions [5,6]. In industry, the Haber–Bosch process combines N_2 and high purity H_2 at high temperatures and pressures over heterogeneous iron- or ruthenium-based catalysts [7–10]. This highly efficient process produces 100 million tons of ammonia per year but is the largest energy-consuming process in the modern world today; the need for direct activation and functionalisation of N_2 under mild conditions is a clear goal. Accessing appropriately reactive phosphorus building blocks presents a different set of challenges based on the sustainability and efficiency of chemical transformations required; from phosphate rock minerals which are mined globally on a 225 million ton scale per year (2013) [11], phosphate fertilisers derived from phosphoric acid are the major products with the remainder used for elemental phosphorus production. Organophosphorus compounds are generally derived from PCl_3, obtained by the chlorination of P_4, and subsequent multi-step procedures [12–14]. Attention has turned to direct and selective activation of elemental phosphorus under mild conditions; this approach is more atom-efficient (which is important given the limited accessible deposits of phosphate rock), avoids the need for large scale production of PCl_3 (which is toxic, corrosive and highly reactive), and is both more economically and environmentally sustainable [15].

The area of dinitrogen activation has been reviewed extensively with particular focus on the Haber–Bosch process [16–18], and biological nitrogen fixation [5,6,19–26]. There are reviews on transition metal N_2 activation which cover N_2 binding modes [27–29], multimetallic N_2 activation [30,31], the relevance of metal hydride complexes to N_2 activation [32,33], N_2 cleavage and functionalisation [34,35] (including electrochemical [36] and photolytic N_2 cleavage [37]), and N_2 activation at bare metal atoms [38] and using surface organometallic chemistry [39]. Specific reviews have also focused on activation by group 4 metals [40–42], iron [31,43,44], molybdenum [24,45–47], and the mid-to-late transition metal centres [48]. In terms of rare earth N_2 activation; an account of work from Evans and co-workers to 2004 has been reported [49], and Gardiner more recently reviewed the chemistry of the lanthanides with dinitrogen and reduced derivatives [50]. The area of actinide N_2 activation has been discussed in the context of small molecule activation by trivalent uranium complexes [51,52].

Transition metal-mediated white phosphorus activation has been previously reviewed [53–56], with specific reviews on both early transition metal complexes [57,58] and late transition metal complexes [59]. More broadly, reviews of P_4 activation by *p*-block compounds have also been reported [60–64].

This review seeks to cover all examples of molecular pnictogen activation (dinitrogen, white phosphorus, yellow arsenic) by both rare earth and actinide complexes to date, focusing on synthetic methodology and the structure and bonding of the resulting complexes. Only well-defined homogeneous complexes will be discussed; heterogeneous and surface chemistry lie beyond the scope of this review.

2. Dinitrogen Activation by Rare Earth Complexes

2.1. Complexes Containing a Formal $N_2{}^{2-}$ Ligand

The majority of rare earth complexes that activate dinitrogen (N_2) result in its formal reduction to the $N_2{}^{2-}$ anion and the formation of bimetallic complexes of the general form $[A_2(thf)_xLn]_2(\mu-\eta^2{:}\eta^2-N_2)$ where there is side-on binding of N_2 ("A is defined as a group that exists as an anion in LnA_3 and provides reductive reactivity in combination with an alkali metal or the equivalent") [65]. The work of Evans and co-workers has led to the development of three key methodologies to access these species: (i) salt metathesis reactions of divalent lanthanide halides with alkali metal salts; (ii) combination of trivalent Ln complexes with alkali metals (LnA_3/M or LnA_2A'/M method); (iii) photochemical activation of LnA_2A' systems (Figure 1).

For reference, N–N and M–N(N_2) bond lengths obtained from single crystal X-ray diffraction experiments, N–N stretching frequencies (obtained by IR or Raman spectroscopy) and $^{14/15}$N-NMR spectroscopic data are summarised in Table 1.

From divalent Ln	From trivalent Ln	
Direct N₂ reduction	*Photochemical activation*	*LnA₃/M or LnA₂A'/M*

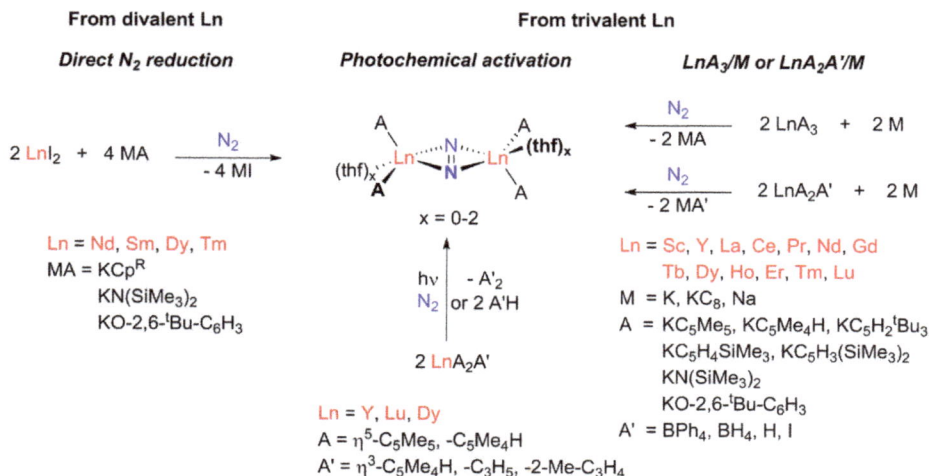

$$2\ LnI_2\ +\ 4\ MA\ \xrightarrow[-\ 4\ MI]{N_2}$$

Ln = Nd, Sm, Dy, Tm
MA = KCp^R
KN(SiMe₃)₂
KO-2,6-^{t}Bu-C₆H₃

x = 0–2

hν | – A'₂
N₂ | or 2 A'H

2 LnA₂A'

Ln = Y, Lu, Dy
A = η⁵-C₅Me₅, -C₅Me₄H
A' = η³-C₅Me₄H, -C₃H₅, -2-Me-C₃H₄

$$\xleftarrow[-\ 2\ MA]{N_2}\ 2\ LnA_3\ +\ 2\ M$$

$$\xleftarrow[-\ 2\ MA']{N_2}\ 2\ LnA_2A'\ +\ 2\ M$$

Ln = Sc, Y, La, Ce, Pr, Nd, Gd
Tb, Dy, Ho, Er, Tm, Lu
M = K, KC₈, Na
A = KC₅Me₅, KC₅Me₄H, KC₅H₂^{t}Bu₃
KC₅H₄SiMe₃, KC₅H₃(SiMe₃)₂
KN(SiMe₃)₂
KO-2,6-^{t}Bu-C₆H₃
A' = BPh₄, BH₄, H, I

Figure 1. Routes to N₂²⁻ complexes using rare earth metals.

Table 1. Summary of rare earth N₂²⁻ complexes.

Complex (#) [Reference]	N–N Bond Length (Å)	Ln–N(N₂) Bond Lengths (Å)	N–N Frequency (cm⁻¹)	¹⁴/¹⁵N-NMR Spectroscopy (ppm) [a]
N₂	1.0975 [66]	-	2331 [67]	−75 [68]
[(η⁵-C₅Me₄H)₂Sc]₂(μ-η²:η²-N₂) (**1**) [69]	1.239(3)	2.216(1) 2.220(1)	-	-
[(η⁵-C₅Me₄H)₂Sc]₂(μ-η²:η²-N₂) (**1'**) [70]	1.229(3)	2.197(2) 2.179(2)	-	385
[(η⁵-C₅Me₅)₂Y]₂(μ-η²:η²-N₂) (**2**) [71]	1.172(6)	2.279(3) 2.292(3)	-	496
[(η⁵-1,2,4-^{t}Bu–C₅H₂)₂Nd]₂(μ-η²:η²-N₂) (**3**) [67]	1.226(12)	2.495(2) 2.497(2)	1622 (¹⁴N₂) 1569 (¹⁵N₂)	-
[(η⁵-C₅Me₅)₂Sm]₂(μ-η²:η²-N₂) [b] (**4**) [72]	1.088(12)	2.348(6) 2.367(6)	-	−117 (263 K) −161 (203 K)
[(η⁵-C₅Me₅)₂Dy]₂(μ-η²:η²-N₂) (**5**) [73]	-	-	-	-
[(η⁵-SiMe₃–C₅H₄)₂Dy]₂(μ-η²:η²-N₂) (**6**) [74]	Connectivity only			-
[(η⁵-C₅Me₅)₂Tm]₂(μ-η²:η²-N₂) (**7**) [75]	Connectivity only			-
[(η⁵-1,3-SiMe₃–C₅H₃)₂Tm]₂(μ-η²:η²-N₂) (**8**) [75]	1.259(4)	2.273(2) 2.272(2)	-	-
[(η⁵-C₅Me₅)₂Lu]₂(μ-η²:η²-N₂) (**9**) [71]	Connectivity only			527
[(η⁵-C₅Me₅)(η⁵-C₅Me₄H)Lu]₂(μ-η²:η²-N₂) (**10**) [76]	1.275(3)	2.291(3) 2.295(3)	1736 (¹⁴N₂) 1678 (¹⁵N₂)	-
[(η⁵-C₅Me₄H)₂Y(thf)]₂(μ-η²:η²-N₂) [c] (**11**) [77]	1.252(5)	2.338(3) 2.370(3)	-	468
[(η⁵-SiMe₃–C₅H₄)₂Y(thf)]₂(μ-η²:η²-N₂) [c] (**12**) [78]	1.244(2)	2.3214(14) 2.3070(14)	-	-
[(η⁵-C₅Me₅)₂La(thf)]₂(μ-η²:η²-N₂) [d] (**13**) [79]	1.233(5)	2.537(4) 2.478(4)	-	569
[(η⁵-C₅Me₄H)₂La(thf)]₂(μ-η²:η²-N₂) [c] (**14**) [79]	1.243(4)	2.457(2) 2.503(2)	-	495
[(η⁵-C₅Me₅)₂Ce(thf)]₂(μ-η²:η²-N₂) [c] (**15**) [68]	1.258(9)	2.4548(15) 2.542(2)	-	871
[(η⁵-C₅Me₄H)₂Ce(thf)]₂(μ-η²:η²-N₂) [d] (**16**) [68]	1.235(6)	2.428(3) 2.475(3)	-	1001

Table 1. *Cont.*

Complex (#) [Reference]	N–N Bond Length (Å)	Ln–N(N$_2$) Bond Lengths (Å)	N–N Frequency (cm^{-1})	$^{14/15}$N-NMR Spectroscopy (ppm) [a]
[(η^5-C$_5$Me$_5$)$_2$Pr(thf)]$_2$(μ-η^2:η^2-N$_2$) [d] (**17**) [68]	1.242(9)	2.4459(14) 2.512(2)	-	2231
[(η^5-C$_5$Me$_4$H)$_2$Pr(thf)]$_2$(μ-η^2:η^2-N$_2$) [c] (**18**) [68]	1.235(7) [e]	2.418(4) [e] 2.455(3) [e]	-	2383
[(η^5-C$_5$Me$_4$H)$_2$Nd(thf)]$_2$(μ-η^2:η^2-N$_2$) [c] (**19**) [79]	1.241(5) [e]	2.404(3) [e] 2.451(2) [e]	-	-
[(η^5-SiMe$_3$–C$_5$H$_4$)$_2$Tm(thf)]$_2$(μ-η^2:η^2-N$_2$) [c] (**20**) [75]	1.236(8)	2.274(4) 2.302(4)	-	-
[(η^5-C$_5$Me$_4$H)$_2$Lu(thf)]$_2$(μ-η^2:η^2-N$_2$) [c] (**21**) [80]	1.243(12)	2.290(6) 2.311(6)	-	521
[{(Me$_3$Si)$_2$N}$_2$Y(thf)]$_2$(μ-η^2:η^2-N$_2$) [c] (**22**) [81]	1.274(3)	2.297(2) 2.308(2)	1425 (^{14}N$_2$) 1377 (^{15}N$_2$)	+513 (t)
[{(Me$_3$Si)$_2$N}$_2$La(thf)]$_2$(μ-η^2:η^2-N$_2$) (**23**) [81]	-	-	-	516
[{(Me$_3$Si)$_2$N}$_2$Nd(thf)]$_2$(μ-η^2:η^2-N$_2$) [c] (**24**) [81]	1.258(3)	2.3758(16) 2.3938(16)	-	-
[{(Me$_3$Si)$_2$N}$_2$Gd(thf)]$_2$(μ-η^2:η^2-N$_2$) [c] (**25**) [81]	1.278(4)	2.326(2) 2.353(2)	-	-
[{(Me$_3$Si)$_2$N}$_2$Tb(thf)]$_2$(μ-η^2:η^2-N$_2$) [c] (**26**) [81]	1.271(4)	2.301(2) 2.328(2)	-	-
[{(Me$_3$Si)$_2$N}$_2$Dy(thf)]$_2$(μ-η^2:η^2-N$_2$) [c] (**27**) [82]	1.305(6)	2.287(3) 2.312(3)	-	-
[{(Me$_3$Si)$_2$N}$_2$Ho(thf)]$_2$(μ-η^2:η^2-N$_2$) [c] (**28**) [83]	1.264(4)	2.296(2) 2.315(2)	-	-
[{(Me$_3$Si)$_2$N}$_2$Er(thf)]$_2$(μ-η^2:η^2-N$_2$) [c] (**29**) [81]	1.276(5)	2.271(3) 2.302(3)	-	-
[{(Me$_3$Si)$_2$N}$_2$Tm(thf)]$_2$(μ-η^2:η^2-N$_2$) [c] (**30**) [82]	1.261(4)	2.271(2) 2.296(2)	-	-
[{(Me$_3$Si)$_2$N}$_2$Lu(thf)]$_2$(μ-η^2:η^2-N$_2$) [c] (**31**) [83]	1.285(4)	2.241(2) 2.272(2)	1451 (^{14}N$_2$)	557
[{(Me$_3$Si)$_2$N}$_2$Y(PhCN)]$_2$(μ-η^2:η^2-N$_2$) [c] (**32**) [84]	1.258(2)	2.2848(13) 2.3092(13)	-	-
[{(Me$_3$Si)$_2$N}$_2$Y(C$_5$H$_5$N)]$_2$(μ-η^2:η^2-N$_2$) [c] (**33**) [84]	1.255(3)	2.2917(16) 2.3107(17)	-	-
[{(Me$_3$Si)$_2$N}$_2$Y(4-NMe$_2$–C$_5$H$_4$N)]$_2$(μ-η^2:η^2-N$_2$) [c] (**34**) [84]	1.259(2)	2.2979(12) 2.3132(12)	-	-
[{(Me$_3$Si)$_2$N}$_2$Y(Ph$_3$PO)]$_2$(μ-η^2:η^2-N$_2$) [c] (**35**) [84]	1.262(2)	2.3000(14) 2.3022(14)	-	-
[{(Me$_3$Si)$_2$N}$_2$Y(Me$_3$NO)]$_2$(μ-η^2:η^2-N$_2$) [c] (**36**) [84]	1.198(3)	2.2925(17) 2.2941(18)	-	-
[(2,6-^{t}Bu–C$_6$H$_3$O)$_2$Nd(thf)$_2$]$_2$(μ-η^2:η^2-N$_2$) (**37**) [82]	1.242(7)	2.397(4) 2.401(3)	-	-
[(2,6-^{t}Bu–C$_6$H$_3$O)$_2$Dy(thf)$_2$]$_2$(μ-η^2:η^2-N$_2$) (**38**) [82]	1.257(7) [f] 1.256(9) [g]	2.328(4) [f] 2.340(4) [f] 2.336(5) [g] 2.336(5) [g]	1526 (^{14}N$_2$)	-
[Na$_4$(thf)$_8$][(η^5:η^1:η^5:η^1-^{Et2}calix[4]pyrrole)Pr]$_2$(μ-η^2:η^2-N$_2$) (**39**) [85]	-	-	-	-
[Na$_4$(dme)$_5$][(η^5:η^1:η^5:η^1-^{Et2}calix[4]pyrrole)Pr]$_2$(μ-η^2:η^2-N$_2$) (**40**) [85]	1.254(7)	2.414(5) 2.457(5)	-	-
[Na$_4$(thf)$_8$][(η^5:η^1:η^5:η^1-^{Et2}calix[4]pyrrole)Nd]$_2$(μ-η^2:η^2-N$_2$) (**41**) [85]	-	-	-	-
[Na$_4$(dioxane)$_6$][(η^5:η^1:η^5:η^1-^{Et2}calix[4]pyrrole)Nd]$_2$(μ-η^2:η^2-N$_2$) (**42**) [85]	1.234(8)	2.511(4) 2.508(4)	-	-
[{HB(3-^{t}Bu-5-Me–pz)}Tm{NH(2,5-^{t}Bu–C$_6$H$_3$)}]$_2$(μ-η^2:η^2-N$_2$) (**43**) [86]	1.215(10)	2.274(8) 2.286(9)	-	-

[a] Referenced to CH$_3$^{15}NO; [b] In equilibrium with [(η^5-C$_5$Me$_5$)$_2$Sm]$_2$; [c] *trans* arrangement of donor solvent; [d] *cis* arrangement of donor solvent; [e] The authors indicate poor data quality or significant disorder in these structures; [f] thf solvent of crystallisation in unit cell; [g] toluene solvent of crystallisation in unit cell.

2.1.1. Cyclopentadienyl Ancillary Ligands

The first isolated, structurally characterised dinitrogen complex of an f-element metal was reported by Evans and co-workers [72]. $[(\eta^5\text{-}C_5Me_5)_2Sm]_2(\mu\text{-}\eta^2\text{:}\eta^2\text{-}N_2)$ (**4**) was isolated by slow crystallisation of a toluene solution of the bent metallocene $[(\eta^5\text{-}C_5Me_5)_2Sm]_2$ under an N_2 atmosphere (Figure 2). **4** exists in dynamic equilibrium with the metallocene starting material involving reversible Sm^{II}/Sm^{III} interconversion. In the solid state, **4** displays tetrahedral coordination around each Sm centre with gearing of the $[Sm(C_5Me_5)_2]$ units and the first example of a co-planar M_2N_2 diamond core for any metal. The bridging, side-on bound N_2 has a short N–N distance of 1.088(12) Å (free N_2: 1.0975 Å [66]) and does not imply reduction to $N_2{}^{2-}$; however, recent studies by Arnold and co-workers have shown that N–N bond lengths determined using X-ray diffraction experiments can be underestimated and so may not provide the best way of assessing the level of dinitrogen reduction [87,88]. Both the Sm–N/C bond lengths and the ^{13}C-NMR spectral data support formulation of the complex as $[Sm^{III}]_2(N_2{}^{2-})$. Maron and co-workers have reported calculations on the interaction of N_2 with $[(\eta^5\text{-}C_5Me_5)_2Ln]$ (Ln = Sm, Eu, Yb) [89].

$Cp^R =$

Ln = Sc	C_5Me_4H (**1**), (**1'**)
Y	C_5Me_5 (**2**)
Nd	1,2,4-$^tBu\text{-}C_5H_2$ (**3**)
Sm	C_5Me_5 (**4**)
Dy	C_5Me_5 (**5**), $SiMe_3\text{-}C_5H_4$ (**6**)
Tm	C_5Me_5 (**7**), 1,3-$SiMe_3\text{-}C_5H_3$ (**8**)
Lu	C_5Me_5 (**9**)
Lu	$Cp^R{}_2 = (C_5Me_5)(C_4Me_4H)$ (**10**)

$Cp^R =$

Ln = Y	C_5Me_4H (**11**), $SiMe_3\text{-}C_5H_4$ (**12**)
La	C_5Me_5 (**13**), C_5Me_4H (**14**)
Ce	C_5Me_5 (**15**), C_5Me_4H (**16**)
Pr	C_5Me_5 (**17**), C_5Me_4H (**18**)
Nd	C_5Me_4H (**19**)
Tm	$SiMe_3\text{-}C_5H_4$ (**20**)
Lu	C_5Me_4H (**21**)

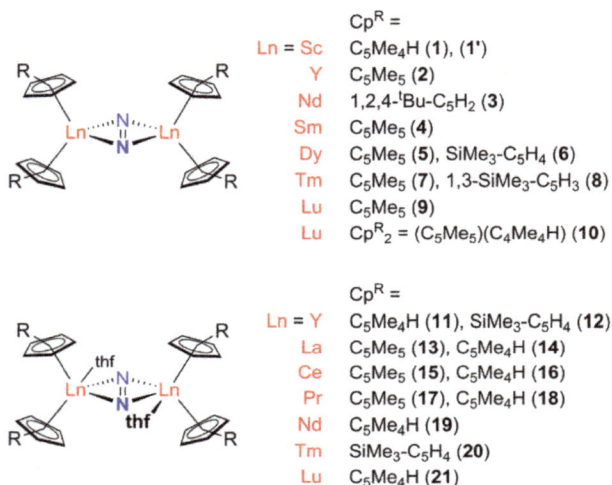

Figure 2. Rare earth complexes with cyclopentadienyl ligands resulting from N_2 activation to $N_2{}^{2-}$.

Since this landmark discovery, the methodology of using reducing divalent rare earth metal complexes to activate N_2 has resulted in analogous cyclopentadienyl complexes of Dy (**5**, **6**) [74] and Tm (**7**, **20**) [75]. Structurally, these complexes all demonstrate a common planar Ln_2N_2 core (Ln–N–N–Ln dihedral angle = 0°), with the arrangement of the cyclopentadienyl ligands being dependent on the metal centre and the nature of the ligand itself (Figure 3).

The number of dinitrogen complexes has been expanded significantly by the report that combination of trivalent lanthanide complexes $LnCp^R{}_3$ or $LnCp^R{}_2A'$ with an alkali metal can also reduce dinitrogen affording side-on bound N_2 complexes $[(\eta^5\text{-}Cp^R)_2Ln]_2(\mu\text{-}\eta^2\text{:}\eta^2\text{-}N_2)$ (Ln = Sc (**1**), Y (**2**), Nd (**3**), Dy (**5**, **6**), Lu (**9**)) [67,69–71,76] or $[(\eta^5\text{-}Cp^R)_2Ln(thf)]_2(\mu\text{-}\eta^2\text{:}\eta^2\text{-}N_2)$ (Ln = Y (**11**, **12**), La (**13**, **14**), Ce (**15**, **16**), Pr (**17**, **18**), Nd (**19**), Lu (**21**)) [68,78–80,83]. The generality of this method has been demonstrated by the wide range of metals utilised as well as the use of both homo- and heteroleptic trivalent lanthanide starting materials with a variety of cyclopentadienyl, amide, aryloxide, hydride, halide and borohydride ligands.

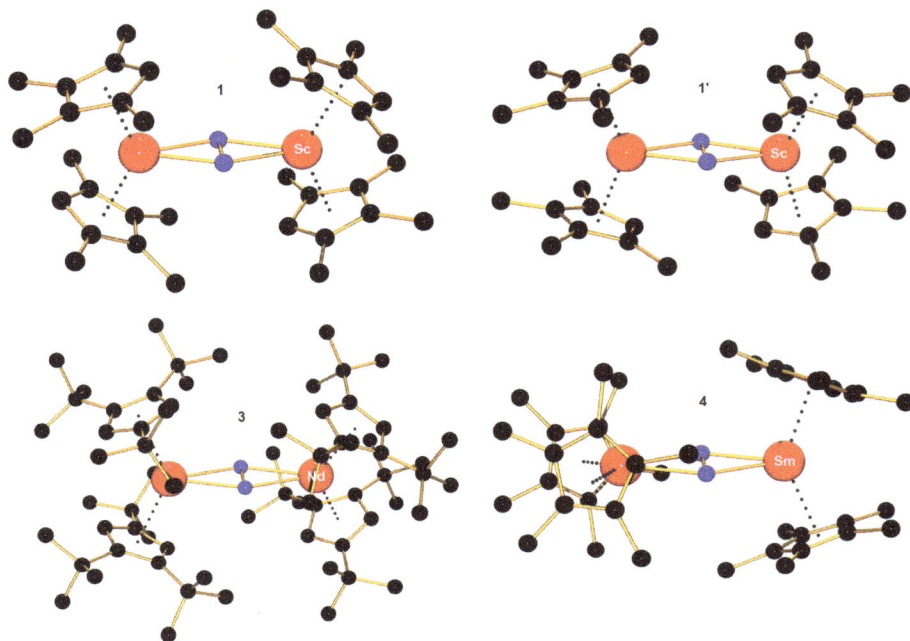

Figure 3. Structural variations in the solid state structures of $[(\eta^5\text{-Cp}^R)_2\text{Ln}]_2(\mu\text{-}\eta^2\text{:}\eta^2\text{-N}_2)$ (Ln = Sc (**1**) and (**1'**); Nd (**3**); Sm (**4**)).

Complexes **11–21** (with donor solvent bound) also have planar Ln_2N_2 cores. However, only **13** (La), **15** (Ce) and **17** (Pr) have a *cis* arrangement of thf molecules and an asymmetrically-bound $\text{N}_2{}^{2-}$ ligand (as a result of crystallographically non-equivalent N atoms). This is most clearly seen though the difference in Ln–N–Ln' angles (**11**: 145.77(16) and 157.33(18)°; **13**: 156.9(3) and 144.7(3)°; **15**: 157.0(3) and 145.1(3)°) (Figure 4).

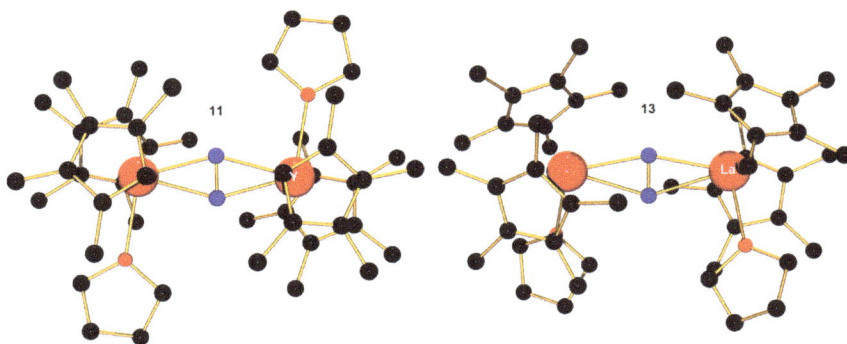

Figure 4. Structural variations in the solid state structures of $[(\eta^5\text{-C}_5\text{Me}_4\text{H})_2\text{Ln(thf)}]_2(\mu\text{-}\eta^2\text{:}\eta^2\text{-N}_2)$ (**11**, **14**, **16**, **18**, **19**, **21**) (**11** depicted, left) and $[(\eta^5\text{-C}_5\text{Me}_5)_2\text{Ln(thf)}]_2(\mu\text{-}\eta^2\text{:}\eta^2\text{-N}_2)$ (**13**, **15**, **17**) (**13** depicted, right). One C_5Me_4 ring in **11** is disordered.

In terms of bonding, calculations were carried out on **1** and the Sc–N (N_2) bonding interaction was found to be a polar covalent two-electron four-centre bond resulting from donation from a filled Sc 3d orbital into an empty N_2 π_g antibonding orbital in the Sc_2N_2 plane. The lowest unoccupied molecular orbital (LUMO) of **1** is an unperturbed antibonding π_g orbital based on N_2. This bonding scheme can likely be extended for all compounds **1–21**, with the donor nd orbital varying based on the nature of the metal ion [69].

Complexes **4** and **15–18** were characterised by ^{15}N-NMR spectroscopy; the first reported examples of such spectra for paramagnetic N_2 complexes [68]. Trivalent Sm, Ce and Pr were chosen due to the low magnetic susceptibility of these ions ($4f^5$ Sm^{III}, μ = 0.84 μ_B; $4f^1$ Ce^{III}, μ = 2.54 μ_B; $4f^2$ Pr^{III}, μ = 3.58 μ_B). Broad singlets at high frequency were observed for **15** (871 ppm), **16** (1001 ppm), **17** (2231 ppm) and **18** (2383 ppm). Consistent with the reversible N_2 coordination to **4**, only a singlet at −75 ppm corresponding to free N_2 is observed at 298 K [68]. Cooling resulted in a new resonance at −117 ppm at 263 K which shifted linearly to −161 ppm at 203 K and accounts for bound N_2. In the context of pioneering NMR spectroscopic characterisation of organometallic complexes, the solid state ^{15}N- and ^{139}La-NMR spectra of **14**–^{15}**N$_2$** have also been reported [90].

Most recently, it has been demonstrated by Evans and co-workers that photochemical activation of the closed shell LnA$_2$A′ complexes Ln(η^5-C$_5$Me$_5$)$_2$(η^3-C$_5$Me$_4$H) or Ln(η^5-C$_5$Me$_5$) (η^5-C$_5$Me$_4$H)(η^3-C$_5$Me$_4$H), which feature a novel η^3 binding mode of a cyclopentadienyl ligand, yields side-on dinitrogen complexes [(η^5-C$_5$Me$_5$)$_2$M]$_2$(μ-η^2:η^2-N$_2$) (M = Y (**2**), Dy (**5**), Lu (**9**)) and [(η^5-C$_5$Me$_5$)(η^5-C$_5$Me$_4$H)Lu]$_2$(μ-η^2:η^2-N$_2$) (**10**), with concomitant formation of (C$_5$Me$_4$H)$_2$ [73]. These reactions typically take place in under 5 h but, in the absence of photochemical activation, normally require a number of weeks. Full conversion to **2** and **9** is achieved via this methodology; this is notable given that other synthetic methods afforded yields of 26%–51% and 49%–59% respectively. Sterically induced reduction, typified by bulky M(η^5-C$_5$Me$_5$)$_3$ complexes, does not account for this process since the less sterically hindered [C$_5$Me$_4$H]$^-$ ligand acts as the reductant [91]. Calculations support a mechanism involving electron transfer from the [η^3-C$_5$Me$_4$H]$^-$ ligand into an empty 4d$_{z^2}$ orbital on the metal centre. This affords a [C$_5$Me$_4$H]· radical which dimerises, and excited [Cp$_2$M]* nd^1 fragment which reduces N_2. Similarly, the allyl complexes Ln(C$_5$Me$_5$)$_2$(η^3-C$_3$RH$_4$) (R = H or Me) can be photochemically activated to yield **2** and **9**; in this case, propene and isobutene are observed as by-products due to H-atom abstraction from solvent rather than radical dimerisation [92].

2.1.2. Amide Ancillary Ligands

The simple silylamide ligand [N(SiMe$_3$)$_2$]$^-$ has also proved suitable to prepare related complexes of the form [{(Me$_3$Si)$_2$N}$_2$Ln(thf)]$_2$(μ-η^2:η^2-N$_2$) (Ln = Y (**22**), La (**23**), Nd (**24**), Gd (**25**), Tb (**26**), Dy (**27**), Ho (**28**), Er (**29**), Tm (**30**), Lu (**31**)) (Figure 5, left) [65,81–83,93–95]. All complexes can be prepared using the LnA$_3$/M or LnA$_2$A′/M method and in addition, **24**, **27** and **30** can be prepared directly from the divalent starting materials NdI$_2$, DyI$_2$ and TmI$_2$(thf)$_3$ respectively (though **24** is notable in that it is only isolated in 4% yield using this synthetic approach) [82].

Lewis base coordination of yttrium complex **22** was investigated through a series of substitution reactions to afford [{(Me$_3$Si)$_2$N}$_2$Ln(L)]$_2$(μ-η^2:η^2-N$_2$) (L = PhCN (**32**), C$_5$H$_5$N (**33**), 4-NMe$_2$–C$_5$H$_4$N (**34**), Ph$_3$PO (**35**), Me$_3$NO (**36**)) (Figure 5, right) [84]. Varying the donor ligand had little effect on the planar Ln$_2$N$_2$ structural core of **22** and **32–35**, though the N–N distance in **36** was unexpectedly short at 1.198(3) Å. Similarly for complexes **1–21**, calculations on [{(Me$_3$Si)$_2$N}$_2$Y(L)]$_2$(μ-η^2:η^2-N$_2$) (**22**) indicate that the Y–N (N_2) bonding interaction involves donation from a filled Y 4d orbital into an antibonding N_2 π_g orbital in the Y$_2$N$_2$ plane. The LUMO is an unperturbed antibonding N_2 π_g orbital [84,93]. UV-Vis spectra of **22** and **31–35** all contained a low energy, low intensity absorption around 700 nm which corresponds to the formally electric-dipole forbidden HOMO–LUMO (a$_g$–a$_g$) transition and act as a fingerprint of the electronic structure of the Y$_2$N$_2$ core.

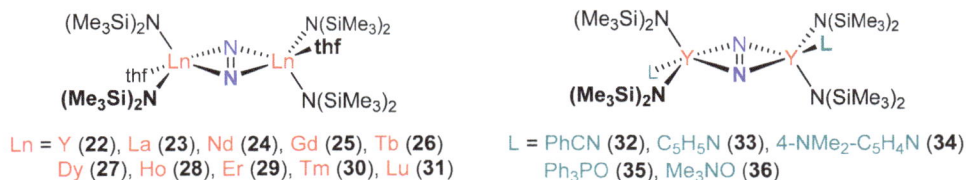

Figure 5. Rare earth complexes with amide ligands resulting from N_2 activation to N_2^{2-}.

In terms of bonding, the closed shell $4f^{14}$ Lu^{III} ion in **31** provides an interesting contrast to yttrium complex **22** with $4f^0$ Y^{III} ions. Calculations support that the bonding is described in analogy with **22**, but using higher energy, radially diffuse 5d orbitals which have a good energy match with the N_2 antibonding π_g orbital and are the correct symmetry for overlap [65]. Hughbanks and co-workers have also reported calculations on the $4f^7$ Gd^{III} complex **33** to analyse magnetic coupling [96].

2.1.3. Aryloxide Ancillary Ligands

Aryloxide ancillary ligands have been used to prepare side-on N_2 complexes of the form $[(2,6-^{t}Bu-C_6H_3O)_2Ln(thf)_2]_2(\mu-\eta^2:\eta^2-N_2)$ (Ln = Nd (**37**) and Dy (**38**)) which now contain two molecules of coordinating solvent per metal (Figure 6) [65,82,93]. Calculations on **38**, which contains open shell $4f^9$ Dy^{III} ions, indicate that Dy–N_2 bonding is derived from a $5d-\pi_g$ interaction in the Dy_2N_2 plane [65].

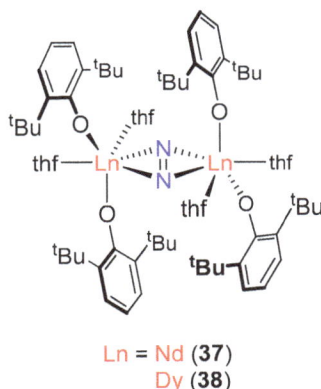

Ln = Nd (**37**)
Dy (**38**)

Figure 6. Rare earth complexes with aryloxide ligands resulting from N_2 activation to N_2^{2-}.

2.1.4. Multidentate Ancillary Ligands

Floriani and co-workers reported N_2 complexes of Pr^{III} (**39**, **40**) and Nd^{III} (**41**, **42**) using a calix[4]pyrrole ligand; these complexes were obtained as single crystals suitable for X-ray diffraction studies and no isolated yields were reported (Figure 7) [85]. $[Na_4(thf)_8][(\eta^5:\eta^1:\eta^5:\eta^1-^{Et2}$ calix[4]pyrrole)Ln]_2(\mu-\eta^2:\eta^2-N_2)$ (Ln = Pr (**39**), Nd (**41**)) were prepared by reduction of $[Na(thf)_2]$ $[(\eta^5:\eta^1:\eta^5:\eta^1-^{Et2}$ calix[4]pyrrole)Ln(thf)]$ (Ln = Pr or Nd) using sodium metal with catalytic napthalene. Addition of dimethoxyethane (dme) to **39** led to solvent exchange to afford $[Na_4(dme)_5]$ $[(\eta^5:\eta^1:\eta^5:\eta^1-^{Et2}$ calix[4]pyrrole)Ln]_2(\mu-\eta^2:\eta^2-N_2)$ (**40**). Similarly, addition of dioxane to **41** affords $[Na_4(dioxane)_6][(\eta^5:\eta^1:\eta^5:\eta^1-^{Et2}$ calix[4]pyrrole)Ln]_2(\mu-\eta^2:\eta^2-N_2)$ (**42**). The solid state structures of **40** and **42** revealed different coordination modes of the sodium counterions and feature N–N bond distances of 1.254(7) and 1.234(8) Å which are consistent with reduction to N_2^{2-}. Measured magnetic moments are also consistent with Ln^{III} oxidation state for both metal centres.

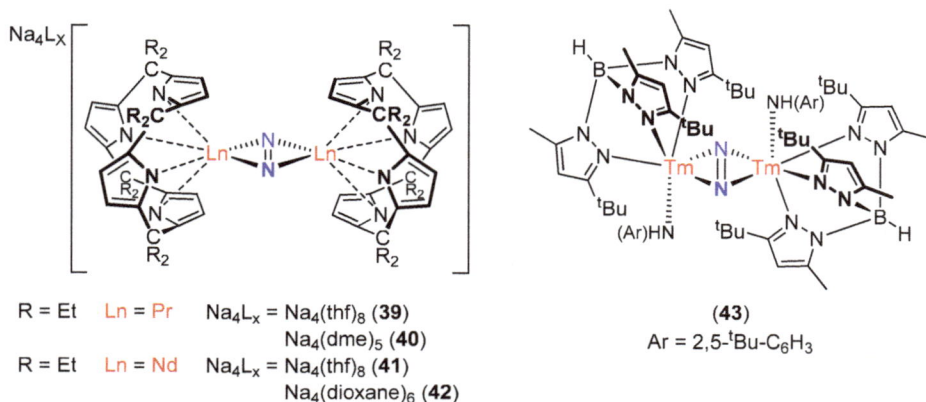

Figure 7. Rare earth complexes with multidentate ligands resulting from N_2 activation to N_2^{2-}.

Takats and co-workers reported the scorpionate complex [{HB(3-^{t}Bu-5-Me–pz)}Tm{NH(2,5-^{t}Bu–C$_6$H$_3$)}]$_2$(μ-η^2:η^2-N$_2$) (**43**) (pz = C$_3$HN$_2$ = pyrazolyl) which was prepared by the protonolysis reaction of 2,5-^{t}Bu–C$_6$H$_3$NH$_2$ with the heteroleptic Tm^{II} hydrocarbyl compound {HB(3-^{t}Bu-5-Me–pz)}Tm{CH(SiMe$_3$)$_2$} [86]. The N–N bond distance of 1.215(10) Å is consistent with reduction to N_2^{2-} and the Ln$_2$N$_2$ core is slightly bent with an Ln–N–N–Ln torsion angle of 0.37°.

2.2. Complexes Containing a Formal N_2^{3-} Ligand

2.2.1. Amide Ancillary Ligands

For reference, N–N and M–N(N$_2$) bond lengths obtained from single crystal X-ray diffraction experiments and N–N stretching frequencies (obtained by IR or Raman spectroscopy) are summarised in Table 2.

The first definitive evidence for an N_2^{3-} reduction product of dinitrogen was demonstrated by Evans and co-workers [93]. The LnA$_3$/M system of Y{N(SiMe$_3$)$_2$}$_3$ with KC$_8$ in thf afforded a mixture of [{(Me$_3$Si)$_2$N}$_2$Y(thf)]$_2$(μ-η^2:η^2-N$_2$) (**22**), [K(thf)$_6$][{(Me$_3$Si)$_2$N}$_2$Y(thf)]$_2$(μ-η^2:η^2-N$_2$) (**44**) and [K][{(Me$_3$Si)$_2$N}$_2$Y(thf)]$_2$(μ_3-η^2:η^2:η^2-N$_2$) (**47**) from which **44** and **47** could be isolated (Figure 8). The EPR spectrum of **44**–^{15}N$_2$ has a 9-line pattern consistent with a triplet of triplets due to two ^{15}N and two ^{89}Y nuclei and has a hyperfine coupling constant of 8.2 G implying a N-centred radical, while **47**–^{15}N$_2$ shows extra coupling to potassium; all spectra indicate the presence of the N_2^{3-} ion. The N–N bond distances are 1.401(6) and 1.405(3) Å respectively and are intermediate between N–N single bonds (1.47 Å in N$_2$H$_4$) and N=N double bonds (1.25 Å in PhN=NPh) [98]. The N–N vibrational stretching frequency in **47** is 989 cm^{-1}, significantly reduced from 1425 cm^{-1} for **22**. Similarly to complexes **22–36**, the Y–N$_2$ bonding interaction in these complexes can be described by the donation from a filled Y 4d orbital into an antibonding N$_2$ π_g orbital (HOMO). However, the orthogonal antibonding N$_2$ π_g orbital is now also occupied by a single electron.

Following this remarkable report, these types of complexes have been extended to more rare earth metal centres; [K(thf)$_6$][{(Me$_3$Si)$_2$N}$_2$Ln(thf)]$_2$(μ-η^2:η^2-N$_2$) (Ln = La (**45**), Lu (**46**)) and [K][{(Me$_3$Si)$_2$N}$_2$Ln(thf)]$_2$(μ_3-η^2:η^2:η^2-N$_2$) (Ln = Gd (**48**), Tb (**49**), Dy (**50**)). Additionally, the solvated counterion can also be varied to include 18-crown-6 in [K(18c6)(thf)$_2$][{(Me$_3$Si)$_2$N}$_2$Ln(thf)]$_2$(μ-η^2:η^2-N$_2$) (Ln = Y (**51**), Gd (**52**), Tb (**53**), Dy (**54**), Ho (**55**), Er (**56**)) or can be exchanged for sodium in [Na(thf)$_6$][{(Me$_3$Si)$_2$N}$_2$Ln(thf)]$_2$(μ-η^2:η^2-N$_2$) (Ln = Y (**57**), Er (**58**)) [65,94,95,97].

Table 2. Summary of rare earth N_2^{3-} complexes.

Complex (#) [Reference]	N–N Bond Length (Å)	Ln–N (N_2) Bond Lengths (Å)	Ln–N–N–Ln Torsion Angle (°)	N–N Frequency (cm^{-1})
N_2	1.0975 [66]	-	-	2331 [67]
[K(thf)$_6$][{(Me$_3$Si)$_2$N}$_2$Y(thf)]$_2$(μ-η^2:η^2-N$_2$) ^{b} (**44**) [93]	1.401(6)	2.194(3) 2.218(3)	0	1002 (^{14}N$_2$) (calculated)
	1.401(6) ^{a}	2.190(3) ^{a} 2.213(3) ^{a}		
[K(thf)$_6$][{(Me$_3$Si)$_2$N}$_2$La(thf)]$_2$(μ-η^2:η^2-N$_2$) (**45**) [65]	-	-	0	-
[K(thf)$_6$][{(Me$_3$Si)$_2$N}$_2$Lu(thf)]$_2$(μ-η^2:η^2-N$_2$) ^{b} (**46**) [65]	1.414(8)	2.163(4) 2.180(4)	0	979 (^{14}N$_2$)
[K][{(Me$_3$Si)$_2$N}$_2$Y(thf)]$_2$(μ_3-η^2:η^2:η^2-N$_2$) ^{b} (**47**) [93]	1.405(3)	2.225(2) 2.242(2)	14.22	989 (^{14}N$_2$) 956 (^{15}N$_2$)
[K][{(Me$_3$Si)$_2$N}$_2$Gd(thf)]$_2$(μ_3-η^2:η^2:η^2-N$_2$) ^{b} (**48**) [97]	1.395(3)	2.248(2) 2.274(2)	13.64	-
[K][{(Me$_3$Si)$_2$N}$_2$Tb(thf)]$_2$(μ_3-η^2:η^2:η^2-N$_2$) ^{b} (**49**) [97]	1.401(3)	2.235(2) 2.260(2)	16.12	-
[K][{(Me$_3$Si)$_2$N}$_2$Dy(thf)]$_2$(μ_3-η^2:η^2:η^2-N$_2$) ^{b} (**50**) [97]	1.404(5)	2.229(4) 2.242(4)	15.27	-
[K(18c6)(thf)$_2$][{(Me$_3$Si)$_2$N}$_2$Y(thf)]$_2$(μ-η^2:η^2-N$_2$) ^{b} (**51**) [97]	1.396(3)	2.1909(17) 2.2136(16)	0	-
[K(18c6)(thf)$_2$][{(Me$_3$Si)$_2$N}$_2$Gd(thf)]$_2$(μ-η^2:η^2-N$_2$) ^{b} (**52**) [94]	1.401(4)	2.224(2) 2.249(2)	0	-
[K(18c6)(thf)$_2$][{(Me$_3$Si)$_2$N}$_2$Tb(thf)]$_2$(μ-η^2:η^2-N$_2$) ^{b} (**53**) [94]	1.394(3)	2.2056(15) 2.2345(15)	0	-
[K(18c6)(thf)$_2$][{(Me$_3$Si)$_2$N}$_2$Dy(thf)]$_2$(μ-η^2:η^2-N$_2$) ^{b} (**54**) [94]	1.393(7)	2.199(4) 2.213(4)	0	-
[K(18c6)(thf)$_2$][{(Me$_3$Si)$_2$N}$_2$Ho(thf)]$_2$(μ-η^2:η^2-N$_2$) ^{b} (**55**) [95]	1.404(4)	2.188(2) 2.210(2)	0	-
[K(18c6)(thf)$_2$][{(Me$_3$Si)$_2$N}$_2$Er(thf)]$_2$(μ-η^2:η^2-N$_2$) ^{b} (**56**) [65]	1.409(4)	2.178(2) 2.204(2)	0	-
[Na(thf)$_6$][{(Me$_3$Si)$_2$N}$_2$Y(thf)]$_2$(μ-η^2:η^2-N$_2$) ^{b} (**57**) [65]	1.393(7)	2.199(4) 2.213(4)	0	-
[Na(thf)$_6$][{(Me$_3$Si)$_2$N}$_2$Er(thf)]$_2$(μ-η^2:η^2-N$_2$) ^{b} (**58**) [65]	1.403(4)	2.1817(19) 2.2019(19)	0	-
[K(thf)$_6$][(2,6-^{t}Bu–C$_6$H$_3$O)$_2$Dy(thf)]$_2$(μ-η^2:η^2-N$_2$) (**59**) [93]	1.396(7)	2.197(3) 2.203(4)	0	962 (^{14}N$_2$)
[K(thf)][(2,6-^{t}Bu–C$_6$H$_3$O)$_2$Dy(thf)]$_2$(μ_3-η^2:η^2:η^2-N$_2$) (**60**) [93]	1.402(7)	2.235(5) 2.209(5)	6.59	-

^{a} Second independent molecule in unit cell; ^{b} *trans* arrangement of thf.

The solid state structures of **44**–**46** and **51**–**58** which contain outer-sphere counterions have the common Ln$_2$N$_2$ planar core like those of rare earth N_2^{2-} complexes **1**–**42** which have been structurally characterised. However, complexes **47**–**50**, which have inner-sphere K$^+$ ions, display bent Ln$_2$N$_2$ cores with torsion angles of 14.22° (**47**), 13.64° (**48**), 16.12° (**49**) and 15.27° (**50**).

Complexes **48**–**50** (inner-sphere K$^+$) and **52**–**56** (outer-sphere K$^+$) all display interesting magnetic properties; **48** and **52** have the strongest magnetic exchange couplings in a Gd^{III} complex with exchange constants of -27 cm^{-1}, and **49**, **50** and **53**–**56** demonstrate single-molecule-magnet behaviour [94,95,99,100]. Combination of rare earth complexes which demonstrate both high anisotropy and strong exchange coupling potentially provides a route to single-molecule magnets with high blocking temperatures. The diffuse nature of the N_2^{3-} radical facilitates strong coupling in these systems by overlap of the Ln 4f orbitals with the bridging dinitrogen ligand. **53** and **54** exhibit magnetic hysteresis up to record blocking temperatures of 13.9 K (0.9 mTs^{-1} sweep rate) and 8.3 K (0.08 Ts^{-1} sweep rate) respectively. Competing Ln^{III}–Ln^{III} antiferromagnetic coupling is observed in complexes at low temperatures in **48**–**50**, which have a non-zero Ln–N–N–Ln dihedral angle, demonstrating the importance of geometry of the Ln$_2$N$_2$ unit to magnetic behaviour.

Ln = Y (**44**), La (**45**), Lu (**46**)   Ln = Y (**47**), Gd (**48**), Tb (**49**), Dy (**50**)

Ln = Y (**51**), Gd (**52**), Tb (**53**), Dy (**54**), Ho (**55**), Er (**56**)

Ln = Y (**57**), Er (**58**)

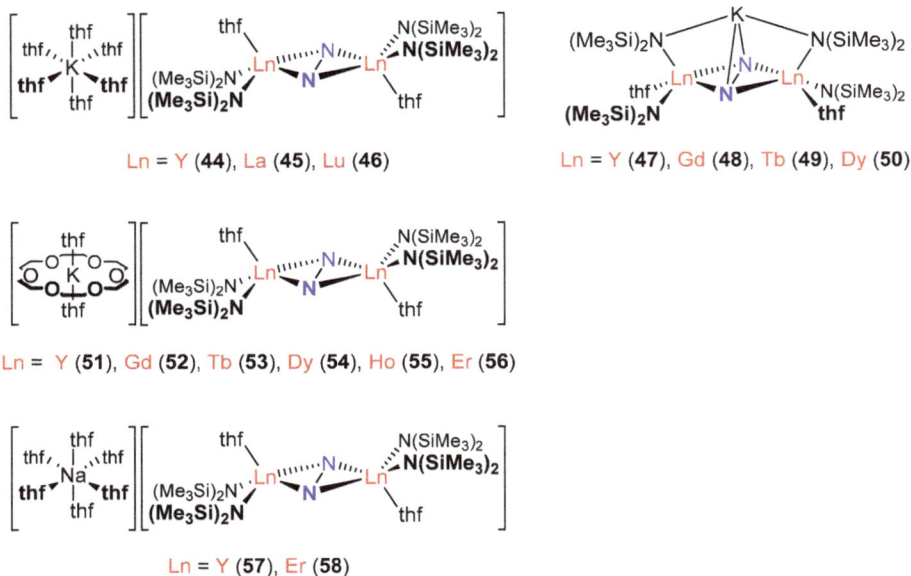

Figure 8. Rare earth complexes with amide ligands resulting from N_2 activation to N_2^{3-}.

2.2.2. Aryloxide Ancillary Ligands

As for yttrium complexes **44** and **47**, the dysprosium aryloxide complexes [K(thf)$_6$][(2,6-^{t}Bu–C$_6$H$_3$O)$_2$Dy(thf)]$_2$(μ-η^2:η^2-N$_2$) (**59**) and [K(thf)][[(2,6-^{t}Bu–C$_6$H$_3$O)$_2$Dy(thf)]$_2$(μ_3-η^2:η^2:η^2-N$_2$) (**60**) were first isolated from reaction of DyI$_2$ with KO-2,6-^{t}Bu–C$_6$H$_3$ (Figure 9) [65,93]. Reoxidation of **59** with AgBPh$_4$ affords the N$_2$$^{2-}$ complex [(2,6-^{t}Bu–C$_6$H$_3$O)$_2$Dy(thf)$_2$]$_2$(μ-η^2:η^2-N$_2$) (**38**). From the solid state structures the N–N bond lengths are 1.396(7) (**59**) and 1.402(7) Å (**60**), which fall in the range of reported N$_2$$^{3-}$ complexes (Table 2). As anticipated, **59** has a planar Ln$_2$N$_2$ core in the solid state whereas it is bent in complex **60** with a Ln–N–N–Ln torsion angle of 6.59°.

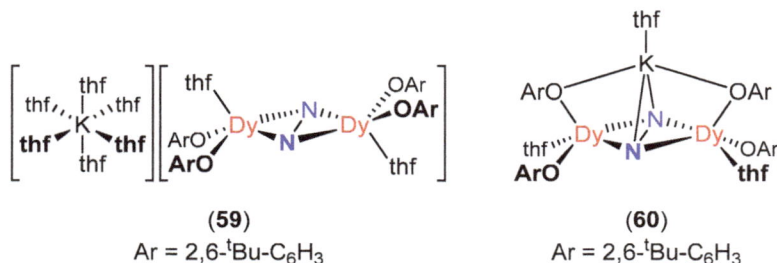

(**59**)   (**60**)
Ar = 2,6-^{t}Bu-C$_6$H$_3$ Ar = 2,6-^{t}Bu-C$_6$H$_3$

Figure 9. Rare earth complexes with aryloxide ligands resulting from N_2 activation to N_2^{3-}.

2.3. Complexes Containing a Formally N_2^{4-} Ligand

For reference, N–N and M–N(N$_2$) bond lengths obtained from single crystal X-ray diffraction experiments are summarised in Table 3.

Table 3. Summary of rare earth N_2^{4-} complexes.

Complex (#) [Reference]	N–N Bond Length (Å)	Ln–N (N_2) Bond Lengths (Side-on) (Å)	Ln–N (N_2) Bond Lengths (End-on) (Å)
N_2	1.0975 [66]	-	-
$[Li(thf)_2]_2[(^{Et2}calix[4]pyrrole)Sm]_2(N_2Li_4)$ (61) [101]	1.525(4)	2.357(2) 2.342(2)	-
$[\{Ph_2C(C_4H_3N)_2\}Sm(thf)]_4(\mu_4\text{-}\eta^1:\eta^1:\eta^2:\eta^2\text{-}N_2)$ (62) [102]	1.412(17)	2.327(3) 2.327(3)	2.177(8) 2.177(8)
$[\{CyC(C_4H_3N)_2\}Sm(thf)]_4(\mu_4\text{-}\eta^1:\eta^1:\eta^2:\eta^2\text{-}N_2)$ (63) [103]	1.392(16)	2.339(3) 2.339(3)	2.160(8) 2.160(8)
$[\{Et_2C(C_4H_3N)_2\}Sm(thf)]_4(\mu_4\text{-}\eta^1:\eta^1:\eta^2:\eta^2\text{-}N_2)$ (64) [104]	1.415(4)	2.328(3) 2.342(3)	2.145(3)
$[\{Ph(Me)C(C_4H_3N)_2\}Sm(dme)]_4(\mu_4\text{-}\eta^1:\eta^1:\eta^2:\eta^2\text{-}N_2)$ (65) [104]	1.42(2)	2.316(13) 2.316(12)	2.149(11)
$[Na(thf)]_2[\{CyC(C_4H_3N)_2\}Sm(thf)]_4(\mu_6\text{-}\eta^1:\eta^1:\eta^1:\eta^1:\eta^2:\eta^2\text{-}N_2)$ (66) [103]	1.371(16)	2.332(11) 2.324(11)	2.178(10)
$[\{Li(thf)\}_3(\mu_3\text{-}Cl)][(^{Cy}calix[4]pyrrole)Sm]_2(\mu\text{-}\eta^2:\eta^2\text{-}N_2)$ (67) [105]	1.08(3)	2.880(18) 2.974(18)	-
$[(Li(thf)_2][(^{Cy}calix[4]pyrrole)_2Sm_3Li_2](\mu_5\text{-}\eta^1:\eta^2:\eta^2:\eta^2\text{-}N_2)$ (68) [105]	1.502(5)	2.249(4) (Sm(1)–N(1)) 2.253(4) (Sm(1)–N(1)) 2.355(4) (Sm(2)–N(1)) 2.370(4) (Sm(2)–N(1)) 2.398(4) (Sm(3)–N(1)) 2.376(4) (Sm(3)–N(1))	-

To date, all examples of complexes containing an N_2^{4-} ligand, derived from dinitrogen activation by a rare earth metal centre involve multidentate ligands (Figure 10) [101–106]. The first example of a structurally characterised rare earth complex containing an N_2^{4-} ligand was reported by Gambarotta and co-workers [101]. The octametallic complex $[Li(thf)_2]_2[(^{Et2}calix[4]pyrrole)Sm]_2(N_2Li_4)$ (61) was prepared by reaction of $SmCl_3(thf)_3$ with $[Li(thf)]_4[^{Et2}calix[4]pyrrole]$ and subsequent reduction with Li metal under an argon atmosphere, followed by exposure to N_2. The reaction by-products were not determined. In the solid state structure, there is an octahedron of metal ions coordinating to the N_2^{4-} ligand which is η^2-bound to the two Sm^{III} in the apical sites, and both η^2- and η^1-bound to the opposite pairs of the four Li^+ ions in the equatorial plane. The N–N distance of 1.525(4) Å, combined with a total magnetic moment of 2.72 μ_B, is consistent with formulation of the complex as $[Sm^{III}]_2(N_2^{4-})$.

A family of tetrametallic Sm dipyrrolide complexes $[\{R_2C(C_4H_3N)_2\}Sm(L)]_4(\mu_4\text{-}\eta^1:\eta^1:\eta^2:\eta^2\text{-}N_2)$ ($L = thf$, $CR_2 = CPh_2$ (62), CCy (63), CEt_2 (64); $L = dme$, $CR_2 = C(Me)Ph$ (65)) was prepared by reaction of $SmI_2(thf)_2$ with the corresponding alkali metal dipyrrolide salt, and display both side-on and end-on N_2 coordination [102–104]. 62–65 were stable both thermally and *in vacuo*. All compounds were structurally characterised using X-ray diffraction experiments; the N–N bond lengths range from 1.392(16) to 1.42(2) Å, and variable coordination modes of the pyrrolide ligands are observed across all complexes. Reduction of 63 with Na afforded $[Na(thf)]_2[\{R_2C(C_4H_3N)_2\}Sm(thf)]_4(\mu_6\text{-}\eta^1:\eta^1:\eta^1:\eta^1:\eta^2:\eta^2\text{-}N_2)$ (66). The N_2^{4-} ligand is bound end-on to two Sm ions and side-on to the other two, as well as being end-on bound to two Na^+ ions. Assignment of the reduction level of dinitrogen to N_2^{4-} leads to a formal oxidation state of +2.5 for each samarium centre. This is proposed on the basis of the N–N bond distance (1.371(16) Å; slightly shorter than that expected for an N–N single bond), magnetic moment (4.05 μ_B; lower than the analogous divalent complex 63) and short Sm–Sm contacts in the solid state which may promote magnetic couplings.

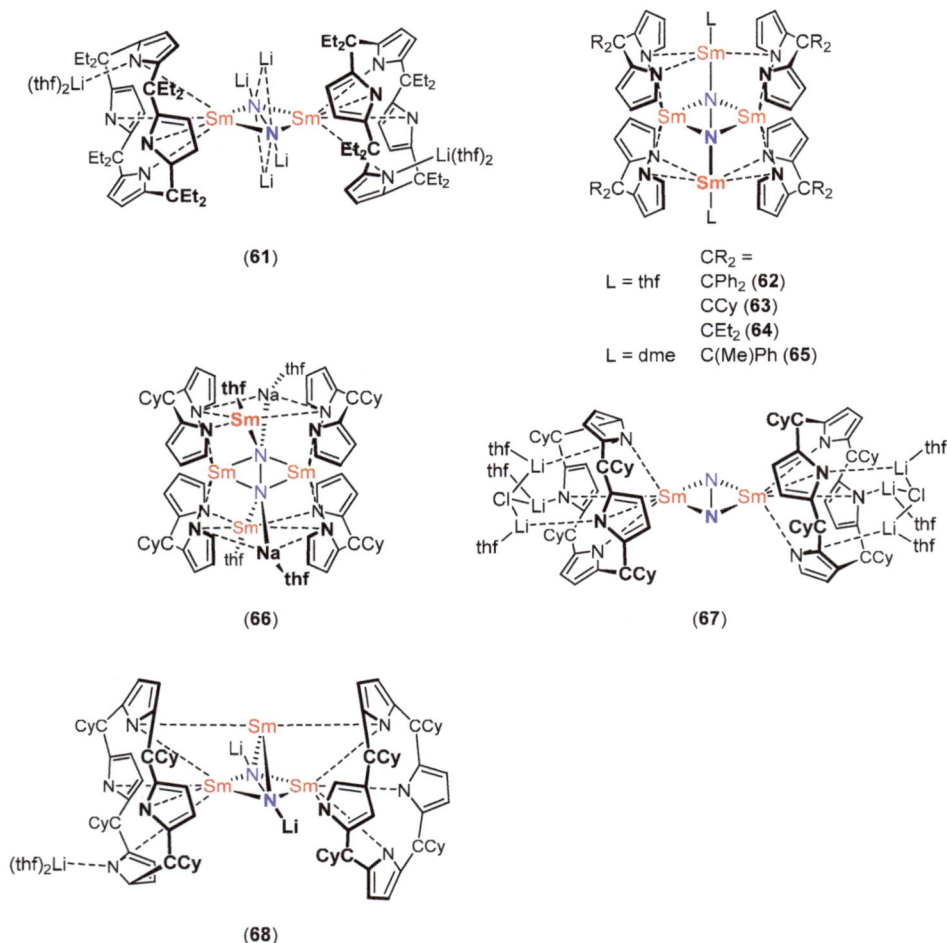

(61)

CR$_2$ =
L = thf   CPh$_2$ (62)
           CCy (63)
           CEt$_2$ (64)
L = dme   C(Me)Ph (65)

(66)

(67)

(68)

Figure 10. Lanthanide complexes with multidentate donors resulting from N$_2$ activation to N$_2^{4-}$.

Tetra-calix-pyrrole complexes [{Li(thf)}$_3$(μ_3-Cl)][(^{Cy}calix[4]pyrrole)Sm]$_2$(μ-η^2:η^2-N$_2$) (67) and the unusual trimetallic [Li(thf)$_2$][(^{Cy}calix[4]pyrrole)$_2$Sm$_3$Li$_2$](μ_5-η^1:η^1:η^2:η^2:η^2-N$_2$) (68) contain formal N$_2^{4-}$ ligands, on the basis of charge neutrality implied from the presence of Sm^{III} centres but have disparate N–N bond lengths of 1.08(3) Å and 1.502(5) Å respectively [105]. It should be noted that the crystallographic data for 67 was only of a good enough quality to obtain structural connectivity.

3. Dinitrogen Activation by Actinide Complexes

3.1. Complexes Containing an Activated N$_2$ Ligand

Perhaps surprisingly, given the number of examples of dinitrogen activation by rare earth complexes, there are very few examples of N$_2$ activation with actinide complexes despite the presence of uranium in early catalysts for the Haber–Bosch process (Figure 11) [107]. Actinide-element bonding in the model system [X$_3$An]$_2$(μ-η^2:η^2-N$_2$) (An = Th–Pu, X = F, Cl, Br, Me, H, OPh) has recently been studied using relativistic DFT calculations [108].

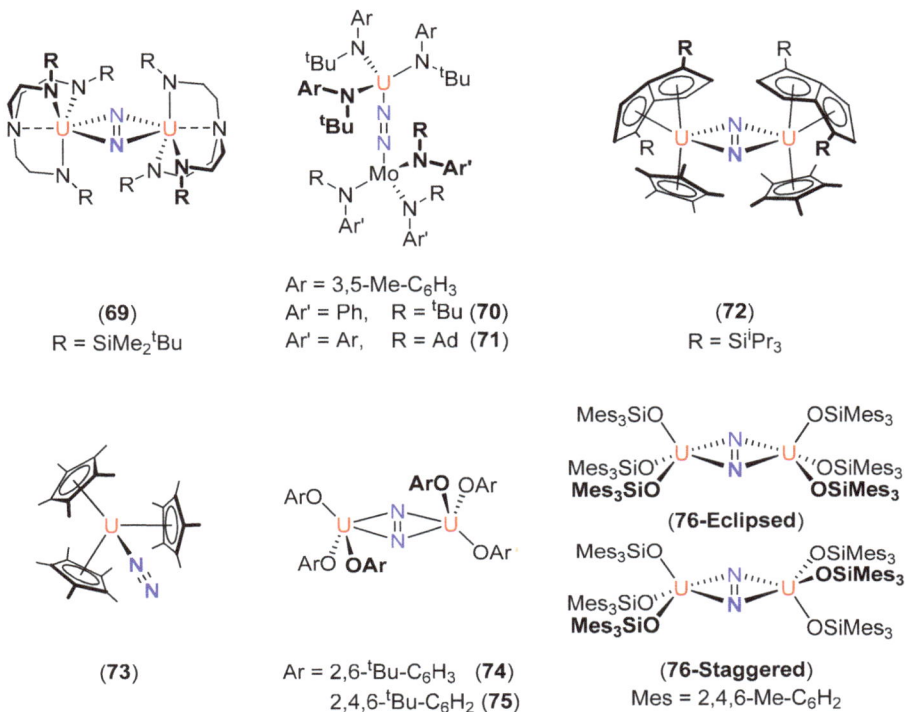

(69)
R = SiMe$_2$^{t}Bu

Ar = 3,5-Me-C$_6$H$_3$
Ar' = Ph, R = ^{t}Bu (70)
Ar' = Ar, R = Ad (71)

(72)
R = Si^{i}Pr$_3$

(73)

Ar = 2,6-^{t}Bu-C$_6$H$_3$ (74)
2,4,6-^{t}Bu-C$_6$H$_2$ (75)

(76-Eclipsed)

(76-Staggered)
Mes = 2,4,6-Me-C$_6$H$_2$

Figure 11. Actinide complexes resulting from N–N activation (no N–N bond cleavage).

For reference, N–N bond lengths obtained from single crystal X-ray diffraction experiments, N–N stretching frequencies (obtained by IR or Raman spectroscopy) and $^{14/15}$N-NMR spectroscopic data of actinide N$_2$ complexes are summarised in Table 4.

Table 4. Summary of actinide N_2 complexes.

Complex (#) [Reference]	Stability	N–N Bond Length (Å)	N–N Frequency (cm^{-1})	$^{14/15}$N-NMR Spectroscopy (ppm) [a]
N_2	-	1.0975 [66]	2331 [67]	−75 [68]
$[\{N(CH_2CH_2NSiMe_2{}^tBu)_3\}U]_2$ $(\mu$-η^2:η^2-$N_2)$ **(69)** [109]	Stable under N_2 (1 atm)	1.109(7)	-	-
	N_2 dissociation *in vacuo*			
$\{Ph({}^tBu)N\}_3Mo(\mu_2$-$\eta^1$:$\eta^1$-$N_2)$ $U\{N({}^tBu)(3,5$-Me$-C_6H_3)\}_3$ **(70)** [110]	Stable *in vacuo* at 25 °C	1.232(11)	- ($^{14}N_2$)	-
	"Thermally stable"		1547 ($^{15}N_2$)	
$\{(3,5$-Me$-C_6H_3)(Ad)N\}_3Mo$ $(\mu_2$-η^1:η^1-$N_2)U\{N({}^tBu)(3,5$-Me$-C_6H_3)\}_3$ **(71)** [110]	Stable *in vacuo* at 25 °C	1.23(2)	1568 ($^{14}N_2$)	-
	"Thermally stable"		1527 ($^{15}N_2$)	
$[(\eta^5$-$C_5Me_5)(\eta^8$-$1,4$-Si${}^iPr_3$-$C_8H_4)U]_2$ $(\mu$-η^2:η^2-$N_2)$ **(72)** [111]	75% conversion to **72** at 50 psi N_2	1.232(10)	-	-
	N_2 dissociation *in vacuo*, in solution and solid state			
$(\eta^5$-$C_5Me_5)_3U(\eta^1$-$N_2)$ **(73)** [112]	Crystallisation at 80 psi N_2	1.120(14)	2207 ($^{14}N_2$)	-
	N_2 dissociation *in vacuo* or in solution under N_2 (1 atm)		2134 ($^{15}N_2$)	
$[(2,6$-${}^tBu$-$C_6H_3O)_3U]_2(\mu$-η^2:η^2-$N_2)$ **(74)** [88]	N_2 dissociation *in vacuo* and in solution at 25 °C	1.163(19) 1.204(17) 1.201(19)	-	-
$[(2,4,6$-${}^tBu$-$C_6H_2O)_3U]_2(\mu$-η^2:η^2-$N_2)$ **(75)** [88]	Stable *in vacuo* at 25 °C	1.236(5)	1451 ($^{14}N_2$)	-
	N_2 dissociation at 80 °C in solution		1404 ($^{15}N_2$)	
$[\{(Mes)_3SiO\}_3U]_2(\mu$-$\eta^2$:$\eta^2$-$N_2)$ **(76)** [87]	Stable in vacuo at 25 °C	1.124(12) (eclipsed)	1437 ($^{14}N_2$)	4213.5
	Slowly forms $U\{OSi(Mes)_3\}_4$ at 100 °C in solution	1.080(11) (staggered)	1372 ($^{15}N_2$)	

[a] Referenced to $CH_3{}^{15}NO$.

The first example of dinitrogen activation by an actinide complex was reported by Scott and co-workers; the trivalent uranium complex $\{N(CH_2CH_2NSiMe_2{}^tBu)_3\}U$ reacts with N_2 (1 atm) to yield $[\{N(CH_2CH_2NSiMe_2{}^tBu)_3\}U]_2(\mu$-$\eta^2$:$\eta^2$-$N_2)$ **(69)** [109]. In solution, the reaction is reversible and **69** converts back to the trivalent uranium starting material when freeze-pump-thaw degassed (Scheme 1). The solid state structure of **69** illustrates the side-on binding mode of N_2 and features an N–N bond length of 1.109(7) Å. Alongside solution magnetic susceptibility measurements of 3.22 μ_B per uranium centre, these data agree with the dimer being formulated as $[U^{III}]_2(N_2{}^0)$.

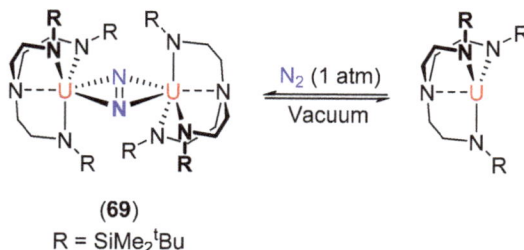

(69)
R = SiMe$_2{}^t$Bu

Scheme 1. Reversible N_2 binding in **69**.

Cummins and co-workers reported thermally stable heterobimetallic U–Mo N_2 complexes featuring the end-on binding mode of N_2; the U^{III} tris(amide) $U\{N(^tBu)Ar\}_3(thf)$ (Ar = 3,5-Me–C_6H_3) reacts with $Mo\{N(R)Ar'\}_3$ under an N_2 atmosphere (1 atm) to yield $\{Ar'(R)N\}_3Mo(\mu_2-\eta^1{:}\eta^1-N_2)U\{N(^tBu)Ar\}_3$ (R = ^tBu, Ar' = Ph (**70**); R = Ad, Ar' = 3,5-Me–C_6H_3 (**71**)) [110]. The solid state molecular structure of **70** shows an N–N bond distance of 1.232(11) Å, consistent with N_2^{2-} and a formal oxidation state of U^{IV}. In principal, U^{III} and Mo^{III} metal centres can provide the 6 electrons required for N_2 bond cleavage and the stability of **70** and **71** should be noted with reference to $Mo\{N(^tBu)Ar'\}_3$ which cleaves N_2 and forms a terminal nitride (Mo≡N) under mild conditions [113–115].

Reversible side-on binding and N_2 activation was demonstrated by Cloke *et al.*, using a mixed sandwich U^{III} pentalene complex. $(\eta^5-C_5Me_5)(\eta^8-1,4-Si^iPr_3-C_8H_4)U$ reacts with N_2 to yield $[(\eta^5-C_5Me_5)(\eta^8-1,4-Si^iPr_3-C_8H_4)U]_2(\mu-\eta^2{:}\eta^2-N_2)$ (**72**) [111]. The solid state structure features an N–N bond length of 1.232(10) Å which is consistent with reduction to N_2^{2-}; regardless of the formal reduction level, the relief of steric crowding in **72** likely drives the facile loss of N_2 when there is no overpressure.

To date, the only example of a monometallic f-element complex of N_2 was reported by Evans *et al.*; sterically crowded $U(\eta^5-C_5Me_5)_3$ reacts with N_2 (80 psi) to afford $(\eta^5-C_5Me_5)_3U(\eta^1-N_2)$ (**73**) [112]. N_2 binding is reversible and lowering the pressure results in N_2 dissociation. The solid state structure of **73** shows the N_2 ligand is bound end-on and linearly (U–N–N = 180°) and the N–N distance is 1.120(14) Å which is statistically equivalent with free N_2.

Arnold and co-workers reported that the trivalent uranium aryloxides $U(OAr)_3$ (Ar = $2,6-^tBu-C_6H_3$ or $2,4,6-^tBu-C_6H_2$) bind N_2 (1 atm) to form the side-on bound N_2 adducts $[(ArO)_3U]_2(\mu-\eta^2{:}\eta^2-N_2)$ (**74** and **75** respectively) [88]. Though **74** was obtained as a minor product, the more sterically hindered **75** was formed in quantitative yield and was stable under dynamic vacuum and in the presence of coordinating solvents and polar small molecules (CO and CO_2) under ambient conditions. N_2 loss was observed when **75** was heated to 80 °C in a toluene solution. The solid state N–N bond lengths are 1.163(19), 1.204(17), 1.201(19) Å (**74**) and 1.236(5) Å (**75**) which indicate significant N_2 reduction by the electron rich U^{III} metal centres. Consistent with this, Raman spectroscopy performed on **75** showed a strong band at 1451 cm^{-1} for the N–N stretch (1404 cm^{-1} in the **75**–$^{15}N_2$) which is significantly lower than in free N_2 (2331 cm^{-1}) [67]. DFT calculations indicate a ^5A_g ground state which agrees with the $[U^{IV}]_2(N_2^{2-})$ description of **75**. Significantly, the U–N (N_2) interaction derives from two occupied MOs showing π backbonding from uranium f orbitals into an N_2 antibonding π_g orbitals and that the interaction is strongly polarised. The bonding description is very similar to that in previously calculated models for **69** (formally N_2^0) [116,117] and **72** (N_2^{2-}) [118] which display very different N–N bond lengths. While experimental N–N bond distances determined by X-ray diffraction experiments are undeniably useful for quick comparisons, the bond length is likely underestimated since the data is based on electron density rather than atomic positions and thus may not reflect the level of dinitrogen reduction. In these studies, N–N stretching wavenumbers were more accurately reproduced by calculation than bond length and it is proposed that this would be a more suitable measurement for probing N_2 reduction.

The most robust actinide N_2 complex prepared to date is $[\{(Mes)_3SiO\}_3U]_2(\mu-\eta^2{:}\eta^2-N_2)$ (**76**) (Mes = 2,4,6-Me–C_6H_2) and is stable both *in vacuo* and in toluene solution up to 100 °C, at which point $U\{OSi(Mes)_3\}_4$ is slowly formed as the major product (52% conversion after 18 h) [87]. **76** is isolated from the reaction of $U\{N(SiMe_3)_2\}_3$ with 3 equivalents of $HOSi(Mes)_3$ under an N_2 atmosphere (1 atm). Raman spectroscopy shows a peak at 1437 cm^{-1} assigned to the N–N stretching mode, indicating a significant level of reduction with respect to free dinitrogen (2331 cm^{-1}) and comparing well with 1451 cm^{-1} recorded for **75** where reduction to N_2^{2-} was assigned. The N–N distances in the solid state are 1.124(12) Å (**76-Eclipsed**) and 1.080(11) Å (**76-Staggered**), which are statistically equivalent to that of free N_2; the disparity in implied reduction of N_2 from Raman spectroscopy and X-ray diffraction experiments again highlighting that the latter may not be best suited for assigning reduction in these systems.

3.2. Complexes Resulting from N_2 Cleavage

Tetra-calix-pyrrole ligands bound to Sm^{II} centres have been demonstrated to activate N_2 by Gambarotta and co-workers [101]. With U^{III}, an unprecedented example of N–N bond cleavage using an molecular f-element complex was observed; when [K(dme)][(^{Et2}calix[4]pyrrole)U(dme)] is treated with potassium naphthalenide under an atmosphere of N_2, N–N bond cleavage occurs to afford [K(dme)$_4$][{K(dme)(^{Et2}calix[4]pyrrole)U}$_2$(μ-NK)$_2$] (**77**) (Scheme 2) [119]. **77** contains two bridging nitrides (U–N: 2.076(6) and 2.099(5) Å) which have contacts with potassium ions (N–K: 2.554(6) Å) that bridge two pyrrolide units on separate ligands. It was postulated that **77** is a Class 1 U^{IV}–U^{V} mixed valence complex on the basis of an absorption at 1247 nm in the near-IR spectrum which is characteristic of U^{V}. The paramagnetism of **77** resulted in NMR silence in both ^{15}N- and ^{14}N-NMR spectra.

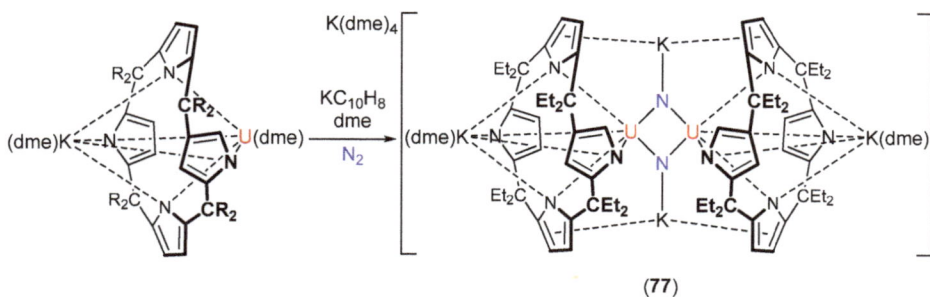

Scheme 2. N–N bond cleavage using a U^{III} complex to form **71**.

Reduction of the thorium bisphenolate complex [K(dme)$_2$][{(2-^{t}Bu-4-Me–C$_6$H$_2$O)$_2$-6-CH$_2$}$_2$ ThCl(dme)] (**A**) with potassium naphthalenide under an atmosphere of dinitrogen unexpectedly resulted in the amide complex [K(dme)$_4$][{(2-^{t}Bu-4-Me–C$_6$H$_2$O)$_2$-6-CH$_2$}$_2$Th(NH$_2$)(dme)] **78** which is the first example of N_2 functionalisation using an f-element complex (Scheme 3). The parent [NH$_2$]$^-$ amide ligand is confirmed through ^{15}N-NMR spectroscopy which shows a triplet at 155.01 ppm ($^1J_{NH}$ = 57.2 Hz) [120]. The proposed mechanism of this transformation involves formation of a formally zero-valent thorium intermediate which contains two bound [C$_{10}$H$_8$K(18c6)] fragments (identified through a single crystal X-ray diffraction experiment). This intermediate can then react with the starting material **A** leading to N_2 activation, cleavage and hydrogenation as a result of H atom abstraction from solvent molecules.

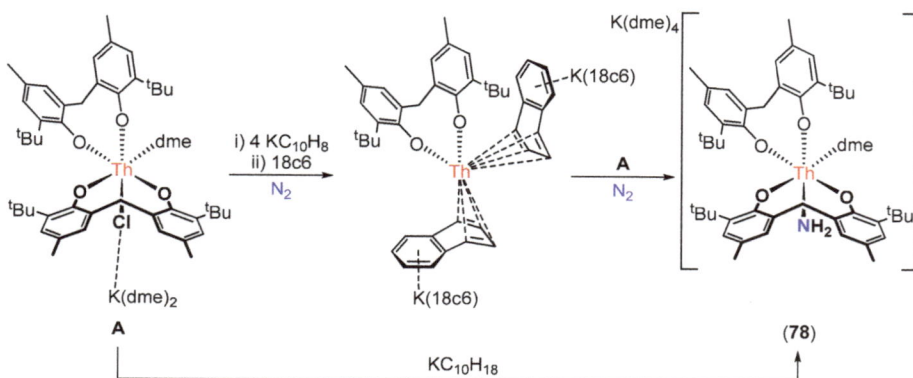

Scheme 3. N_2 cleavage and hydrogenation by a thorium complex.

4. White Phosphorus Activation by Rare Earth Complexes

Rare earth complexes resulting from P_4 activation are illustrated in Figure 12. For reference, average P–P and Ln–P bond lengths obtained from single crystal X-ray diffraction experiments, and ^{31}P-NMR spectroscopic resonances are summarised in Table 5. The structural cores of complexes **79–84** are shown in Figure 13 for clarity.

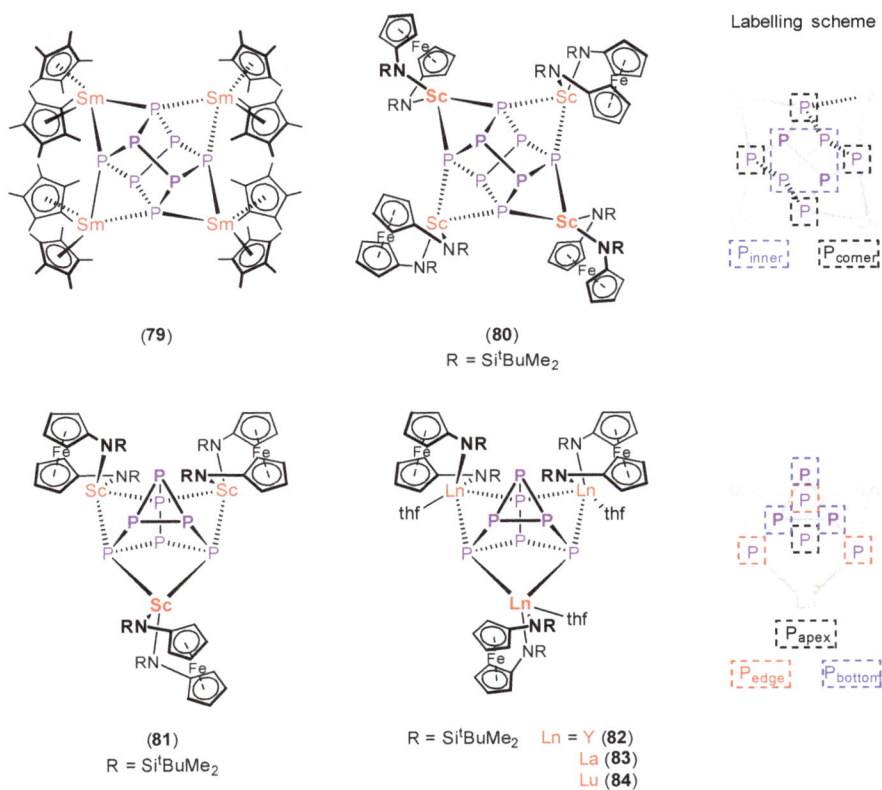

Figure 12. Rare earth complexes resulting from P_4 activation.

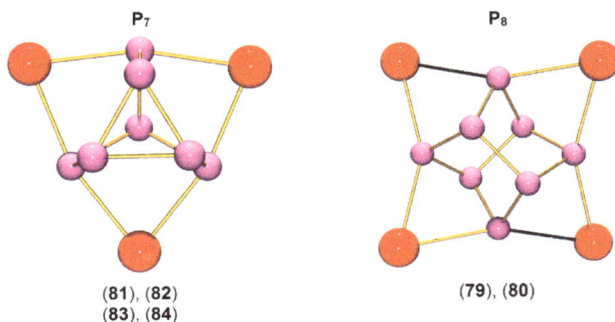

Figure 13. Overview of the Ln_xP_n structural cores resulting from P_4 activation by rare earth complexes.

Table 5. Summary of rare earth P_4 activation complexes.

Complex (#) [Reference]	Average P–P Bond Lengths (Å)	Average M–P Bond Length (Å)	^{31}P-NMR Spectroscopy (298 K) (ppm) [a]
P_4	2.21 [121]	-	−488 to −527 [122]
$[(\eta^5\text{-}C_5Me_5)_2Sm]_4(\mu_4\text{-}\eta^2{:}\eta^2{:}\eta^2{:}\eta^2\text{-}P_8)$ (**79**) [123]	2.195 (P_{corner}–P_{inner}) 2.291 (P_{inner}–P_{inner})	3.047	-
$[\{Fe(1\text{-}NSi^tBuMe_2\text{-}C_5H_4)_2\}Sc](\mu_4\text{-}\eta^2{:}\eta^2{:}\eta^2{:}\eta^2\text{-}P_8)$ (**80**) [124]	2.204 (P_{corner}–P_{inner}) 2.308 (P_{inner}–P_{inner})	2.768	+45.7 +96.2
$[\{Fe(1\text{-}NSi^tBuMe_2\text{-}C_5H_4)_2\}Sc]_3(\mu_3\text{-}\eta^2{:}\eta^2{:}\eta^2\text{-}P_7)$ (**81**) [124]	2.229 (P_{bottom}–P_{bottom}) 2.197 (P_{edge}–P_{bottom}) 2.201 (P_{apex}–P_{edge})	2.750	+23.1 −118.9 −131.4
$[\{Fe(1\text{-}NSi^tBuMe_2\text{-}C_5H_4)_2\}Y(thf)]_3(\mu_3\text{-}\eta^2{:}\eta^2{:}\eta^2\text{-}P_7)$ (**82**) [124]	2.238 (P_{bottom}–P_{bottom}) 2.176 (P_{edge}–P_{bottom}) 2.188 (P_{apex}–P_{edge})	2.950	−21.1 −82.4 −130.3
$[\{Fe(1\text{-}NSi^tBuMe_2\text{-}C_5H_4)_2\}La(thf)]_3(\mu_3\text{-}\eta^2{:}\eta^2{:}\eta^2\text{-}P_7)$ (**83**) [125]	2.258 (P_{bottom}–P_{bottom}) 2.161 (P_{edge}–P_{bottom}) 2.191 (P_{apex}–P_{edge})	3.120	−75
$[\{Fe(1\text{-}NSi^tBuMe_2\text{-}C_5H_4)_2\}Lu(thf)]_3(\mu_3\text{-}\eta^2{:}\eta^2{:}\eta^2\text{-}P_7)$ (**84**) [125]	2.233 (P_{bottom}–P_{bottom}) 2.181 (P_{edge}–P_{bottom}) 2.183 (P_{apex}–P_{edge})	2.893	+0.8 −96.8 −133.3

[a] Referenced to 85% H_3PO_4.

Roesky and co-workers reported the first example of a molecular polyphosphide of the rare earth elements, $[(\eta^5\text{-}C_5Me_5)_2Sm]_4(\mu_4\text{-}\eta^2{:}\eta^2{:}\eta^2{:}\eta^2\text{-}P_8)$ (**79**) [123]. The samarocene $(\eta^5\text{-}C_5Me_5)_2Sm$ activates P_4 to yield a P_8^{4-} fragment with a realgar-type structure, a process proposed to be driven by the one-electron oxidation of the divalent samarium metal centre. **79** has molecular D_{2d} symmetry and the [Cp*$_2$Sm] units bridge the P_8^{4-} cage with Sm–P distances in the range of 2.997(2) to 3.100(2) Å. DFT calculations support the strongly ionic character of the Sm–P bonds.

Activation of P_4 by group 3 metal centres was first reported by Diaconescu and co-workers [124]. Reaction of the scandium arene inverse-sandwich complexes $[\{Fe(1\text{-}NSi^tBuMe_2\text{-}C_5H_4)_2\}Sc]_2(\mu\text{-arene})$ (arene = $C_{10}H_8$ or $C_{14}H_{10}$) with P_4 resulted in displacement of the neutral arene and formation of a mixture of the tetrametallic $[\{Fe(1\text{-}NSi^tBuMe_2\text{-}C_5H_4)_2\}Sc]_4(\mu_4\text{-}\eta^2{:}\eta^2{:}\eta^2{:}\eta^2\text{-}P_8)$ (**80**) and trimetallic $[\{Fe(1\text{-}NSi^tBuMe_2\text{-}C_5H_4)_2\}Sc]_3(\mu_3\text{-}\eta^2{:}\eta^2{:}\eta^2\text{-}P_7)$ (**81**). The mixtures were readily separated and the product distribution could be controlled by the stoichiometry of P_4 and the nature of the arene starting material. **80** possesses a realgar-type P_8^{4-} unit whereas **81** contains a Zintl-type P_7^{3-} unit [126], the first example of its formation in the absence of strong alkali metal reducing agents. Solution phase ^{31}P-NMR spectroscopy demonstrates a diagnostic AA′A″MM′M″X spin system. The analogous yttrium arene inverse-sandwich complex $[\{Fe(1\text{-}NSi^tBuMe_2\text{-}C_5H_4)_2\}Y(thf)]_2(\mu\text{-}C_{10}H_8)$ activates P_4 to yield $[\{Fe(1\text{-}NSi^tBuMe_2\text{-}C_5H_4)_2\}Y(thf)]_3(\mu_3\text{-}\eta^2{:}\eta^2{:}\eta^2\text{-}P_7)$ (**82**) as the sole product where the larger coordination sphere of yttrium is saturated with an additional thf molecule [127]. Importantly, in the context of functionalisation of white phosphorus to organophosphorus compounds, both **81** and **82** were shown to react with 3 equivalents of Me_3SiI to yield $P_7(SiMe_3)_3$ and $\{Fe(1\text{-}NSi^tBuMe_2\text{-}C_5H_4)_2\}MI$ (Scheme 4).

This chemistry was later extended to lanthanum and lutetium using the same methodology, forming $[\{Fe(1\text{-}NSi^tBuMe_2\text{-}C_5H_4)_2\}Ln(thf)]_3(\mu_3\text{-}\eta^2{:}\eta^2{:}\eta^2\text{-}P_7)$ (Ln = La (**83**), Lu (**84**)) [125]. The valence tautomerisation of P_7^{3-} in **83**, which occurs at a similar temperature to Li_3P_7 but does not require donor solvents, was proposed to take place by a lanthanum-assisted mechanism involving simultaneous formation and breaking of 4 La–P bonds.

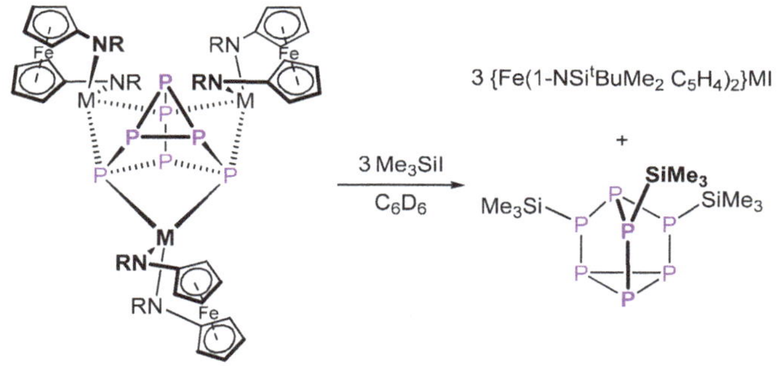

R = SiMe$_2$^{t}Bu M = Sc (**81**)
Y(thf) (**82**)

Scheme 4. P_7^{3-} functionalisation with Me_3SiI.

5. White Phosphorus Activation by Actinide Complexes

Actinide complexes resulting from P_4 activation are illustrated in Figure 14. For reference, average P–P and An–P bond lengths obtained from single crystal X-ray diffraction experiments, and ^{31}P-NMR spectroscopic resonances are summarised in Table 6. The structural cores of complexes **85–91** are shown in Figure 15 for clarity.

Table 6. Summary of actinide P_4 activation complexes.

Complex (#) [Reference]	Average P–P Bond Lengths (Å)	Average An–P Bond Length (Å)	^{31}P-NMR Spectroscopy (ppm) ^{a}
P_4	2.21 [121]	-	−488 to −527 [122]
[(η^5-1,3-^{t}Bu–C$_5$H$_3$)$_2$Th](μ-η^3:η^3-*cyclo*-P$_3$)[(η^5-1,3-^{t}Bu–C$_5$H$_3$)$_2$ThCl] (**85**) [128]	2.185	2.913	−75.7 (293 K)
[(η^5-1,3-^{t}Bu–C$_5$H$_3$)$_2$Th]$_2$(μ-η^3:η^3-P$_6$) (**86**) [128]	2.234	2.904 (Th-η^2-P) 2.844 (Th-η^1-P)	+125.4 (293 K) −41.9 (293 K)
[{(3,5-Me–C$_6$H$_3$)(^{t}Bu)N}$_3$U](μ-η^4:η^4-*cyclo*-P$_4$) (**87**) [129]	2.160	3.127	+794
[{(3,5-Me–C$_6$H$_3$)(Ad)N}$_3$U](μ-η^4:η^4-*cyclo*-P$_4$) (**88**) [129]	2.159	3.124	+803
[(η^5-C$_5$Me$_5$)(η^8-1,4-Si^{i}Pr$_3$–C$_8$H$_6$)U]$_2$(μ-η^2:η^2-*cyclo*-P$_4$) (**89**) [130]	2.150	2.977	+718
[HC(SiMe$_2$N-4-Me–C$_6$H$_4$)$_3$U]$_3$(μ$_3$-η^2:η^2:η^2-P$_7$) (**90**) [130]	2.249 (P$_{bottom}$–P$_{bottom}$) 2.187 (P$_{edge}$–P$_{bottom}$) 2.209 (P$_{apex}$–P$_{edge}$)	2.990	-
[{N(CH$_2$CH$_2$NSi^{i}Pr$_3$)$_3$}U]$_2$(μ-η^5:η^5-*cyclo*-P$_5$) (**91**) [131]	2.006	3.280	-

^{a} Referenced to 85% H_3PO_4.

The first report of P_4 activation by an actinide complex came from Scherer *et al.*, using thorium [128]. The butadiene complex (η^5-1,3-^{t}Bu–C$_5$H$_3$)$_2$Th(η^4-C$_4$H$_6$) reacts with P_4 at 100 °C in the presence of MgCl$_2$(OEt$_2$) to yield [(η^5-1,3-^{t}Bu–C$_5$H$_3$)$_2$Th](μ-η^3:η^3-*cyclo*-P$_3$)[(η^5-1,3-^{t}Bu–C$_5$H$_3$)$_2$ThCl] (**85**). In the absence of MgCl$_2$(OEt$_2$), [(η^5-1,3-^{t}Bu–C$_5$H$_3$)$_2$Th]$_2$(μ-η^3:η^3-P$_6$) (**86**) was afforded as a consequence of P_4 fragmentation and subsequent catenation. **85** features a *cyclo*-P_3^{3-} unit and formal Th^{IV} centres. In the solid state structure, the coordination environment about the thorium centres is trigonal planar and tetrahedral (Cl bound) with Th–P distances ranging from 2.809(6) to 2.974(8) Å. The P–P distances are 2.171(9), 2.192(9) and 2.192(8) Å. Only a single broad resonance is observed in the ^{31}P-NMR spectrum at 293 K but cooling to 193 K leads to splitting and the observation of an A_2B system; the barrier to rotation of the *cyclo*-P_3 unit was estimated to be *ca.* 44 kJ·mol^{-1}. **86** contains a P_6^{4-} bicycle with thorium metal centres capping the five-membered rings. Th–P bond lengths in the solid state structure range from 2.840(7) to 2.919(7) Å.

(85)	**(86)**
	1,3-^tBu groups on Cp
	rings omitted for clarity

Ar = 3,5-Me–C$_6$H$_3$ R = ^{t}Bu (**87**)
Ad (**88**)

(89)
R = Si^{i}Pr$_3$

(90)
Ar = *p*-tolyl

(91)
R = Si^{i}Pr$_3$

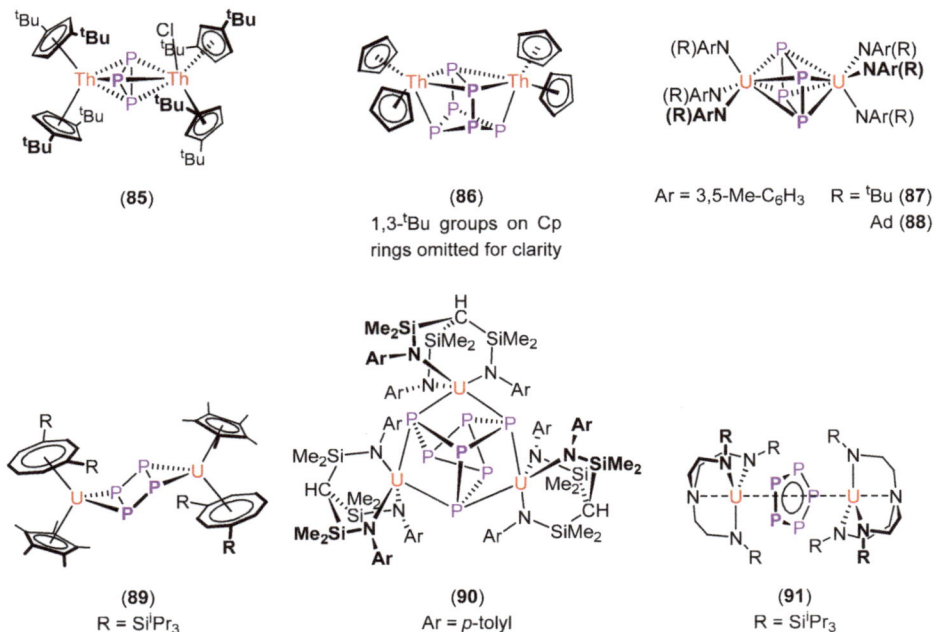

Figure 14. Actinide complexes resulting from P$_4$ activation.

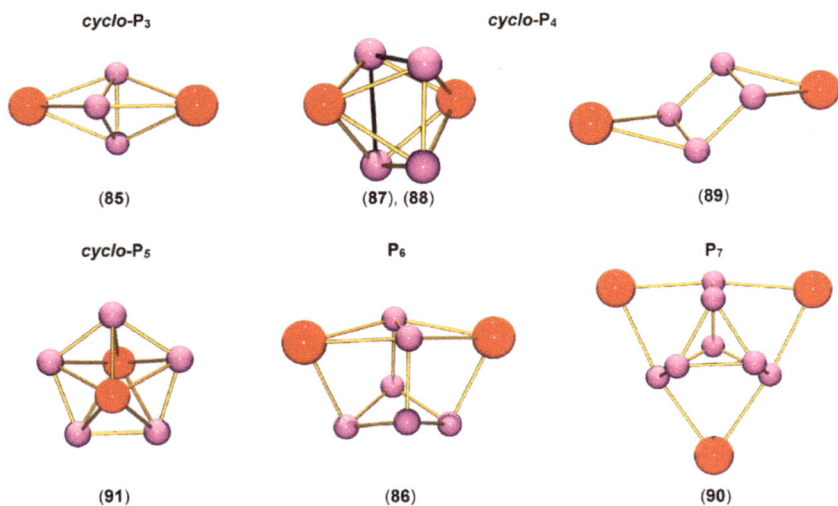

cyclo-P$_3$

cyclo-P$_4$

(85)

(87), (88)

(89)

cyclo-P$_5$

P$_6$

P$_7$

(91)

(86)

(90)

Figure 15. Overview of the An$_x$P$_n$ structural cores resulting from P$_4$ activation by actinide complexes.

The following report of P$_4$ activation came over a decade later and was the first using a uranium complex [129]. The U^{III} tris(amide) U{N(R)Ar}$_3$(thf) (R = ^{t}Bu or Ad, Ar = 3,5-Me–C$_6$H$_3$) reacts with 0.5 equivalents of P$_4$ to yield [{Ar(R)N}$_3$U](μ-η^4:η^4-*cyclo*-P$_4$) (R = ^{t}Bu (**87**), Ad (**88**)) which contains a *cyclo*-P$_4$$^{2-}$ unit and where the metal centres have been formally oxidised to U^{IV}. In the solid state structures, the average P–P bond distance is 2.159 Å and the P–P–P angle is 90°; both statistically equivalent across the two structures. Resonances at 794 and 803 ppm were observed for **87** and **88** respectively in the ^{31}P-NMR spectrum. Computational studies implied that the U–P bonding character is largely ionic with the presence of a weak δ-bonding interaction between filled U df hybrid orbitals and the P_4^{2-} LUMO.

Cloke and co-workers described the related *cyclo*-P$_4$ example [(η^5-C$_5$Me$_5$)(η^8-1,4-Si^{i}Pr$_3$–C$_8$H$_6$)U]$_2$ (μ-η^2:η^2-*cyclo*-P$_4$) (**89**) which was prepared from (η^5-C$_5$Me$_5$)(η^8-1,4-Si^{i}Pr$_3$–C$_8$H$_6$)U(thf) and 0.5 equivalents of P$_4$ [130]. This was the first example of the μ-η^2:η^2-P$_4$ coordination mode [132,133] and DFT studies on a model system [(η^5-C$_5$H$_5$)(η^8-C$_8$H$_8$)U](μ-η^2:η^2-*cyclo*-P$_4$) support the formulation of the dimer with P$_4^{2-}$ and U^{IV} oxidation states. The tilted *cyclo*-P$_4^{2-}$ unit leads to U–P bonding interactions involving both σ and π orbitals. The wedge shaped nature of the sterically demanding (η^5-C$_5$Me$_5$)(η^8-1,4-Si^{i}Pr$_3$–C$_8$H$_6$)U fragment likely results in the slipped μ-η^2:η^2 coordination mode.

Following on from the rare earth inverse sandwich complexes that resulted in P$_8^{4-}$ and P$_7^{3-}$ clusters, and P$_7^{3-}$ functionalisation [124,125], Liddle and co-workers reported the reaction of [HC(SiMe$_2$NAr)$_3$U]$_2$(μ-η^6-η^6-C$_6$H$_5$CH$_3$) (Ar = 4-Me–C$_6$H$_4$) with 1.1 equivalents of P$_4$ which afforded the first actinide Zintl complex [HC(SiMe$_2$NAr)$_3$U]$_3$(μ$_3$-η^2:η^2:η^2-P$_7$) (**90**) [134]. U–P bonding was determined to be essentially ionic. Interestingly, reaction of **90** with a number of electrophiles under ambient conditions led to functionalisation of the P$_7^{3-}$ unit and liberation of P$_7$R$_3$ (R = SiMe$_3$, Me, Ph, Li(tmeda)) after P–Si, P–C or P–Li bond formation. Though not catalytic, **90** could be regenerated from this reaction mixture and two turnovers achieved demonstrating a significant step towards controlled P$_4$ activation under mild conditions.

Very recently, Liddle and co-workers described the first example of a *cyclo*-P$_5$ complex resulting from activation of P$_4$ by an f-block complex [131]. [{N(CH$_2$CH$_2$NSi^{i}Pr$_3$)$_3$}U]$_2$(μ-η^5:η^5-*cyclo*-P$_5$) (**91**) was prepared by reaction of {N(CH$_2$CH$_2$NSi^{i}Pr$_3$)$_3$}U with 0.25 equivalents of P$_4$. Spectroscopic and magnetic measurements support oxidation to afford U^{IV} centres and charge transfer resulting in a formal P$_5^{2-}$ ligand in this inverse sandwich complex. Despite the isolobal analogy of *cyclo*-P$_5$ with the cyclopentadienyl anion, which bonds to metal centres using primarily σ- and π-bonding, calculations on **91** suggest that the principal U–P interactions involve polarised δ-bonding and this can be attributed to the energetically available uranium 5f orbitals of correct δ-symmetry.

Compared to both rare earth and actinide complexes, the activation of P$_4$ by transition metal complexes have proven to result in a wide variety of activation products [58,59]; with notable examples including fragmentation resulting in terminal and bridging P$_1$ ligands [135–137], P$_2$ ligands [138], *cyclo*-P$_3$ ligands [56,139], fragmentation to other P$_4$ ligands [140], coordination of P$_4$ tetrahedra [141,142], and expansion to P$_n$ (n = 5–14) ligands [143–146]. More significantly, functionalisation of these phosphorus units has also been observed.

6. Arsenic, Antimony and Bismuth Activation by Rare Earth and Actinide Complexes

There is only a single example of molecular arsenic activation by a rare earth or actinide metal complex. Reaction of the thorium butadiene complex (η^5-1,3-^{t}Bu–C$_5$H$_3$)$_2$Th(η^4-C$_4$H$_6$) with As$_4$ (yellow arsenic) in boiling xylene affords [(η^5-1,3-^{t}Bu–C$_5$H$_3$)$_2$Th]$_2$(μ-η^3:η^3-As$_6$) in analogy to the previously reported P$_4$ chemistry of Scherer *et al.* (Figure 16) [147]. In the solid state molecular structure; the average Th–(η^2-As$_6$) bonds are 3.027 Å, average Th–(η^1-As) bonds are 2.922 Å and average As–As bonds are 2.459 Å.

In contrast, activation of molecular arsenic by homogeneous transition metal complexes is considerably more diverse; formation of *cyclo*-As$_n$ ligands (n = 3–6, 8) [54,148], metal arsenic clusters [149], coordination of intact As$_4$ tetrahedra to metal ions [133,150,151], fragmentation into As$_2$ and other As$_4$ ligands [152–154], catenation to As$_{10}$ and As$_{12}$ ligands [155], reactions to form P$_n$As$_m$ ligands [156], and full As$_4$ fragmentation resulting in terminal M≡As arsenide bonds [157] have all been reported.

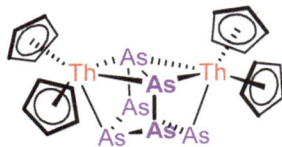

(92)

1,3-^{t}Bu groups on Cp
rings omitted for clarity

Figure 16. Actinide complexes resulting from As$_4$ activation.

Beyond arsenic in group 15 are antimony and bismuth. While activation of molecular forms of these elements is unlikely, it is worth noting that Scheer and co-workers reported a tungsten terminal stibido complex {N(CH$_2$CH$_2$NSiMe$_3$)$_3$}$_3$W≡Sb prepared from reaction of {N(CH$_2$CH$_2$NSiMe$_3$)$_3$}$_3$WCl with LiSb(H){CH(SiMe$_3$)$_2$} [158], while Breunig *et al.*, have reported [(η^5-C$_5$Me$_5$)$_2$Mo(CO)$_2$](*cyclo*-Sb$_3$) and [(η^5-C$_5$H$_5$)$_2$Mo(CO)$_2$](*cyclo*-Sb$_3$), which are a result of reaction of [(η^5-Cp^R)$_2$Mo(CO)$_3$]$_2$ with (^{t}BuSb)$_4$ [2]. In terms of rare earth complexes; [(η^5-C$_5$Me$_5$)$_2$Sm]$_2$(μ$_3$-η^1:η^2:η^2-Sb$_3$){(η^5-C$_5$Me$_5$)$_2$(thf)Sm} and [(η^5-C$_5$Me$_5$)$_2$Sm]$_2$(η^2:η^2-Bi$_2$) were prepared by reaction of [(η^5-C$_5$Me$_5$)$_2$Sm]$_2$ with SbPh$_3$ and BiPh$_3$ respectively [3,4].

7. Conclusions and Perspectives

To date, a wide range of rare earth dinitrogen complexes have been prepared (**1–68**), including group 3 metal ions and 4f elements at both ends of the periodic table, despite the limited radial extension of the 4f orbitals and the trivalent oxidation state being the most prevalent. In fact, apart from the very first f-element dinitrogen complex (**4**), all of the other complexes are air-sensitive but stable to N$_2$ dissociation *in vacuo*. Reduction of N$_2$ to N$_2$$^{2-}$ has most commonly been achieved with the [A$_2$(thf)$_x$Ln]$_2$(μ-η^2:η^2-N$_2$) structural motif (**1–38**) whereas examples of reduction to N$_2$$^{4-}$ have all involved more complex multidentate ligands (**61–68**). The nature of the bonding in complexes of the form [A$_2$(thf)$_x$Ln]$_2$(μ-η^2:η^2-N$_2$) has been extensively studied for both group 3, and closed and open shell 4f^n metal ions demonstrating that the Ln–N (N$_2$) bonding is based on a Ln nd–N$_2$ π* interaction. These systems have allowed for the first definitive characterisation of the N$_2$$^{3-}$ radical reduction product of dinitrogen (**44, 47**) and this has now been extended to many of the rare earth elements (**45, 46, 48–60**). Isolation of the N$_2$$^{3-}$ radical in homogeneous complexes is significant since it is likely to be a transient species in other transition metal systems and may also have a role in biological N$_2$ fixation.

In terms of reactivity, the rare earth N$_2$ complexes prepared thus far react with N$_2$ dissociation rather than N$_2$ cleavage and functionalisation [77,159–161]. Related to this, the understanding of the nature of bonding of Ln–N multiple bonds is of fundamental interest and the isolation of terminal imido (Ln=NR) complexes has only recently been reported [162–164]. It is also significant to consider the reactivity of Ln–N$_2$ complexes in the context of other low electron count early transition metal complexes; here, N$_2$ cleavage or functionalisation can be achieved through ligand induced reductive cleavage [165–173], and more generally, N–N bond scission of reduced N$_2$ derivatives can be attained through metal-ligand cooperativity [174,175].

Actinide dinitrogen complexes are much rarer, with just 8 examples of well-defined molecular uranium complexes (**69–76**). Of these, only the heterobimetallic U–Mo end-on N$_2$ complexes (**70, 71**) and recently prepared U^{IV} aryloxide and siloxide side-on N$_2$ complexes (**75, 76**) are thermally robust and stable when exposed to vacuum. Key steps forward have been made in the understanding of U–N (N$_2$) bonding in the side-on N$_2$ complexes as a polar covalent U 5f-N$_2$ π* interaction, and in the rationalisation of the differences between solid state N–N bond lengths and the overall electronic structure of these complexes. This understanding, in combination with well-designed ligand sets may lead the way in the preparation of other isolable actinide dinitrogen complexes for further study.

Though the isolated actinide N_2 complexes, like the rare earth N_2 complexes, tend to react with N_2 loss rather than N_2 functionalisation or cleavage, there are two reports of such reactivity. Importantly, in both cases, the putative $An(N_2)$ complex was not observed. Cleavage of N_2 by a uranium tetra-calix-pyrrole complex results in a bimetallic complex with bridging nitrides (**78**) whereas a thorium bisphenolate complex activates and functionalises N_2 to a parent amide ligand $[NH_2]^-$ (**79**) through an unknown mechanism. Both reactions occur in the presence of an external reductant. Examples of isolable terminal uranium nitrides (U≡N), derived from NaN_3, have only recently been reported [176–178], but it has already been demonstrated that these systems are capable of nitride functionalisation. **78** and **79** remain standout examples in demonstrating that actinide complexes can both cleave and functionalise N_2, but also highlight how much more remains to be understood in this field.

P_4 activation by rare earth complexes has led to both P_8^{4-} ions with realgar-type structures (**79**, **80**), and P_7^{3-} ions (**81–84**) using cyclopentadienyl and amido ancillary ligands. Promisingly, functionalisation of the P_7^{3-} unit in **81** and **82** was found possible using Me_3SiI to afford $P_7(SiMe_3)_3$. This is an interesting prospect for the synthesis of organophosphorus compounds from a P_4 building block. Actinide P_4 activation results in a more diverse array of phosphorus ligands; P_3^{3-} (**85**), P_6^{4-}, (**86**), *cyclo*-P_4^{2-} (**87–89**), P_7^{3-} (**90**), and *cyclo*-P_5^{2-} (**91**). Similar to rare earth chemistry, the P_7^{3-} ions in **90** could be functionalised by P–Si, P–C or P–Li bond formation to afford P_7R_3 units and this reaction cycle could be completed with two turnovers. **91** is the first example of an f-element being able to fragment and catenate P_4 to *cyclo*-P_5^{2-}. Despite the parallels with the cyclopentadienyl ligand, calculations suggest the U–P (*cyclo*-P_5) interaction to be based on polarised δ-bonding and electronic structure in these systems can be described as $[U^{IV}]_2(P_5^{2-})$. Putting this area into perspective, transition metals have already been shown to activate molecular phosphorus and, in limited examples, to result in further functionalisation.

There remains only a lone example of arsenic activation by a thorium butadiene complex leading to a P_6 cage (**92**) and no examples using rare earth metals. The first examples of crystallographically characterised uranium arsenide (U–AsH$_2$), arsenidene (U=AsH) and arsenido (U≡AsK) complexes have only recently been reported, using $KAsH_2$ as a source of arsenic [179]. These compounds raise the question of the diverse reactivity that actinide complexes could be expected to show with molecular arsenic and whether the formation of an unsupported, terminal actinide arsenide bond (M≡As) is accessible.

It is clear that molecular pnictogen activation by rare earth and actinide metal complexes is an exciting field of study which remains underdeveloped with respect to transition metals and main group elements. These unique metals offer the potential of new reactivity and functionalisation chemistry with the pnictogen elements, while the fundamental study of M–pnictogen bonds remains important.

Acknowledgments: Zoë R. Turner thanks Prof. Dermot O'Hare (University of Oxford), SCG Chemicals for financial support and a SCG Research Fellowship, and Trinity College for a Junior Research Fellowship.

Conflicts of Interest: The author declares no conflict of interest.

References

1. Kaltsoyannis, N. Does covalency increase or decrease across the actinide series? Implications for minor actinide partitioning. *Inorg. Chem.* **2013**, *52*, 3407–3413. [CrossRef] [PubMed]
2. Breunig, H.J.; Rösler, R.; Lork, E. Complexes with Sb_2 and *cyclo*-Sb_3 ligands: The tetrahedranes $[\{C_5H_5(CO)_2Mo\}_2Sb_2]$, $[C_5H_5(CO)_2MoSb_3]$, and $[C_5Me_5(CO)_2MoSb_3]$. *Angew. Chem. Int. Ed.* **1997**, *36*, 2819–2821. [CrossRef]
3. Evans, W.J.; Gonzales, S.L.; Ziller, J.W. The utility of $(C_5Me_5)_2Sm$ in isolating crystallographically characterizable zintl ions. X-ray crystal structure of a samarium complex of $(Sb_3)^{3-}$. *J. Chem. Soc. Chem. Commun.* **1992**, 1138–1139. [CrossRef]

4.   Evans, W.J.; Gonzales, S.L.; Ziller, J.W. Organosamarium-mediated synthesis of bismuth–bismuth bonds: X-ray crystal structure of the first dibismuth complex containing a planar $M_2(\mu\text{-}\eta^2{:}\eta^2\text{-Bi}_2)$ unit. *J. Am. Chem. Soc.* **1991**, *113*, 9880–9882. [CrossRef]

5.   Hoffman, B.M.; Dean, D.R.; Seefeldt, L.C. Climbing nitrogenase: Toward a mechanism of enzymatic nitrogen fixation. *Acc. Chem. Res.* **2009**, *42*, 609–619. [CrossRef] [PubMed]

6.   Hu, Y.; Ribbe, M.W. Decoding the nitrogenase mechanism: The homologue approach. *Acc. Chem. Res.* **2010**, *43*, 475–484. [CrossRef] [PubMed]

7.   Hellman, A.; Baerends, E.J.; Biczysko, M.; Bligaard, T.; Christensen, C.H.; Clary, D.C.; Dahl, S.; van Harrevelt, R.; Honkala, K.; Jonsson, H.; et al. Predicting catalysis: Understanding ammonia synthesis from first-principles calculations. *J. Phys. Chem. B* **2006**, *110*, 17719–17735. [CrossRef] [PubMed]

8.   Schrock, R.R. Reduction of dinitrogen. *Proc. Natl. Acad. Sci. USA* **2006**, *103*, 17087. [CrossRef] [PubMed]

9.   Smil, V. *Enriching the Earth: Fritz Haber, Carl Bosch and the Transformation of World Food Production*; Massachusetts Institute of Technology Press: Cambridge, MA, USA, 2001.

10. Erisman, J.W.; Sutton, M.A.; Galloway, J.; Klimont, Z.; Winiwarter, W. How a century of ammonia synthesis changed the world. *Nat. GeoSci.* **2008**, *1*, 636–639. [CrossRef]

11. *U.S. Geographical Survey, Mineral Commodity Surveys*; U.S. Department of the Interior: Washington, DC, USA, 2015.

12. Corbridge, D. *Phosphorus: An Outline of its Chemistry, Biochemistry, and Technology*, 5th ed.; Elsevier: New York, NY, USA, 1994.

13. Quin, L.D. *A Guide to Organophosphorus Chemistry*; Wiley: New York, NY, USA, 2000.

14. Engel, R. *Synthesis of Carbon Phosphorus Bonds*, 2nd ed.; CRC Press: Boca Raton, FL, USA, 2004.

15. Withers, P.J.A.; Elser, J.J.; Hilton, J.; Ohtake, H.; Schipper, W.J.; van Dijk, K.C. Greening the global phosphorus cycle: How green chemistry can help achieve planetary P sustainability. *Green Chem.* **2015**, *17*, 2087–2099. [CrossRef]

16. Liu, H. *Catalytic Ammonia Synthesis*; Plenum Press: New York, NY, USA, 1991.

17. Schlögl, R. Catalytic synthesis of ammonia—A "Never-Ending Story"? *Angew. Chem. Int. Ed.* **2003**, *42*, 2004–2008. [CrossRef] [PubMed]

18. Leigh, G.J. Haber–Bosch and Other Industrial Processes. In *Catalysts for Nitrogen Fixation*; Smith, B., Richards, R., Newton, W., Eds.; Springer: Houten, The Netherlands, 2004; Volume 1, pp. 33–54.

19. Studt, F.; Tuczek, F. Theoretical, spectroscopic, and mechanistic studies on transition-metal dinitrogen complexes: Implications to reactivity and relevance to the nitrogenase problem. *J. Comput. Chem.* **2006**, *27*, 1278–1291. [CrossRef] [PubMed]

20. MacKay, B.A.; Fryzuk, M.D. Dinitrogen coordination chemistry: On the biomimetic borderlands. *Chem. Rev.* **2004**, *104*, 385–401. [CrossRef] [PubMed]

21. Barrière, F. Model Complexes of the Active Site of Nitrogenases: Recent Advances. In *Bioinspired Catalysis*; Wiley-VCH Verlag GmbH & Co. KGaA: Weinheim, Germany, 2014; pp. 225–248.

22. Rolff, M.; Tuczek, F. Nitrogenase and Nitrogen Activation. In *Comprehensive Inorganic Chemistry II*, 2nd ed.; Poeppelmeier, J.R., Ed.; Elsevier: Amsterdam, The Netherlands, 2013; pp. 593–618.

23. Ribbe, M.W. *Nitrogen Fixation: Methods and Protocols*; Humana Press: New York, NY, USA, 2011.

24. Barrière, F. Modeling of the molybdenum center in the nitrogenase FeMo-cofactor. *Coord. Chem. Rev.* **2003**, *236*, 71–89. [CrossRef]

25. Smith, B.E.; Durrant, M.C.; Fairhurst, S.A.; Gormal, C.A.; Grönberg, K.L.C.; Henderson, R.A.; Ibrahim, S.K.; le Gall, T.; Pickett, C.J. Exploring the reactivity of the isolated iron-molybdenum cofactor of nitrogenase. *Coord. Chem. Rev.* **1999**, *185–186*, 669–687. [CrossRef]

26. Rehder, D. Vanadium nitrogenase. *J. Inorg. Biochem.* **2000**, *80*, 133–136. [CrossRef]

27. MacLachlan, E.A.; Fryzuk, M.D. Synthesis and reactivity of side-on-bound dinitrogen metal complexes. *Organometallics* **2006**, *25*, 1530–1543. [CrossRef]

28. Fryzuk, M.D. Side-on end-on bound dinitrogen: An activated bonding mode that facilitates functionalizing molecular nitrogen. *Acc. Chem. Res.* **2009**, *42*, 127–133. [CrossRef] [PubMed]

29. Poveda, A.; Perilla, I.C.; Pérez, C.R. Some considerations about coordination compounds with end-on dinitrogen. *J. Coord. Chem.* **2001**, *54*, 427–440. [CrossRef]

30. Gambarotta, S.; Scott, J. Multimetallic cooperative activation of N_2. *Angew. Chem. Int. Ed.* **2004**, *43*, 5298–5308. [CrossRef] [PubMed]

31. McWilliams, S.F.; Holland, P.L. Dinitrogen binding and cleavage by multinuclear iron complexes. *Acc. Chem. Res.* **2015**, *48*, 2059–2065. [CrossRef] [PubMed]

32. Ballmann, J.; Munha, R.F.; Fryzuk, M.D. The hydride route to the preparation of dinitrogen complexes. *Chem. Commun.* **2010**, *46*, 1013–1025. [CrossRef] [PubMed]

33. Jia, H.-P.; Quadrelli, E.A. Mechanistic aspects of dinitrogen cleavage and hydrogenation to produce ammonia in catalysis and organometallic chemistry: Relevance of metal hydride bonds and dihydrogen. *Chem. Soc. Rev.* **2014**, *43*, 547–564. [CrossRef] [PubMed]

34. Sivasankar, C.; Baskaran, S.; Tamizmani, M.; Ramakrishna, K. Lessons learned and lessons to be learned for developing homogeneous transition metal complexes catalyzed reduction of N_2 to ammonia. *J. Organomet. Chem.* **2014**, *752*, 44–58. [CrossRef]

35. Tanabe, Y.; Nishibayashi, Y. Developing more sustainable processes for ammonia synthesis. *Coord. Chem. Rev.* **2013**, *257*, 2551–2564. [CrossRef]

36. Van der Ham, C.J.M.; Koper, M.T.M.; Hetterscheid, D.G.H. Challenges in reduction of dinitrogen by proton and electron transfer. *Chem. Soc. Rev.* **2014**, *43*, 5183–5191. [CrossRef] [PubMed]

37. Rebreyend, C.; de Bruin, B. Photolytic N_2 splitting: A road to sustainable NH_3 production? *Angew. Chem. Int. Ed.* **2015**, *54*, 42–44. [CrossRef] [PubMed]

38. Himmel, H.-J.; Reiher, M. Intrinsic dinitrogen activation at bare metal atoms. *Angew. Chem. Int. Ed.* **2006**, *45*, 6264–6288. [CrossRef] [PubMed]

39. Chow, C.; Taoufik, M.; Quadrelli, E.A. Ammonia and dinitrogen activation by surface organometallic chemistry on silica-grafted tantalum hydrides. *Eur. J. Inorg. Chem.* **2011**, *2011*, 1349–1359. [CrossRef]

40. Ohki, Y.; Fryzuk, M.D. Dinitrogen activation by group 4 metal complexes. *Angew. Chem. Int. Ed.* **2007**, *46*, 3180–3183. [CrossRef] [PubMed]

41. Chirik, P.J. Dinitrogen functionalization with bis(cyclopentadienyl) complexes of zirconium and hafnium. *Dalton Trans.* **2007**, *1*, 16–25. [CrossRef] [PubMed]

42. Kuganathan, N.; Green, J.C.; Himmel, H.-J. Dinitrogen fixation and activation by Ti and Zr atoms, clusters and complexes. *New J. Chem.* **2006**, *30*, 1253–1262. [CrossRef]

43. Crossland, J.L.; Tyler, D.R. Iron–dinitrogen coordination chemistry: Dinitrogen activation and reactivity. *Coord. Chem. Rev.* **2010**, *254*, 1883–1894. [CrossRef]

44. Hazari, N. Homogeneous iron complexes for the conversion of dinitrogen into ammonia and hydrazine. *Chem. Soc. Rev.* **2010**, *39*, 4044–4056. [CrossRef] [PubMed]

45. Schrock, R.R. Catalytic reduction of dinitrogen to ammonia at a single molybdenum center. *Acc. Chem. Res.* **2005**, *38*, 955–962. [CrossRef] [PubMed]

46. Nishibayashi, Y. Molybdenum-catalyzed reduction of molecular dinitrogen into ammonia under ambient reaction conditions. *Comptes Rendus Chim.* **2015**, *18*, 776–784. [CrossRef]

47. Schrock, R.R. Catalytic Reduction of Dinitrogen to Ammonia by Molybdenum. In *Catalysis without Precious Metals*; Wiley-VCH Verlag GmbH & Co. KGaA: Weinheim, Germany, 2010; pp. 25–50.

48. Khoenkhoen, N.; de Bruin, B.; Reek, J.N.H.; Dzik, W.I. Reactivity of dinitrogen bound to mid- and late-transition-metal centers. *Eur. J. Inorg. Chem.* **2015**, *2015*, 567–598. [CrossRef]

49. Evans, W.J.; Lee, D.S. Early developments in lanthanide-based dinitrogen reduction chemistry. *Can. J. Chem.* **2005**, *83*, 375–384. [CrossRef]

50. Gardiner, M.G.; Stringer, D.N. Dinitrogen and related chemistry of the lanthanides: A review of the reductive capture of dinitrogen, as well as mono- and di-aza containing ligand chemistry of relevance to known and postulated metal mediated dinitrogen derivatives. *Materials* **2010**, *3*, 841–862. [CrossRef]

51. Gardner, B.M.; Liddle, S.T. Small-molecule activation at uranium(III). *Eur. J. Inorg. Chem.* **2013**, *2013*, 3753–3770. [CrossRef]

52. Liddle, S.T. The renaissance of non-aqueous uranium chemistry. *Angew. Chem. Int. Ed.* **2015**, *54*, 8604–8641. [CrossRef] [PubMed]

53. Scherer, O.J. Complexes with substituent-free acyclic and cyclic phosphorus, arsenic, antimony, and bismuth ligands. *Angew. Chem. Int. Ed.* **1990**, *29*, 1104–1122. [CrossRef]

54. Scherer, O.J. P_n and As_n ligands: A novel chapter in the chemistry of phosphorus and arsenic. *Acc. Chem. Res.* **1999**, *32*, 751–762. [CrossRef]

55. Peruzzini, M.; Gonsalvi, L.; Romerosa, A. Coordination chemistry and functionalization of white phosphorus via transition metal complexes. *Chem. Soc. Rev.* **2005**, *34*, 1038–1047. [CrossRef] [PubMed]

56. Vaira, M.D.; Sacconi, L. Transition metal complexes with *cyclo*-triphosphorus (η^3-P$_3$) and *tetrahedro*-tetraphosphorus (η^1-P$_4$) ligands. *Angew. Chem. Int. Ed.* **1982**, *21*, 330–342. [CrossRef]

57. Figueroa, J.S.; Cummins, C.C. A niobaziridine hydride system for white phosphorus or dinitrogen activation and N- or P-atom transfer. *Dalton Trans.* **2006**, 2161–2168. [CrossRef] [PubMed]

58. Cossairt, B.M.; Piro, N.A.; Cummins, C.C. Early-transition-metal-mediated activation and transformation of white phosphorus. *Chem. Rev.* **2010**, *110*, 4164–4177. [CrossRef] [PubMed]

59. Caporali, M.; Gonsalvi, L.; Rossin, A.; Peruzzini, M. P$_4$ activation by late-transition metal complexes. *Chem. Rev.* **2010**, *110*, 4178–4235. [CrossRef] [PubMed]

60. Giffin, N.A.; Masuda, J.D. Reactivity of white phosphorus with compounds of the p-block. *Coord. Chem. Rev.* **2011**, *255*, 1342–1359. [CrossRef]

61. Balázs, G.; Seitz, A.; Scheer, M. Activation of White Phosphorus (P$_4$) by Main Group Elements and Compounds. In *Comprehensive Inorganic Chemistry II*, 2nd ed.; Poeppelmeier, J.R., Ed.; Elsevier: Amsterdam, The Netherlands, 2013; pp. 1105–1132.

62. Khan, S.; Sen, S.S.; Roesky, H.W. Activation of phosphorus by group 14 elements in low oxidation states. *Chem. Commun.* **2012**, *48*, 2169–2179. [CrossRef] [PubMed]

63. Johnson, B.P.; Balázs, G.; Scheer, M. Low-coordinate E$_1$ ligand complexes of group 15 elements—A developing area. *Coord. Chem. Rev.* **2006**, *250*, 1178–1195. [CrossRef]

64. Scheer, M.; Balázs, G.; Seitz, A. P$_4$ activation by main group elements and compounds. *Chem. Rev.* **2010**, *110*, 4236–4256. [CrossRef] [PubMed]

65. Fang, M.; Bates, J.E.; Lorenz, S.E.; Lee, D.S.; Rego, D.B.; Ziller, J.W.; Furche, F.; Evans, W.J. (N$_2$)$^{3-}$ radical chemistry via trivalent lanthanide salt/alkali metal reduction of dinitrogen: New syntheses and examples of (N$_2$)$^{2-}$ and (N$_2$)$^{3-}$ complexes and density functional theory comparisons of closed shell Sc^{3+}, Y^{3+}, and Lu^{3+} *versus* 4f^9 Dy^{3+}. *Inorg. Chem.* **2011**, *50*, 1459–1469. [CrossRef] [PubMed]

66. Fryzuk, M.D.; Johnson, S.A. The continuing story of dinitrogen activation. *Coord. Chem. Rev.* **2000**, *200–202*, 379–409. [CrossRef]

67. Jaroschik, F.; Momin, A.; Nief, F.; LeGoff, X.-F.; Deacon, G.B.; Junk, P.C. Dinitrogen reduction and C–H activation by the divalent organoneodymium complex [(C$_5$H$_2$^{t}Bu$_3$)$_2$Nd(μ-I)K([18]crown-6)]. *Angew. Chem. Int. Ed.* **2009**, *48*, 1117–1121. [CrossRef] [PubMed]

68. Evans, W.J.; Rego, D.B.; Ziller, J.W. Synthesis, structure, and ^{15}N-NMR studies of paramagnetic lanthanide complexes obtained by reduction of dinitrogen. *Inorg. Chem.* **2006**, *45*, 10790–10798. [CrossRef] [PubMed]

69. Demir, S.; Lorenz, S.E.; Fang, M.; Furche, F.; Meyer, G.; Ziller, J.W.; Evans, W.J. Synthesis, structure, and density functional theory analysis of a scandium dinitrogen complex, [(C$_5$Me$_4$H)$_2$Sc]$_2$(μ-η^2:η^2-N$_2$). *J. Am. Chem. Soc.* **2010**, *132*, 11151–11158. [CrossRef] [PubMed]

70. Demir, S.; Siladke, N.A.; Ziller, J.W.; Evans, W.J. Scandium and yttrium metallocene borohydride complexes: Comparisons of (BH$_4$)$^{1-}$ *vs.* (BPh$_4$)$^{1-}$ coordination and reactivity. *Dalton Trans.* **2012**, *41*, 9659–9666. [CrossRef] [PubMed]

71. Schmiege, B.M.; Ziller, J.W.; Evans, W.J. Reduction of dinitrogen with an yttrium metallocene hydride precursor, [(C$_5$Me$_5$)$_2$YH]$_2$. *Inorg. Chem.* **2010**, *49*, 10506–10511. [CrossRef] [PubMed]

72. Evans, W.J.; Ulibarri, T.A.; Ziller, J.W. Isolation and X-ray crystal structure of the first dinitrogen complex of an f-element metal, [(C$_5$Me$_5$)$_2$Sm]$_2$N$_2$. *J. Am. Chem. Soc.* **1988**, *110*, 6877–6879. [CrossRef]

73. Fieser, M.E.; Bates, J.E.; Ziller, J.W.; Furche, F.; Evans, W.J. Dinitrogen reduction via photochemical activation of heteroleptic tris(cyclopentadienyl) rare-earth complexes. *J. Am. Chem. Soc.* **2013**, *135*, 3804–3807. [CrossRef] [PubMed]

74. Evans, W.J.; Allen, N.T.; Ziller, J.W. Expanding divalent organolanthanide chemistry: The first organothulium(II) complex and the *in situ* organodysprosium(II) reduction of dinitrogen. *Angew. Chem. Int. Ed.* **2002**, *41*, 359–361. [CrossRef]

75. Evans, W.J.; Allen, N.T.; Ziller, J.W. Facile dinitrogen reduction via organometallic Tm(II) chemistry. *J. Am. Chem. Soc.* **2001**, *123*, 7927–7928. [CrossRef] [PubMed]

76. Mueller, T.J.; Fieser, M.E.; Ziller, J.W.; Evans, W.J. (C$_5$Me$_4$H)$^{1-}$-based reduction of dinitrogen by the mixed ligand tris(polyalkylcyclopentadienyl) lutetium and yttrium complexes, (C$_5$Me$_5$)$_{3-x}$(C$_5$Me$_4$H)$_x$Ln. *Chem. Sci.* **2011**, *2*, 1992–1996. [CrossRef]

77. Lorenz, S.E.; Schmiege, B.M.; Lee, D.S.; Ziller, J.W.; Evans, W.J. Synthesis and reactivity of bis(tetramethylcyclopentadienyl) yttrium metallocenes including the reduction of Me_3SiN_3 to $[(Me_3Si)_2N]^-$ with $[(C_5Me_4H)_2Y(THF)]_2(\mu\text{-}\eta^2{:}\eta^2\text{-}N_2)$. *Inorg. Chem.* **2010**, *49*, 6655–6663. [CrossRef] [PubMed]

78. MacDonald, M.R.; Ziller, J.W.; Evans, W.J. Synthesis of a crystalline molecular complex of Y^{2+}, $[(18\text{-crown-}6)K][(C_5H_4SiMe_3)_3Y]$. *J. Am. Chem. Soc.* **2011**, *133*, 15914–15917. [CrossRef] [PubMed]

79. Evans, W.J.; Lee, D.S.; Lie, C.; Ziller, J.W. Expanding the LnZ_3/alkali-metal reduction system to organometallic and heteroleptic precursors: Formation of dinitrogen derivatives of lanthanum. *Angew. Chem. Int. Ed.* **2004**, *43*, 5517–5519. [CrossRef] [PubMed]

80. Evans, W.J.; Lee, D.S.; Johnston, M.A.; Ziller, J.W. The elusive $(C_5Me_4H)_3Lu$: Its synthesis and $LnZ_3/K/N_2$ reactivity. *Organometallics* **2005**, *24*, 6393–6397. [CrossRef]

81. Evans, W.J.; Lee, D.S.; Rego, D.B.; Perotti, J.M.; Kozimor, S.A.; Moore, E.K.; Ziller, J.W. Expanding dinitrogen reduction chemistry to trivalent lanthanides via the LnZ_3/Alkali metal reduction system: Evaluation of the generality of forming $Ln_2(\mu\text{-}\eta^2{:}\eta^2\text{-}N_2)$ complexes via LnZ_3/K. *J. Am. Chem. Soc.* **2004**, *126*, 14574–14582. [CrossRef] [PubMed]

82. Evans, W.J.; Zucchi, G.; Ziller, J.W. Dinitrogen reduction by Tm(II), Dy(II), and Nd(II) with simple amide and aryloxide ligands. *J. Am. Chem. Soc.* **2003**, *125*, 10–11. [CrossRef] [PubMed]

83. Evans, W.J.; Lee, D.S.; Ziller, J.W. Reduction of dinitrogen to planar bimetallic $M_2(\mu\text{-}\eta^2{:}\eta^2\text{-}N_2)$ complexes of Y, Ho, Tm, and Lu using the $K/Ln[N(SiMe_3)_2]_3$ reduction system. *J. Am. Chem. Soc.* **2004**, *126*, 454–455. [CrossRef] [PubMed]

84. Corbey, J.F.; Farnaby, J.H.; Bates, J.E.; Ziller, J.W.; Furche, F.; Evans, W.J. Varying the Lewis base coordination of the Y_2N_2 core in the reduced dinitrogen complexes $\{[(Me_3Si)_2N]_2(L)Y\}_2(\mu\text{-}\eta^2{:}\eta^2\text{-}N_2)$ (L = benzonitrile, pyridines, triphenylphosphine oxide, and trimethylamine N-oxide). *Inorg. Chem.* **2012**, *51*, 7867–7874. [CrossRef] [PubMed]

85. Campazzi, E.; Solari, E.; Floriani, C.; Scopelliti, R. The fixation and reduction of dinitrogen using lanthanides: Praseodymium and neodymium *meso*-octaethylporphyrinogen-dinitrogen complexes. *Chem. Commun.* **1998**, 2603–2604. [CrossRef]

86. Cheng, J.; Takats, J.; Ferguson, M.J.; McDonald, R. Heteroleptic Tm(II) complexes: One more success for Trofimenko's scorpionates. *J. Am. Chem. Soc.* **2008**, *130*, 1544–1545. [CrossRef] [PubMed]

87. Mansell, S.M.; Farnaby, J.H.; Germeroth, A.I.; Arnold, P.L. Thermally stable uranium dinitrogen complex with siloxide supporting ligands. *Organometallics* **2013**, *32*, 4214–4222. [CrossRef]

88. Mansell, S.M.; Kaltsoyannis, N.; Arnold, P.L. Small molecule activation by uranium tris(aryloxides): Experimental and computational studies of binding of N_2, coupling of CO, and deoxygenation insertion of CO_2 under ambient conditions. *J. Am. Chem. Soc.* **2011**, *133*, 9036–9051. [CrossRef] [PubMed]

89. Perrin, L.; Maron, L.; Eisenstein, O.; Schwartz, D.J.; Burns, C.J.; Andersen, R.A. Bonding of H_2, N_2, ethylene, and acetylene to bivalent lanthanide metallocenes: Trends from DFT calculations on Cp_2M and Cp^*_2M (M = Sm, Eu, Yb) and experiments with Cp^*_2Yb. *Organometallics* **2003**, *22*, 5447–5453. [CrossRef]

90. Hamaed, H.; Lo, A.Y.; Lee, D.S.; Evans, W.J.; Schurko, R.W. Solid-state ^{139}La- and ^{15}N-NMR spectroscopy of lanthanum-containing metallocenes. *J. Am. Chem. Soc.* **2006**, *128*, 12638–12639. [CrossRef] [PubMed]

91. Evans, W.J.; Davis, B.L. Chemistry of tris(pentamethylcyclopentadienyl) f-element complexes, $(C_5Me_5)_3M$. *Chem. Rev.* **2002**, *102*, 2119–2136. [CrossRef] [PubMed]

92. Fieser, M.E.; Johnson, C.W.; Bates, J.E.; Ziller, J.W.; Furche, F.; Evans, W.J. Dinitrogen reduction, sulfur reduction, and isoprene polymerization via photochemical activation of trivalent bis(cyclopentadienyl) rare-earth-metal allyl complexes. *Organometallics* **2015**, *34*, 4387–4393. [CrossRef]

93. Evans, W.J.; Fang, M.; Zucchi, G.; Furche, F.; Ziller, J.W.; Hoekstra, R.M.; Zink, J.I. Isolation of dysprosium and yttrium complexes of a three-electron reduction product in the activation of dinitrogen, the $(N_2)^{3-}$ radical. *J. Am. Chem. Soc.* **2009**, *131*, 11195–11202. [CrossRef] [PubMed]

94. Rinehart, J.D.; Fang, M.; Evans, W.J.; Long, J.R. Strong exchange and magnetic blocking in N_2^{3-}-radical-bridged lanthanide complexes. *Nat. Chem.* **2011**, *3*, 538–542. [CrossRef] [PubMed]

95. Rinehart, J.D.; Fang, M.; Evans, W.J.; Long, J.R. A N_2^{3-} radical-bridged terbium complex exhibiting magnetic hysteresis at 14 K. *J. Am. Chem. Soc.* **2011**, *133*, 14236–14239. [CrossRef] [PubMed]

96. Roy, L.E.; Hughbanks, T. Magnetic coupling in dinuclear Gd complexes. *J. Am. Chem. Soc.* **2006**, *128*, 568–575. [CrossRef] [PubMed]

97. Meihaus, K.R.; Corbey, J.F.; Fang, M.; Ziller, J.W.; Long, J.R.; Evans, W.J. Influence of an inner-sphere K^+ ion on the magnetic behavior of N_2^{3-} radical-bridged dilanthanide complexes isolated using an external magnetic field. *Inorg. Chem.* **2014**, *53*, 3099–3107. [CrossRef] [PubMed]

98. Allen, F.H.; Kennard, O.; Watson, D.G.; Brammer, L.; Orpen, A.G.; Taylor, R. Tables of bond lengths determined by X-ray and neutron diffraction. Part 1. Bond lengths in organic compounds. *J. Chem. Soc. Perkin Trans.* **1987**, *2*, S1–S19. [CrossRef]

99. Zhang, Y.-Q.; Luo, C.-L.; Wang, B.-W.; Gao, S. Understanding the magnetic anisotropy in a family of N_2^{3-} radical-bridged lanthanide complexes: Density functional theory and *ab initio* calculations. *J. Phys. Chem. A* **2013**, *117*, 10873–10880. [CrossRef] [PubMed]

100. Rajeshkumar, T.; Rajaraman, G. Is a radical bridge a route to strong exchange interactions in lanthanide complexes? A computational examination. *Chem. Commun.* **2012**, *48*, 7856–7858. [CrossRef] [PubMed]

101. Jubb, J.; Gambarotta, S. Dinitrogen reduction operated by a samarium macrocyclic complex. Encapsulation of dinitrogen into a Sm_2Li_4 metallic cage. *J. Am. Chem. Soc.* **1994**, *116*, 4477–4478. [CrossRef]

102. Dubé, T.; Conoci, S.; Gambarotta, S.; Yap, G.P.A.; Vasapollo, G. Tetrametallic reduction of dinitrogen: Formation of a tetranuclear samarium dinitrogen complex. *Angew. Chem. Int. Ed.* **1999**, *38*, 3657–3659. [CrossRef]

103. Dubé, T.; Ganesan, M.; Conoci, S.; Gambarotta, S.; Yap, G.P.A. Tetrametallic divalent samarium cluster hydride and dinitrogen complexes. *Organometallics* **2000**, *19*, 3716–3721. [CrossRef]

104. Bérubé, C.D.; Yazdanbakhsh, M.; Gambarotta, S.; Yap, G.P.A. Serendipitous isolation of the first example of a mixed-valence samarium tripyrrole complex. *Organometallics* **2003**, *22*, 3742–3747. [CrossRef]

105. Guan, J.; Dubé, T.; Gambarotta, S.; Yap, G.P.A. Dinitrogen labile coordination *versus* four-electron reduction, THF cleavage, and fragmentation promoted by a (calix-tetrapyrrole)Sm(II) complex. *Organometallics* **2000**, *19*, 4820–4827. [CrossRef]

106. Ganesan, M.; Gambarotta, S.; Yap, G.P.A. Highly reactive Sm^{II} macrocyclic clusters: Precursors to N_2 reduction. *Angew. Chem. Int. Ed.* **2001**, *40*, 766–769. [CrossRef]

107. Fox, A.R.; Bart, S.C.; Meyer, K.; Cummins, C.C. Towards uranium catalysts. *Nature* **2008**, *455*, 341–349. [CrossRef] [PubMed]

108. Huang, Q.-R.; Kingham, J.R.; Kaltsoyannis, N. The strength of actinide-element bonds from the quantum theory of atoms-in-molecules. *Dalton Trans.* **2015**, *44*, 2554–2566. [CrossRef] [PubMed]

109. Roussel, P.; Scott, P. Complex of dinitrogen with trivalent uranium. *J. Am. Chem. Soc.* **1998**, *120*, 1070–1071. [CrossRef]

110. Odom, A.L.; Arnold, P.L.; Cummins, C.C. Heterodinuclear uranium/molybdenum dinitrogen complexes. *J. Am. Chem. Soc.* **1998**, *120*, 5836–5837. [CrossRef]

111. Cloke, F.G.N.; Hitchcock, P.B. Reversible binding and reduction of dinitrogen by a uranium(III) pentalene complex. *J. Am. Chem. Soc.* **2002**, *124*, 9352–9353. [CrossRef] [PubMed]

112. Evans, W.J.; Kozimor, S.A.; Ziller, J.W. A monometallic f element complex of dinitrogen: $(C_5Me_5)_3U(\eta^1\text{-}N_2)$. *J. Am. Chem. Soc.* **2003**, *125*, 14264–14265. [CrossRef] [PubMed]

113. Laplaza, C.E.; Cummins, C.C. Dinitrogen cleavage by a three-coordinate molybdenum(III) complex. *Science* **1995**, *268*, 861–863. [CrossRef] [PubMed]

114. Mindiola, D.J.; Meyer, K.; Cherry, J.-P.F.; Baker, T.A.; Cummins, C.C. Dinitrogen cleavage stemming from a heterodinuclear niobium/molybdenum N_2 complex: New nitridoniobium systems including a niobazene cyclic trimer. *Organometallics* **2000**, *19*, 1622–1624. [CrossRef]

115. Curley, J.J.; Cook, T.R.; Reece, S.Y.; Müller, P.; Cummins, C.C. Shining light on dinitrogen cleavage: Structural features, redox chemistry, and photochemistry of the key intermediate bridging dinitrogen complex. *J. Am. Chem. Soc.* **2008**, *130*, 9394–9405. [CrossRef] [PubMed]

116. Kaltsoyannis, N.; Scott, P. Evidence for actinide metal to ligand π backbonding. Density functional investigations of the electronic structure of $[\{(NH_2)_3(NH_3)U\}_2(\mu^2\text{-}\eta^2\text{:}\eta^2\text{-}N_2)]$. *Chem. Commun.* **1998**, 1665–1666. [CrossRef]

117. Roussel, P.; Errington, W.; Kaltsoyannis, N.; Scott, P. Back bonding without σ-bonding: A unique π-complex of dinitrogen with uranium. *J. Organomet. Chem.* **2001**, *635*, 69–74. [CrossRef]

118. Cloke, F.G.N.; Green, J.C.; Kaltsoyannis, N. Electronic structure of $[U_2(\mu^2\text{-}N_2)(\eta^5\text{-}C_5Me_5)_2(\eta^8\text{-}C_8H_4(SiPr^i_3)_2)_2]$. *Organometallics* **2004**, *23*, 832–835. [CrossRef]

119. Korobkov, I.; Gambarotta, S.; Yap, G.P.A. A highly reactive uranium complex supported by the calix[4]tetrapyrrole tetraanion affording dinitrogen cleavage, solvent deoxygenation, and polysilanol depolymerization. *Angew. Chem. Int. Ed.* **2002**, *41*, 3433–3436. [CrossRef]

120. Korobkov, I.; Gambarotta, S.; Yap, G.P.A. Amide from dinitrogen by *in situ* cleavage and partial hydrogenation promoted by a transient zero-valent thorium synthon. *Angew. Chem. Int. Ed.* **2003**, *42*, 4958–4961. [CrossRef] [PubMed]

121. Corbridge, D.E.C.; Lowe, E.J. Structure of white phosphorus: Single crystal X-ray examination. *Nature* **1952**, *170*, 629–629. [CrossRef]

122. Kühl, O. *Phosphorus-31 NMR Spectroscopy: A Concise Introduction for the Synthetic Organic and Organometallic Chemist*; Springer-Verlag GmbH: Berlin, Germany; Heidelberg, Germany, 2008.

123. Konchenko, S.N.; Pushkarevsky, N.A.; Gamer, M.T.; Köppe, R.; Schnöckel, H.; Roesky, P.W. [{(η^5-C$_5$Me$_5$)$_2$Sm}$_4$P$_8$]: A molecular polyphosphide of the rare-earth elements. *J. Am. Chem. Soc.* **2009**, *131*, 5740–5741. [CrossRef] [PubMed]

124. Huang, W.; Diaconescu, P.L. P$_4$ activation by group 3 metal arene complexes. *Chem. Commun.* **2012**, *48*, 2216–2218. [CrossRef] [PubMed]

125. Huang, W.; Diaconescu, P.L. P$_4$ Activation by lanthanum and lutetium naphthalene complexes supported by a ferrocene diamide ligand. *Eur. J. Inorg. Chem.* **2013**, *2013*, 4090–4096. [CrossRef]

126. Turbervill, R.S.P.; Goicoechea, J.M. From clusters to unorthodox pnictogen sources: Solution-phase reactivity of [E$_7$]$^{3-}$ (E = P–Sb) anions. *Chem. Rev.* **2014**, *114*, 10807–10828. [CrossRef] [PubMed]

127. Shannon, R. Revised effective ionic radii and systematic studies of interatomic distances in halides and chalcogenides. *Acta Crystallogr. Sect. A Found. Crystallogr.* **1976**, *32*, 751–767. [CrossRef]

128. Scherer, O.J.; Werner, B.; Heckmann, G.; Wolmershäuser, G. Bicyclic P$_6$ as complex ligand. *Angew. Chem. Int. Ed.* **1991**, *30*, 553–555. [CrossRef]

129. Stephens, F.H. Activation of White Phosphorus by Molybdenum and Uranium tris-Amides. Ph.D. Thesis, Massachusetts Institute of Technology, Cambridge, MA, USA, May 2004.

130. Frey, A.S.P.; Cloke, F.G.N.; Hitchcock, P.B.; Green, J.C. Activation of P$_4$ by U(η^5-C$_5$Me$_5$)(η^8-C$_8$H$_6$ (Si^{i}Pr$_3$)$_2$-1,4)(THF); the X-ray structure of [U(η^5-C$_5$Me$_5$)(η^8-C$_8$H$_6$(Si^{i}Pr$_3$)$_2$-1,4)]$_2$(μ-η^2:η^2-P$_4$). *New J. Chem.* **2011**, *35*, 2022–2026. [CrossRef]

131. Gardner, B.M.; Tuna, F.; McInnes, E.J.L.; McMaster, J.; Lewis, W.; Blake, A.J.; Liddle, S.T. An inverted-sandwich diuranium μ-η^5:η^5-cyclo-P$_5$ complex supported by U–P$_5$ δ-bonding. *Angew. Chem. Int. Ed.* **2015**, *54*, 7068–7072. [CrossRef] [PubMed]

132. Forfar, L.C.; Clark, T.J.; Green, M.; Mansell, S.M.; Russell, C.A.; Sanguramath, R.A.; Slattery, J.M. White phosphorus as a ligand for the coinage metals. *Chem. Commun.* **2012**, *48*, 1970–1972. [CrossRef] [PubMed]

133. Spitzer, F.; Sierka, M.; Latronico, M.; Mastrorilli, P.; Virovets, A.V.; Scheer, M. Fixation and release of intact E$_4$ tetrahedra (E = P, As). *Angew. Chem. Int. Ed.* **2015**, *54*, 4392–4396. [CrossRef] [PubMed]

134. Patel, D.; Tuna, F.; McInnes, E.J.L.; Lewis, W.; Blake, A.J.; Liddle, S.T. An actinide zintl cluster: A tris(triamidouranium)μ_3-η^2:η^2:η^2-heptaphosphanortricyclane and its diverse synthetic utility. *Angew. Chem. Int. Ed.* **2013**, *52*, 13334–13337. [CrossRef] [PubMed]

135. Laplaza, C.E.; Davis, W.M.; Cummins, C.C. A molybdenum–phosphorus triple bond: Synthesis, structure, and reactivity of the terminal phosphido (P^{3-}) complex [Mo(P)(NRAr)$_3$]. *Angew. Chem. Int. Ed.* **1995**, *34*, 2042–2044. [CrossRef]

136. Cherry, J.-P.F.; Stephens, F.H.; Johnson, M.J.A.; Diaconescu, P.L.; Cummins, C.C. Terminal phosphide and dinitrogen molybdenum compounds obtained from pnictide-bridged precursors. *Inorg. Chem.* **2001**, *40*, 6860–6862. [CrossRef] [PubMed]

137. Chisholm, M.H.; Folting, K.; Pasterczyk, J.W. A phosphido-capped tritungsten alkoxide cluster: W$_3$(μ_3-P)(μ-OCH$_2$-*t*-Bu)$_3$(OCH$_2$-*t*-Bu)$_6$ and speculation upon the existence of a reactive (*t*-BuCH$_2$O)$_3$W≡P intermediate. *Inorg. Chem.* **1988**, *27*, 3057–3058. [CrossRef]

138. Scherer, O.J.; Sitzmann, H.; Wolmershäuser, G. Umsetzung von P$_4$ mit (η^5-C$_5$H$_5$)(CO)$_2$Mo≡Mo(CO)$_2$(η^5-C$_5$H$_5$) zu den tetraedrischen molybdänkomplexen P$_n$[Mo(CO)$_2$(η^5-C$_5$H$_5$)]$_{4-n}$ (*n* = 2,3). *J. Organomet. Chem.* **1984**, *268*, C9–C12. [CrossRef]

139. Di Vaira, M.; Ghilardi, C.A.; Midollini, S.; Sacconi, L. *cyclo*-Triphosphorus (δ-P$_3$) as a ligand in cobalt and nickel complexes with 1,1,1-tris(diphenylphosphinomethyl)ethane. Formation and structures. *J. Am. Chem. Soc.* **1978**, *100*, 2550–2551. [CrossRef]

140. Chirik, P.J.; Pool, J.A.; Lobkovsky, E. Functionalization of elemental phosphorus with [Zr(η^5-C$_5$Me$_5$) (η^5-C$_5$H$_4$^{t}Bu)H$_2$]$_2$. *Angew. Chem. Int. Ed.* **2002**, *41*, 3463–3465. [CrossRef]

141. Gröer, T.; Baum, G.; Scheer, M. Complexes with a monohapto bound phosphorus tetrahedron and phosphaalkyne. *Organometallics* **1998**, *17*, 5916–5919. [CrossRef]

142. Peruzzini, M.; Marvelli, L.; Romerosa, A.; Rossi, R.; Vizza, F.; Zanobini, F. Synthesis and characterisation of *tetrahedro*-tetraphosphorus complexes of rhenium—Evidence for the first bridging complex of white phosphorus. *Eur. J. Inorg. Chem.* **1999**, *1999*, 931–933. [CrossRef]

143. Urnėžius, E.; Brennessel, W.W.; Cramer, C.J.; Ellis, J.E.; Schleyer, P.V.R. A carbon-free sandwich complex [(P$_5$)$_2$Ti]$^{2-}$. *Science* **2002**, *295*, 832–834.

144. Scherer, O.J.; Sitzmann, H.; Wolmershäuser, G. Hexaphosphabenzene as complex ligand. *Angew. Chem. Int. Ed.* **1985**, *24*, 351–353. [CrossRef]

145. Scherer, O.J.; Berg, G.; Wolmershäuser, G. P$_8$ and P$_{12}$ as complex ligands. *Chem. Ber.* **1996**, *129*, 53–58. [CrossRef]

146. Scherer, M.; Deng, S.; Scherer, O.J.; Sierka, M. Tetraphosphacyclopentadienyl and triphosphaallyl ligands in iron complexes. *Angew. Chem. Int. Ed.* **2005**, *44*, 3755–3758. [CrossRef] [PubMed]

147. Scherer, O.J.; Schulze, J.; Wolmershäuser, G. Bicyclisches As$_6$ als komplexligand. *J. Organomet. Chem.* **1994**, *484*, C5–C7. [CrossRef]

148. Di Vaira, M.; Midollini, S.; Sacconi, L.; Zanobini, F. *cyclo*-Triarsenic as μ,η-ligand in transition-metal complexes. *Angew. Chem. Int. Ed.* **1978**, *17*, 676–677. [CrossRef]

149. Scherer, O.J.; Kemény, G.; Wolmershäuser, G. [Cp$_4$Fe$_4$(E$_2$)$_2$] clusters with triangulated dodecahedral Fe$_4$E$_4$ skeletons (E = P, As). *Chem. Ber.* **1995**, *128*, 1145–1148. [CrossRef]

150. Schwarzmaier, C.; Timoshkin, A.Y.; Scheer, M. An end-on-coordinated As$_4$ tetrahedron. *Angew. Chem. Int. Ed.* **2013**, *52*, 7600–7603. [CrossRef] [PubMed]

151. Schwarzmaier, C.; Sierka, M.; Scheer, M. Intact As$_4$ tetrahedra coordinated side-on to metal cations. *Angew. Chem. Int. Ed.* **2013**, *52*, 858–861. [CrossRef] [PubMed]

152. Scherer, O.J.; Sitzmann, H.; Wolmershäuser, G. (E$_2$)$_2$-einheiten (E = P, As) als clusterbausteine. *J. Organomet. Chem.* **1986**, *309*, 77–86. [CrossRef]

153. Spinney, H.A.; Piro, N.A.; Cummins, C.C. Triple-bond reactivity of an AsP complex intermediate: Synthesis stemming from molecular arsenic, As$_4$. *J. Am. Chem. Soc.* **2009**, *131*, 16233–16243. [CrossRef] [PubMed]

154. Heinl, S.; Scheer, M. Activation of group 15 based cage compounds by [Cp^{BIG}Fe(CO)$_2$] radicals. *Chem. Sci.* **2014**, *5*, 3221–3225. [CrossRef]

155. Gra; Bodensteiner, M.; Zabel, M.; Scheer, M. Synthesis of arsenic-rich As$_n$ ligand complexes from yellow arsenic. *Chem. Sci.* **2015**, *6*, 1379–1382.

156. Schwarzmaier, C.; Bodensteiner, M.; Timoshkin, A.Y.; Scheer, M. An approach to mixed P$_n$As$_m$ ligand complexes. *Angew. Chem. Int. Ed.* **2014**, *53*, 290–293. [CrossRef] [PubMed]

157. Curley, J.J.; Piro, N.A.; Cummins, C.C. A terminal molybdenum arsenide complex synthesized from yellow arsenic. *Inorg. Chem.* **2009**, *48*, 9599–9601. [CrossRef] [PubMed]

158. Balázs, G.; Sierka, M.; Scheer, M. Antimony–tungsten triple bond: A stable complex with a terminal antimony ligand. *Angew. Chem. Int. Ed.* **2005**, *44*, 4920–4924. [CrossRef] [PubMed]

159. Evans, W.J.; Fang, M.; Bates, J.E.; Furche, F.; Ziller, J.W.; Kiesz, M.D.; Zink, J.I. Isolation of a radical dianion of nitrogen oxide (NO)$^{2-}$. *Nat. Chem.* **2010**, *2*, 644–647. [CrossRef] [PubMed]

160. Corbey, J.F.; Fang, M.; Ziller, J.W.; Evans, W.J. Cocrystallization of (μ-S$_2$)$^{2-}$ and (μ-S)$^{2-}$ and formation of an [η^2-S$_3$N(SiMe$_3$)$_2$] ligand from chalcogen reduction by (N$_2$)$^{2-}$ in a bimetallic yttrium amide complex. *Inorg. Chem.* **2015**, *54*, 801–807. [CrossRef] [PubMed]

161. Evans, W.J.; Lee, D.S.; Ziller, J.W.; Kaltsoyannis, N. Trivalent [(C$_5$Me$_5$)$_2$(THF)Ln]$_2$(μ-η^2:η^2-N$_2$) complexes as reducing agents including the reductive homologation of CO to a ketene carboxylate, (μ-η^4-O$_2$CCCO)$_2$. *J. Am. Chem. Soc.* **2006**, *128*, 14176–14184. [CrossRef] [PubMed]

162. Lu, E.; Li, Y.; Chen, Y. A scandium terminal imido complex: Synthesis, structure and DFT studies. *Chem. Commun.* **2010**, *46*, 4469–4471. [CrossRef] [PubMed]

163. Rong, W.; Cheng, J.; Mou, Z.; Xie, H.; Cui, D. Facile preparation of a scandium terminal imido complex supported by a phosphazene ligand. *Organometallics* **2013**, *32*, 5523–5529. [CrossRef]

164. Schädle, D.; Meermann-Zimmermann, M.; Schädle, C.; Maichle-Mössmer, C.; Anwander, R. Rare-earth metal complexes with terminal imido ligands. *Eur. J. Inorg. Chem.* **2015**, *2015*, 1334–1339. [CrossRef]

165. Sobota, P.; Janas, Z. Formation of a nitrogen–carbon bond from N_2 and CO. Influence of $MgCl_2$ on the N_2 reduction process in the system $TiCl_4$/Mg. *J. Organomet. Chem.* **1984**, *276*, 171–176. [CrossRef]

166. Semproni, S.P.; Margulieux, G.W.; Chirik, P.J. Di- and tetrametallic hafnocene oxamidides prepared from CO-induced N_2 bond cleavage and thermal rearrangement to hafnocene cyanide derivatives. *Organometallics* **2012**, *31*, 6278–6287. [CrossRef]

167. Knobloch, D.J.; Lobkovsky, E.; Chirik, P.J. Carbon monoxide-induced dinitrogen cleavage with group 4 metallocenes: Reaction scope and coupling to N–H bond formation and CO deoxygenation. *J. Am. Chem. Soc.* **2010**, *132*, 10553–10564. [CrossRef] [PubMed]

168. Semproni, S.P.; Milsmann, C.; Chirik, P.J. Structure and reactivity of a hafnocene μ-nitrido prepared from dinitrogen cleavage. *Angew. Chem. Int. Ed.* **2012**, *51*, 5213–5216. [CrossRef] [PubMed]

169. Semproni, S.P.; Chirik, P.J. Synthesis of a base-free hafnium nitride from N_2 cleavage: A versatile platform for dinitrogen functionalization. *J. Am. Chem. Soc.* **2013**, *135*, 11373–11383. [CrossRef] [PubMed]

170. MacKay, B.A.; Johnson, S.A.; Patrick, B.O.; Fryzuk, M.D. Functionalization and cleavage of coordinated dinitrogen via hydroboration using primary and secondary boranes. *Can. J. Chem.* **2005**, *83*, 315–323. [CrossRef]

171. Fryzuk, M.D.; MacKay, B.A.; Johnson, S.A.; Patrick, B.O. Hydroboration of coordinated dinitrogen: A new reaction for the N_2 ligand that results in its functionalization and cleavage. *Angew. Chem. Int. Ed.* **2002**, *41*, 3709–3712. [CrossRef]

172. Fryzuk, M.D.; MacKay, B.A.; Patrick, B.O. Hydrosilylation of a dinuclear tantalum dinitrogen complex: Cleavage of N_2 and functionalization of both nitrogen atoms. *J. Am. Chem. Soc.* **2003**, *125*, 3234–3235. [CrossRef] [PubMed]

173. Spencer, L.P.; MacKay, B.A.; Patrick, B.O.; Fryzuk, M.D. Inner-sphere two-electron reduction leads to cleavage and functionalization of coordinated dinitrogen. *Proc. Natl. Acad. Sci. USA* **2006**, *103*, 17094–17098. [CrossRef] [PubMed]

174. Margulieux, G.W.; Turner, Z.R.; Chirik, P.J. Synthesis and ligand modification chemistry of a molybdenum dinitrogen complex: Redox and chemical activity of a bis(imino)pyridine ligand. *Angew. Chem. Int. Ed.* **2014**, *53*, 14211–14215. [CrossRef] [PubMed]

175. Milsmann, C.; Turner, Z.R.; Semproni, S.P.; Chirik, P.J. Azo N=N bond cleavage with a redox-active vanadium compound involving metal–ligand cooperativity. *Angew. Chem. Int. Ed.* **2012**, *51*, 5386–5390. [CrossRef] [PubMed]

176. King, D.M.; Tuna, F.; McInnes, E.J.L.; McMaster, J.; Lewis, W.; Blake, A.J.; Liddle, S.T. Synthesis and structure of a terminal uranium nitride complex. *Science* **2012**, *337*, 717–720. [CrossRef] [PubMed]

177. King, D.M.; Tuna, F.; McInnes, E.J.L.; McMaster, J.; Lewis, W.; Blake, A.J.; Liddle, S.T. Isolation and characterization of a uranium(VI)–nitride triple bond. *Nat. Chem.* **2013**, *5*, 482–488. [CrossRef] [PubMed]

178. Cleaves, P.A.; King, D.M.; Kefalidis, C.E.; Maron, L.; Tuna, F.; McInnes, E.J.L.; McMaster, J.; Lewis, W.; Blake, A.J.; Liddle, S.T. Two-electron reductive carbonylation of terminal uranium(V) and uranium(VI) nitrides to cyanate by carbon monoxide. *Angew. Chem. Int. Ed.* **2014**, *53*, 10412–10415. [CrossRef] [PubMed]

179. Gardner, B.M.; Balázs, G.; Scheer, M.; Tuna, F.; McInnes, E.J.L.; McMaster, J.; Lewis, W.; Blake, A.J.; Liddle, S.T. Triamidoamine uranium(IV)–arsenic complexes containing one-, two- and threefold U–As bonding interactions. *Nat. Chem.* **2015**, *7*, 582–590. [CrossRef] [PubMed]

inorganics

[MDPI]

Article

New Lanthanide Alkynylamidinates and Diiminophosphinates

Farid M. Sroor [1], Cristian G. Hrib [2] and Frank T. Edelmann [2,*]

[1] Organometallic and Organometalloid Chemistry Department, National Research Centre, 12622 Cairo, Egypt; faridsroor@gmx.de

[2] Chemisches Institut der Otto-von-Guericke-Universität Magdeburg, 39106 Magdeburg, Germany; cristian.hrib@ovgu.de

* Correspondence: frank.edelmann@ovgu.de; Tel.: +49-391-67-58327; Fax: +49-391-67-12933.

Academic Editors: Stephen Mansell and Steve Liddle

Received: 7 October 2015; Accepted: 27 October 2015; Published: 5 November 2015

Abstract: This contribution reports the synthesis and structural characterization of several new lithium and lanthanide alkynylamidinate complexes. Treatment of PhC≡CLi with N,N'-diorganocarbodiimides, R–N=C=N–R (R = ^{i}Pr, Cy (cyclohexyl)), in THF or diethyl ether solution afforded the lithium-propiolamidinates Li[Ph–C≡C–C(NCy)$_2$] S (**1**: R = ^{i}Pr, S = THF; **2**: R = Cy, S = THF; **3**: R = Cy, S = Et$_2$O). Single-crystal X-ray diffraction studies of **1** and **2** showed the presence of typical ladder-type dimeric structures in the solid state. Reactions of anhydrous LnCl$_3$ (Ln = Ce, Nd, Sm or Ho) with **2** in a 1:3 molar ratio in THF afforded a series of new homoleptic lanthanide tris(propiolamidinate) complexes, [Ph–C≡C–C(NCy)$_2$]$_3$Ln (**4**: Ln = Ce; **5**: Ln = Nd; **6**: Ln = Sm; **7**: Ln = Ho). The products were isolated in moderate to high yields (61%–89%) as brightly colored, crystalline solids. The chloro-functional neodymium(III) bis(cyclopropylethynylamidinate) complex [{c-C$_3$H$_5$–C≡C–C(N^{i}Pr)$_2$}$_2$Ln(μ-Cl)(THF)]$_2$ (**8**) was prepared from NdCl$_3$ and two equiv. of Li[c-C$_3$H$_5$–C≡C–C(N^{i}Pr)$_2$] in THF and structurally characterized. A new monomeric Ce(III)-diiminophosphinate complex, [Ph$_2$P(NSiMe$_3$)$_2$]$_2$Ce(μ-Cl)$_2$Li(THF)$_2$ (**9**), has also been synthesized in a similar manner from CeCl$_3$ and two equiv. of Li[Ph$_2$P(NSiMe$_3$)$_2$]. Structurally, this complex resembles the well-known "ate" complexes (C$_5$Me$_5$)$_2$Ln(μ-Cl)$_2$Li(THF)$_2$. Attempts to oxidize compound **9** using trityl chloride or phenyliodine(III) dichloride did not lead to an isolable cerium(IV) species.

Keywords: amidinate; alkynylamidinate; diiminophosphinate; lithium; lanthanide complexes; crystal structure

1. Introduction

Monoanionic N,N'-chelating ligands like amidinates, [RC(NR')$_2$]$^-$, guanidinates, [R$_2$NC(NR')$_2$]$^-$, and diiminophosphinates, [R$_2$P(NR')$_2$]$^-$, are commonly regarded as steric equivalents of the omnipresent cyclopentadienyl ligands [1–7]. In the case of rare-earth metals, mono-, di- and trisubstituted lanthanide amidinate and guanidinate complexes are all accessible, just like the mono-, di- and tricyclopentadienyl complexes. Over the past *ca.* 25 years, lanthanide amidinates have undergone an impressive transformation from laboratory curiosities to highly efficient homogeneous catalysts as well as versatile precursors in materials science. Lanthanide amidinates have been demonstrated to be efficient homogeneous catalysts e.g., for ring-opening polymerization reactions of lactones, the guanylation of amines, the addition of terminal alkynes to carbodiimides, as well as hydroamination and hydroamination/cyclization reactions [8,9]. Moreover, homoleptic alkyl-substituted lanthanide tris(amidinate) complexes often exhibit high volatility and can be employed as promising precursors for ALD (atomic layer deposition) and MOCVD (metal-organic chemical vapor deposition) processes, e.g., for the deposition of lanthanide oxide (Ln$_2$O$_3$) or lanthanide nitride (LnN) thin layers [10–19].

What makes the amidinate and guanidinate ligands extremely versatile is the possibility of varying the substituents at all three atoms of the N–C–N unit. Attachment of an alkynyl substituent to the central carbon atom of amidines leads to alkynylamidines, R–C≡C–C(=NR′)(NHR′). Alkynylamidines are of significant interest due to their diverse applications in organic synthesis [20–23] as well as in biological and pharmacological systems [24–27]. Alkynylamidinate complexes of various transition metals and lanthanides have been shown to be efficient and versatile catalysts e.g., for C–C and C–N bond formation, the addition of C–H, N–H and P–H bonds to carbodiimides as well as ε-caprolactone polymerization [28–33]. Previously used alkynylamidinate ligands include e.g., the propiolamidine derivatives [Ph–C≡C–C(NR)$_2$]$^-$ (R = ^{i}Pr, ^{t}Bu) [16,28,34] and the trimethylsilylacetylene-derived anions [Me$_3$Si–C≡C–C(NR)$_2$]$^-$ (R = cyclohexyl (Cy), ^{i}Pr) [35]. In 2008, the compound Li[Ph–C≡C–C(N^{i}Pr)$_2$] has been reported to be suitable precursor for the preparation of *d*-transition metal complexes containing bridging alkynylamidinate ligand [36]. Moreover, the same ligand has been used in the synthesis of the unsolvated homoleptic Ce(III) complex [Ph–C≡C–C(N^{i}Pr)$_2$]$_3$Ce [16].

In the course of our ongoing investigation of lanthanide amidinates we recently started a project to study lanthanide alkynylamidinates derived from phenylacetylene and cyclopropylacetylene. The cyclopropyl group was chosen because of the well-known electron-donating ability of this substituent to an adjacent electron-deficient center [37–43]. This would provide a rare chance to electronically influence the amidinate ligand system rather than altering only its steric demand. In the course of this study, we first reported the synthesis and full characterization of a series of lithium-cyclopropylethinylamidinates [44]. These precursors are readily available on a large scale using commercially available starting materials (cyclopropylacetylene, N,N′-diorganocarbodiimides). In subsequent contributions, we described the first trivalent rare-earth metal complexes comprising the new cyclopropylethynylamidinate ligands and their use as guanylation, hydroamination, and hydroacetylenation catalysts [45–47].

We now report the synthesis and structural characterization of two lithium-propiolamidinates and their use as precursors for new homoleptic lanthanide(III)-tris(propiolamidinate) complexes. Moreover, the synthesis and crystal structures of a dimeric neodymium(III) bis(cyclopropylethynylamidinate) complex and a monomeric Ce(III)-diiminophosphinate complex are reported. Scheme 1 shows the anionic N,N′-chelating ligands employed in this study.

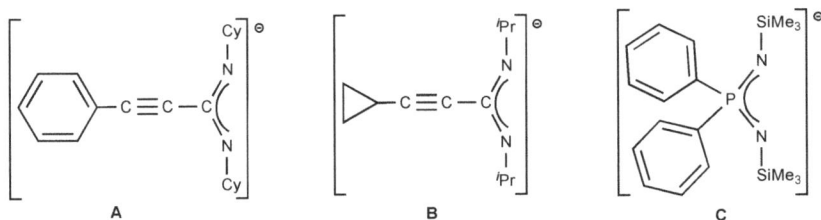

Scheme 1. Anionic N,N′-chelating ligands employed in this study: N,N′-Dicyclohexylpropiolamidinate (**A**); N,N′-diisopropyl-cyclopropylethynylamidinate (**B**); and N,N′-bis(trimethylsilyl)-P,P-diphenyldiiminophosphinate (**C**).

2. Results and Discussion

2.1. Lithium-Propiolamidinate Precursors

While the N,N′-isopropyl-substituted lithium-propiolamidinate salt Li[Ph–C≡C–C(N^{i}Pr)$_2$] THF (**1**) was already known from previous studies [16,28,34], we prepared the cyclohexyl derivatives **2** and **3** in a straightforward manner according to Scheme 2 by *in situ* deprotonation of phenylacetylene with ^{n}BuLi at −20 °C in THF or Et$_2$O followed by addition of N,N′-dicyclohexylcarbodiimide. After 10 min, the reaction mixtures were warmed to room temperature and stirred for two hours. The solvent was removed under vacuum, affording white solids of Li[Ph–C≡C–C(NCy)$_2$]·S (**2**: S = THF, **3**: S = Et$_2$O) in good yields (**2**: 88%, **3**: 76%).

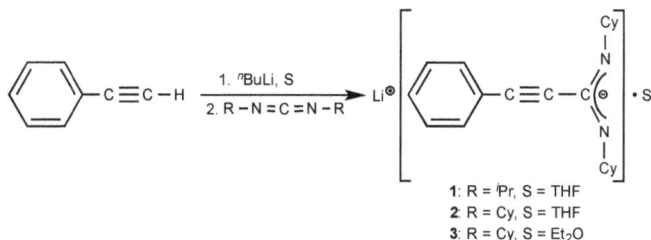

Scheme 2. Synthesis of the lithium-propiolamidinates **1–3**.

Both alkynylamidinates **2** and **3** have been fully characterized by spectroscopic methods and elemental analysis. The IR spectrum of **2** shows a medium band at 2217 cm^{-1}, which could be assigned to the C≡C triple bond, while in the spectrum of **3** it appears as a weak band at 2211 cm^{-1}. The C=N stretching vibrations of the NCN unit of amidinate moieties is observed at 1610 cm^{-1} as a very strong band in the spectrum of **2** and at 1592 cm^{-1} as strong band in that of **3** [16]. The NMR spectra were recorded in toluene-d_8 and C$_6$D$_6$ for **2** and **3**, respectively. The ^{1}H and ^{13}C NMR analyses were in good agreement with the formation of Li[Ph–C≡C–C(NCy)$_2$]. Resonances of all protons and carbons were observed except for the carbon atom of the NCN unit and one C atom of the two acetylenic carbon atoms in **3**.

Single-crystal X-ray diffraction studies were carried out on the lithium-propiolamidinates **1** and **2**. In both cases, X-ray quality, block-like single-crystals were obtained by cooling saturated solutions in THF to 5 °C for a few days. Figures 1 and 2 depict the molecular structures of **1** and **2** along with selected bond lengths and angles, while crystallographic details are listed in Table 1. Full crystallographic data can be found in the Supplementary Information. The crystal structure determinations of Li[Ph–C≡C–C(N^{i}Pr)$_2$] THF (**1**) and Li[Ph–C≡C–C(NCy)$_2$] THF (**2**) in both cases revealed the presence of a ladder-type dimeric structure, which is the most characteristic structural motif of most previously characterized lithium amidinates and guanidinates [48–58]. The dimers are centrosymmetric with the central Li$_2$N$_2$ ring being strictly planar. In the solid state, both lithium propiolamidinates crystallize in the triclinic space group P-1 with one dimeric molecule in the unit cell. The bond lengths and bond angles are in good agreement with similar structures of amidinate ligands [48–58]. The N–C distances in the amidinate N–C–N unit are 1.321(2) and 1.330(2) Å in **1** and 1.3201(13) and 1.3407(12) Å in **2**. These values clearly indicate uniform π-delocalization within the N–C–N moieties. The distances Li–N distances 1.990(3) and 2.261(3) Å in **1** and 2.031(2) and 2.188(2) Å in **2**, while the C≡C bond lengths are virtually identical with 1.189(2) Å (**1**) and 1.2018(18) Å (**2**). The bond angles N–C–N and N–Li–N are 119.13(13)° and 64.67(10)° in the isopropyl derivative **1** and 118.89(8)° and 65.67(6)°, respectively, in the cyclohexyl-substituted derivative **2**.

2.2. Homoleptic Lanthanide(III)-tris(Propiolamidinate) Complexes

A series of homoleptic lanthanide(III)-tris(propiolamidinates) were prepared in a straightforward manner according to Scheme 3. The reaction of anhydrous LnCl$_3$ (Ln = Ce, Nd, Sm or Ho) with **2** in a 1:3 molar ratio using THF as solvent afforded a series of new homoleptic lanthanide tris(propiolamidinate) complexes, [Ph–C≡C–C(NCy)$_2$]$_3$Ln (Ln = Ce (**4**), Nd (**5**), Sm (**6**), Ho (**7**)). The products were isolated in moderate to high yields (**4**: 61%, **5**: 56%, **6**: 59%, **7**: 89%) yields as unsolvated complexes in the form of brightly colored crystals (**4**: orange, **5**: pale green, **6**: yellow, **7**: bright yellow).

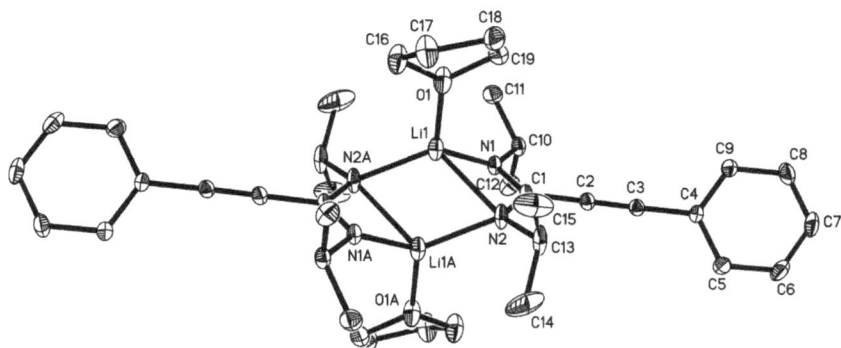

Figure 1. Molecular structure of **1** in the crystal, showing the atom-labeling scheme. Displacement ellipsoids are drawn at the 50% probability level. Selected bond lengths (Å) and angles (°): O(1)–Li(1) 1.908(3), Li(1)–N(1) 1.990(3), Li(1)–N(2) 2.261(3), N(1)–C(1) 1.321(2), N(2)–C(1) 1.330(2), C(1)–C(2) 1.470(2), C(2)–C(3) 1.189(2), C(3)–C(4) 1.439(2), O(1)–Li(1)–N(1) 118.80(16), O(1)–Li(1)–N(2) 113.49(14), N(1)–Li(1)–N(2) 64.67(10), N(1)–C(1)–N(2) 119.13(13), N(1)–C(1)–C(2) 120.72(14), N(2)–C(1)–C(2) 120.15(14), C(3)–C(2)–C(1) 179.05(18), C(2)–C(3)–C(4) 178.56(17).

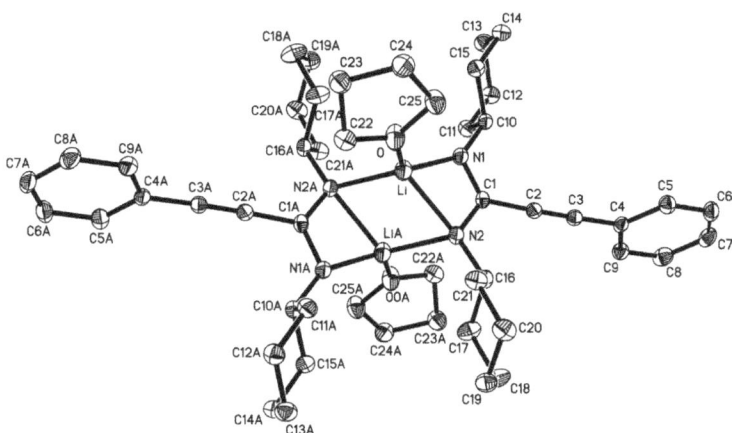

Figure 2. Molecular structure of **2** in the crystal, showing the atom-labeling scheme. Displacement ellipsoids are drawn at the 50% probability level. Selected bond lengths (Å) and angles (°): O–Li 1.9518(19), N(1)–C(1) 1.3201(13), N(1)–Li 2.031(2), N(2)–Li 2.188(2), N(2)–C(1) 1.3407(12), C(1)–C(2) 1.4681(13), C(2)–C(3) 1.2018(14), C(3)–C(4) 1.4366(13), O–Li–N(1) 126.92(10), O–Li–N(2) 122.23(9), N(1)–Li–N(2) 65.67(6), N(1)–C(1)–N(2) 118.89(8), N(1)–C(1)–C(2) 120.49(9), N(2)–C(1)–C(2) 120.58(9), C(3)–C(2)–C(1) 175.39(10), C(2)–C(3)–C(4) 178.27(11).

The new unsolvated lanthanide-tris(propiolamidinate) complexes, [Ph–C≡C–C(NCy)$_2$]$_3$Ln (**4–7**), have been fully characterized by spectroscopic methods and elemental analysis. Unfortunately, attempted recrystallization of complexes **4–7** from various solvents such as toluene, pentane, THF or diethyl ether did not provide single-crystals suitable for X-ray diffraction. Only on one occasion, well-formed crystals of **7** obtained from *n*-pentane could be successfully subjected to X-ray diffraction, but the crystal quality was too poor to allow full refinement of the crystal structure. The NMR spectra were in good agreement with those of similar unsolvated homoleptic lanthanide amidinate complexes, [Ph–C≡C–C(N^{i}Pr)$_2$]$_3$Ce [16], [Ph–C≡C–C(N^{i}Pr)$_2$]$_3$Y and [Ph–C(NCy)$_2$]$_3$Ln (Ln = Pr, Nd or Sm) [59].

Due to the strongly paramagnetic nature of Ho^{3+} ion, it was impossible to obtain NMR spectra for **7**. According to the ^1H–^{13}C correlation (HSQC) technique, the protons of CH in the cyclohexyl group observed at three different positions, $\delta = 9.49, 3.76$ and 3.56 ppm in **4** (Figure 3) and at $\delta = 18.33, 3.77$ and 3.57 ppm in **5** (Figure S1) are in agreement with the CH-protons in the complex [Ph–C(NCy)$_2$]$_3$Ln (Ln = Pr or Nd), whereas the CH-protons of c-C$_6$H$_{11}$ in **6** (Figure S2) appear at δ 3.71 and 3.30 ppm [34]. The significantly different chemical shifts of the CH proton resonances originate from the different pramagnetism of the Ce^{3+}, Nd^{3+}, and Sm^{3+} ions, respectively. The protons of the phenyl group appear in the range of $\delta = 7.40$–9.41 ppm [16,34].

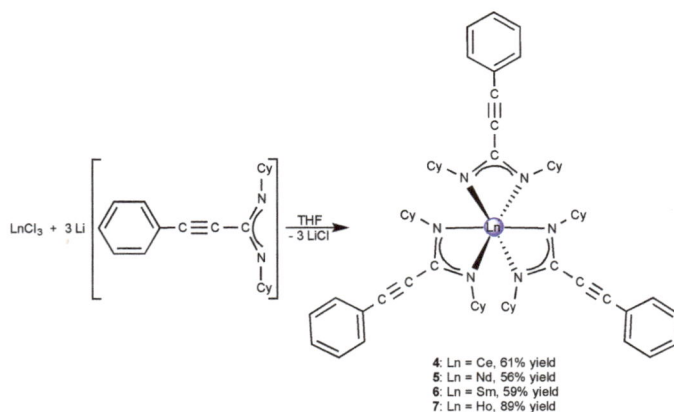

Scheme 3. Synthesis of the unsolvated, homoleptic lanthanide-tris(propiolamidinates) **4–7**.

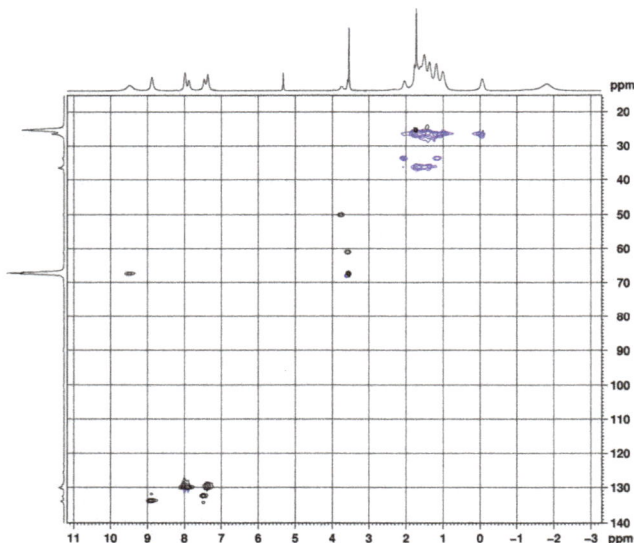

Figure 3. HSQC spectrum (400 MHz, THF-d_8, 25 °C) of [Ph–C≡C–C(NCy)$_2$]$_3$Ce (**4**).

The ^1H NMR spectra reflected the stronger paramagnetic properties for the complexes of cerium **4** and neodymium **5** as compared to the less paramagnetic samarium complex **6**. On the other hand, the $^{13}C\{^1H\}$ NMR spectra showed that the signals of the phenyl carbon atoms are observed in the range of

δ = 125–140 ppm in all three cases. The carbon atoms of the CH of the cyclohexyl groups are observed at δ = 67.1, 61.1 and 56.3 ppm in **4**, and at δ = 78.5, 61.1 and 50.0 ppm in **5**, whereas in **6** they appear at δ = 59.5 and 57.4 ppm. The carbon atoms of the CH_2 in the cyclohexyl group appear in the same range at δ = 25.7–36.5 ppm for all of three complexes. All data clearly showed that the new complexes **4–7** are unsolvated, homoleptic lanthanide(III)-tris(propiolamidinates) like the previously reported [Ph–C≡C–C(N^{i}Pr)$_2$]$_3$Ce [16] and [Ph–C≡C–C(N^{i}Pr)$_2$]$_3$Y [34].

2.3. A Neodymium(III)-bis(Cyclopropylethinylamidinate)

We recently reported that reactions between anhydrous lanthanide trichlorides LnCl$_3$ (Ln = Ce, Pr, Nd) with two equiv. of the lithium-cyclopropylethynylamidinates Li[*c*-C$_3$H$_5$–C≡C–C(NR)$_2$] (R = ^{i}Pr, Cy) in THF solution afford the chloro-functional lanthanide(III) bis(cyclopropylethynylamidinate) complexes [{*c*-C$_3$H$_5$–C≡C–C(NR)$_2$}$_2$Ln(μ-Cl)(THF)]$_2$ (Ln = Ce, Nd; R = ^{i}Pr, Cy). In the course of this study, we carried out an X-ray diffraction study of the neodymium derivative [{*c*-C$_3$H$_5$–C≡C–C(N^{i}Pr)$_2$}$_2$Nd(μ-Cl)(THF)]$_2$ (**8**). This complex was prepared according to Scheme 4 by treatment of anhydrous NdCl$_3$ with two equiv. of Li[{*c*-C$_3$H$_5$–C≡C–C(N^{i}Pr)$_2$] [44] following our synthetic protocol [45].

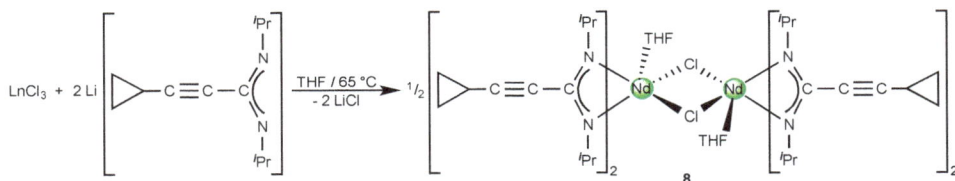

Scheme 4. Preparation of **8**.

Dark green, needle-like single crystals of **8** suitable for X-ray diffraction were obtained by slow cooling of saturated solutions in *n*-pentane to −30 °C. The single-crystal X-ray diffraction study clearly established the presence of a chloro-bridged dimer of the composition [{*c*-C$_3$H$_5$–C≡C–C(NR)$_2$}$_2$Nd(μ-Cl)(THF)]$_2$. The molecular structure of **8** is depicted in Figure 4; the crystal and structural refinement data are summarized in Table 1. Complete structural data of **8** can be found in the Supplementary Information. The molecule is a centrosymmetric dimer of the type [L$_2$Ln(μ-Cl)(THF)]$_2$ comprising a planar four-membered Nd$_2$Cl$_2$ ring. With angles of 106.85(5)° (Nd–Cl–Nd) and 73.15(5)° (Cl–Nd–Cl), the central Nd$_2$Cl$_2$ moiety is rhomb-shaped. The cyclopropyl-ethinylamidinate ligands are coordinated in the "classical" *N,N'*-chelating coordination mode, and the chloro ligands act as bridging ligands. The coordination sphere around the large Nd^{3+} ion still leaves room for an additional THF ligand, resulting in a formal coordination number of seven around neodymium. It has been pointed out earlier [45], that in this respect the cyclopropylethynylamidinates differ from the related dimeric yttrium, ytterbium, and lutetium bis(guanidinato) complexes such as [{(Me$_3$Si)$_2$NC(N^{i}Pr)$_2$}$_2$Y(μ-Cl)]$_2$ [60] and [{(Me$_3$Si)$_2$NC(NCy)$_2$}$_2$Ln(μ-H)]$_2$ [61]. In these compounds, the higher steric demand of the supporting guanidinate ligands prevents the formation of THF adducts, resulting in six-coordinate lanthanide ions. This shows that the steric demand of the cyclopropylethynylamidinate anions is significantly lower than that of the *N,N*-bis(trimethylsilyl)-substituted guanidinate anion [(Me$_3$Si)$_2$NC(NCy)$_2$]$^-$. Within the N–C–N groups the C–N distances are equal (average C–N = 1.335 Å), indicating full π-electron delocalization within the chelating amidinate units. The bond lengths of the triple bonds in the cyclopropylethynyl units are C(2)–C(3) 1.186(9) Å and C(22)–C(23) 1.188(9) Å.

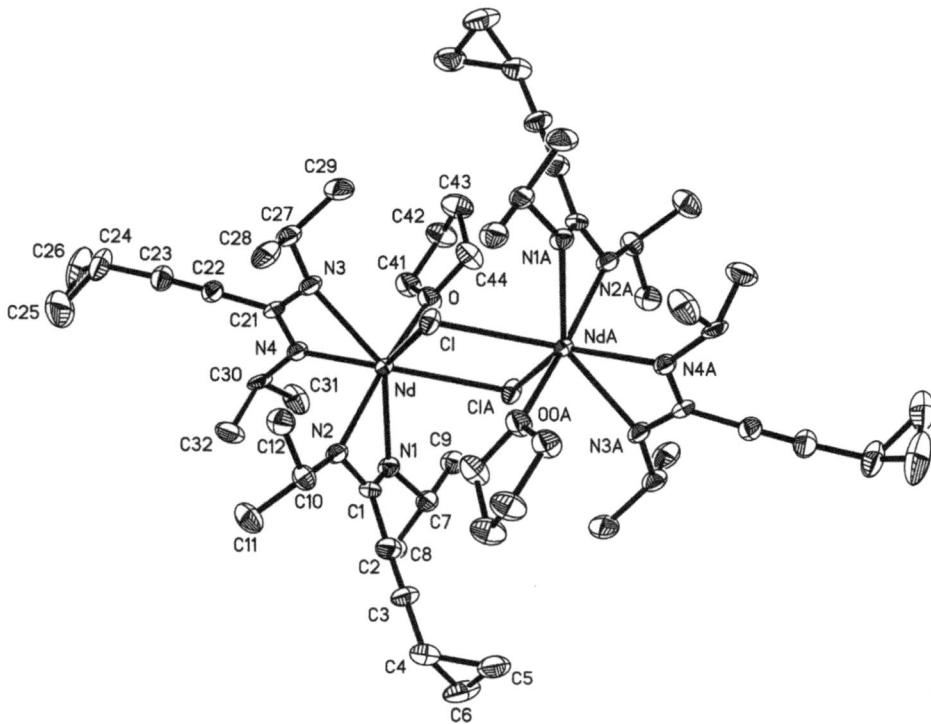

Figure 4. Molecular structure of **8** in the crystal, showing the atom-labeling scheme. Displacement ellipsoids are drawn at the 50% probability level. Selected bond lengths (Å) and angles (°): Cl–Nd 2.7961(16), Nd–N(1) 2.478(5), Nd–N(2) 2.423(5), Nd–N(4) 2.464(5), Nd–N(3) 2.484(5), Nd–O 2.504(5), N(1)–C(1) 1.338(8), N(2)–C(1) 1.332(7), N(3)–C(21) 1.321(8), N(4)–C(21) 1.345(7), N(4)–C(30) 1.456(8), C(1)–C(2) 1.444(8), C(2)–C(3) 1.186(9), C(3)–C(4) 1.425(10), C(21)–C(22) 1.458(8), C(22)–C(23) 1.188(9), C(23)–C(24) 1.439(9), Nd–Cl–NdA 106.85(5), Cl–Nd–ClA 73.15(5), N(2)–Nd–N(4) 91.56(17), N(2)–Nd–N(1) 54.95(17), N(4)–Nd–N(3) 54.74(16), N(2)–C(1)–N(1) 115.9(5), N(3)–C(21)–N(4) 117.1(5).

2.4. A Cerium(III)-bis(Diiminophosphinate) Complex

Rare-earth metal complexes comprising related N–P–N chelating diiminophosphinate ligands [$R_2P(NR')_2$]⁻ were first reported in the literature in 1989 [62], although the number of well-characterized examples remains small [63,64]. However, there is renewed interest in this class of compounds. A recent report by Cui *et al.* demonstrated that new diiminophosphinato-supported lanthanide dialkyl complexes are valuable catalysts for the 3,4-polymerization of isoprene. The NPN-bidentate diiminophosphinate-ligated rare-earth metal complexes provided both high *syndio*- and *iso*-3,4-selectivities (3,4 > 99%, *rr* = 66%, *mmmm* = 96%), depending on the frameworks, steric environment and geometry of the ligands. The regio- and stereoselective mechanisms proceeding in these systems were explicated by DFT simulation [65]. A new monomeric Ce(III)-diiminophosphinate complex, [$Ph_2P(NSiMe_3)_2$]$_2$Ce(μ-Cl)$_2$Li(THF)$_2$ (**9**), has been synthesized in the course of this work. The starting material Li[$Ph_2P(NSiMe_3)_2$] was prepared according to the literature method [66,67]. According to Scheme 5, a suspension of anhydrous CeCl$_3$ in THF was added to a solution of Li[$Ph_2P(NSiMe_3)_2$] in THF. The reaction mixture was stirred over night at room temperature. The product **9** was extracted as a golden-yellow solution in *n*-pentane. Complex **9** was isolated as an exceedingly air- and moisture-sensitive bright yellow, block-like crystals at 5 °C in 78% yield.

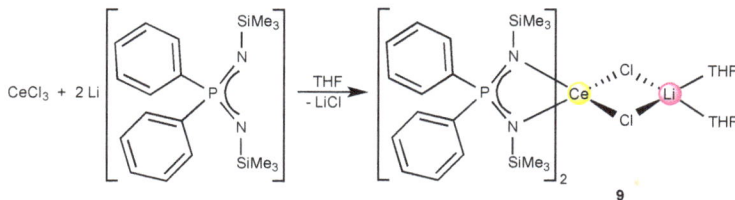

Scheme 5. Preparation of **9**.

The new compound $[Ph_2P(NSiMe_3)_2]_2Ce(\mu-Cl)_2Li(THF)_2$ (**9**) has been fully characterized by spectroscopic methods, elemental analysis and single-crystal X-ray diffraction. The molecular structure of **9** is in good agreement with related structures such as $[Ph_2P(NSiMe_3)_2]_2Sm(\mu-I)_2Li(THF)_2$ [63]. The IR spectrum showed medium bands at 1180 and 1116 cm^{-1}, which can be assigned to the PNSi unit. The strong bands that appear at 1246, 933 and 840 cm^{-1} can be attributed to the SiMe$_3$ groups. The ^{1}H NMR spectrum of **9** shows sets of signals assignable to the phenyl groups and the coordinated THF molecules. Due to the paramagnetic nature of the Ce^{3+} ion, the protons of the Si(CH$_3$)$_3$ groups appear as singlets at high magnetic field at $\delta = -6.50$ ppm. The X-ray study revealed the presence of the "ate" complex $[Ph_2P(NSiMe_3)_2]_2Ce(\mu-Cl)_2Li(THF)_2$ (**9**). The cerium(III) ion is coordinated with four nitrogen atoms of the diiminophosphinate ligands and two bridging chloride ions, giving a formal coordination number of six, as shown in Figure 5. The phenomenon of "ate" complex formation via retention of alkali metal halides in the products is quite common in organolanthanide chemistry [68]. It can be traced back to the strong tendency of the large Ln^{3+} ions to adopt high coordination numbers.

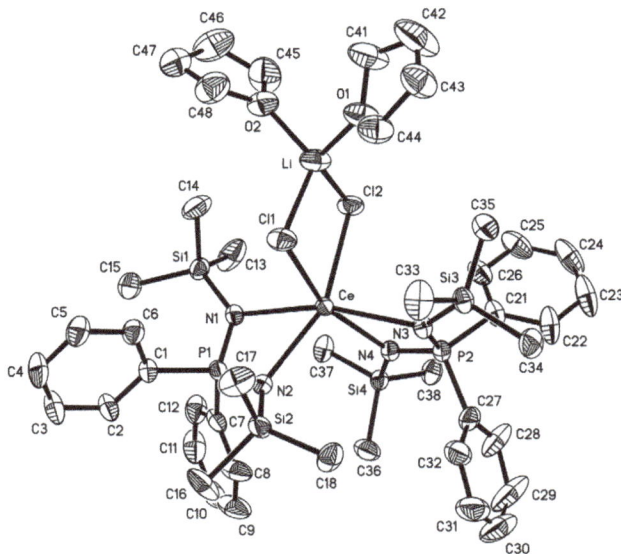

Figure 5. Molecular structure of **9** in the crystal, showing the atom-labeling scheme. Displacement ellipsoids are drawn at the 50% probability level. Selected bond lengths (Å) and angles (°): Li–O(1) 1.938(6), Li–O(2) 1.948(6), Li–Cl(1) 2.359(6), Li–Cl(2) 2.385(6), Ce–N(1) 2.555(2), Ce–N(2) 2.499(2), Ce–N(3) 2.557(2)Ce–N(4) 2.483(2), Ce–Cl(1) 2.8095(8), Ce–Cl(2) 2.8069(8), P(1)–N(1) 1.593(2), P(1)–N(2) 1.594(2), P(2)–N(4) 1.593(2), P(2)–N(3) 1.598(2), O(1)–Li–O(2) 99.1(3), Cl(1)–Li–Cl(2) 97.7(2), N(2)–Ce–N(1) 60.96(7), N(4)–Ce–N(3) 60.80(7), Cl(2)–Ce–Cl(1) 79.00(3), N(1)–P(1)–N(2) 107.10(11), N(4)–P(2)–N(3) 106.18(12).

Compound **9** crystallizes in the monoclinic space group $P2_1/n$ with four molecules of the complex in the unit cell. The average Ce–N bond distance (2.551 Å) is almost identical compared to the Ce–N distances in cerium(III) amidinate complexes such as [Ph–C≡C–C(N^{i}Pr)$_2$]$_3$Ce [16] and [{c-C$_3$H$_5$–C≡C–C(NR)$_2$}$_2$Ce(μ-Cl)(THF)]$_2$ (R = ^{i}Pr, Cy) [45]. Likewise, the Ce–Cl distance (average 2.810 Å) is very similar. The bond lengths P1–N1 (1.593(2) Å), P1–N2 (1.594(2) Å), P2–N3 (1.597(2) Å) and P2–N3 (1.593(2) Å) are almost equal, confirming the delocalization of the negative charge in the N1–P1–N2 and N3–P2–N4 units, respectively. The bond angles N1–Ce–N2 (60.96(7)°) and N3–Ce–N4 (60.80(7)°) in **9** are somewhat larger than those observed in [{c-C$_3$H$_5$–C≡C–C(NR)$_2$}$_2$Ce(μ-Cl)(THF)]$_2$ (R = ^{i}Pr, Cy) [45]. The bond angles Ce–Cl1–Li, Cl1–Li–Cl2, and Cl1–Ce–Cl2 are 39.80(10)°, 97.7(2)° and 79.01(3)°, respectively, to form the rhomb-shaped Ce–Cl1–Li–Cl2 unit. Two THF ligands complete the distorted tetrahedral coordination sphere around lithium.

2.5. Attempted Oxidation of **9** to A Cerium(IV) Complex

The preparation and characterization of well-defined molecular cerium(IV) complexes remains a considerable synthetic challenge. However, significant progress has been made in the field of cerium(IV) amides and related species. Various Ce(IV) amide complexes such as heteroleptic Ce[N{CH$_2$CH$_2$N(SiMe$_2$^{t}Bu)}$_3$]I [69], Ce[N(SiMe$_3$)$_2$]$_3$X (X = Cl, Br) [70] and [(Me$_3$Si)$_2$N]$_3$Ce(OC$_6$H$_4$O)Ce[N(SiMe$_3$)$_2$]$_3$ [71] as well as homoleptic Ce(N^{i}Pr$_2$)$_4$ [72] Ce(NCy$_2$)$_4$ [73] and Ce[N(SiHMe$_2$)$_2$]$_4$ [74] have been successfully synthesized by oxidation of cerium(III) precursors. Suitable oxidizing agents are e.g., TeCl$_4$, trityl chloride (Ph$_3$CCl), hexachloroethane (C$_2$Cl$_6$) or phenyliodine(III) dichloride (PhICl$_2$). The first cerium(IV) amidinate complex, [p-MeOC$_6$H$_4$C(NSiMe$_3$)$_2$]$_3$CeCl, was obtained in 61% yield upon oxidation of the cerium(III) precursor [p-MeOC$_6$H$_4$C(NSiMe$_3$)$_2$]$_3$Ce(N≡CC$_6$H$_4$OMe-p) with PhICl$_2$ in n-pentane [75]. Thus it seemed reasonable to attempt a similar oxidation reaction of [Ph$_2$P(NSiMe$_3$)$_2$]$_2$Ce(μ-Cl)$_2$Li(THF)$_2$ (**9**). Compound **9** was treated with calculated amounts of either trityl chloride or PhICl$_2$ in different solvents (n-pentane, toluene, THF). In all cases, intensely colored (dark green or green-brown) solutions were obtained, but we were unable to isolate the tentative cerium(IV) complex [Ph$_2$P(NSiMe$_3$)$_2$]$_2$CeCl$_2$ under any circumstances.

Table 1. Crystallographic Data for **1**, **2**, **8** and **9**.

Parameter	1	2	8	9
Empirical formula	C$_{38}$H$_{54}$Li$_2$N$_4$O$_2$	C$_{50}$H$_{70}$Li$_2$N$_4$O$_2$	C$_{56}$H$_{92}$Cl$_2$N$_8$Nd$_2$O$_2$	C$_{44}$H$_{72}$CeCl$_2$LiN$_4$O$_2$P$_2$Si$_4$
a(Å)	9.6687(19)	9.8768(3)	10.049(2)	11.117(2)
b(Å)	9.948(2)	10.2082(3)	11.337(2)	36.425(7)
c(Å)	19.611(4)	11.6883(4)	14.471(3)	13.538(3)
α(°)	93.77(3)	104.144(3)	82.53(3)	90
β(°)	91.33(3)	90.248(3)	83.02(3)	96.70(3)
γ(°)	90.96(3)	104.209(3)	73.74(3)	90
V(Å^3)	1881.3(6)	1105.18(6)	1562.9(5)	5444.3(19)
Z	2	1	1	4
formula weight	612.73	772.98	1268.76	1081.32
space group	P-1	P-1	P-1	P2$_1$/n
T (°C)	−120	−173	−120	−120
λ (Å)	0.71073	1.54184	0.71073	0.71073
D_{calcd} (g cm^{-3})	1.082	1.161	1.348	1.319
μ (mm^{-1})	0.066	0.530	1.771	1.118
F(000)	664	420	654	2244
data/restraints/parameters	7395/0/469	4628/0/262	8340/0/324	12,841/0/569
goodness-of-fit on F^2	1.013	1.027	1.219	0.977
R (Fo or Fo^2)	0.0520	0.0392	0.0547	0.0363
R$_w$ (Fo or Fo^2)	0.1347	0.0974	0.1663	0.0859

3. Experimental Section

3.1. General Procedures

All experiments were carried out under an inert atmosphere of dry argon employing standard Schlenk and glovebox techniques (<1 ppm O_2, <1 ppm H_2O). *n*-Pentane and THF were distilled from sodium/benzophenone under nitrogen atmosphere prior to use. All glassware was oven-dried at 120 °C for at least 24 h, assembled while hot, and cooled under high vacuum prior to use. The starting materials, anhydrous $LnCl_3$ (Ln = Ce, Nd) [76] and $Li[Ph_2P(NSiMe_3)_2]$ [66,67], and compound **8** [45] were prepared according to the literature methods. *^{n}BuLi*, phenylacetylene, and *N*,*N'*-dicyclohexylcarbodiimide were purchased from Aldrich and used as received. ^{1}H-NMR (400 MHz) and ^{13}C-NMR (100.6 MHz) were recorded in C_6D_6 or $CDCl_3$ solutions on a Bruker DPX 400 spectrometer at 25 °C. Chemical shifts were referenced to TMS. Assignment of signals was made from ^{1}H–^{13}C HSQC 2D NMR experiments. IR spectra were recorded using KBr pellets on a Perkin Elmer FT-IR spectrometer system 2000 (Perkin Elmer, Rodgau, Germany) between 4000 and 400 cm^{-1}. Microanalyses of the compounds were performed using a LECO CHNS 932 apparatus (LECO Corp., St. Joesph, MI, USA).

3.2. Syntheses

$Li[Ph-C\equiv C-C(NCy)_2]\cdot THF$ (**2**). A solution of phenylacetylene (3.30 mL, 30 mmol) in THF (80 mL) was cooled to −20 °C and treated slowly with *n*-butyllithium (18.85 mL, 1.6 M solution in *n*-hexane). The solution was stirred for 15 min, before *N*,*N'*-dicyclohexylcarbodiimide (6.20 g, 30 mmol) was added. The reaction mixture was stirred for 10 min at −20 °C, and then warmed to room temperature and stirred for another 2 h. The solution was reduced to a small volume under vacuum (25 mL) and stored at −25 °C in a freezer to obtain single-crystals of **2**. Yield: 10.20 g, 88%. Elemental analysis for $C_{25}H_{35}LiN_2O$ (386.51 g·mol^{-1}): C, 77.69%; H, 9.13%; N, 7.25%; found C, 77.15%; H, 9.33%; N, 7.08%. ^{1}H NMR (400 MHz, toluene-d_8, 25 °C): δ (ppm) 7.52 (d, *J* = 6.6 Hz, 2H, C_6H_5), 6.99 (m, 3H, C_6H_5), 3.96 (m, 2H, *CH*, Cy), 3.65 (m, 4H, THF), 1.44 (m, 4H, THF), 1.31–2.12 (m, 20H, CH_2, Cy); ^{13}C{^{1}H} NMR (100.6 MHz, toluene-d_8, 25 °C): δ (ppm) 159.3 (NCN), 158.5 (C≡C–C), 132.2 (C_6H_5), 123.8 (C_6H_5), 94.4 (C_6H_5), 82.4 (C_6H_5–C≡C), 68.2 (THF), 59.8 (CH, Cy), 37.6 (CH_2, Cy), 26.8 (CH_2, Cy), 25.6 (THF), 23.0 (CH_2, Cy).

$Li[Ph-C\equiv C-C(NCy)_2]\cdot Et_2O$ (**3**). The compound was prepared by following the procedure for **2** using diethyl ether as solvent. The solvent was removed under vacuum affording **2** as white solid. Yield: 8.85 g, 76%. Elemental analysis for $C_{25}H_{37}LiN_2O$ (388.52 g·mol^{-1}): C, 77.29%; H, 9.60%; N, 7.21%; found C, 76.06%; H, 9.52%; N, 7.60%. ^{1}H NMR (400 MHz, C_6D_6, 25 °C): δ (ppm) 7.56 (d, *J* = 6.8 Hz, 2H, Ph), 6.95 (m, 3H, Ph), 4.01(s, br, 2H, *CH*, Cy), 3.25 (q, 4H, CH_2, Et_2O), 2.40 (m, 4H, CH_2, Cy), 1.93 (m, 6H, CH_2, Cy), 1.24–1.80 (m, 10H, CH_2, Cy), 1.15 (t, 6H, CH_3, Et_2O); ^{13}C{^{1}H} NMR (100.6 MHz, C_6D_6, 25 °C): δ (ppm) 132.3 (Ph), 123.8 (Ph), 94.3 (Ph), 83.1 (Ph–C≡C), 65.9 (Et_2O), 59.8 (CH, Cy), 37.7 (CH_2, Cy), 26.9 (CH_2, Cy), 26.5 (CH_2, Cy), 15.3 (Et_2O).

$[Ph-C\equiv C-C(NCy)_2]_3Ce$ (**4**). A solution of anhydrous $CeCl_3$ (1.0 g, 4 mmol) in 20 mL of THF was added to a solution of **2** (4.6 g, 12 mmol) in 60 mL of THF. The reaction mixture was stirred over night at the room temperature. The solvent was removed under vacuum to dryness followed by extraction with *n*-pentane (3 × 10 mL) to give a clear, orange solution. The filtrate was evaporated under vacuum to afford **4** as yellow solid. Yield: 3.4 g, 81%. Elemental analysis for $C_{63}H_{81}CeN_6$ (1062.50 g·mol^{-1}): C, 71.22%; H, 7.68%; N, 7.91%; found C, 71.36%; H, 7.64%; N, 7.90%. ^{1}H NMR (400 MHz, THF-d_8, 25 °C): δ (ppm) 9.49 (s, br, 2H, *CH*, Cy), 8.81 (s, 3H, Ph), 7.8–8.05 (m, 6H, Ph), 7.29–7.56 (m, 6H, Ph), 3.76 (s, br, 2H, *CH*, Cy), 3.56 (m, 2H, *CH*, Cy), 0.93–2.12 (m, 50H, CH_2, Cy), −0.05 (s, br, 4H, CH_2, Cy), −1.82 (s, br, 6H, CH_2, Cy); ^{13}C{^{1}H} NMR (100.6 MHz, THF-d_8, 25 °C): δ (ppm) 160.7 (NCN), 140.5 (Ph), 134 (Ph), 132.6 (Ph), 130.2 (Ph), 129.4 (Ph), 100.5 (C≡C–C), 90.1 (Ph–C≡C), 67.1 (CH, Cy), 61.1 (CH, Cy), 56.3 (CH, Cy), 36.4 (CH_2, Cy), 34.1 (CH_2, Cy), 26.2 (CH_2, Cy).

[Ph–C≡C–C(NCy)$_2$]$_3$Nd (**5**). A solution of anhydrous NdCl$_3$ (0.5 g, 2 mmol) in 20 mL of THF was added to a solution of **2** (2.3 g, 6 mmol) dissolved in 60 mL of THF. Following the procedure described for **4**, compound **5** was isolated as a pale green solid. Yield: 1.5 g, 73%. Elemental analysis for C$_{63}$H$_{81}$N$_6$Nd (1066.63 g·mol^{-1}): C, 70.94%; H, 7.65%; N, 7.88%; found C, 68.88%; H, 7.41%, N, 7.79%. ^{1}H NMR (400 MHz, THF-d_8, 25 °C): δ (ppm) 18.33 (s, 2H, CH, Cy), 9.40 (m, 3H, Ph), 8.25 (m, 4H, Ph), 8.05 (m, 2H, Ph), 7.48 (m, 2H, Ph), 7.38 (m, 4H, Ph), 3.77 (s, 2H, CH, Cy), 3.57 (s, 2H, CH, Cy), 0.56–2.10 (m, br, 50H, CH$_2$, Cy), −0.66 (s, 4H, CH$_2$, Cy), −3.55 (s, 6H, CH$_2$, Cy); ^{13}C{^{1}H} NMR (100.6 MHz, THF-d_8, 25 °C): δ (ppm) 140.5 (Ph), 135.5 (Ph), 132.6 (Ph), 131.0 (Ph), 130.4 (Ph), 129.8 (Ph), 125.3 (Ph), 102.0 (C≡C–C), 90.0 (Ph–C≡C), 78.5 (CH, Cy), 61.1 (CH, Cy), 50.0 (CH, Cy), 36.0 (CH$_2$, Cy), 33.7 (CH$_2$, Cy), 27.2 (CH$_2$, Cy), 25.7 (CH$_2$, Cy).

[Ph–C≡C–C(NCy)$_2$]$_3$Sm (**6**). Reaction of anhydrous SmCl$_3$ (0.5 g, 2 mmol) with **2** (2.3 g, 6 mmol) following the procedure described for **4** afforded **6** as pale yellow solid. Yield: 1.3 g, 64%. Elemental analysis for C$_{63}$H$_{81}$N$_6$Sm (1072.74 g·mol^{-1}): C, 70.54%; H, 7.61%; N, 7.83%; found C, 69.62%; H, 7.70%; N, 7.60%. ^{1}H NMR (400 MHz, THF-d_8, 25 °C): δ (ppm) 8.23 (d, 3H, Ph), 8.04 (s, br, 3H, Ph), 7.30–7.64 (m, 9H, Ph), 3.71 (s, br, 3H, CH, Cy), 3.30 (m, 3H, CH, Cy), 0.00–0.16 (m, 4H, CH$_2$, Cy), 0.30–2.06 (m, br, 50H, CH$_2$, Cy), −1.70 (s, br, 6H, CH$_2$, Cy); ^{13}C{^{1}H} NMR (100.6 MHz, THF-d_8, 25 °C): δ (ppm) 183.3 (NCN), 133.1 (Ph), 130.1(Ph), 129.6 (Ph), 124.4 (Ph), 97.7 (C≡C–C), 84.1 (Ph–C≡C), 59.5 (CH, Cy), 57.4 (CH, Cy), 36.6 (CH$_2$, Cy), 33.7 (CH$_2$, Cy), 26.9 (CH$_2$, Cy).

[Ph–C≡C–C(NCy)$_2$]$_3$Ho (**7**). Reaction of anhydrous HoCl$_3$ (1.0 g, 3.7 mmol) with **2** (4.3 g, 11.1 mmol). Following the procedure described for **4**, the reaction afforded **7** as pale yellow solid. Yield: 3.6 g, 91%. Elemental analysis for C$_{63}$H$_{81}$HoN$_6$ (1087.31 g·mol^{-1}): C, 69.59%; H, 7.51%; N, 7.73%; found C, 69.45%; H, 7.41%; N, 7.73%.

[Ph$_2$P(NSiMe$_3$)$_2$]$_2$Ce(μ-Cl)$_2$Li(THF)$_2$ (**9**). A solution of anhydrous CeCl$_3$ (1.0 g, 4.1 mmol) in 30 mL of THF was added to a solution of Li[Ph$_2$P(NSiMe$_3$)$_2$] (3.0 g, 8.2 mmol) in 60 mL of THF. The reaction mixture was stirred over night at room temperature. The solution color changed to golden yellow. The solvent was removed under vacuum followed by extraction with *n*-pentane (30 mL) to give a clear, bright yellow solution. The filtrate was concentrated under vacuum to *ca.* 10 mL. Crystallization at 5 °C afforded **9** in the form of bright yellow, block-like crystals. Yield: 3.4 g, 78%. Elemental analysis for C$_{44}$H$_{72}$CeCl$_2$LiN$_4$O$_2$P$_2$Si$_4$ (1081.32 g·mol^{-1}): C, 48.82%; H, 6.65%; N, 5.17%; found C, 42.10%; H, 6.70%; N, 5.30%. ^{1}H NMR (400 MHz, THF-d_8, 25 °C): δ (ppm) 10.22 (m, 8H, C$_6$H$_5$), 9.50 (s, br, 4H, C$_6$H$_5$), 7.70 (m, 8H, C$_6$H$_5$), 1.44 (m, 8H, THF), −4.80 (s, br, 3H, Si(CH$_3$)$_3$), −5.83 (s, br, 33H, Si(CH$_3$)$_3$); ^{13}C{^{1}H} NMR (100.6 MHz, THF-d_8, 25 °C): δ (ppm) 144.0 (C$_6$H$_5$), 133.2 (C$_6$H$_5$), 132.5 (C$_6$H$_5$), 131.2 (C$_6$H$_5$), 64.7 (THF), 24.1 (THF), −1.3 (Si(CH$_3$)$_3$); ^{29}Si NMR (80 MHz, THF-d_8, 25 °C): δ (ppm) −13.90 (Si(CH$_3$)$_3$), −21.24 (Si(CH$_3$)$_3$); ^{31}P NMR (162 MHz, THF-d_8, 25 °C): δ (ppm) −4.94.

3.3. X-ray Crystallography

The intensity data of **1**, **2**, **8**, and **9** were collected on a Stoe IPDS 2T diffractometer with MoK$_\alpha$ radiation. The data were collected with the Stoe XAREA [77] program using ω-scans. The space groups were determined with the XRED32 [78] program. Absorption corrections were applied using the multi-scan method. The structures were solved by direct methods (SHELXS-97) [79] and refined by full matrix least-squares methods on F2 using SHELXL-97 [80]. Data collection parameters are given in Table 1. CCDC 1419935 (**1**), 1419937 (**2**), 1400418 (**8**), and 1419934 (**9**) contain the supplementary crystal data for this article. These data can be obtained free of charge from the Cambridge Crystallographic Data Centre via www.ccdc.cam.ac.uk/data_request/cif.

4. Conclusions

In summarizing the results reported here, we extended the series of homoleptic lanthanide tris(propiolamidinate) complexes by preparing four new representatives, [Ph–C≡C–C(NCy)$_2$]$_3$Ln (**4**: Ln = Ce; **5**: Ln = Nd; **6**: Ln = Sm; **7**: Ln = Ho). Two lithium propiolamidinate precursors, Li[Ph-C≡C-C(NCy)$_2$] S (**1**: R = ^{i}Pr, S = THF; **2**: R = Cy, S = THF), have been structurally characterized by X-ray diffraction. Moreover, crystal structure determinations of [{c-C_3H_5–C≡C–C(N^{i}Pr)$_2$}$_2$Ln(μ-Cl)(THF)]$_2$ (**8**) and the new monomeric Ce(III)-diiminophosphinate complex, [Ph$_2$P(NSiMe$_3$)$_2$]$_2$Ce(μ-Cl)$_2$Li(THF)$_2$ (**9**), have been carried out. Attempts to oxidize compound **9** using trityl chloride or phenyliodine(III) dichloride did not lead to an isolable cerium(IV) species.

Supplementary: Supplementary materials can be found at http://www.mdpi.com/2304-6740/3/4/0429/s1.

Acknowledgments: Generous support of this work through a research grant by the Deutsche Forschungsgemeinschaft DFG (Grant No. AN 238/16 "Organocer(IV)-Chemie") is gratefully acknowledged. This work was also financially supported by the Otto-von-Guericke-Universität Magdeburg. Farid M. Sroor is grateful to the Ministry of Higher Educational Scientific Research (MHESR), Egypt, and the Germany Academic Exchange Service (DAAD), Germany, for a Ph.D. scholarship within the German Egyptian Research Long-Term Scholarship (GERLS) program.

Author Contributions: Farid M. Sroor carried out the experimental work. Cristian G. Hrib did the crystal structure determinations. Frank T. Edelmann conceived and supervised the experiments and wrote the paper.

Conflicts of Interest: The authors declare no conflict of interest.

References

1. Edelmann, F.T. Advances in the coordination chemistry of amidinate and guanidinate ligands. *Adv. Organomet. Chem.* **2008**, *57*, 183–352.
2. Coles, M.P. Bicyclic-guanidines, -guanidinates and -guanidinium salts: Wide ranging applications from a simple family of molecules. *Chem. Commun.* **2009**, 3659–3676. [CrossRef] [PubMed]
3. Jones, C. Bulky guanidinates for the stabilization of low oxidation state metallacycles. *Coord. Chem. Rev.* **2010**, *254*, 1273–1289. [CrossRef]
4. Trifonov, A.A. Guanidinate and amidopyridinate rare-earth complexes: Towards highly reactive alkyl and hydrido species. *Coord. Chem. Rev.* **2010**, *254*, 1327–1347. [CrossRef]
5. Mohamed, A.A.; Abdou, H.E., Jr.; Fackler, J.P. Coordination chemistry of gold(II) with amidinate, thiolate and ylide ligands. *Coord. Chem. Rev.* **2010**, *254*, 1253–1259. [CrossRef]
6. Collins, S. Polymerization catalysis with transition metal amidinate and related complexes. *Coord. Chem. Rev.* **2011**, *255*, 118–138. [CrossRef]
7. Edelmann, F.T. Recent progress in the chemistry of metal amidinates and guanidinates: Syntheses, catalysis and materials. *Adv. Organomet. Chem.* **2013**, *61*, 55–374.
8. Edelmann, F.T. Lanthanide Amidinates and guanidinates: From laboratory curiosities to efficient homogeneous catalysts and precursors for rare-earth oxide thin films. *Chem. Soc. Rev.* **2009**, *38*, 2253–2268. [CrossRef] [PubMed]
9. Edelmann, F.T. Lanthanide amidinates and guanidinates in catalysis and materials science: A continuing success story. *Chem. Soc. Rev.* **2012**, *41*, 7657–7672. [CrossRef] [PubMed]
10. Fujita, H.; Endo, R.; Aoyama, A.; Ichii, T. Propiolamidines I. syntheses of *N,N'*-disubstituted phenylpropiolamidines and new routes to 5-*N*-substituted amino-3-phenylisoxazoles and 5-*N*-substituted amino-1,3-diphenylpyrazoles. *Bull. Chem. Soc. Jpn.* **1972**, *45*, 1846–1852. [CrossRef]
11. Himbert, G.; Feustel, M.; Jung, M. Aminoethinylierungen, 2. Synthese von (Aminoethinyl)ketonen über β-stannylierte Inamine. *Liebigs Ann. Chem.* **1981**, 1907–1927. [CrossRef]
12. Himbert, G.; Schwickerath, W. (Aminoethinyl)metallierungen, 13. Synthese und reaktionen von 3-aminopropiolamidinen. *Liebigs Ann. Chem.* **1984**, 85–97. [CrossRef]
13. Schmidt, G.F.; Süss-Fink, G. Katalytische C–C– und C–N–kupplungsreaktionen von carbodiimiden durch übergangsmetallcluster. *J. Organomet. Chem.* **1988**, *356*, 207–211. [CrossRef]
14. Ong, T.-G.; O'Brien, J.S.; Korobkov, I.; Richeson, D.S. Facile and atom-efficient amidolithium-catalyzed C–C and C–N formation for the construction of substituted guanidines and propiolamidines. *Organometallics* **2006**, *25*, 4728–4730. [CrossRef]
15. Xu, X.; Gao, J.; Cheng, D.; Li, J.; Qiang, G.; Guo, H. Copper-catalyzed highly efficient multicomponent reactions of terminal alkynes, acid chlorides, and carbodiimides: Synthesis of functionalized propiolamidine derivatives. *Adv. Synth. Catal.* **2008**, *350*, 61–64. [CrossRef]

16. Dröse, P.; Hrib, C.G.; Edelmann, F.T. Synthesis and structural characterization of a homoleptic cerium(III) propiolamidinate. *J. Organomet. Chem.* **2010**, *695*, 1953–1956. [CrossRef]

17. Weingärtner, W.; Kantlehner, W.; Maas, G. A convenient synthesis and some characteristic reactions of novel propiolamidinium salts. *Synthesis* **2011**, *42*, 265–272.

18. Weingärtner, W.; Maas, G. Cycloaddition reactions of propiolamidinium salts. *Eur. J. Org. Chem.* **2012**, *2012*, 6372–6382. [CrossRef]

19. Devi, A. "Old Chemistries" for new applications: Perspectives for development of precursors for MOCVD and ALD applications. *Coord. Chem. Rev.* **2013**, *257*, 3332–3384. [CrossRef]

20. Fujita, H.; Endo, R.; Murayama, K.; Ichii, T. A novel cleavage of carbon–carbon triple bond. *Bull. Chem. Soc. Jpn.* **1972**, *45*, 1581. [CrossRef]

21. Ried, W.; Wegwitz, M. Synthese von speziell substituierten 1,3-Dithiol-2-thionen und 1,3-Thiaselenol-2-thionen. *Liebigs Ann. Chem.* **1975**, *1975*, 89–94. [CrossRef]

22. Ried, W.; Schweitzer, R. Neue cyclisierungsprodukte aus *N,N'*-diarylsubstituierten propiolamidinen. *Chem. Ber.* **1976**, *109*, 1643–1649. [CrossRef]

23. Ried, W.; Winkler, H. Cyclisierungsprodukte aus *N*-arylsubstituierten propiolamidinen. *Chem. Ber.* **1979**, *112*, 384–388. [CrossRef]

24. Sienkiewich, P.; Bielawski, K.; Bielawska, A.; Palka, J. Inhibition of collagen and DNA biosynthesis by a novel amidine analogue of chlorambucil is accompanied by deregulation of β1-integrin and IGF-I receptor signaling in MDA-MB 231 cells. *Environ. Toxicol. Pharmacol.* **2005**, *20*, 118–124. [CrossRef] [PubMed]

25. Sielecki, T.M.; Liu, J.; Mousa, S.A.; Racanelli, A.L.; Hausner, E.A.; Wexler, R.R.; Olson, R.E. Synthesis and pharmacology of modified amidine isoxazoline glycoprotein IIb/IIIa receptor antagonists. *Bioorg. Med. Chem. Lett.* **2001**, *11*, 2201–2204. [CrossRef]

26. Stephens, C.E.; Tanious, E.; Kim, S.; Wilson, D.W.; Schell, W.A.; Perfect, J.R.; Franzblau, S.G.; Boykin, D.W. Diguanidino and "reversed" diamidino 2,5-diarylfurans as antimicrobial agents. *J. Med. Chem.* **2001**, *44*, 1741–1748. [CrossRef] [PubMed]

27. Rowley, C.N.; DiLabio, G.A.; Barry, S.T. Theoretical and synthetic investigations of carbodiimide insertions into Al–CH$_3$ and Al–N(CH$_3$)$_2$ bonds. *Inorg. Chem.* **2005**, *44*, 1983–1991. [CrossRef] [PubMed]

28. Zhang, W.-X.; Nishiura, M.; Hou, Z. Catalytic addition of terminal alkynes to carbodiimides by half-sandwich rare earth metal complexes. *J. Am. Chem. Soc.* **2005**, *127*, 16788–16789. [CrossRef] [PubMed]

29. Zhou, S.; Wang, S.; Yang, G.; Li, Q.; Zhang, L.; Yao, Z.; Zhou, Z.; Song, H.-B. Synthesis, structure, and diverse catalytic activities of [ethylenebis(indenyl)]lanthanide(III) amides on N–H and C–H addition to carbodiimides and ε-caprolactone polymerization. *Organometallics* **2007**, *26*, 3755–3761. [CrossRef]

30. Zhang, W.-X.; Hou, Z. Catalytic addition of alkyne C–H, amine N–H, and phosphine P–H bonds to carbodiimides: An efficient route to propiolamidines, guanidines, and phosphaguanidines. *Org. Biomol. Chem.* **2008**, *6*, 1720–1730. [CrossRef] [PubMed]

31. Du, Z.; Li, W.; Zhu, X.; Xu, F.; Shen, Q. Divalent lanthanide complexes: Highly active precatalysts for the addition of N–H and C–H bonds to carbodiimides. *J. Org. Chem.* **2008**, *73*, 8966–8972. [CrossRef] [PubMed]

32. Rowley, C.N.; Ong, T.-G.; Priem, J.; Richeson, D.S.; Woo, T.K. Analysis of the critical step in catalytic carbodiimide transformation: Proton transfer from amines, phosphines, and alkynes to guanidinates, phosphaguanidinates, and propiolamidinates with Li and Al catalysts. *Inorg. Chem.* **2008**, *47*, 12024–12031. [CrossRef] [PubMed]

33. Wu, Y.; Wang, S.; Zhang, L.; Yang, G.; Zhu, X.; Zhou, Z.; Zhu, H.; Wu, S. Cyclopentadienyl-free rare-earth metal amides [{(CH$_2$SiMe$_2$){(2,6-^{i}Pr$_2$C$_6$H$_3$)N}$_2$}Ln{N(SiMe$_3$)$_2$}(THF)] as highly efficient versatile catalysts for C–C and C–N bond formation. *Eur. J. Org. Chem.* **2010**, *41*, 326–332. [CrossRef]

34. Xu, L.; Wang, Y.-C.; Zhang, W.-X.; Xi, Z. Half-sandwich bis(propiolamidinate) rare-earth metal complexes: Synthesis, structure and dissociation of the cyclopentadienyl ligand via competition with an amidinate. *Dalton Trans.* **2013**, *42*, 16466–16469. [CrossRef] [PubMed]

35. Seidel, W.W.; Dachtler, W.; Pape, T. Synthesis, structure, and reactivity of Ru^{II} complexes with trimethylsilylethynylamidinate ligands. *Z. Anorg. Allg. Chem.* **2012**, *638*, 116–121. [CrossRef]

36. Brown, D.J.; Chisholm, M.H.; Gallucci, J.C. Amidinate-carboxylate complexes of dimolybdenum and ditungsten: M$_2$(O$_2$CR)$_2$((N^{i}Pr)$_2$CR')$_2$. Preparations, molecular and electronic structures and reactions. *Dalton Trans.* **2008**, 1615–1624. [CrossRef] [PubMed]

37. Fischer, P.; Kurtz, W.; Effenberger, F. NMR-spektroskopische untersuchung der konformation und reaktivität von cyclopropylbenzolen. *Chem. Ber.* **1973**, *106*, 549–559. [CrossRef]

38. Wilcox, C.F.; Loew, L.M.; Hoffman, R. Why a cyclopropyl group is good at stabilizing a cation but poor at transmitting substituent effects. *J. Am. Chem. Soc.* **1973**, *95*, 8192–8193. [CrossRef]

39. De Meijere, A. Bonding properties of cyclopropane and their chemical consequences. *Angew. Chem. Int. Ed.* **1979**, *18*, 809–826. [CrossRef]

40. Brown, H.C.; Periasamy, M.; Perumal, P.T. Structural effects in solvolytic reactions. 47. Effects of *p*-alkyl and *p*-cycloalkyl groups on the carbon-13 NMR shifts of the cationic carbon center in *p*-alkyl-*tert*-cumyl cations. *J. Org. Chem.* **1984**, *49*, 2754–2757. [CrossRef]

41. Roberts, D.D. Electron-donating ability of the cyclopropyl substituent in the solvolyses of an α-CF$_3$-substituted secondary-alkyl tosylate. *J. Org. Chem.* **1991**, *56*, 5661–5665. [CrossRef]

42. Sauers, R.R. Cyclopropanes: Calculation of NMR spectra by *ab initio* methodology. *Tetrahedron* **1998**, *54*, 337–348. [CrossRef]

43. De Meijere, A.; Kozhushkov, S.I. Small rings in focus: Recent routes to cyclopropylgroup containing carbo- and heterocycles. *Mendeleev Commun.* **2010**, *20*, 301–311. [CrossRef]

44. Sroor, F.M.A.; Hrib, C.G.; Hilfert, L.; Edelmann, F.T. Lithium-cyclopropylethinylamidinates. *Z. Anorg. Allg. Chem.* **2013**, *639*, 2390–2394. [CrossRef]

45. Sroor, F.M.; Hrib, C.G.; Hilfert, L.; Jones, P.G.; Edelmann, F.T. Lanthanide(III)-bis (cyclopropylethinylamidinates): Synthesis, structure, and catalytic activity. *J. Organomet. Chem.* **2015**, *785*, 1–10. [CrossRef]

46. Sroor, F.M.; Hrib, C.G.; Hilfert, L.; Busse, S.; Edelmann, F.T. Synthesis and catalytic activity of homoleptic lanthanide-tris(cyclopropylethinyl)amidinates. *New J. Chem.* **2015**, *39*, 7595–7601. [CrossRef]

47. Sroor, F.M.; Hrib, C.G.; Hilfert, L.; Hartenstein, L.; Roesky, P.W.; Edelmann, F.T. Synthesis and structural characterization of new bis(alkynylamidinato)lanthanide(III)-amides. *J. Organomet. Chem.* **2015**, *799–800*, 160–165. [CrossRef]

48. Stalke, D.; Wedler, M.; Edelmann, F.T. Dimere Alkalimetallbenzamidinate: Einfluß des metallions auf die struktur. *J. Organomet. Chem.* **1992**, *431*, C1–C5. [CrossRef]

49. Barker, J.; Barr, D.; Barnett, N.D.R.; Clegg, W.; Cragg-Hine, I.; Davidson, M.G.; Davies, R.P.; Hodgson, S.M.; Howard, J.A.K.; Kilner, M.; *et al.* Lithiated amidines: Syntheses and structural characterisations. *Dalton Trans.* **1997**, 951–955. [CrossRef]

50. Hagadorn, J.R.; Arnold, J. Ferrocene-substituted amidinate derivatives: Syntheses and crystal structures of lithium, iron(II), and cobalt(II) complexes. *Inorg. Chem.* **1997**, *36*, 132–133. [CrossRef]

51. Caro, C.F.; Hitchcock, P.B.; Lappert, M.F.; Layh, M. Neutral isonitrile adducts of alkali and alkaline earth metals. *Chem. Commun.* **1998**, 1297–1298. [CrossRef]

52. Knapp, C.; Lork, E.; Watson, P.G.; Mews, R. Lithium fluoroarylamidinates: Syntheses, structures, and reactions. *Inorg. Chem.* **2002**, *41*, 2014–2025. [CrossRef] [PubMed]

53. Boyd, C.L.; Tyrrell, B.R.; Mountford, P. Bis(μ-*N*-*n*-propyl-*N*-trimethylsilylbenz-amidinato)bis[(diethyl ether-*O*)lithium. *Acta Crystallogr. Sect. E* **2002**, *58*, m597–m598. [CrossRef]

54. Aharonovich, S.; Kapon, M.; Botoshanski, M.; Eisen, M.S. *N,N'*-bis-silylated lithium aryl amidinates: Synthesis, characterization, and the gradual transition of coordination mode from σ toward π originated by crystal packing interactions. *Organometallics* **2008**, *27*, 1869–1877. [CrossRef]

55. Nevoralová, J.; Chlupaty, T.; Padelková, Z.; Ruzicka, A. Quest for lithium amidinates containing adjacent amino donor group at the central carbon atom. *J. Organomet. Chem.* **2013**, *745–746*, 186–189.

56. Hong, J.; Zhang, L.; Wang, K.; Chen, Z.; Wu, L.; Zhou, X. Synthesis, structural characterization, and reactivity of mono(amidinate) rare-earth-metal bis(aminobenzyl) complexes. *Organometallics* **2013**, *32*, 7312–7322. [CrossRef]

57. Snaith, R.; Wright, D.S. *Lithium Chemistry, A Theoretical and Experimental Overview*; Sapse, A., von R. Schleyer, P., Eds.; Wiley: New York, NY, USA, 1995.

58. Downard, A.; Chivers, T. Applications of the laddering principle—A two-stage approach to describe lithium heterocarboxylates. *Eur. J. Inorg. Chem.* **2001**, *2001*, 2193–2221. [CrossRef]

59. Richter, J.; Feiling, J.; Schmidt, H.-G.; Noltemeyer, M.; Brüser, W.; Edelmann, F.T. Lanthanide(III) amidinates with six- and seven-coordinate metal atoms. *Z. Anorg. Allg. Chem.* **2004**, *630*, 1269–1275. [CrossRef]
60. Yao, Y.; Luo, Y.; Chen, J.; Zhang, Z.; Zhang, Y.; Shen, Q. Synthesis and characterization of bis(guanidinate)lanthanide diisopropylamido complexes—New highly active initiators for the polymerizations of ε-caprolactone and methyl methacrylate. *J. Organomet. Chem.* **2003**, *679*, 229–237. [CrossRef]
61. Lyubov, D.M.; Bubnov, A.M.; Fukin, G.K.; Dolgushin, F.M.; Antipin, M.Y.; Pelcé, O.; Schappacher, M.; Guillaume, S.M.; Trifonov, A.A. Hydrido complexes of yttrium and lutetium supported by bulky guanidinato ligands [Ln(μ-H){(Me₃Si)₂NC(NCy)₂}₂]₂ (Ln = Y, Lu): Synthesis, structure, and reactivity. *Eur. J. Inorg. Chem.* **2008**, *2008*, 2090–2098. [CrossRef]
62. Recknagel, A.; Witt, M.; Edelmann, F.T. Synthese von metallacyclophosphazenen mit praseodym, neodym, uran und thorium. *J. Organomet. Chem.* **1989**, *371*, C40–C44. [CrossRef]
63. Recknagel, A.; Steiner, A.; Noltemeyer, M.; Brooker, S.; Stalke, D.; Edelmann, F.T. Diiminophosphinate des lithiums, samariums und ytterbiums: Molekülstrukturen von Li[Ph₂P(NSiMe₃)₂](THF)₂ und [Ph₂P(NSiMe₃)₂]₂Sm(μ-I)₂Li(THF)₂. *J. Organomet. Chem.* **1991**, *414*, 327–335. [CrossRef]
64. Schumann, H.; Winterfeld, J.; Hemling, H.; Hahn, F.E.; Reich, P.; Brzezinka, K.-W.; Edelmann, F.T.; Kilimann, U.; Schäfer, M.; Herbst-Irmer, R. (Cyclooctatetraenyl)[N,N′-bis(trimethylsilyl)benzamidinato]- und -[P,P-diphenyl-bis-(trimethylsilylimino)phosphinato]-Komplexe der Seltenen Erden: Röntgenstrukturanalyse von (C₈H₈)Tm[PhC(NSiMe₃)₂](THF), (C₈H₈)Lu[4-MeOC₆H₄C(NSiMe₃)₂](THF) und (C₈H₈)Nd[Ph₂P(NSiMe₃)₂](THF). *Chem. Ber.* **1995**, *128*, 395–404.
65. Liu, B.; Li, L.; Sun, G.; Liu, J.; Wang, M.; Li, S.; Cui, D. 3,4-Polymerization of Isoprene by using NSN- and NPN-ligated rare earth metal precursors: Switching of stereo selectivity and mechanism. *Macromolecules* **2014**, *47*, 4971–4978. [CrossRef]
66. Schmidbaur, H.; Schwirten, K.; Pickel, H. Kleine anorganische ringe, II. ein cyclisches alumophosphazan und seine gallium- und Indiumanalogen. *Chem. Ber.* **1969**, *102*, 564–567. [CrossRef]
67. Wolfsberger, W.; Hager, W. Phosphinimino- und arsinimino-phosphoniumhalogenide. *Z. Anorg. Allg. Chem.* **1977**, *433*, 247–254. [CrossRef]
68. Edelmann, F.T. *Comprehensive Organometallic Chemistry III, Vol. 3, Complexes of Scandium, Yttrium and Lanthanide Elements*; Crabtree, R.H., Mingos, D.M.P., Eds.; Elsevier: Oxford, UK, 2006.
69. Morton, C.; Alcock, N.W.; Lees, M.R.; Munslow, I.J.; Sanders, C.J.; Scott, P. Stabilization of cerium(IV) in the presence of an iodide ligand: Remarkable effects of lewis acidity on valence state. *J. Am. Chem. Soc.* **1992**, *121*, 11255–11256. [CrossRef]
70. Eisenstein, O.; Hitchcock, P.B.; Hulkes, A.G.; Lappert, M.F.; Maron, L. Cerium masquerading as a group 4 element: Synthesis, structure and computational characterisation of [CeCl{N(SiMe₃)₂}₃]. *Chem. Commun.* **2001**, *17*, 1560–1561. [CrossRef]
71. Hitchcock, P.B.; Hulkes, A.G.; Lappert, M.F. Oxidation in nonclassical organolanthanide chemistry: Synthesis, characterization, and X-ray crystal structures of cerium(III) and -(IV) amides. *Inorg. Chem.* **2003**, *43*, 1031–1038. [CrossRef] [PubMed]
72. Werner, D.; Deacon, G.B.; Junk, P.C.; Anwander, R. Cerium(III/IV) formamidinate chemistry, and a stable cerium(IV) diolate. *Chem. Eur. J.* **2014**, *20*, 4426–4438. [CrossRef] [PubMed]
73. Schneider, D.; Spallek, T.; Maichle-Mössmer, C.; Törnroos, K.W.; Anwander, R. Cerium tetrakis(diisopropylamide)—A useful precursor for cerium(IV) chemistry. *Chem. Commun.* **2014**, *50*, 14763–14766. [CrossRef] [PubMed]
74. Hitchcock, P.B.; Lappert, M.F.; Protchenko, A.V. Facile formation of a homoleptic Ce(IV) amide *via* aerobic oxidation. *Chem. Commun.* **2006**, 3546–3548. [CrossRef] [PubMed]
75. Crozier, A.R.; Bienfait, A.M.; Maichle-Mössmer, C.; Törnroos, K.W.; Anwander, R. A homoleptic tetravalent cerium silylamide. *Chem. Commun.* **2013**, *49*, 87–89. [CrossRef] [PubMed]
76. Freeman, J.H.; Smith, M.L. The preparation of anhydrous inorganic chlorides by dehydration with thionyl chloride. *J. Inorg. Nucl. Chem.* **1958**, *7*, 224–227. [CrossRef]
77. *XAREA program for X-ray crystal data collection*; XRED32; Stoe: Darmstadt, Germany, 2002.
78. *SMART, SAINT and SADABS*; Bruker AXS Inc.: Madison, WI, USA, 1998.

79. Sheldrick, G.M. *SHELXS-97 Program for Crystal Structure Solution*; Universität Göttingen: Göttingen, Germany, 1997.
80. Sheldrick, G.M. *SHELXL-97 Program for Crystal Structure Refinement*; Universität Göttingen: Göttingen, Germany, 1997.

![inorganics logo] *inorganics*

MDPI

Article

Dinuclear Lanthanide (III) Coordination Polymers in a Domino Reaction

Edward Loukopoulos [1], Kieran Griffiths [1], Geoffrey R. Akien [2,*], Nikolaos Kourkoumelis [3], Alaa Abdul-Sada [1] and George E. Kostakis [1,*]

[1] Department of Chemistry, School of Life Sciences, University of Sussex, Brighton BN1 9QJ, UK; E.Loukopoulos@sussex.ac.uk (E.L.); K.Griffiths@sussex.ac.uk (K.G.); a.abdul-sada@sussex.ac.uk (A.A.-S.)
[2] Department of Chemistry, Lancaster University, Lancaster LA1 4YB, UK
[3] Department of Medical Physics, School of Health Sciences, University of Ioannina, 45110 Ioannina, Greece; nkourkou@cc.uoi.gr
* Correspondence: g.akien@lancaster.ac.uk (G.R.A.); g.kostakis@sussex.ac.uk (G.E.K.); Tel.: +44-1273-877-339 (G.E.K); +44-1524-593-790 (G.R.A.); Fax: +44-1273-876-687 (G.E.K.).

Academic Editors: Stephen Mansell and Steve Liddle
Received: 13 October 2015; Accepted: 29 October 2015; Published: 6 November 2015

Abstract: A systematic study was performed to further optimise the catalytic room-temperature synthesis of *trans*-4,5-diaminocyclopent-2-enones from 2-furaldehyde and primary or secondary amines under a non-inert atmosphere. For this purpose, a series of dinuclear lanthanide (III) coordination polymers were synthesised using a dianionic Schiff base and their catalytic activities were investigated.

Keywords: coordination polymers; lanthanide; single crystal; [89]Y-NMR; catalysis; domino reaction

1. Introduction

The interest in lanthanide (4f) coordination chemistry has grown considerably during the last decade. From a synthetic point of view, the high coordination number that 4f elements favour and the steric demand of the organic ligands used govern the size and the nuclearity of the resulting coordination clusters (CCs) [1]. Unprecedented and aesthetically pleasant structures, such as the "bottlebrush" Ln_{19} [2], the peptoid-based Ln_{15} [3], the two dimensional (2D) coordination polymers based on Ln_{36} units [4], and the world record four-shell Keplerates Ln_{104} [5], have been recently reported. However, the popularity of 4f coordination chemistry is being exploited rapidly due to their association in diverse research areas, such as molecular magnetism [6,7], molecular bioimaging [8], optics [9,10], catalysis [11–13], and the possibility to simultaneously exhibit more than one property [14–18].

Schiff base ligands, due to their low cost, ease of access and several coordination modes, have received a large amount of attention, especially in the last five years, and thus have been widely employed in the synthesis of polynuclear CCs [19–37]. Using (*E*)-2-(2-hydroxy-3-methoxybenzylideneamino)phenol (H_2L1, Scheme 1 left), we recently initiated a project towards the synthesis of 3d/4f CCs that would display catalytic properties [38]. We showed that three 3d/Dy(III) CCs remain intact into solution and display homogeneous catalytic behaviour in the room-temperature synthesis of *trans*-4,5-diaminocyclopent-2-enones from 2-furaldehyde and primary or secondary amines under a non-inert atmosphere [38]. Our studies indicated that the catalytic activity is owed to the Dy(III) ion, highly dependent on its coordination environment, and required an order of magnitude less loading in comparison with the parent catalyst $Dy(OTf)_3$ [39].

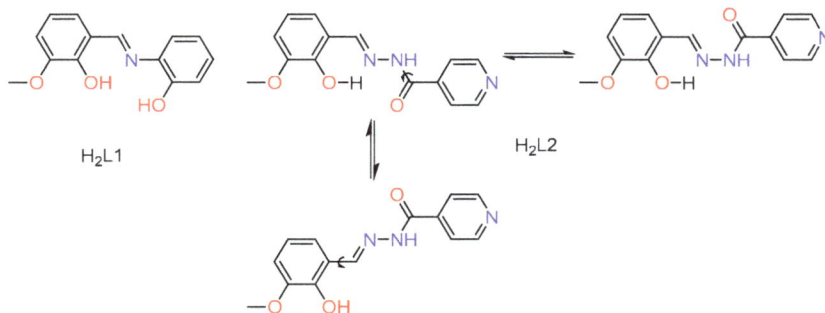

Scheme 1. The organic ligands discussed in the text. In DMSO-d_6, ligand H_2L2 exhibits two rotamers in an 11.5:1 ratio with observable NOESY exchange peaks between the two, and their identity confirmed by ^{15}N-HMBC experiments giving $^1J_{NH}$ of 93.6 and 91.6 Hz for the major and minor amide protons, respectively. Since there are two plausible bonds with restricted rotation it is not possible to unambiguously assign their identities. Nevertheless the distinctly different chemical shifts for the phenol OH's (10.71 ppm for the major and 9.43 for the minor), imply that one of the two rotamers possesses an intramolecular H-bond.

With this in mind, we decided to employ a pure 4f CC as catalyst in the aforementioned domino reaction. The employment of H_2L1 in 4f chemistry affords tetranuclear compounds that possess a defect dicubane motif [26], but the coordination sites of all four 4f ions are occupied by the organic moieties and most importantly do not retain their structure into solution. Therefore, we addressed to employ another Schiff base ligand that exhibits more steric demands and thus we concluded that (E)-N'-(2-hydroxy-3-methoxybenzylidene)isonicotinohydrazide (H_2L2, Scheme 1 right) can serve this purpose. This ligand has been used for the synthesis of a series of Mn compounds [40] as well two pseudopolymophic [41–43] compounds formulated as [Dy$_2$(L2)$_2$(NO$_3$)$_2$(MeOH)$_2$] and [Dy$_2$(L2)$_2$(NO$_3$)$_2$(MeOH)$_2$]·MeCN (**1-Dy** MeCN) [44]. Interestingly, compound **1-Dy** forms a 2D network via the pyridine group and each Dy centre coordinates to one solvent molecule (Figure 1).

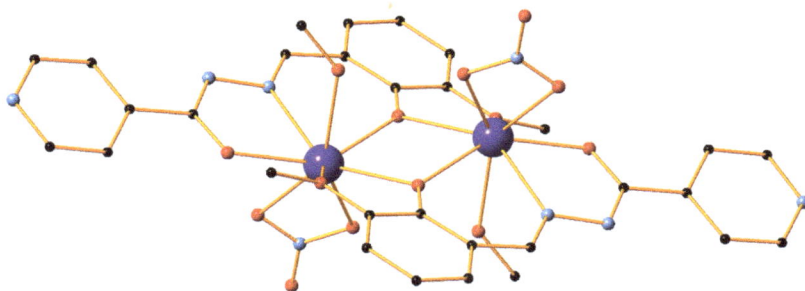

Figure 1. The structure of **1-Dy** as reported by Murugesu *et al.* [44]. H atoms and solvent molecules are omitted for clarity. Colour code Dy (purple), C (black), N (blue), O (red).

Being motivated by our initial catalytic results [38], in this work, we decided to investigate the possibility of using **1-Dy** as catalyst in the aforementioned domino reaction. Moreover, having in mind that (a) the catalytic efficiency is highly dependent on the coordination environment of the metal centre, and (b) the imperative to seek lower cost materials, we decided to perform a systematic study incorporating other lanthanides and thus we present herein the synthesis and characterization of the following compounds [Dy$_2$(L2)$_2$(NO$_3$)$_2$(MeOH)$_2$]·MeCN (**1-Dy**·MeCN), [Gd$_2$(L2)$_2$(NO$_3$)$_2$(MeOH)$_2$]·MeCN·MeOH (**1-Gd**),

Inorganics **2015**, *3*, 82–97

$[Y_2(L2)_2(NO_3)_2(EtOH)_{0.32}(H_2O)_{0.68}]\cdot0.68EtOH$ **(1-Y)**, $[Dy_2(HL2)_3Cl_2]\cdot Cl\cdot(H_2O)_6\cdot(MeOH)$ **(2-Dy)**, $[Gd_2(HL2)_3Cl_2]\cdot Cl\cdot(H_2O)_6$ **(2-Gd)**, $[Y_2(HL2)_3Cl_2]\cdot Cl\cdot(H_2O)_6\cdot(MeOH)$ **(2-Y)**, $[Dy(HL2)(NO_3)_2(MeOH)_2]$ **(3-Dy)** and $[Y(HL2)_2(MeOH)_2]\cdot Cl\cdot MeCN$ **(4-Y)**. The catalytic performance of compounds **1-Y** and **2-Y** is further monitored by solution NMR studies (^{1}H and ^{89}Y NMR).

2. Results and Discussion

Having chosen to study this metal:ligand system for catalytic purposes, we initiated our study by employing the dinuclear Dy$_2$ complex (**1-Dy**), which was reported by Murugesu *et al.* [44]. We then expanded the system by employing a series of lanthanides, Ln(NO$_3$)$_3$ and LnCl$_3$, while keeping the synthetic method similar, then exploring their catalytic potential. This approach was adopted for the following reasons: (a) this would account for a fully systematic study and a better understanding of the role that the lanthanide and the anion play during the catalytic process; (b) the proposed synthetic method was easy and also ideal for affording high yields of the product; (c) while the Ln(NO$_3$)$_3$ sources are generally cheaper, affording the products using LnCl$_3$ sources and the aforementioned synthetic method has the least cost overall.

As such, in our efforts to synthesise an analogue of **1-Dy** using DyCl$_3$ as the metal source and a similar synthetic method, we obtained compound **2-Dy**. The compound is also a Dy (III) dimer (Figure 2A) which crystallises in the monoclinic P2$_1$/n space group (Z = 4). The structure contains three chlorine atoms, out of which two act as terminal ligands, and three ligand molecules. Despite the unsatisfactory crystallographic data, TGA analysis also confirms the presence of one methanol and six water molecules in the lattice. Each of the organic ligands is only deprotonated in the phenoxide group (HL2$^-$) to account for the total charge balance. This is further confirmed by the related C–O distances (Table S1), which are indicative of a double bond. Dy1 is bonded to a {ClN$_2$O6} coordination, including a pyridyl N atom of a neighbouring HL2$^-$ ligand (Scheme S1). As a result, the compound forms a one-dimensional zig-zag network which extends to the *a0c* plane (Figure 2B). Using SHAPE software [45], the geometry of Dy1 may be best described as a tridiminished icosahedron (J63). Dy2 is also coordinated to nine atoms in a {ClN$_2$O6} environment and possesses a muffin geometry. The two metal centres are bridged by the phenoxide oxygen atoms of all three ligands, with the Dy–O–Dy angles ranging from 95.92(17)° to 99.73(19)°. Out of the respective Dy–O distances, Dy1–O6 was the longest at 2.570(5) Å, while Dy1–O8 was the shortest at 2.277(4) Å. Similarly, for the respective Dy–N distances, the mean Dy1–N3 distance was the longest at 2.692(6) Å, while Dy1–N7 was the shortest at 2.495(7) Å. Selected bond lengths and angles are given in Table S2.

We then attempted to test the catalytic properties of **1-Dy** and **2-Dy** to determine whether Dy(III) behaves in a similar fashion to the reported Dy(OTf)$_3$ [39] and our initial results [38]. In the following protocol, the reactions were performed in acetonitrile in air, at room temperature using variable catalytic loadings ranging from 10%–1%. Before going into detail, it is crucial to mention that a solubility test was carried out to identify the behaviour of **1-Dy** and **2-Dy** in CH$_3$CN solution; despite the polymeric character of **1-Dy** and **2-Dy**, 2D and 1D, respectively, both compounds are slightly soluble in CH$_3$CN solutions and, thus, the catalytic process takes place in a homogeneous phase. Under these conditions, the reaction between 2-furaldehyde and morpholine led to the quantitative formation of (4*S*,5*R*)-dimorpholinocyclopent-2-enone at 10% loading with both complexes (Table 1, Entries 1 + 5). With the reduction of loading there are small decreases in yields, however even the lowest of these is acceptable. Noticeably, the nitrate analogue **1-Dy** performs marginally better than the chloride analogue (**2-Dy**) (6%–7%) at 5% and 2.5% loading (Table 1, Entries 2 + 3 and 6 + 7, respectively).

Our next step was to test the efficiency of lanthanides lighter than Dy. Compound **1-Gd** was synthesised in an analogous manner and was confirmed to be pseudopolymorphic [41–43] to **1-Dy** through IR and thermal studies. As such, it is a dinuclear Gd$_2$ complex containing the same Ln$_2$(HL2)$_2$(NO$_3$)$_2$(MeOH)$_2$ core and extending to two dimensions through the pyridine group. In our efforts to synthesise the chloride-based Gd analogue, we obtained compound **2-Gd**. X-ray and thermal studies, as well as IR (Figure S1) reveal that it is a dinuclear Gd$_2$ complex, pseudopolymorphic to **2-Dy** and therefore also a 1D coordination polymer. As shown in Table 1, Entries 9 and 10, both Gd(III) analogues show a reduced activity in comparison to the Dy(III) compounds.

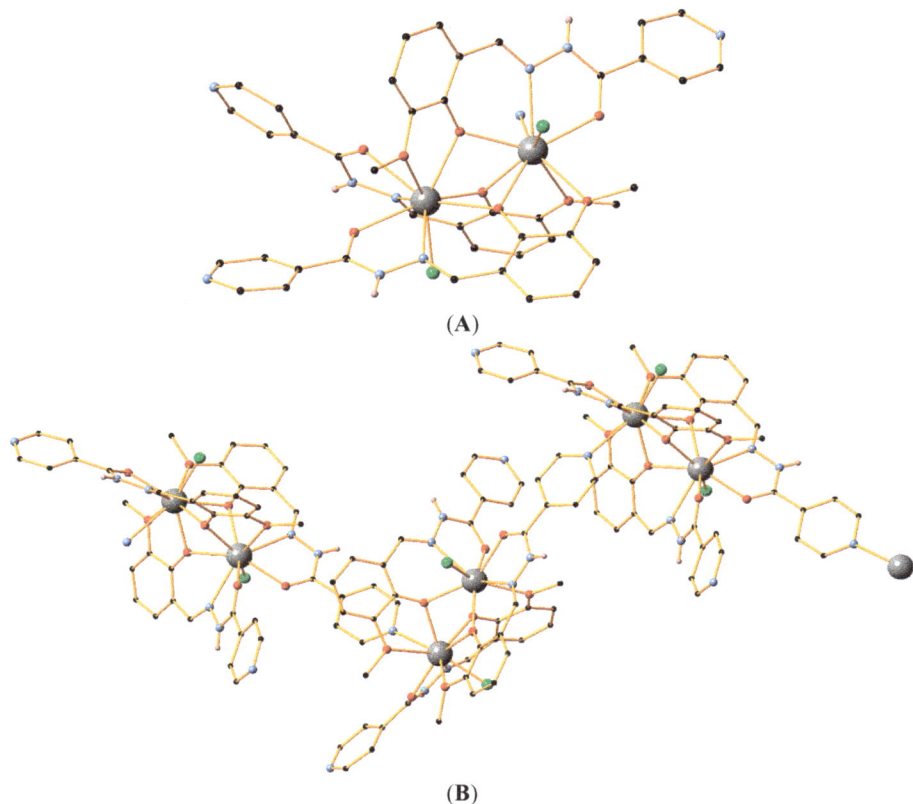

Figure 2. (**A**) The structure of compounds **2-Dy**, **2-Gd** and **2-Y**. X = Dy, Gd, Y; (**B**) Part of the 1D framework of compounds **2-Dy**, **2-Gd** and **2-Y**. Certain H atoms and all solvent molecules are omitted for clarity. Colour code X (grey), Cl (green), C (black), N (light blue), O (red).

We then proceeded to explore the attributes of Y, which has similar ionic radius to Dy. We first employed Y(NO$_3$)$_3$ as the metal source to obtain compound **1-Y**. The complex forms a 2D network via the pyridine group and each Y centre coordinates to one solvent molecule (Figure S2). The crystal structure was found to be pseudopolymorphic to **1-Dy** and as a result it will not be described in further detail. However, its measured diffusion coefficient in DMSO-d_6 was of a similar order of magnitude to that of **2-Y** (1.98 versus 2.15 × 10^{-10} m$^2 \cdot$s^{-1}), so it is assumed to dissociate on dissolution. The imine proton was broadened to an extent consistent with a $^3J_{YH}$ 0.9 Hz doublet (lineshape fitting), and this narrowed significantly on the application of ^{89}Y-decoupling. While it was possible to observe a strong 3-bond correlation between the imine H and the Y centre (92 ppm) via ^{89}Y-HMBC (S/N 275 in 20 min experiment time, optimised for J_{YH} 2 Hz), it was not possible to detect any 3-bond correlation between the methoxy and Y atom, even with many more scans. It is possible that this is a normal feature of such complexes, but it is certainly suggestive of the methoxy being uncoordinated in the solution state.

Compound **2-Y** was synthesised using YCl$_3$ and the same synthetic method as above for **2-Dy**. IR and thermal studies confirm that the compound is isostructural to **2-Dy** and contains the same

$Ln_2(HL_2)_3Cl_2$ core. Similarly as with **1-Y**, it was possible to measure a strong 3-bond correlation between the imine H and Y (111 ppm), but again, no correlation to the methoxy group. The ^{15}N-HMBC correlation between the imine H and the hydrazide N (321.7 ppm) was also intense enough that it was possible to acquire a highly-aliased high-resolution spectrum sufficient to determine the first ever $^1J_{YN}$ coupling of 9.1 Hz. It was not possible to do this for **1-Y**, at the same concentration (20 mg/0.5 mL). Due to the similarities of the ^1H and ^{13}C spectra for **1-Y** and **2-Y**, a mixture of 9.9 mg **1-Y** and 5.8 mg **2-Y** were mixed and dissolved in DMSO-d_6. Only one set of peaks could be observed by ^1H NMR, and it was not possible to measure any correlations by ^{89}Y-HMBC at all. A plausible explanation is due to intermediate exchange between the two species in solution, but since the ^1H and ^{13}C chemical shifts are so similar, there is negligible broadening in their respective 1D spectra. However, the ^{13}C resonances with the largest differences between **1-Y** and **2-Y** ($\Delta0.55$ ppm) was the quaternary carbon adjacent to the imine, and these were significantly broadened.

Regarding their catalytic activity, the efficiency of the Y_2 analogues (Table 1, Entries 11–16), with either chloride or nitrate ion, out-perform any all counterpart Ln(III) analogues including Dy(III). These analogues retain good catalytic activity when the loading is decreased from 2.5% to 1% with yields above 89% (entries 12 and 15), however decrease significantly when lowered to 0.5%. The Dy(III)and Y(III) ions have a similar ionic radius, which could relate to comparable activities, with 4f orbital effects influencing the difference.

Table 1. Comparison of catalytic activity for compounds **1-Dy**, **2-Dy**, **1-Gd**, **2-Gd**, **1-Y** and **2-Y**

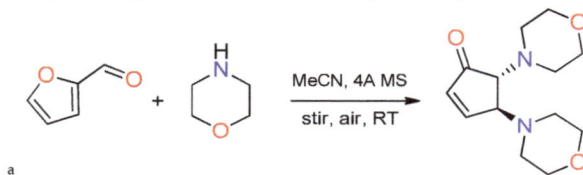

Entry	Catalyst	Loading/% [b]	Time/h	Yield/%
1	**1-Dy**	10	2	100
2	**1-Dy**	5	2	96
3	**1-Dy**	2.5	2	95
4	**1-Dy**	1	2	80
5	**2-Dy**	10	2	100
6	**2-Dy**	5	2	90
7	**2-Dy**	2.5	2	88
8	**2-Dy**	1	2	79
9	**1-Gd**	2.5	2	80
10	**2-Gd**	2.5	2	54
11	**1-Y**	2.5	2	100
12	**1-Y**	1	2	89
13	**1-Y**	0.5	2	68
14	**2-Y**	2.5	2	100
15	**2-Y**	1	2	85
16	**2-Y**	0.5	2	38

[a] Reaction conditions: morpholine, 1 mmol; 2-furaldehyde, 0.5 mmol; 4A MS, 100 mg; anhydrous MeCN, 4 mL; room temperature; [b] Catalyst loading calculated per equivalent of Ln[III]. Yields calculated by ^1H NMR spectroscopy.

A closer look in the ESI-MS studies (Section 2.1) reveals that the core in all aforementioned compounds behave similarly in solution. All complexes exhibit a peak that corresponds to the $[Ln(L)(MeOH)_2]^{1+}$ monocationic fragment, indicating that this species contributes to the catalysis. Having this in mind, in order to better study the system we attempted to synthesise a mononuclear analogue, focusing on Dy(III) and Y(III) compounds which provided the best catalytic yields. As a result, we obtained complexes **3-Dy** and **4-Y**. The former is a Dy(III) monomer, with two molecules in the asymmetric unit (Z = 4). The monomer consists of two nitrate and two methanol molecules that act as terminal ligands, as well as one organic ligand molecule which is deprotonated only in the phenoxide group (Figure 3A). Each metal centre possesses a {N_1O_8} coordination environment and their geometry may be best described as a capped square antiprism (J10) [45]. The compound crystallises in the triclinic $P\bar{1}$ space group and is stabilised by strong O–H···O, N–H···O and O–H···N hydrogen bonds through participation of the ligated methanols, the nitrate anions and the pyridine nitrogen atom (Table S5). Table S3 contains selected bond distances for this structure. **4-Y** is an Y(III) monomer which crystallises in the monoclinic P2$_1$/c space group (Z = 4). The structure consists of two ligand molecules, two methanol solvent molecules which act as terminal ligands, one acetonitrile solvent molecule in the lattice and one chlorine counter ion (Figure 3B). The organic ligands are once again only deprotonated in the phenoxide group, as evidenced by the C–O bond distances (Table S1). The metal centre has a {N_2O_6} coordination environment and its geometry may be best described as triangular dodecahedron [45]. Each ligand coordinates to the Y(III) centre through the pocket which consists of the phenoxide, imine and carbonyl atoms. The Y–O distances range from 2.210(2) Å to 2.402(2) Å, while the Y–N distances are longer, ranging from 2.482(3) Å to 2.509(3) Å. Selected bond lengths may be found in Table S4. The structure is further stabilised by various strong intermolecular O–H···N and intramolecular N–H···Cl hydrogen bonds, which also account for the formation of a two-dimensional architecture at a supramolecular level. Characteristics for these bonds may be found in Table S6.

The **1-Y** analogue, which demonstrated the highest efficiency of all the characterised compounds (Table 1, Entries 11 and 12) was chosen to optimise and explore the scope of this for a variety of secondary and primary amines as substrates. Due to **1-Y** showing a decrease in activity from quantitative, when loadings are reduced to 1% (Table 1, Entry 12), which is not observed in the previously reported procedure [38], a higher loading of catalyst (2.5%) was employed in an attempt to reduce the reaction time for quantitative conversion.

In all cases the reaction proceeded to the expected products (Table 2) and demonstrated the same ability as previously reported catalyst [38] to form deprotonated Stenhouse salts from the primary amine substrates. The increased % loading of catalyst, in 8 h (From 16 h) with primary amine substrates affords a higher yield of Stenhouse salt products (Table 2, **5a**–**5b**) than previously reported conditions. The time for secondary amine substrate conversion was reduced from 16 to 4 h, with similar conversion percents for products **5a**, **5b** and **5c**. However, the bulky substrates **5d** and **5f**, with the same catalyst loading (2.5%), had a far reduced activity in the 4-h time period. When this is increased to 18 h, the improvement is only marginal (93%, 85%), suggesting the **1-Y** is more subject to steric effects than the previously reported 3d/4f catalysts [38].

In an effort to understand the catalysis further, we monitored the reaction *in situ*. **1-Y** is only slightly soluble in MeCN-d_3, so we opted to run the reaction in DMSO-d_6. A solution of 6.9 mg **1-Y** (~11.5 mol. %) in 0.55 mL DMSO-d_6 gives a Y chemical shift of 95.9 ppm changing slightly to 97.4 ppm on the addition of 0.125 mmol morpholine, and all the ^{1}H peaks broadening slightly. After the addition of 0.0625 mmol 2-furaldehyde, the reaction at room temperature begins immediately, with *ca.* 50% of the furaldehyde forming the corresponding imine from morpholine (most notable, H-1 as a singlet at 5.13 ppm, Scheme 2), and only 4% of the intended product. Whether the imine is a necessary intermediate or merely a resting state is not yet clear—this part of the reaction occurs in the absence of a catalyst. There was no difference in the Y chemical shift with time, and it was not possible to observe any Y catalytic intermediates. After *ca.* 17 h, 5% furfural remained, <1% of the imine adduct, *ca.* 75% of the desired product, and 20% of a second species which slowly increased in concentration, although so far it has not been possible to unambiguously characterise its structure. In addition, <5% of morpholine remained (due to overlap with other species it is difficult to be certain to a lower limit of detection), and the ^{1}H peak shape had improved such that the smaller *J*-couplings in furfural are observable, as they are for pure solutions of furfural in DMSO-d_6.

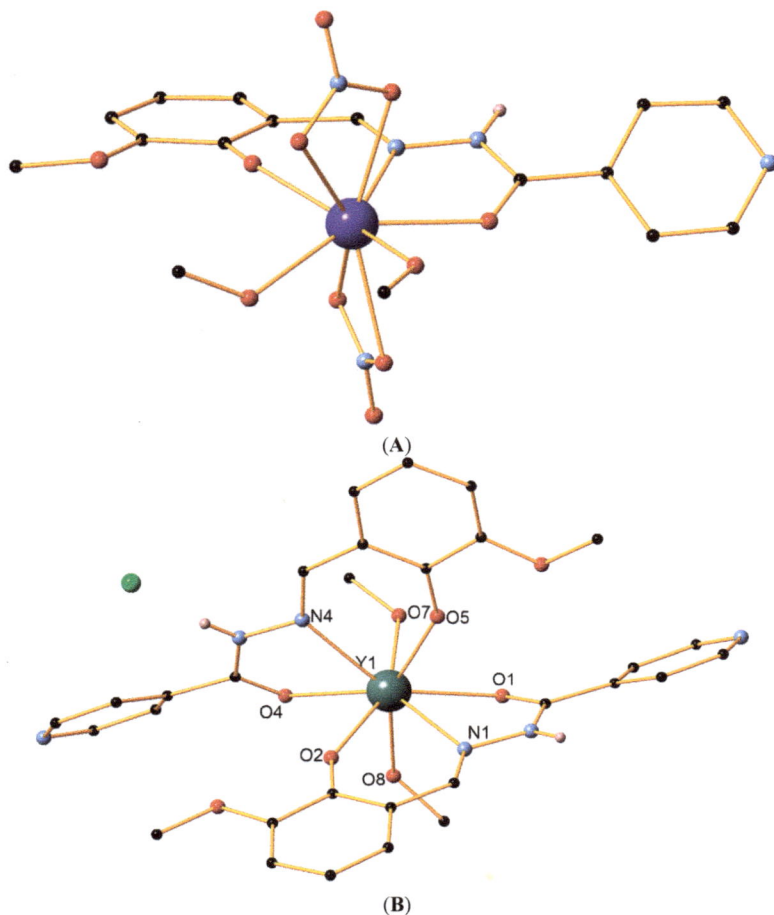

Figure 3. (**A**) The structure of the **3-Dy** monomer; (**B**) The structure of compound **4-Y**. Certain H atoms and all lattice solvent molecules are omitted for clarity. Colour code Dy (purple), Y (dark green), Cl (green), C (black), N (light blue), O (red).

Table 2. Y(III)-catalysed condensation/ring-opening/cyclisation of secondary amines with 2-furaldehyde and condensation/ring opening of primary amines with 2-furaldehyde. Reaction conditions: amine, 1 mmol; 2-furaldehyde, 0.5 mmol; 4 A MS, 100 mg; anhydrous MeCN, 4 mL; room temperature. Yields calculated by isolation of product via column chromatography.

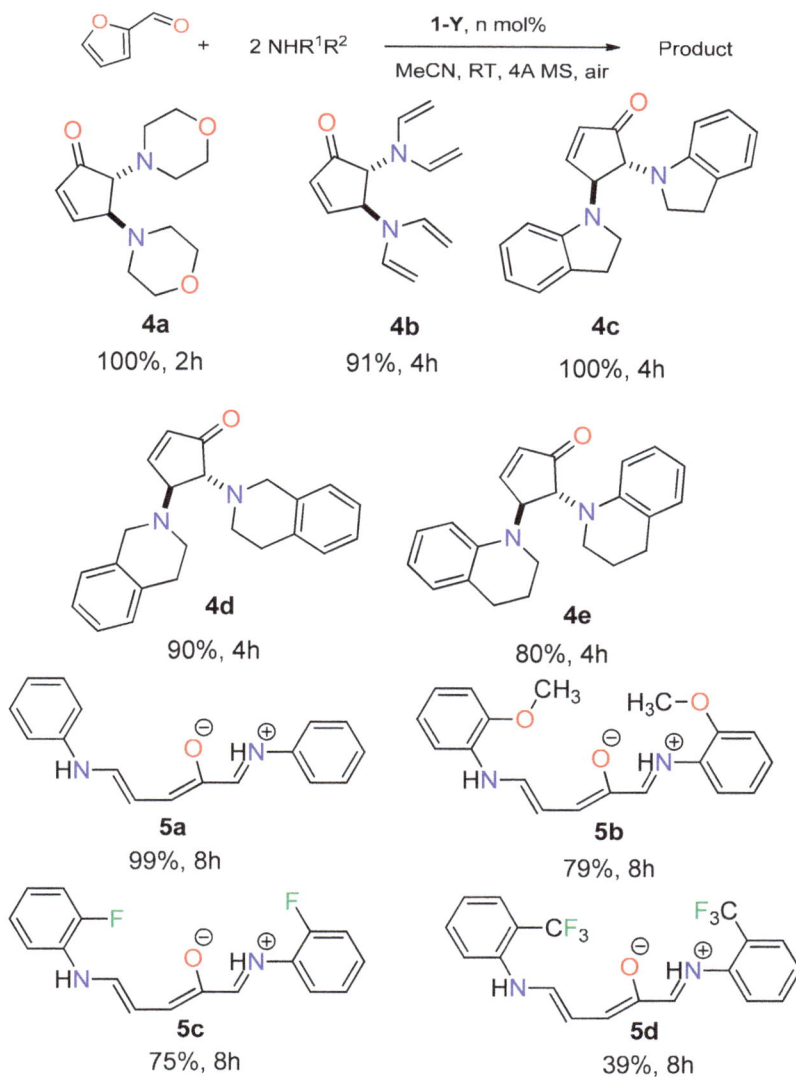

4a
100%, 2h

4b
91%, 4h

4c
100%, 4h

4d
90%, 4h

4e
80%, 4h

5a
99%, 8h

5b
79%, 8h

5c
75%, 8h

5d
39%, 8h

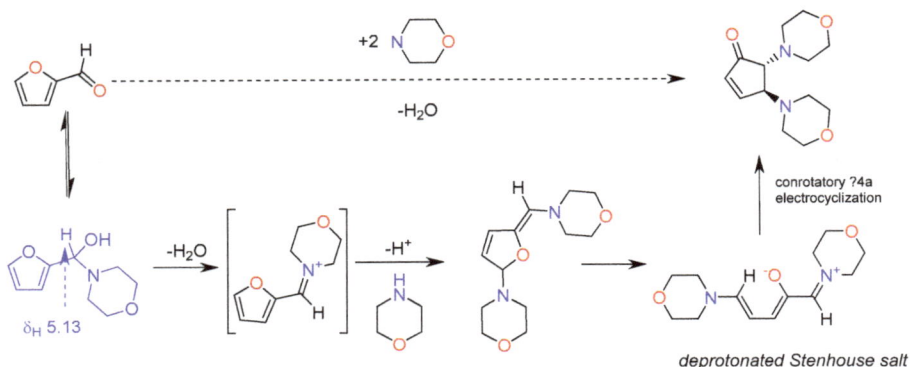

Scheme 2. Modified mechanism for the Ln^{III}-catalysed synthesis of *trans*-4,5-Dimorpholin-4-yl-cyclopent-2-enone. Possible imine intermediate step shown in blue.

2.1. Solution Studies

Electrospray ionization mass spectrometry (ESI-MS) was performed to confirm the identity and behaviour of the compounds in solution. Representative spectra for the compounds are presented in Figure S3, Figure S4, Figure S5 and Figure S6. The MS (positive-ion mode) for the Dy (III) complexes **1-Dy** and **2-Dy** shows one main peak at 497.06 *m*/*z* which perfectly corresponds to the respective $[Dy(L)(MeOH)_2 - 2H]^{1+}$ monocationic fragment. The respective isostructural complexes **1-Gd** and **2-Gd** also behave in a similar fashion, showing one main peak at 491.05 *m*/*z* that corresponds to the $[Gd(L)(MeOH)_2 - 2H]^{1+}$ monocationic fragment. Compounds **1-Y** and **2-Y** exhibit common peaks at 422.04, 629.08 and 986.07 *m*/*z*, corresponding to the $[Y(L)(MeOH)_2 - H]^{1+}$, $[Y(L)_2]^{1+}$ and $[Y_2(L)_3 - 2H]^{1+}$ fragments, respectively. **1-Y** shows extra main peaks at 760.96, 823.97 *m*/*z*, corresponding to the respective $[Y_2(L)_2MeCN]^{1+}$, $[Y_2(L)_2(MeCN)(MeOH)_2]^{1+}$ fragments. For **2-Y** two additional peaks appear at 900.18 and 1257.16 *m*/*z* that match the respective $[Y(L)_3 + H]^{1+}$ and $[Y_2(L)_4 - H]^{1+}$ monocationic fragments. These results indicate that compounds behave in a similar way. Uniquely for the Y(III) compounds, the ESI-MS studies suggest that both dimer and monomer forms are present into the solution.

2.2. Thermal and Powder XRD Studies

In order to confirm phase purity of the most effective catalysts **1-Dy** and **1-Y**, these were further characterised through Powder XRD studies, indicating that both compounds are formed in high purity (Figure S7). In order to examine the thermal stability of all complexes, TGA was carried out between room temperature and 1000 °C under N$_2$ atmosphere. The TGA curves are presented in Figure 4.

Regarding the nitrate-based compounds, the TGA graphs for **1-Dy** and **1-Gd** are very similar, showing a mass loss at ~150 °C, which corresponds to the loss of respective lattice solvent molecules. The remaining core is stable until the region of 250–300 °C, after which gradual decomposition begins, leaving metal oxide as the final residue. The method was also used to confirm the identity of **1-Gd**, which could not be obtained as crystalline material; the first observed mass loss (10.78%) at 161 °C corresponds to the loss of two acetonitrile and one methanol molecules (theoretical weight loss: 9.87%). The second mass loss occurs at 242 °C, corresponding to the gradual decomposition and was measured at 57.88%, within a reasonable range to the expected value of 55.21%.

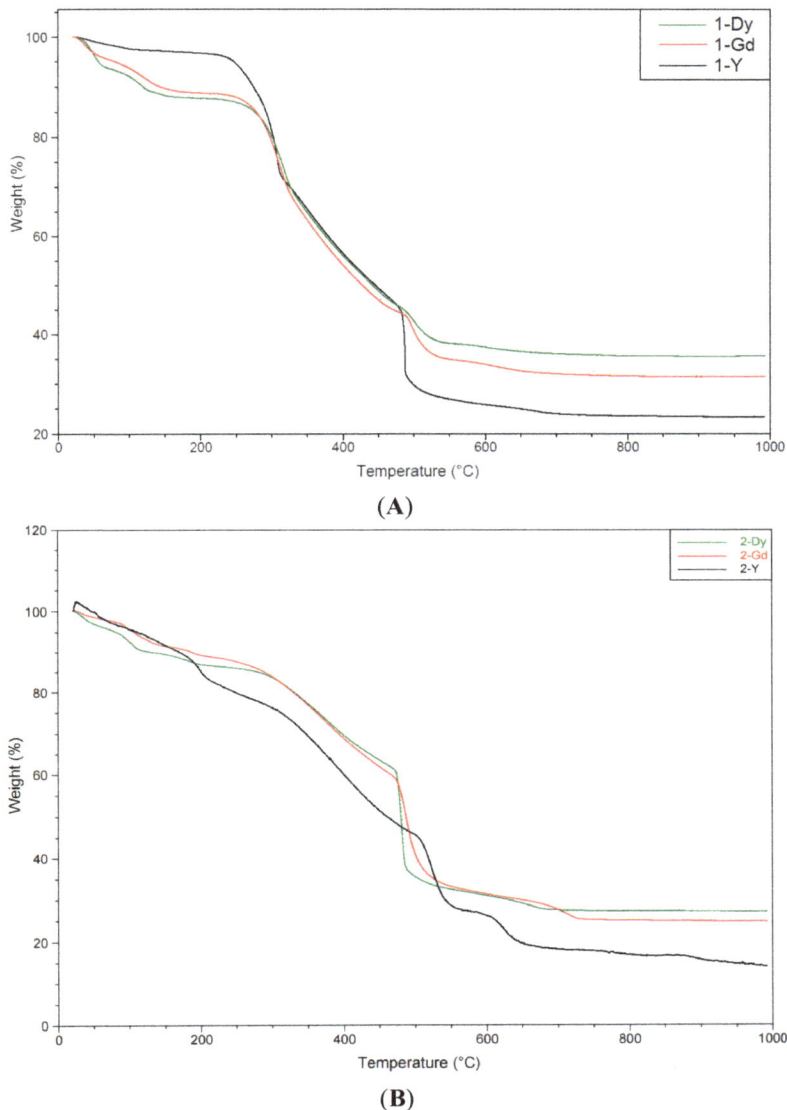

Figure 4. (A) TGA overlay of **1-Dy**, **1-Gd**, **1-Y**; (B) TGA overlay of **2-Dy**, **2-Gd**, **2-Y**.

TGA analysis was further used to confirm the structure of compound **2-Dy** and identify the correct number of solvent molecules, which could not be determined through X-ray crystallography due to bad data. The first observed mass loss (9.96%) at 132 °C corresponds to the loss of six water and one methanol molecules (theoretical weight loss: 10.13%). The second mass loss occurs from 132 to 222 °C and corresponds to the loss of the lattice chlorine atom (calculated: 3.29%, theoretical: 2.53%). Finally, the third loss which begins at 222 °C, attributed to gradual decomposition was measured at 59.61%, in good agreement to the expected value of 60.17%. A similar identification took place for the pseudopolymorph **2-Gd**. A continuous mass loss occurs from room temperature up to 196 °C, corresponding in good agreement to the loss of the six water molecules and the lattice chlorine atom (calc.: 10.66%, theor.: 10.71%). Almost instantly (200 °C) gradual decomposition takes place, with the calculated (64.38%) and the theoretical (62.24%) values being in a reasonable range. The bulk material of **2-Y** was identified likewise. For **2-Y**, the first mass loss is again continuous, occurring from room temperature to 198 °C and corresponds to the loss of all lattice components (six water molecules, one methanol molecule and one chlorine atom, calc.: 14.49%, theor.: 14.22%). The second mass loss begins almost instantly at 200 °C and corresponds to the gradual decomposition of the remaining core (calc.: 68.45%, theor.: 67.5%).

3. Experimental Section

3.1. Materials

Chemicals (reagent grade) were purchased from Sigma Aldrich (St Louis, MO, USA), Acros Organics (Morris Plains, NJ, USA) and Alfa Aesar (Ward Hill, MA, USA). All experiments were performed under aerobic conditions using materials and solvents as received.

3.2. Instrumentation

NMR spectra were measured on a Varian VNMRS solution-state spectrometer (Bruker BioSpin, Rheinstetten, Germany) at 500 MHz at 30 °C using residual isotopic solvent (DMSO-d_5, δ_H = 2.50 ppm) as an internal chemical shift reference. ^{1}H DOSY (ledbpgp2s), ^{15}N-HMBC and ^{89}Y-HMQC spectra were acquired at 298.0 K on a Bruker Avance III 400 instrument (Bruker BioSpin) equipped with a broadband observe probe (BBO). Chemical shifts are quoted in ppm. Coupling constants (*J*) are recorded in Hz. IR spectra of the samples were recorded over the range of 4000–650 cm^{-1} on a Perkin Elmer Spectrum One FT-IR spectrometer (PerkinElmer, Waltham, MA, USA) fitted with an UATR polarization accessory (PerkinElmer). ESI-MS was performed on a VG Autospec Fissions instrument (EI at 70 eV). TGA analysis was performed on a TA Instruments Q-50 model (TA, Surrey, UK) under nitrogen and at a scan rate of 10 °C/min. X-ray powder diffraction patterns were recorded using a Bruker-AXS D8-Advance diffractometer (Bruker AXS GmbH, Karlsruhe, Germany) equipped with a Cu sealed-tube radiation source (λ = 1.5418 Å) and a secondary beam graphite monochromator. Data were collected from 4° to 50° in the 2θ mode at 5 s/step.

3.3. Crystallography

Data for **2-Dy**, **2-Gd** and **1-Y** were collected (ω-scans) at the University of Sussex using an Agilent Xcalibur Eos Gemini Ultra diffractometer (Agilent Technologies, Palo Alto, CA, USA) with CCD plate detector under a flow of nitrogen gas at 173(2) K using Mo Kα radiation (λ = 0.71073 Å). CRYSALIS CCD and RED software was used respectively for data collection and processing. Reflection intensities were corrected for absorption by the multi-scan method. Data for **4-Y** were collected at the National Crystallography Service, University of Southampton [46] using a Rigaku Saturn 724+ area detector mounted at the window of an FR-E+ rotating anode generator with a Mo anode (λ = 0.71075Å) under a flow of nitrogen gas at 100(2) K. All structures were determined using Olex2 [47], solved using either Superflip [48] or SHELXT [49,50] and refined with SHELXL-2014 [51]. All non-H atoms were refined with anisotropic thermal parameters, and H-atoms were introduced at calculated positions and allowed to ride on their carrier atoms. For **2-Dy** and **2-Gd**, attempts to model the lattice solvents were unsuccessful despite multiple data collections. Therefore, the solvent mask function of Olex2 [47] was employed to remove the contribution of the electron density associated with those molecules from the intensity data. The presence of those solvent molecules in each structure was confirmed through TGA and/or elemental analysis. Geometric/crystallographic calculations for all structures were performed using PLATON [52], Olex2 [47], and WINGX [49] packages; graphics were prepared with Crystal Maker [53]. CCDC 1430641–1430645.

3.4. Ligand Synthesis

The synthesis of H_2L2 has been carried out according to the reported synthetic procedure [40], although the authors did not report data for the minor rotamer. Due to peak overlap and low S/N, only a partial reporting of ^{13}C (via HSQC) and ^{15}N chemical shifts was possible. Rotamer A. ^1H NMR (400 MHz, DMSO-d_6) δ 12.26 (br. s, 1H), 10.71 (br. S, 1H), 8.81 (d, J = 4.5 Hz, 2H), 8.71 (s, 1H), 7.85 (d, J = 4.5 Hz), 7.21 (m, 1H)., 7.06 (m, 1H), 6.88 (m, 1H), 3.83 (s, 3H); ^{13}C NMR (100 MHz, DMSO-d_6) δ 161.8, 150.8, 148.5, 147.6, 140.5, 122.0, 119.6, 119.5, 114.4, 56.3; ^{15}N NMR (40.5 MHz, DMSO-d_6) δ 326.0, 303.2, 172.0. Rotamer B. ^1H NMR (400 MHz, DMSO-d_6) δ 12.02 (br. s, 1H), 9.43 (br. S, 1H), 8.71 (m), 8.41 (s, 1H), 7.64 (m), 7.00 (m, 1H), 6.97 (m, 1H), 6.77 (m, 1H), 3.79 (s, 3H); ^{13}C NMR (100 MHz, DMSO-d_6) δ 143.5, 123.2, 119.7, 118.9, 56.3; ^{15}N NMR (40.5 MHz, DMSO-d_6) δ 312.1, 179.7.

Synthesis of **1-Dy**. The reported synthetic procedure was followed [44].

Synthesis of $[Dy_2(HL2)_3Cl_2] \cdot Cl \cdot (H_2O)_6 \cdot (MeOH)$ (**2-Dy**). 0.25 mmol (0.068 g) of L and 0.37 mmol (0.099 g) of $DyCl_3$ were dissolved in MeCN/MeOH (15:5mL) while stirring to produce an orange solution, which was further treated with 1 mmol pyridine (81.0 µL). The resulting yellow solution was stirred for a further 15 min, filtrated, then left to evaporate slowly. Orange block crystals were obtained after 15 days. Yield: 46%. Selected IR peaks (cm^{-1}): 3115 (w), 1600 (s), 1566 (m), 1521 (w), 1454 (s), 1382 (m), 1318 (w), 1279 (m), 1222 (s), 1170 (w), 1108 (w), 1066 (m), 1007 (w), 960 (m), 908 (w), 851 (m), 785 (w), 741 (s), 694 (s). Elemental analysis for $C_{43}H_{52}Cl_3Dy_2N_9O_{16}$: C 37.36, H 3.79, N 9.12; found C 35.49, H 3.60, N 8.75. This result corresponds to the loss of one methanol molecule and the presence of ten water molecules: $C_{42}H_{56}Cl_3Dy_2N_9O_{19}$: C 35.47, H 3.97, N 8.86.

Synthesis of $[Dy(HL2)(NO_3)_2(MeOH)_2]$ (**3-Dy**). A solution of 0.3 mmol (0.133 g) of $Dy(NO_3)_3 \cdot 5H_2O$ in 12.5 mL MeOH was added to a solution of 0.15 mmol (0.041 g) of L in 12.5 mL MeOH while stirring. The resulting orange solution was stirred for a further 15 min, filtrated, then kept in a stored vial. Yellow/orange block crystals were obtained after 5 days. Yield: 21%. Selected IR peaks (cm^{-1}): 3087 (w), 1607 (s), 1549 (m), 1458 (s), 1430 (m), 1412 (m), 1394 (m), 1301 (w), 1245 (m), 1221 (s), 1169 (w), 1082 (w), 1063 (m), 1025 (w), 1008 (w), 967 (m), 912 (w), 867 (w), 850 (m), 740 (s), 699 (s), 639 (m).

Synthesis of $[Gd_2(L2)_2(NO_3)_2(MeOH)_2] \cdot MeCN \cdot MeOH$ (**1-Gd**). The reported synthetic procedure was followed [44], with $Gd(NO_3)_3 \cdot 6H_2O$ as the metal salt. A yellow powder was obtained and its identity was confirmed through IR and TGA. Selected IR peaks (cm^{-1}): 3135 (w), 1605 (s), 1572 (m), 1521 (w), 1456 (s), 1408 (w), 1380 (m), 1306 (w), 1266 (m), 1241 (w), 1219 (s), 1174 (w), 1105 (w), 1078 (m), 1058 (w), 1009 (w), 960 (m), 913 (w), 849 (m), 786 (w), 740 (s), 692 (s).

Synthesis of $[Gd_2(HL2)_3Cl_2] \cdot Cl \cdot (H_2O)_6$ (**2-Gd**). A synthetic procedure similar to **2-Dy** was followed, with 0.29 mmol (0.107 g) of $GdCl_3 \cdot 6H_2O$ added. Orange block crystals were obtained after 7 days. Yield: 28%. Selected IR peaks (cm^{-1}): 3133 (w), 2944 (w), 2834 (w), 1600 (s), 1547 (m), 1454 (s), 1408 (w), 1380 (m), 1317 (w), 1288 (m), 1221 (s), 1172 (w), 1105 (w), 1078 (w), 1063 (m), 1024 (w), 965 (m), 904 (w), 851 (m), 786 (w), 740 (s), 692 (s).

Synthesis of $[Y_2(L2)_2(NO_3)_2(EtOH)_{0.32}(H_2O)_{0.68}] \cdot 0.68EtOH$ (**1-Y**) The reported synthetic procedure was followed [44], with $Y(NO_3)_3 \cdot 6H_2O$ as the metal salt. A yellow powder was obtained, which was then recrystallised in EtOH/hexane to produce yellow block crystals after 1 day. Yield: 18%. Selected IR peaks (cm^{-1}): 3148 (w), 2953 (w), 1606 (s), 1574 (m), 1521 (s), 1459 (s), 1370 (m), 1320 (m), 1268 (m), 1244 (m), 1220 (s), 1172 (w), 1098 (w), 1080 (m), 1014 (w), 962 (m), 920 (w), 849 (m), 785 (w), 739 (s), 707 (s). A minor impurity was insoluble in DMSO0-d_6, so NMR samples were filtered through a 0.45 µm syringe filter. The ^{15}N-HMBC S/N was too low to identify the second N atom in the hydrazide bridge. ^1H NMR (400 MHz, DMSO-d_6) δ 8.63 (m, 2H), 8.43 (br. s, 1H), 7.90 (m, 2H), 6.88 (dd, J = 7.9, 1.4 Hz, 1H), 6.83 (dd, J = 7.7, 1.4 Hz, 1H), 6.43 (t, J = 7.7 Hz, 1H), 3.73 (s, 3H); ^{13}C NMR (100 MHz, DMSO-d_6) δ 166.9, 157.8 (br.), 157.0 (d, J_{CY} = 2.2 Hz), 151.1, 150.2, 143.9 (br.), 126.1, 122.1,

Inorganics **2015**, *3*, 82–97

121.3, 114.7. 113.4, 55.8; ^{15}N NMR (40.5 MHz, DMSO-d_6) δ 319.7, 314.4; ^{89}Y NMR (19.6 MHz, DMSO-d_6) δ 92.4. D 2.0 × 10^{-10} m^2·s^{-1}.

Synthesis of [Y$_2$(HL2)$_3$Cl$_2$]·Cl·(H$_2$O)$_6$·(MeOH) (**2-Y**). The synthetic procedure for **2-Dy** was followed, with YCl$_3$·6H$_2$O as the metal salt. A yellow powder was obtained and its identity was confirmed through IR and TGA. Selected IR peaks (cm^{-1}): 3145 (w), 2938 (w), 2847 (w), 1599 (s), 1549 (m), 1517 (w), 1451 (s), 1411 (w), 1379 (m), 1288 (m), 1220 (s), 1168 (w), 1063 (m), 1005 (w), 966 (m), 905 (w), 847 (m), 787 (w), 739 (s), 692 (s). ^{1}H NMR (400 MHz, DMSO-d_6) δ 8.61 (m, 2H), 8.41 (br. s, 1H), 7.89 (m, 2H), 6.85 (m, 1H)., 6.83 (m, 1H), 6.43 (m, 1H), 3.73 (s, 3H); ^{13}C NMR (100 MHz, DMSO-d_6) δ 167.1 (d, J_{CY} = 1.3 Hz), 158.0, 156.7 (d, J_{CY} = 2.0 Hz), 151.0, 150.1, 144.3, 126.1, 122.1, 121.5, 114.5. 113.3, 55.8; ^{15}N NMR (40.5 MHz, DMSO-d_6) δ 323.7, 321.7 ($^1J_{YN}$ = 9.1 Hz*), 284.9; ^{89}Y NMR (19.6 MHz, DMSO-d_6) δ 111.0. D 2.2 × 10^{-10} m^2·s^{-1}. *Measured via high-resolution aliased ^{15}N-HMBC spectra.

Synthesis of [Y(HL2)$_2$(MeOH)$_2$]·Cl·MeCN (**4-Y**). A synthetic procedure similar to **2-Dy** was followed, with 0.125 mmol (0.038 g) of YCl$_3$·6H$_2$O added. Small yellow block crystals were obtained after 1 day. Yield: 29%. Selected IR peaks (cm^{-1}): 3140 (w), 2935 (w), 2822 (w), 1596 (s), 1544 (m), 1444 (s), 1415 (w), 1389 (m), 1320 (w), 1218 (s), 1165 (w), 1108 (w), 1065 (m), 1007 (w), 975 (m), 907 (w), 855 (m), 788 (w), 739 (s), 692 (s).

2-Furaldehyde morpholine hemi-aminal. ^{1}H NMR (400 MHz, DMSO-d_6) δ 7.58 (dd, J = 1.9, 0.9 Hz, 1H), 6.40 (dd, 3.2, 1.8 Hz, 1H), 6.32 (dt, 3.2, 0.9 Hz, 1H), 5.13 (s, 1H), 3.55 (m, 4H), 2.51 (m, 4.8H*). *Overlaps with DMSO-d_5; ^{13}C NMR (100 MHz, DMSO-d_6).

4. Conclusions

To summarise, a series of 4f compounds has been obtained and tested for their catalytic activity towards trans-4,5-diaminocyclopent-2-enones from 2-furaldehyde and primary or secondary amines. From this systematic study we have derived that the afforded Y(III) compounds exhibit remarkable activity as homogeneous catalysts and at a lower cost compared to other lanthanides. The monitored performance of those compounds through ^{1}H and ^{89}Y NMR studies has also allowed us to further understand the catalytic reaction; as a result, we have proposed an extra step in its reported mechanism (Scheme 2). However, the presence of two different ligand rotamers in solution, as well as both dimer and monomer forms for the Y(III) compounds make this system a less than ideal candidate for further investment in this field. Despite that, this work certainly demonstrates that the plethora of polynuclear 4f compounds reported to date could exhibit other interesting properties, apart from molecular magnetism.

Supplementary Materials: Supplementary materials can be found at http://www.mdpi.com/2304-6740/3/4/0448/s1.

Acknowledgments: We thank the EPSRC (UK) for funding (grant number EP/M023834/1) and the EPSRC UK National Crystallography Service at the University of Southampton for the collection of the crystallographic data for compound **4-Y**.

Author Contributions: Edward Loukopoulos and George E. Kostakis conceived and designed the experiments; Edward Loukopoulos and Kieran Griffiths performed the experiments; Edward Loukopoulos, Geoffrey R. Akien and George E. Kostakis analysed the data; Alaa Abdul-Sada and Nikolaos Kourkoumelis contributed reagents/materials/analysis tools; Edward Loukopoulos, Geoffrey R. Akien and George E. Kostakis wrote the paper.

Conflicts of Interest: The authors declare no conflict of interest.

References

1. Zheng, Z. Ligand-controlled self-assembly of polynuclear lanthanide-oxo/hydroxo complexes: From synthetic serendipity to rational supramolecular design. *Chem. Commun.* **2001**, 2521–2529. [CrossRef]

2. D'Alessio, D.; Sobolev, A.N.; Skelton, B.W.; Fuller, R.O.; Woodward, R.C.; Lengkeek, N.A.; Fraser, B.H.; Massi, M.; Ogden, M.I. Lanthanoid "bottlebrush" clusters: Remarkably elongated metal-oxo core structures with controllable lengths. *J. Am. Chem. Soc.* **2014**, *136*, 15122–15125. [CrossRef] [PubMed]

3. Thielemann, D.T.; Wagner, A.T.; Lan, Y.; Oña-Burgos, P.; Fernández, I.; Rösch, E.S.; Kölmel, D.K.; Powell, A.K.; Bräse, S.; Roesky, P.W. Peptoid-ligated pentadecanuclear yttrium and dysprosium hydroxy clusters. *Chem. Eur. J.* **2015**, *21*, 2813–2820. [CrossRef] [PubMed]

4. Wu, M.; Jiang, F.; Kong, X.; Yuan, D.; Long, L.; Al-Thabaiti, S.A.; Hong, M. Two polymeric 36-metal pure lanthanide nanosize clusters. *Chem. Sci.* **2013**, *4*, 3104–3109. [CrossRef]

5. Peng, J.-B.; Kong, X.-J.; Zhang, Q.-C.; Orendáč, M.; Prokleška, J.; Ren, Y.-P.; Long, L.-S.; Zheng, Z.; Zheng, L.-S. Beauty, symmetry, and magnetocaloric effect—Four-shell keplerates with 104 lanthanide atoms. *J. Am. Chem. Soc.* **2014**, *136*, 17938–17941. [CrossRef] [PubMed]

6. Woodruff, D.N.; Winpenny, R.E.P.; Layfield, R.A. Lanthanide single-molecule magnets. *Chem. Rev.* **2013**, *113*, 5110–5148. [CrossRef] [PubMed]

7. Zhang, P.; Guo, Y.-N.; Tang, J. Recent advances in dysprosium-based single molecule magnets: Structural overview and synthetic strategies. *Coord. Chem. Rev.* **2013**, *257*, 1728–1763. [CrossRef]

8. Amoroso, A.J.; Pope, S.J.A. Using lanthanide ions in molecular bioimaging. *Chem. Soc. Rev.* **2015**, *44*, 4723–4442. [CrossRef] [PubMed]

9. Bünzli, J.-C.G. On the design of highly luminescent lanthanide complexes. *Coord. Chem. Rev.* **2015**, *293–294*, 19–47. [CrossRef]

10. Eliseeva, S.V.; Bünzli, J.-C.G. Lanthanide luminescence for functional materials and bio-sciences. *Chem. Soc. Rev.* **2010**, *39*, 189–227. [CrossRef] [PubMed]

11. Anastasiadis, N.C.; Granadeiro, C.M.; Klouras, N.; Cunha-Silva, L.; Raptopoulou, C.P.; Psycharis, V.; Bekiari, V.; Balula, S.S.; Escuer, A.; Perlepes, S.P. Dinuclear lanthanide(III) complexes by metal-ion-assisted hydration of di-2-pyridyl ketone azine. *Inorg. Chem.* **2013**, *52*, 4145–4147. [CrossRef] [PubMed]

12. Robinson, J.R.; Gu, J.; Carroll, P.J.; Schelter, E.J.; Walsh, P.J. Exchange processes in Shibasaki's rare earth alkali metal BINOLate frameworks and their relevance in multifunctional asymmetric catalysis. *J. Am. Chem. Soc.* **2015**, *137*, 7135–7144. [CrossRef] [PubMed]

13. Roesky, P.W.; Canseco-Melchor, G.; Zulys, A. A pentanuclear yttrium hydroxo cluster as an oxidation catalyst. Catalytic oxidation of aldehydes in the presence of air. *Chem. Commun.* **2004**, 738–739. [CrossRef] [PubMed]

14. Bi, Y.; Wang, X.-T.; Liao, W.; Wang, X.; Deng, R.; Zhang, H.; Gao, S. Thiacalix[4]arene-supported planar Ln(4) (Ln = Tb^{III}, Dy^{III}) clusters: Toward luminescent and magnetic bifunctional materials. *Inorg. Chem.* **2009**, *48*, 11743–11747. [CrossRef] [PubMed]

15. Alexandropoulos, D.I.; Fournet, A.; Cunha-Silva, L.; Mowson, A.M.; Bekiari, V.; Christou, G.; Stamatatos, T.C. Fluorescent naphthalene diols as bridging ligands in Ln^{III} cluster chemistry: Synthetic, structural, magnetic, and photophysical characterization of Ln$^{III}_8$ "Christmas Stars". *Inorg. Chem.* **2014**, *53*, 5420–5422. [CrossRef] [PubMed]

16. Pointillart, F.; le Guennic, B.; Golhen, S.; Cador, O.; Maury, O.; Ouahab, L. A redox-active luminescent ytterbium based single molecule magnet. *Chem. Commun.* **2013**, *49*, 615–617. [CrossRef] [PubMed]

17. Mazarakioti, E.C.; Poole, K.M.; Cunha-Silva, L.; Christou, G.; Stamatatos, T.C. A new family of Ln$_7$ clusters with an ideal D_{3h} metal-centered trigonal prismatic geometry, and SMM and photoluminescence behaviors. *Dalton Trans.* **2014**, *43*, 11456–11460. [CrossRef] [PubMed]

18. Bag, P.; Rastogi, C.K.; Biswas, S.; Sivakumar, S.; Mereacre, V.; Chandrasekhar, V. Homodinuclear lanthanide {Ln$_2$} (Ln = Gd, Tb, Dy, Eu) complexes prepared from an *o*-vanillin based ligand: Luminescence and single-molecule magnetism behavior. *Dalton Trans.* **2015**, *44*, 4328–4340. [CrossRef] [PubMed]

19. Chandrasekhar, V.; Bag, P.; Colacio, E. Octanuclear {Ln$^{III}_8$}(Ln = Gd, Tb, Dy, Ho) macrocyclic complexes in a cyclooctadiene-like conformation: Manifestation of slow relaxation of magnetization in the Dy^{III} derivative. *Inorg. Chem.* **2013**, *52*, 4562–4570. [CrossRef] [PubMed]

20. Pasatoiu, T.D.; Tiseanu, C.; Madalan, A.M.; Jurca, B.; Duhayon, C.; Sutter, J.P.; Andruh, M. Study of the luminescent and magnetic properties of a series of heterodinuclear [Zn^{II}Ln^{III}] complexes. *Inorg. Chem.* **2011**, *50*, 5879–5889. [CrossRef] [PubMed]

21. Ehama, K.; Ohmichi, Y.; Sakamoto, S.; Fujinami, T.; Matsumoto, N.; Mochida, N.; Ishida, T.; Sunatsuki, Y.; Tsuchimoto, M.; Re, N. Synthesis, structure, luminescent, and magnetic properties of carbonato-bridged $Zn^{II}_2Ln^{III}_2$ complexes $[(\mu_4-CO_3)_2\{Zn^{II}L^nLn^{III}(NO_3)\}_2]$ ($Ln^{III} = Gd^{III}$, Tb^{III}, Dy^{III}; $L^1 =$ *N,N'*-bis(3-methoxy-2-oxybenzylidene)-1,3-propanediaminato, $L^2 = $ *N,N'*-bis(3-ethoxy-2-oxybenzylidene)-1,3-propanediaminato). *Inorg. Chem.* **2013**, *52*, 12828–12841. [PubMed]

22. Goura, J.; Mereacre, V.; Novitchi, G.; Powell, A.K.; Chandrasekhar, V. Homometallic Fe^{III}_4 and heterometallic $\{Fe^{III}_4Ln^{III}_2\}$ (Ln = Dy, Tb) complexes—Syntheses, structures, and magnetic properties. *Eur. J. Inorg. Chem.* **2015**, *2015*, 156–165. [CrossRef]

23. Ke, H.; Zhao, L.; Guo, Y.; Tang, J. A Dy_6 cluster displays slow magnetic relaxation with an edge-to-edge arrangement of two Dy_3 triangles. *Eur. J. Inorg. Chem.* **2011**, *2011*, 4153–4156. [CrossRef]

24. Liao, S.; Yang, X.; Jones, R.A. Self-assembly of luminescent hexanuclear lanthanide salen complexes. *Cryst. Growth Des.* **2012**, *12*, 970–974. [CrossRef]

25. Sarwar, M.; Madalan, A.M.; Tiseanu, C.; Novitchi, G.; Maxim, C.; Marinescu, G.; Luneau, D.; Andruh, M. A new synthetic route towards binuclear 3d-4f complexes, using non-compartmental ligands derived from *o*-vanillin. Syntheses, crystal structures, magnetic and luminescent properties. *New J. Chem.* **2013**, *37*, 2280–2292. [CrossRef]

26. Mondal, K.C.; Kostakis, G.E.; Lan, Y.; Powell, A.K. Magnetic properties of five planar defect dicubanes of $[Ln^{III}_4(\mu_3-OH)_2(L)_4(HL)_2]\cdot2THF$ (Ln = Gd, Tb, Dy, Ho and Er). *Polyhedron* **2013**, *66*, 268–273. [CrossRef]

27. Gómez, V.; Vendier, L.; Corbella, M.; Costes, J.-P. Tetranuclear $[Co-Gd]_2$ complexes: Aiming at a better understanding of the 3d-Gd magnetic interaction. *Inorg. Chem.* **2012**, *51*, 6396–6404. [CrossRef] [PubMed]

28. Athanasopoulou, A.A.; Pilkington, M.; Raptopoulou, C.P.; Escuer, A.; Stamatatos, T.C. Structural aesthetics in molecular nanoscience: A unique Ni_{26} cluster with a "rabbit-face" topology and a discrete Ni_{18} "molecular chain". *Chem. Commun.* **2014**, *50*, 14942–14945. [CrossRef] [PubMed]

29. Tziotzi, T.G.; Tzimopoulos, D.I.; Lis, T.; Inglis, R.; Milios, C.J. Dodecanuclear [MnLn] species: Synthesis, structures and characterization of magnetic relaxation phenomena. *Dalton Trans.* **2015**, *44*, 11696–11699. [CrossRef] [PubMed]

30. Meng, Z.-S.; Guo, F.-S.; Liu, J.-L.; Leng, J.-D.; Tong, M.-L. Heterometallic cubane-like $\{M_2Ln_2\}$ (M = Ni, Zn; Ln =, Gd, Dy) and $\{Ni_2Y_2\}$ aggregates. Synthesis, structures and magnetic properties. *Dalton Trans.* **2012**, *41*, 2320–2329. [CrossRef] [PubMed]

31. Fan, L.-L.; Guo, F.-S.; Yun, L.; Lin, Z.-J.; Herchel, R.; Leng, J.-D.; Ou, Y.-C.; Tong, M.-L. Chiral transition metal clusters from two enantiomeric schiff base ligands. Synthesis, structures, CD spectra and magnetic properties. *Dalton Trans.* **2010**, *39*, 1771–1780. [CrossRef] [PubMed]

32. Thio, Y.; Toh, S.W.; Xue, F.; Vittal, J.J. Self-assembly of a 15-nickel metallamacrocyclic complex derived from the L-glutamic acid Schiff base ligand. *Dalton Trans.* **2014**, *43*, 5998–6001. [CrossRef] [PubMed]

33. Berkoff, B.; Griffiths, K.; Abdul-Sada, A.; Tizzard, G.J.; Coles, S.; Escuer, A.; Kostakis, G.E. A new family of high nuclearity CoII/DyIII coordination clusters possessing robust and unseen topologies. *Dalton Trans.* **2015**, *44*, 12788–12795. [CrossRef] [PubMed]

34. Nemec, I.; Machata, M.; Herchel, R.; Boča, R.; Trávníček, Z. A new family of Fe2Ln complexes built from mononuclear anionic Schiff base subunits. *Dalton Trans.* **2012**, *41*, 14603–14610. [CrossRef] [PubMed]

35. Costes, J.-P.; Duhayon, C. An ionic dysprosium complex made of a hexanuclear Dy_6 cationic cluster and a mononuclear Dy anionic unit. *Eur. J. Inorg. Chem.* **2014**, *2014*, 4745–4749. [CrossRef]

36. Loukopoulos, E.; Berkoff, B.; Abdul-Sada, A.; Tizzard, G.J.; Coles, S.J.; Escuer, A.; Kostakis, G.E. A disk-like $Co^{II}_3Dy^{III}_4$ coordination cluster exhibiting single molecule magnet behavior. *Eur. J. Inorg. Chem.* **2015**, *2015*, 2646–2649. [CrossRef]

37. Andruh, M. The exceptionally rich coordination chemistry generated by Schiff-base ligands derived from *o*-vanillin. *Dalton Trans.* **2015**, *44*, 16633–16653. [CrossRef] [PubMed]

38. Griffiths, K.; Gallop, C.W.D.; Abdul-Sada, A.; Vargas, A.; Navarro, O.; Kostakis, G.E. Heteronuclear 3 d/Dy^{III} coordination clusters as catalysts in a domino reaction. *Chem. Eur. J.* **2015**, *21*, 6358–6361. [CrossRef] [PubMed]

39. Li, S.-W.; Batey, R.A. Mild lanthanide(III) catalyzed formation of 4,5-diaminocyclopent-2-enones from 2-furaldehyde and secondary amines: A domino condensation/ring-opening/electrocyclization process. *Chem. Commun.* **2007**, 3759–3761. [CrossRef] [PubMed]

40. Lin, P.-H.; Gorelsky, S.; Savard, D.; Burchell, T.J.; Wernsdorfer, W.; Clérac, R.; Murugesu, M. Synthesis, characterisation and computational studies on a novel one-dimensional arrangement of Schiff-base Mn$_3$ single-molecule magnet. *Dalton Trans.* **2010**, *39*, 7650–7658. [CrossRef] [PubMed]

41. Monfared, H. H.; Chamayou, A.-C.; Khajeh, S.; Janiak, C. Can a small amount of crystal solvent be overlooked or have no structural effect? Isomorphous non-stoichiometric hydrates (pseudo-polymorphs): The case of salicylaldehyde thiosemicarbazone. *CrystEngComm* **2010**, *12*, 3526–3530. [CrossRef]

42. Nangia, A. Pseudopolymorph: Retain this widely accepted term. *Cryst. Growth Des.* **2006**, *6*, 2–4. [CrossRef]

43. Dokorou, V.N.; Powell, A.K.; Kostakis, G.E. Two pseudopolymorphs derived from alkaline earth metals and the pseudopeptidic ligand trimesoyl-tris-glycine. *Polyhedron* **2013**, *52*, 538–544. [CrossRef]

44. Lin, P.-H.; Burchell, T.J.; Clerac, R.; Murugesu, M. Dinuclear dysprosium(III) single-molecule magnets with a large anisotropic barrier. *Angew. Chem. Int. Ed.* **2008**, *47*, 8848–8851. [CrossRef] [PubMed]

45. Llunell, M.D.; Casanova, J.; Cirera, P.; Alemany, S.A. *SHAPE*, Version 2.0; SHAPE: Barcelona, Spain, 2010.

46. Coles, S.J.; Gale, P.A. Changing and challenging times for service crystallography. *Chem. Sci.* **2012**, *3*, 683–689. [CrossRef]

47. Dolomanov, O.V.; Blake, A.J.; Champness, N.R.; Schröder, M. OLEX: New software for visualization and analysis of extended crystal structures. *J. Appl. Crystallogr.* **2003**, *36*, 1283–1284. [CrossRef]

48. Palatinus, L.; Chapuis, G. SUPERFLIP—A computer program for the solution of crystal structures by charge flipping in arbitrary dimensions. *J. Appl. Crystallogr.* **2007**, *40*, 786–790. [CrossRef]

49. Farrugia, L.J. *WinGX* suite for small-molecule single-crystal crystallography. *J. Appl. Crystallogr.* **1999**, *32*, 837–838. [CrossRef]

50. Sheldrick, G.M. SHELXT—Integrated space-group and crystal-structure determination. *Acta Crystallogr. Sect. A* **2015**, *71*, 3–8. [CrossRef] [PubMed]

51. Sheldrick, G.M. A short history of SHELX. *Acta Crystallogr. Sect. A* **2008**, *64*, 112–122. [CrossRef] [PubMed]

52. Spek, A.L. Single-crystal structure validation with the program PLATON. *J. Appl. Crystallogr.* **2003**, *36*, 7–13. [CrossRef]

53. Macrae, C.F.; Edgington, P.R.; McCabe, P.; Pidcock, E.; Shields, G.P.; Taylor, R.; Towler, M.; van de Streek, J. Mercury: Visualization and analysis of crystal structures. *J. Appl. Crystallogr.* **2006**, *39*, 453–457. [CrossRef]

inorganics

MDPI

Communication

Luminescent Lanthanide Metal Organic Frameworks for *cis*-Selective Isoprene Polymerization Catalysis

Samantha Russell [1], Thierry Loiseau [1], Christophe Volkringer [1,2] and Marc Visseaux [1,*

[1]   Univ. Lille, CNRS, Ecole Nationale Supérieure de Chimie de Lille, UMR 8181-UCCS—Unité de Catalyse et Chimie du Solide, F-59000 Lille, France; sr90@st-andrews.ac.uk (S.R.); thierry.loiseau@ensc-lille.fr (T.L.); christophe.volkringer@ensc-lille.fr (C.V.)
[2]   Institut Universitaire de France, 75231 Paris, France
*   Correspondence: marc.visseaux@ensc-lille.fr; Tel.: +33-320-336-483; Fax: +33-320-436-585

Academic Editors: Stephen Mansell and Steve Liddle
Received: 18 September 2015; Accepted: 29 October 2015; Published: 9 November 2015

Abstract: In this study, we are combining two areas of chemistry; solid-state coordination polymers (or Metal-Organic Framework—MOF) and polymerization catalysis. MOF compounds combining two sets of different lanthanide elements (Nd^{3+}, Eu^{3+}/Tb^{3+}) were used for that purpose: the use of neodymium was required due to its well-known catalytic properties in dienes polymerization. A second lanthanide, europium or terbium, was included in the MOF structure with the aim to provide luminescent properties. Several lanthanides-based MOF meeting these criteria were prepared according to different approaches, and they were further used as catalysts for the polymerization of isoprene. Stereoregular *cis*-polyisoprene was received, which in some cases exhibited luminescent properties in the UV-visible range.

Keywords: MOF; lanthanide; neodymium; isoprene polymerisation; *cis*-selective; luminescence

1. Introduction

Metal-Organic Frameworks (MOFs) are three-dimensional frameworks that are obtained by a reaction between organic O-/N-donor ligands and metallic cationic species [1]. The resulting crystalline structure consists of inorganic units connected via organic linkers that are often porous; their porosity are exploited in many applications, such as gas storage, separation, controlled drug release and catalysis [2]. Luminescence properties can be observed when MOFs materials are built up from active elements like for example light-emitting rare earths europium or terbium [3,4]. We showed recently that Nd carboxylate based MOFs are efficient pre-catalysts for the stereo-selective polymerization of conjugated dienes [5]. In addition, some unreacted MOF compound residues, that had not been involved in the polymerization, were found disseminated into the polymer matrix, so as the result could be considered as a MOF/polymer composite. However, at this stage, despite the presence of Nd element, the resulting material showed no luminescent properties [6].

Typically, organic luminescent materials, such as polymers, suffer poor stability under harsh conditions and have poor long-term reliability [7]. However, they have a greater ductility and processability than inorganic materials, which allows for polymer films to be produced and the material to be molded and shaped [8]. In comparison, inorganic luminescent materials are known to be a lot more durable and have a greater thermal stability. Therefore, the combination of inorganic and organic materials, such as the MOF and polymer respectively, may allow for the production of a durable luminescent hybrid material with enhanced properties in comparison to organic-only or inorganic-only luminescent materials [9]. Previous work on luminescent polymer composites has been published but current limitations are related to the nature of the polymer matrix, until now rather restricted to thermoplastics [10] or hard to process chitosan [11], and with only scarce examples of synthetic elastomers [12–14].

This contribution considered different lanthanide-based MOF assemblies, obtained by three different synthetic methods, in which were included two elements, neodymium, for its catalytic ability, and europium (or terbium), for luminescence. These compounds were then assessed for isoprene polymerization, with the aim to afford a luminescent diene based-rubber.

2. Results and Discussion

There were two main MOF structures that were intended in the present work. The first one was a lanthanide formate $Ln(form)_3$, a MOF-like compound that has been previously synthesized and contains the Ln element (typically neodymium) for the inorganic moieties and formic acid to produce formate groups as organic linkers [6]. With neodymium, the three-dimensional structure consists of NdO_9 neodymium-centred tricapped trigonal prisms that are connected to each other via the formate ligands (Figure 1A). Previous work showed that $Nd(form)_3$ was an efficient pre-catalyst for the polymerization of isoprene when combined with MMAO (modified methylalumoxane), with yields of polyisoprene ranging from 27% to 83% and *cis*-1,4 selectivity ranging from 48% to 88% [6]. The analysis of the synthesized polyisoprene by scanning electron microscopy (SEM) and powder X-ray diffraction (PXRD) revealed that partial fragmentation of the MOF particles had occurred and produced a dispersion of un-activated MOF compound within the polymer matrix. We prepared similarly as their congeners Eu- and Tb-based MOFs having the above-described structures. They were assessed for polymerization catalysis, but no polymer was obtained (*vide supra*), showing that Nd is necessary to catalyze the polymerization. The second MOF structure used was the material MIL-103(Nd) (Figure 1B), a porous neodymium-based MOF [15], which had already previously been shown to act as a successful pre-catalyst for isoprene polymerization [5].

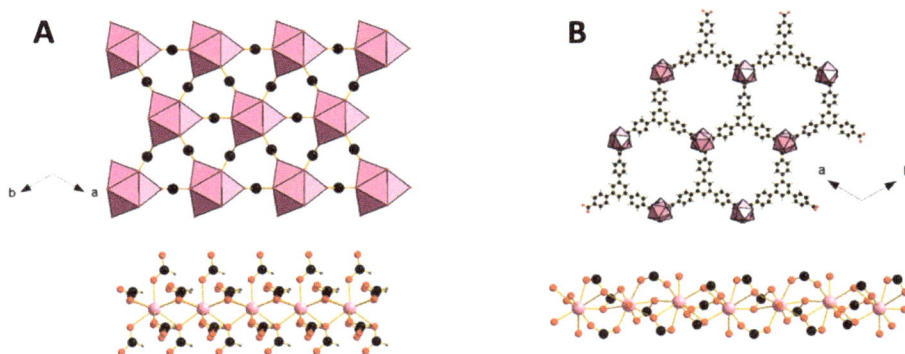

Figure 1. Illustration of the structural organization of $Nd(form)_3$ (**A**) and MIL-103(Nd) (**B**). Pink for Nd, black for C, red for O, grey for H, a and b: crystallographic axes.

Three strategies that are represented in Figure 2 were considered, to have a combination of Nd/Eu(Tb) elements within a MOF structure intended to be further assessed for polymerization: (a) the synthesis of mixed Ln based-MOFs; (b) the use of a mixture of two different pure Ln based-MOF; (c) the inclusion of a second Ln element into a Nd based-MOF.

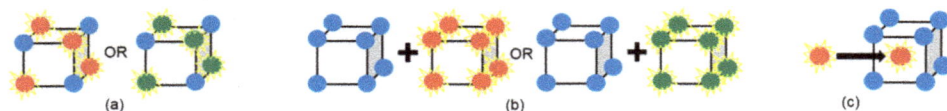

Figure 2. Schematic representations of the expected combined MOF structures: (**a**) mixed Ln based-MOFs; (**b**) mixture of two different pure Ln based-MOF; (**c**) Nd based-MOF including a second Ln element (Nd, blue; Eu, red; Tb, green; yellow circumference represents luminescence).

2.1. Mixed Ln-MOF Synthesis for Ln(form)₃

MOF compounds containing both neodymium and europium (or neodymium and terbium) within the same structure (Figure 2a) were synthesized. The feed molar ratios of the two lanthanides were varied to 1:3, 1:1 and 3:1 in three different MOFs preparations. The true ratios of the two lanthanides within the final MOF structure were determined by Inductively Coupled Plasma-Atomic Emission Spectroscopy (ICP-AES). The results (Experimental Section) revealed a difference between the initial feed ratio and the final quantities of the two lanthanides within the MOF, however the major lanthanide element was the one expected (Nd/Eu: 0.49, 1.59, 5.14, respectively). The powder X-ray diffraction (PXRD) of Ln(form)₃ (Ln = Nd/Eu, Figure 3; Ln = Nd/Tb, Figure S1), showed in both cases a single peak with gradual shift in the 2θ value as the quantity of europium (or terbium) within the MOF increased, proving that the two lanthanides are present within the same crystalline MOF structure.

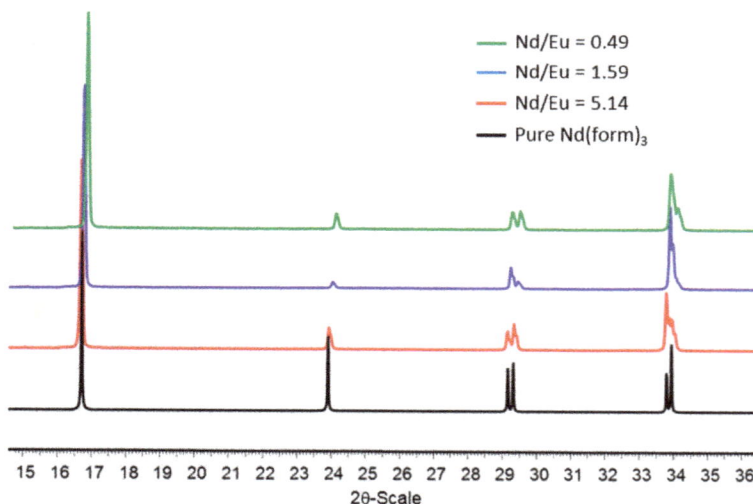

Figure 3. The PXRD pattern of the Nd(form)₃ MOF and the three MOFs with varying Nd/Eu ratios.

Scanning Electron Microscopy (SEM) was used to study the size and shape of the crystals. In pure Ln(form)₃ MOFs, europium-based crystals are smaller than neodymium-based crystals, where the former ones are relatively spherical and the latter ones are longer and more needle shaped (Figure S2). Although the SEM images have shown that there is not a uniform size for all the crystals within one synthesis—some larger needle-shaped crystals due to a slight variation of Nd/Eu ratio are present within every sample—a general trend that smaller crystals are produced when the MOF has a higher europium content may be deduced from observations (Figure 4).

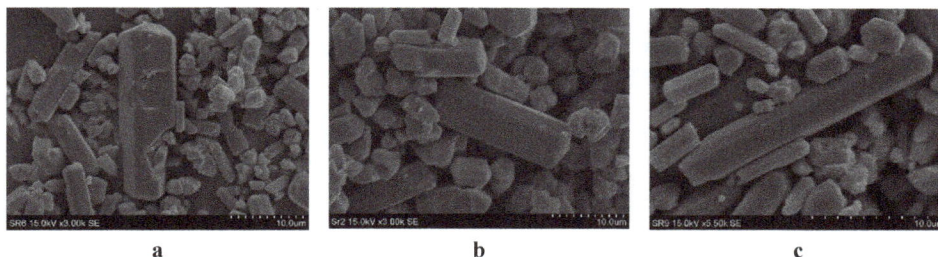

Figure 4. SEM image of Ln(form)$_3$ MOF compounds with Nd/Eu of 0.49 (**a**, average crystals size ≈ 3 μm), 1.59 (**b**, av. crystal size ≈ 5 μm), and 5.14 (**c**, av. crystals size ≈ 6 μm).

2.2. Pure Lanthanide-Based MOFs

A second approach attempted to mix two lanthanides by combining two distinct MOF solids; this mixing took place further during the polymerization. For that purpose, pure MOF compounds containing only one lanthanide of neodymium, europium or terbium were synthesized. The picture representation of this method can be seen in Figure 2b. Powder X-ray diffraction was used to confirm the purity of the three distinct MOF solids. (Figure S3) The resulting pattern showed that the three MOF compounds were pure and they had the same peak pattern. Due to the different lanthanides, there was a slight shift of the Bragg peaks in the 2θ scale. This is due to the well-known contraction effect in the lanthanide series, since ionic radius of neodymium is the largest and that of terbium the smallest of the lanthanides investigated here.

2.3. MIL-103(Nd) with Eu-Insertion

The third method used the material MIL-103(Nd) (Figure 1B), with the aim of inserting europium within the pores (Figure 2c). A dried sample of MIL-103(Nd) was impregnated with a solution of Eu^{3+}, then isolated and dried (see Experimental Section). A luminescence spectrum was recorded, showing the characteristic europium peak at 614 nm, which was a positive indication that the europium was within the pores [16]. However, the rest of the spectrum was severely distorted, signifying that the luminescence was limited (Figure S4). This distortion could be due to a slight quenching from the neodymium within the MIL-103 framework, likely related that the europium and neodymium were now too close in space. Since the result was not concluding in terms of luminescence properties, further characterization of this modified MIL-103 MOF were not conducted.

2.4. Elaboration of cis-Polyisoprene/MOF Composite Materials

The MOFs were studied as pre-catalysts for isoprene polymerization (Scheme 1) in the presence of MMAO (Modified Methyl AluminOxide) as co-catalyst, by varying the percentage of lanthanide present in each MOF structure, and with the expectation to finally produce a polymer material having luminescent properties.

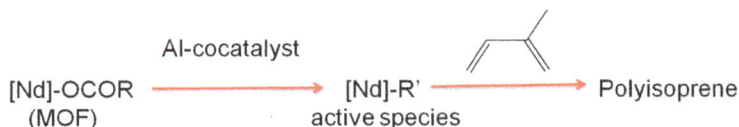

Scheme 1. The activation of the catalyst by the combination of a MOF pre-catalyst and alkyl-aluminum co-catalyst.

We had previously established that homoleptic Nd(form)$_3$ and mixed Nd(2,6-ndc)(form) (2,6-napthalenedicarboxylate ligand) are efficient pre-catalysts towards isoprene polymerization in the presence of an Al co-catalyst [5,6]. This reactivity is connected to the neodymium carboxylate nature of these MOF compounds, which are known to produce efficient catalysts for conjugated dienes [17]. The experimental conditions in the present study were then chosen as adequate to the heterogeneous nature of the catalytic system, *i.e.*, heating (in general 50 °C), long reaction times (>20 h), and large excesses of MMAO (20–50-fold molar compared to the Ln pre-catalyst), in order to produce *cis*-stereoregular polyisoprene, as previously determined [5].

2.4.1. Mixed Nd/Ln(form)$_3$ MOF Polymerization (Ln = Eu, Tb)

The polymerizations were completed in duplicate and the results are shown in Table 1 below.

Table 1. Isoprene polymerization using mixed Nd/Ln(form)$_3$ (Ln = Eu or Tb) as pre-catalysts.

Run [a]	Ratio of Nd/Ln	Yield (%)	M_n (Đ) [b]	Selectivity (%) [c]
				cis-/*trans*-/3,4-
1 [d]	100% Nd	27	64,900 (2.3)	92/2/6
2 [d]	5.14	46	nd	84/8/8
3 [d]	1.59	29	34,400 (5.0)	95/1/4
4 [d]	0.49	11	nd	79/14/7
5 [d]	100% Eu	5	-	-
6 [e]	3.00	40	77,600 (2.4)	92/2/6
7 [e]	1.00	36	49,700 (3.5)	82/1/17
8 [e]	0.33	35	35,600 (3.8)	85/7/8
9 [e]	100% Tb	1	nd	-

[a] Typical conditions: 1 Ln:100 MMAO:500 Isoprene; $V_{toluene}$ = $V_{isoprene}$ = 1 mL; V_{MMAO} = 1 mL (1.84 mmol); t = 24 h; T = 50 °C; [b] Determined by SEC with reference to PS standards; Đ = M_w/M_n; [c] From ^{1}H and ^{13}C NMR; [d] The mass of MOF precatalyst was fixed at 6 mg (*ca.* 20 µmol), where Ln = Nd + Eu; [e] All quantities divided by a factor of 2: mass of MOF precatalyst 3 mg (*ca.* 10 µmol), where Ln = Nd + Tb, $V_{isoprene}$ = $V_{toluene}$ = 0.5 mL. V_{MMAO} = 0.5 mL (0.92 mmol). nd: non determined.

The results of the polymerization show a trend between the ratio of Nd/Eu and the yield, *i.e.*, the larger the quantity of neodymium, the greater the yield (runs 2–4). This result is due to the fact that neodymium is the active species within the polymerization, with the exception of run 1, where the yield of polymer obtained is lower than the yield obtained for the MOF containing lower quantities of neodymium (runs 2,3). A possible reason could be due to insufficient grinding and drying of the MOF prior to the polymerisation as already noticed [6]. Smaller effect of the Nd quantity, though similar, is observed with experiments performed with Nd/Tb(form)$_3$ (based on the initial ratio of the starting materials for the preparation of the mixed MOFs). MOFs containing no Nd were poorly active (run 5) or inactive (run 9). The NMR analyses of the polymers received show, as expected (*vide infra*) a highly *cis*-1,4-polyisoprene selective polymerization (79%–95%), with a particularly stereoregular polymer produced when the compound with Nd/Eu ratio of 1.59 was used as pre-catalyst (run 3, Figure S5). Molecular weights were found in the range 30,000–80,000, and dispersities were rather broad (2.3–5.0), as already noticed and discussed under similar polymerization conditions [5,6]. SEM images of the polymers in film form were recorded to determine if MOF fragments were dispersed throughout the polymer matrix (run 3, Table 1, Figure S6). The results showed that there were relatively small fragments (≈1–50 µm) within the sample, suggesting that some MOF particles remained. Unfortunately, the luminescence tests of all the polymers using the spectrofluorimeter SAFAS FLX-Xenius (equipped with a Xenon lamp) gave spectra that showed no emission. Coming back to the starting mixed MOFs Nd/Ln(form)$_3$ (Ln = Eu, Tb), we observed that the results were also negative in terms of luminescence. This suggested that quenching process, *i.e.*, the excited energy of a center is not emitted as light, but instead transferred to another unit within the system [18] was present within the MOF structure, and the problem did not lie within the transfer of MOF properties to the hybrid MOF/polymer material.

2.4.2. Pure MOF-[Nd(form)$_3$ + Ln(form)$_3$] Polymerization (Ln = Eu, Tb)

A range of molar ratios of [Nd(form)$_3$ + (Eu or Tb)(form)$_3$] mixtures was used in an attempt to find out the best conditions for the polymerization, where the three main factors were the yield, selectivity and luminescence. The volumes of MMAO, isoprene and toluene stayed constant at 0.5 mL throughout the experiments while the masses of the two MOFs were changed, affecting the final ratios of all reagents. The mass of Ln(form)$_3$ was increased to attempt to enhance the luminescence properties within the final polymer. The conditions and results of the polymerizations can be seen in Table 2.

Table 2. Isoprene polymerization using [Nd(form)$_3$ + Ln(form)$_3$] as pre-catalysts.

Run [a]	Ratio of Reagents Nd/Ln/MMAO/Isoprene	Time (h)	Yield (%)	Selectivity (%) [b] cis-/trans-/3,4-	Luminescence
10	1/10Eu/100/500	48	77	87/5/8	yes
11 [c]	10/10Eu/100/500	48	78	93/2/5	yes
12 [d]	50/50Eu/100/500	75	80	92/2/6	yes
13 [d]	50/50Eu/100/500	336	89	86/4/10	yes
14	1/10Tb/100/500	48	77	70/4/26	yes
15 [d]	50/50Tb/100/500	166	64	91/2/7	yes
16 [d]	50/50Tb/100/500	336	85	89/6/5	yes
17 [e]	1/92/500	48	17	88/1/11	no

[a] Typical conditions: [Nd] = 10 μmol; $V_{toluene}$ = $V_{isoprene}$; T = 50 °C; [b] From ^{1}H and ^{13}C NMR; [c] [Nd] = 100 μmol; [d] [Nd] = 500 μmol; [e] ratio of reagents Nd/MMAO/Isoprene, [Nd] = 10 μmol.

All experiments afforded polyisoprene having high *cis*-content. Remarkably, the excess of non-catalyst-active MOF (Eu or Tb) was a drawback neither with regard to the yield, at the condition to extend the reaction time (runs 11–12 and 15–16) nor with regard to the stereo-selectivity. Selected samples were analyzed by SEC, to verify the high M_n values (124,000 and 302,000, runs 11 and 14, respectively), which was in agreement with the slow kinetics of the initiation step of the polymerization, due to the robust nature of the MOF pre-catalyst, as previously observed [5,6]. When using Tb(form)$_3$ instead of Eu(form)$_3$ in association with Nd(form)$_3$, the yields of polymer were a little lower, as well as the *cis*-1,4-polyisoprene selectivity. A powder XRD diagram of polyisoprene was recorded (Figure 5a, run 11). The diagram showed two peaks, which corresponded to Nd(form)$_3$ and to Eu(form)$_3$, indicative of MOF solids still remaining within the final polymer matrix in its original MOF form. SEM images of the polymer materials showed small fragments, most likely the MOF residues detected by PXRD, which were dispersed throughout the polymer matrix (run 11, Table 2, Figure 5b). This result would confirm that the polymerization reaction is most likely occurring on the surface of the MOF.

The luminescence of all hybrid materials thus prepared was this time both visible by eye under the UV lamp, and detected by the spectrophotometer, as shown for selected samples (Figure 6). The luminescence spectrum of pink sample 12 shows three clear peaks, which represent the $^5D_0 \rightarrow {}^7F_1$, $^5D_0 \rightarrow {}^7F_2$ and $^5D_0 \rightarrow {}^7F_4$ transitions. The two weaker transitions of $^5D_0 \rightarrow {}^7F_0$ and $^5D_0 \rightarrow {}^7F_3$ are not as prominent within the spectrum, although both expected transition wavelengths, at 578 and 650 nm respectively, do appear to show very weak broad bands. The green luminescent material received from run 15 has a spectrum showing $^5D_4 \rightarrow {}^7F_6$, $^5D_4 \rightarrow {}^7F_5$, $^5D_4 \rightarrow {}^7F_4$ and $^5D_4 \rightarrow {}^7F_3$ transitions, which provides the evidence that there are Tb^{3+} particles within the polymer sample, but there is also an extra peak at 466 nm, which remained present within all the luminescence scans of the polymer samples synthesized using [Nd(form)$_3$ + Tb(form)$_3$] as pre-catalysts.

a)

b)

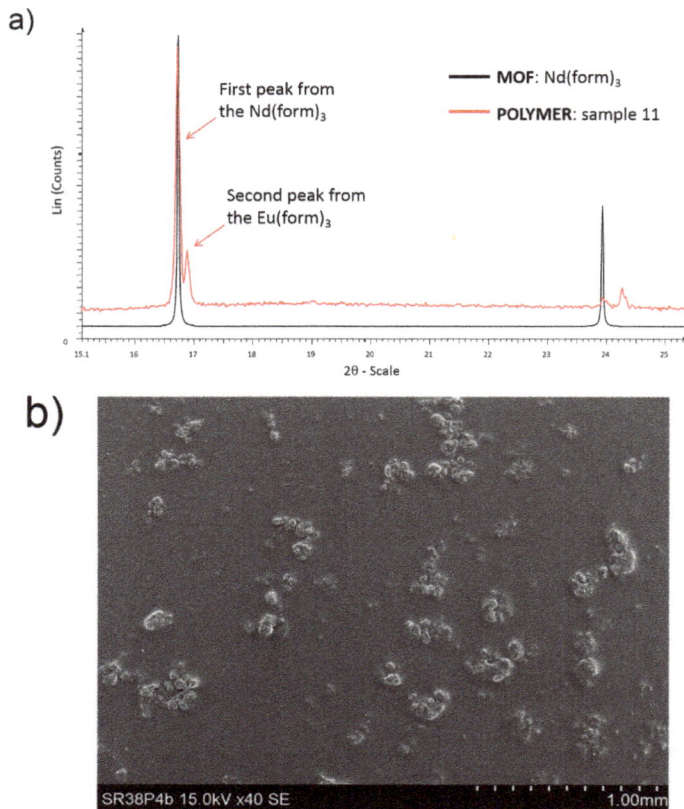

Figure 5. (**a**) The PXRD diagram of polymer sample 11 showing MOF residues of Nd(form)$_3$ and Eu(form)$_3$; (**b**) The SEM image of sample isolated from run 11 showing dispersed MOF particles.

2.4.3. Polymerization with MIL-103(Nd) with Eu-Inserted

The experiment was conducted with 10 μmol of MOF as precatalyst (run 17, Table 2). The resulting polymer showed a highly quenched luminescence spectrum (Figure S7) and no color emission was observed under the UV lamp. Moreover, just regarding the result of this polymerization, *i.e.*, relatively low yield (17% in 48 h) and high percentage of *cis*-1,4-polyisoprene (88%), it can be compared to the previous work [5] done with MIL-103(Nd) as a pre-catalyst for isoprene polymerization. One main difference in experimental is that the authors obtained 33.5% yield in 20 h, when using the porous MIL-103 with empty pores. This would suggest that the filling of the pores of the MOF is detrimental to the activity of the pre-catalyst. This is possibly due to a limitation of active catalytic sites, as filled pores do not allow access to the active metal sites. Previous papers have discussed the advantages of porous material in the use of catalysis and the discussion of a confinement effect due to the controlled polymerization that can take place with porous materials [19].

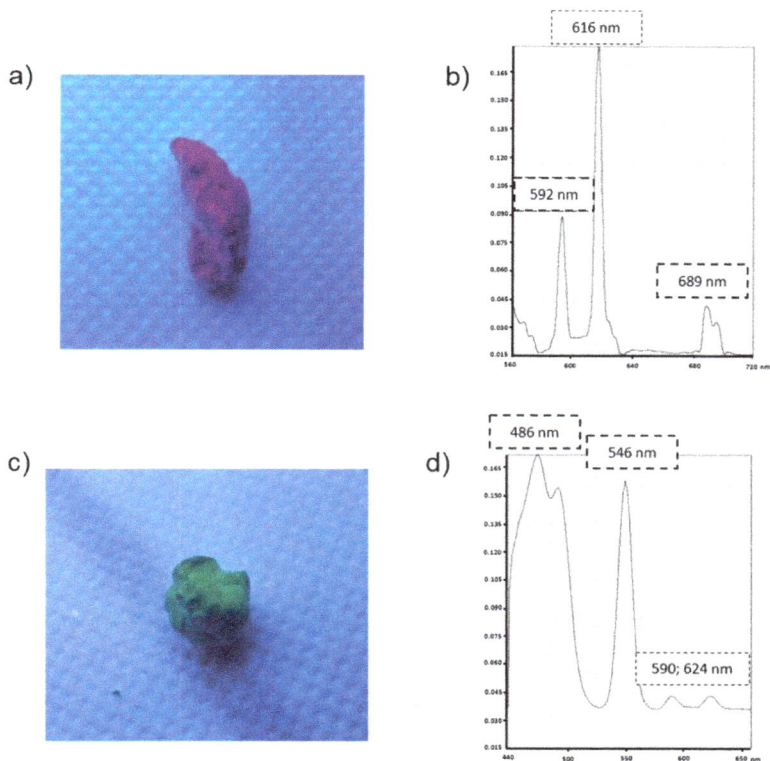

Figure 6. (**a**) The image of sample material obtained from run 12 as seen under the UV lamp and (**b**) the luminescence spectrum of the same sample showing the characteristic peaks of Eu^{3+}; (**c**) the image of sample material obtained from run 15 as seen under the UV lamp and (**d**) the luminescence spectrum of the same sample showing the characteristic peaks of Tb^{3+}.

3. Experimental Section

3.1. MOF Syntheses

3.1.1. Pure MOF Synthesis—Ln(form)₃ (Ln = Nd, Eu, Tb)

The preparation of Nd(form)₃ was recently described in the literature [20]. We used a closely related protocol for its synthesis [6]: A mixture of 0.360 g (1 mmol) of $NdCl_3 \cdot 6H_2O$, 3 mL (79 mmol) of formic acid, and 2 mL (2 mmol) of 1 M KOH was placed in a Parr bomb and then heated statically at 180 °C for 24 h. The solution pH was 1.45 at the end of the reaction. The resulting pink product was then filtered off, washed with water, and dried at room temperature. Elemental analysis, observed (calculated): C, 12.6% (12.9%); H, 0.2% (1.1%). The same procedure was also completed using $EuCl_3 \cdot 6H_2O$ (0.36 g, 1 mmol) and $TbCl_3 \cdot 6H_2O$ (0.37 g, 1 mmol). The MOF was collected via filtration, washed with water and left to dry under air. The purity was analysed by microscopy, powder X-ray diffraction (PXRD), and eventually ICP-AES.

3.1.2. Mixed Ln-MOF Synthesis

Syntheses were completed using two lanthanides with varying ratios; the equivalents of each lanthanide used in each synthesis are detailed in Table 3. $NdCl_3 \cdot 6H_2O$ and $EuCl_3 \cdot 6H_2O$ were weighed on the balance and placed within the Teflon part of an autoclave. Formic acid (3 mL, 25 M, 75 mmol) and sodium hydroxide (2 mL, 4 M, 8 mmol) were then added using syringes. The Teflon part was closed and sealed within the outer steel part. The autoclave reactor was then placed inside the oven for 24 h at 180 °C. The same procedure was also completed using $NdCl_3 \cdot 6H_2O$ and $TbCl_3 \cdot 6H_2O$. Once removed from the oven, the sample was collected via filtration, washed with water and left to dry under air. The purity was determined by two methods, microscopy and powder X-ray diffraction.

Table 3. The reactants used within the combined $Nd/Ln(form)_3$ MOF syntheses (0.5 mmol Nd initial feed).

Structure	Ln Ratio (Feed)	Ln Ratio (ICP-AES)
$Ln(form)_3$	1Nd:1Eu	1.59
$Ln(form)_3$	1Nd:3Eu	0.49
$Ln(form)_3$	3Nd:1Eu	5.14
$Ln(form)_3$	1Nd:1Tb	-
$Ln(form)_3$	1Nd:3Tb	-
$Ln(form)_3$	3Nd:1Tb	-

MIL-103 Synthesis—$Nd(C_{27}H_{15}O_6)(H_2O) \cdot (C_6H_{11}OH)$: the MIL-103 MOF was synthesized using previous literature [15]. H_3BTB (1,3,5-Benzenetrisbenzoic acid, 500 mg, 1.14 mmol) and $Nd(NO_3)_3 \cdot 6H_2O$ (500 mg, 1.14 mmol) were placed inside a 125 mL Teflon-lined steel autoclave. Water (12.5 mL) and NaOH (1 mL, 2M, 2 mmol) were added using syringes and the mixture was then stirred for 2 min. Cyclohexanol (12.5 mL, 0.12 mol) was warmed a little to increase the fluidity of the solution, then added to the Teflon part using a syringe. The solution was stirred for a further 10 min. The autoclave was sealed and placed in the oven for 5 days at 100 °C. Once removed from the oven, the sample was filtered and washed with ethanol (20 mL), water (10 mL), ethanol (10 mL), acetone (10 mL) and then left to dry in the air. The purity was determined using powder X-ray diffraction (PXRD). Procedure for MIL-103 Eu-Insertion: the synthesised MIL-103 (120 mg) was weighed into a vial and dried in the oven at 130 °C for 3 h. Meanwhile, a 1.0 M solution of $EuCl_3 \cdot 6H_2O$ was prepared by placing $EuCl_3 \cdot 6H_2O$ (3.66 g, 1 mmol) in 10 mL of water. The 10 mL solution was added to the dried MIL-103 in the vial and allowed to soak for 24 h. The solution was then removed using centrifugation and the sample washed with 3×10 mL of water, where centrifugation was used to remove the water after each washing. The sample was then placed in the oven at 150 °C for 12 h to dry. To determine if the europium insertion had been successful, a luminescence spectrum was recorded and the sample was examined under the UV lamp.

3.2. Isoprene Polymerisation

The MOF (pre-catalyst) was ground and dried in the oven prior to use to optimize the catalytic activity. The polymerizations were completed in duplicate to ensure reliable results. The polymerization tubes, stirrer bars and MOF were inserted into the Jacomex glove box via the antechamber. The general equivalents of reagents used for the polymerization were as follows: 1 Ln:100 MMAO:500 Isoprene, with $V_{isoprene} = V_{toluene}$. The required mass of MOF was weighed using the balance inside the glove box and was then placed within the polymerisation tube. Next, the toluene (1 mL, 9 mmol) was added using a syringe, followed by the MMAO (1 mL, 1.84 mmol). The sample was then left to stir for 20 min. After this the monomer, isoprene (1 mL, 10 mmol) was added. The lid was placed on the tubes using grease to ensure a tight seal and the tubes were removed from the glove box. The tubes were placed in an oil bath, with stirring, set at the correct temperature (50 °C) and left until the solution increased in viscosity. The polymerization was quenched by toluene containing H^+ ions, which came from diluted hydrochloric acid. Toluene was added to the quenched polymer until the solution became fluid enough to pipette. The polymer solution was then pipetted drop-wise into a stirring beaker of methanol to unravel the chains of polymer. The methanol contained the anti-oxidizing agent 2,6-di-tert-butyl-4-methylphenol as a stabilizer which helps prevent oxidative and thermal degradation of the polymer over time. The methanol was decanted and the polymer collected and dried under air.

3.3. MOF Characterization

3.3.1. Powder X-Ray Diffraction (PXRD)

The purity of the MOFs was determined by powder X-ray diffraction (PXRD). X-ray diffraction is an analysis method, which uses Bragg's equation to determine the scattering of radiation from crystals.

A small quantity of MOF was placed onto the center of a sample holder and flattened using a glass slide until securely held in place. The samples were analyzed by a Siemens D5000 diffractometer (Siemens AG, Munich, Germany) working at the CuKα radiation in the θ–2θ mode. The selected parameters were a scanning range of 5°–50°, a step of 0.6° and a time of 0.2 s/step. The results were compared with computer-generated references.

3.3.2. Scanning Electron Microscopy (SEM)

SEM images were used to determine the size and shape of the crystals within the MOF powder. A small quantity of crystals were placed on a sample holder and sprayed with gold. Gold is a conductive metal and is used to prevent the collecting of electrons on the non-conductive sample, which could affect the electron beam and result in blurred images. The samples were analyzed using a Hitachi S-3400N scanning electron microscope (Hitachi High-Technologies, Krefeld, Germany).

3.3.3. ICP-AES

Inductively coupled plasma atomic emission spectroscopy (ICP-AES, Vista-Pro VARIAN, Agilent tech, Santa Clara, CA, USA) was used to accurately determine the concentrations of the two lanthanides within the final MOF structure.

The ICP-AES samples were prepared by initially creating the oxide of the MOF, this was achieved by heating the sample up to 700 °C. The oxide (3 mg) was then dissolved in concentrated nitric acid (1 mL, 16 M, 16 mmol). The sample was diluted by a factor of 500 by placing 0.2 mL into a 100 mL volumetric flask and filling with water. A small part of the solution was removed for the analysis (*ca.* 10 mL).

3.4. Polymer Material Characterization

3.4.1. Polymer Yield

The final mass of polyisoprene obtained and the volume of isoprene used in the polymer synthesis were used to determine the final yield of polymer. The calculation was completed as follows:

$$\% \; YIELD = (MASS \; OF \; POLYMER \div MASS \; OF \; MONOMER) \times 100 \tag{1}$$

3.4.2. NMR Analysis

The selectivity of the polymer was determined by ^1H NMR recorded on an AC 300 Bruker (Bruker Biospin, Wissembourg, France) at 300 MHz, with assignation as published [21]. Five to ten mg of polymer was placed into an NMR tube and approximately 0.6 mL of the solvent, chloroform-D, was added. When necessary, ^{13}C NMR analysis with quantitative Bruker sequences was performed.

3.4.3. Luminescence

There were two methods used for determining the luminescent properties of the polymer. The first was the use of a UV lamp in a dark environment to observe any colour emission. The second used a SAFAS Xenius XC spectrofluorimeter (SAFAS, Monaco) with a xenon lamp to record the emission spectrum of an excited compound. Europium-based compounds were excited at 394 nm and the emission spectrum was recorded between 560 and 720 nm. Terbium based compounds were excited at 283 nm and the emission spectrum recorded between 440 and 660 nm.

3.4.4. Preparation of Polymer Films–Spin Coater

Before examining the polymers by SEM, they were prepared into films using a SPIN150 Wafer Spinner (SPS, Antwerp, Belgium) to provide clearer images of any MOF dispersion within the matrix. A small quantity of polymer (≈50 mg) was dissolved in toluene (≈5 mL) to produce a viscous solution. Five drops of the sample were placed onto the center of a glass slide, which was placed onto the center of the spin coater. The vacuum was activated to keep the slide in place, the lid closed and the spin coater started. The result was a thin circular polymer film on a glass slide.

4. Conclusions

Lanthanide based-MOF compounds having luminescent properties and polymerization catalysis capability were prepared and assessed to the stereoregular polymerization of isoprene. This was supposed to allow for any properties exhibited from the lanthanide-based MOF, to be transferred to the resulting polymer material, and hence give an elastomer composite having luminescent properties. The results of the polymerizations showed that the combination of the two MOF-like compounds, Nd(form)$_3$ + Ln(form)$_3$, when added together at the beginning of the polymerization, successfully produced luminescent *cis*-1,4 polyisoprenes. Two alternative strategies involving mixed (Nd, Eu) and (Nd, Tb) lanthanide MOFs and porous MIL-103(Nd) with Eu included, as a pre-catalysts for isoprene polymerization, produced polyisoprene with good *cis*-selectivity, but they failed to afford a luminescent material due to quenching phenomena.

Supplementary Materials: Supplementary materials can be found at http://www.mdpi.com/2304-6740/3/4/0467/s1.

Acknowledgments: The authors thank the Region Nord-Pas-De-Calais-Picardie and the Fédération Chevreul FR 2638 for financial support (MULTI-NANO-MOF project) and Hao Wu (ISP Master Student) for his participation to this work. Erasmus Program is also acknowledged for funding to Samantha Russell.

Author Contributions: Samantha Russell: MOFs syntheses and polymerizations, SEM, PXRD, luminescence measurements; Thierry Loiseau: discussion of the results; Christophe Volkringer: MOFs syntheses, PXRD measurements, SEM characterization, luminescence, discussion of the results; Marc Visseaux: Writing the manuscript and discussion of the results, NMR measurements.

Conflicts of Interest: The authors declare no conflict of interest.

References

1. Ferey, G. Hybrid porous solids: Past, present, future. *Chem. Soc. Rev.* **2008**, *37*, 191–214. [CrossRef] [PubMed]
2. Special Issue—Metal-Organic Frameworks. Available online: http://pubs.acs.org/toc/chreay/112/2 (accessed on 5 August 2015).
3. Binnemans, K. Lanthanide-based luminescent hybrid materials. *Chem. Rev.* **2009**, *109*, 4283–4374. [CrossRef] [PubMed]
4. Cui, Y.; Yue, Y.; Qian, G.; Chen, B. Luminescent functional metal-organic frameworks. *Chem. Rev.* **2012**, *112*, 1126–1162. [CrossRef] [PubMed]
5. Vitorino, M.; Devic, T.; Tromp, M.; Ferey, G.; Visseaux, M. Lanthanide metal-organic frameworks as Ziegler–Natta catalysts for the selective polymerisation of isoprene. *Macromol. Chem. Phys.* **2009**, *210*, 1923–1932. [CrossRef]

6. Rodrigues, I.; Mihalcea, I.; Volkringer, C.; Loiseau, T.; Visseaux, M. Water-free neodymium 2,6-naphthalenedicarboxylates coordination complexes and their application as catalysts for isoprene polymerisation. *Inorg. Chem.* **2012**, *51*, 483–490. [CrossRef] [PubMed]

7. Escribano, P.; Julian-Lopez, B.; Planelles-Arago, J.; Cordoncillo, E.; Viana, B.; Sanchez, C. Photonic and nanobiophotonic properties of luminescent lanthanide-doped hybrid organic–inorganic materials. *J. Mater. Chem.* **2008**, *18*, 23–40. [CrossRef]

8. Hou, S.; Chan, W.K. Preparation of functionalized polystyrene-block-polyisoprene copolymers and their luminescence properties. *Macromolecules* **2002**, *35*, 850–856. [CrossRef]

9. Otsuka, T.; Chujo, Y. Highly stabilized luminescent polymer nanocomposites: Fluorescence emission from metal quinolate complexes with inorganic nanocrystals. *J. Mater. Chem.* **2010**, *20*, 10688–10695. [CrossRef]

10. Yang, C.; Sun, Z.; Liu, L.; Zhang, L. Preparation and luminescence performance of rare earth agriculture-used light transformation composites. *J. Mater. Sci.* **2008**, *43*, 1681–1687. [CrossRef]

11. Liu, F.; Carlos, L.D.; Ferreira, R.A.S.; Rocha, J.; Ferro, M.C.; Tourrette, A.; Quignard, F.; Robitzer, M. Synthesis, texture, and photoluminescence of lanthanide-containing chitosan-silica hybrids. *J. Phys. Chem. B* **2010**, *114*, 77–83. [CrossRef] [PubMed]

12. Agostini, G. Self-luminescent pneumatic tire. U.S. Patent S7234498 B2, 26 June 2004.

13. Zuo, Y.; Lu, H.; Xue, L.; Wang, X.; Wu, L.; Feng, S. Polysiloxane-based luminescent elastomers prepared by thiol-ene "click" chemistry. *Chem. Eur. J.* **2014**, *20*, 12924–12932. [CrossRef] [PubMed]

14. Bao, S.; Li, J.; Lee, K.I.; Shao, S.; Hao, J.; Fei, B.; Xin, J.H. Reversible mechanochromism of a luminescent elastomer. *Appl. Mater. Interfaces* **2013**, *5*, 4625–4631. [CrossRef] [PubMed]

15. Devic, T.; Wagner, V.; Guillou, N.; Vimont, A.; Haouas, M.; Pascolini, M.; Serre, C.; Marrot, J.; Daturi, M.; Taulelle, F.; *et al.* Synthesis and characterization of a series of porous lanthanide tricarboxylates. *Microporous Mesoporous Mater.* **2011**, *140*, 25–33. [CrossRef]

16. An, J.; Shade, C.M.; Chengelis-Czegan, D.A.; Petoud, S.; Rosi, N.L. Zinc-Adeninate metal-organic framework for aqueous encapsulation and sensitization of near-infrared and visible emitting lanthanide cations. *J. Am. Chem. Soc.* **2011**, *133*, 1220–1223. [CrossRef] [PubMed]

17. Friebe, L.; Nuyken, O.; Obrecht, W. Neodymium-based Ziegler/Natta catalysts and their application in diene polymerisation. *Adv. Polym. Sci.* **2006**, *204*, 1–154.

18. Kautsky, H. Quenching of luminescence by oxygen. *Trans. Faraday Soc.* **1939**, *35*, 216–219. [CrossRef]

19. Uemura, T.; Hiramatsu, D.; Kubota, Y.; Takata, M.; Kitagawa, S. Topotactic linear radical polymerization of divinylbenzenes in porous coordination polymers. *Angew. Chem. Int. Ed.* **2007**, *46*, 4987–4990. [CrossRef] [PubMed]

20. Lin, J.M.; Guan, Y.F.; Wang, D.Y.; Dong, W.; Wang, X.T.; Gao, S. Syntheses, structures and properties of seven isomorphous 1D Ln^{3+} complexes $Ln(BTA)(HCOO)(H_2O)_3$ (H2BTA = bis(tetrazoly)amine, Ln = Pr, Gd, Eu, Tb, Dy, Er, Yb) and two 3D Ln^{3+} complexes $Ln(HCOO)_3$ (Ln = Pr, Nd). *Dalton Trans.* **2008**, 6165–6169. [CrossRef] [PubMed]

21. Martins, N.; Bonnet, F.; Visseaux, M. Highly efficient *cis*-1,4 polymerisation of isoprene using simple homoleptic amido rare earth-based catalysts. *Polymer* **2014**, *55*, 5013–5016. [CrossRef]

![inorganics logo] *inorganics*

MDPI

Article

Assessing Covalency in Cerium and Uranium Hexachlorides: A Correlated Wavefunction and Density Functional Theory Study

Reece Beekmeyer [1] and Andrew Kerridge [1,2,*]

[1]   Department of Chemistry, University College London, 20 Gordon Street, London WC1H 0AJ, UK;
      reece.beekmeyer.13@ucl.ac.uk
[2]   Department of Chemistry, Lancaster University, Lancaster LA1 4YW, UK
*   Correspondence: a.kerridge@lancaster.ac.uk; Tel.: +441-524-594-770

Academic Editors: Stephen Mansell and Steve Liddle
Received: 14 September 2015; Accepted: 30 October 2015; Published: 9 November 2015

Abstract: The electronic structure of a series of uranium and cerium hexachlorides in a variety of oxidation states was evaluated at both the correlated wavefunction and density functional (DFT) levels of theory. Following recent experimental observations of covalency in tetravalent cerium hexachlorides, bonding character was studied using topological and integrated analysis based on the quantum theory of atoms in molecules (QTAIM). This analysis revealed that M–Cl covalency was strongly dependent on oxidation state, with greater covalency found in higher oxidation state complexes. Comparison of M–Cl delocalisation indices revealed a discrepancy between correlated wavefunction and DFT-derived values. Decomposition of these delocalisation indices demonstrated that the origin of this discrepancy lay in *ungerade* contributions associated with the f-manifold which we suggest is due to self-interaction error inherent to DFT-based methods. By all measures used in this study, extremely similar levels of covalency between complexes of U and Ce in the same oxidation state was found.

Keywords: covalency; cerium; uranium; CASSCF; electron density; QTAIM; DFT

1. Introduction

The question of covalency in f-element bonding is challenging to both experimentalists and theorists alike. Complexes of the f-elements typically exhibit strong relativistic effects, substantial dynamical electron correlation and weak crystal fields. These phenomena result in highly complicated electronic structures and, as such, theoretical measures of covalency based on different premises can lead to qualitatively different conclusions [1]: In particular, the strong deviation from an independent particle approximation in these strongly-correlated systems can lead to consistent, but apparently contradictory, orbital-based descriptions of the electronic structure [2–5]. This ambiguity can be avoided by instead turning to analytical methods based on the experimentally observable electron density. Such approaches are appealing since they are not directly influenced by the theoretical methodology employed in the obtention of the density to be analysed: in fact, such analyses can be applied to experimental densities derived from low temperature X-ray diffraction (XRD) data [6]. The most popular of these density-based approaches is the Quantum Theory of Atoms in Molecules (QTAIM) approach of Bader [7], which provides the theoretical framework for an unambiguous and transferable analytical tool with which to probe the nature of electronic structure and bonding in chemical systems. This approach has been applied to several problems in f-element chemistry [4,6,8–16] and is able to provide quantitative measures of covalency via both topological and integrated properties of the electron-density.

Recently, X-ray absorption spectroscopy (XAS) has emerged as an extremely powerful experimental technique for the characterisation of bonding in organometallic and inorganic complexes of the f-elements [17–19]. Intriguingly, this approach has provided compelling evidence of covalent interactions in Ce(IV) hexachloride [20]. Furthermore, the degree of covalency has been shown to be comparable to that found in the uranium analogue, although the relative contribution from d- and f-shells differs. These results motivated us to perform a theoretical study of the bonding in uranium and cerium hexachlorides, combining state of the art multiconfigurational quantum chemical simulations with the density-based analyses discussed above. These simulations consider a variety of oxidation states and support the experimentally-based assertion of non-negligible covalent character, while highlighting the sensitivity of this phenomenon to the quantum chemical methodology employed.

2. Results

2.1. Structural Characterisation

Structural optimisation of the MCl_6^{n-} complexes was performed numerically at the DFT level of theory using the PBE and B3LYP functionals, as well as using the Restricted Active Space Self Consistent Field with 2nd Order Perturbation Theory (RASPT2) methodology. For the formally closed shell U(VI) and Ce(IV) complexes, a ^1A_g ground state was assumed, whereas for open shell systems the lowest energy state of each irreducible representation was calculated in order to identify the ground state. Optimisations were performed by varying the M–Cl bond length in increments of 0.01 Å until an energetic minimum was obtained. Table 1 reports the ground state symmetries and spin-multiplicities, along with optimal bond-lengths, for each level of theory. For comparison, literature data of experimentally derived bond lengths are also included. Where experimental data exists [20,21], M–Cl bond lengths appear to be ~0.02 Å longer in the aqueous phase than in the solid state and, bearing in mind that a continuum solvation model is employed here, it might be expected that theoretical results would be in better agreement with aqueous phase data.

Table 1. Ground states and calculated optimal bond-lengths for all complexes considered in this study.

Complex	State	$r_{M–Cl}$ (Å)				
		CASPT2	RASPT2	B3LYP	PBE	Experimental
$[U(VI)Cl_6]$	^1A_g	2.45	2.48	2.47	2.48	2.47 [a]
$[U(V)Cl_6]^-$	^2A_u	2.53	2.54	2.55	2.55	2.52 [b]
$[U(IV)Cl_6]^{2-}$	$^3B_{1g}/^3B_{2g}/^3B_{3g}$	2.65	2.69	2.66	2.65	2.62 [c], 2.65 [d]
$[U(III)Cl_6]^{3-}$	^4A_u	2.86	2.89	2.83	2.80	-
$[Ce(IV)Cl_6]^{2-}$	^1A_g	2.62	2.64	2.67	2.65	2.60 [e], 2.62 [f]
$[Ce(III)Cl_6]^{3-}$	$^2B_{1u}/^2B_{2u}/^2B_{3u}$	2.85	2.86	2.82	2.79	2.77 [e], 2.79 [f]

[a] XRD [22]; [b] XRD [23]; [c] XRD [21]; [d] EXAFS [21]; [e] XRD [20]; [f] EXAFS [20].

All methodologies reproduce XRD-derived bond lengths with high accuracy in the closed shell U(VI) complex, whereas there is slight overestimation in the U(V) complex. This overestimation is also present in the RASPT2 bond lengths of the U(IV) complex, although CASPT2 results are in excellent agreement with both DFT and experiment. To the authors' knowledge, no experimental data exists for the U(III) complex, but the U(III) ionic radii is some 0.14 Å greater than that of U(IV), in line with the increased bond lengths found theoretically. The significantly longer U(III)–Cl bond found at the RASPT2 level may indicate that the large active space employed has reduced the effect of the PT2 correction, which would normally correct for the bond length overestimation found in the absence of dynamical correlation. Again, CASPT2 gives a shorter U–Cl bond.

When considering the closed shell Ce(IV) complex, CASPT2 best reproduces the EXAFS-derived bond lengths but both wavefunction-based methods overestimate those of the Ce(III) complex, whereas

DFT simulations, particularly those employing the PBE exchange-correlation functional, give much better agreement with experiment.

In order to validate the numerical optimisations performed here, analytical optimisations were performed for all systems at the B3LYP level of theory using the TURBOMOLE quantum chemistry code. In all cases agreement was obtained to within 0.02 Å, with the greatest deviation occurring for the more highly charged systems. It can be seen in Table 1 that deviation from experiment is greatest when the overall charge of the system is high. Gas phase optimisations were also performed (See Table S1 of electronic supporting information (ESI)) and showed that the presence of the continuum solvation model leads to a significant reduction in bond lengths, particularly for more highly charged systems.

2.2. Natural Orbital Occupancies

In this section the natural orbital [24] occupancies (NOOs) of active space orbitals in each of the systems studied are considered. The natural orbitals (NOs) provide a basis for the most compact CI expansion of the exact wavefunction and their associated occupation numbers therefore provide a measure of multiconfigurational character [25]. Typically, this multiconfigurational character manifests itself in terms of "strongly occupied" orbitals with NOOs close to two and "weakly occupied" correlating orbitals with NOOs close to zero. Often the strongly and weakly occupied orbitals have bonding and antibonding character, respectively. Another potential origin of multiconfigurational character can manifest itself in two or more singly-occupied metal-based orbitals having NOOs deviating significantly from one due to the fact that configurations in which different combinations of these orbitals are occupied are near-degenerate.

Of the seven orbitals available for f-electron occupation, three have formally σ-antibonding character, three δ-antibonding character and one non-bonding character, although it should be borne in mind that the degree of interaction with the ligands may be minimal: hereafter, these orbitals will be labelled as f_σ, f_δ, and f_{NB}, respectively. Table 2 summarises the natural occupancies of these f-orbitals for each system under consideration.

Table 2. f-Orbital occupations for each complex considered in this study, calculated at the Complete/Restricted Active Space Self-Consistent Field (CAS/RASSCF) level of theory.

Complex	f-Orbital Occupation	
	CASSCF	RASSCF
$[U(VI)Cl_6]$	f^0	f^0
$[U(V)Cl_6]^-$	$0.996\ f^1_{NB}$	$1.001\ f^1_{NB}$
$[U(IV)Cl_6]^{2-}$	$0.998\ f^1_{NB}\ 0.998\ f^1_\sigma$	$1.000\ f^1_{NB}\ 1.000\ f^1_\sigma$
$[U(III)Cl_6]^{3-}$	$0.915\ f^3_\delta\ (0.091\ f^1_\sigma)$	$0.999\ f^3_\delta\ (0.007\ f^1_\sigma)$
$[Ce(IV)Cl_6]^{2-}$	f^0	f^0
$[Ce(III)Cl_6]^{3-}$	$0.999\ f^1_\delta$	$1.000\ f^1_\delta$

Typically, f-orbital occupation is uncomplicated, with NOOs close to unity in all cases, however the CASSCF simulation of the U(III) complex presents deviation from this behaviour. Whilst to a first degree of approximation, the system would be described as having a $5f^3_\delta$ subconfiguration, the NOO for each orbital is just 0.915. Correspondingly, each of the three f_σ orbitals have NOOs of 0.091. Occupation numbers deviating from integer values by 0.1 or greater are indicative of considerable multiconfigurational character [25] and so, by this definition, the U(III) system should be considered as multiconfigurational.

Table 3 summarises the NOOs of orbitals with either M–Cl σ- or δ-bonding/antibonding character at the both the RASSCF and CASSCF levels of theory with active spaces as defined in the computational details. Here, a different perspective on the multiconfigurational character of these systems can be found. A trend for decreasing multiconfigurational character with decreasing oxidation state can be seen in the uranium complexes, this being most pronounced at the CASSCF level of theory. This multiconfigurational character is reasonably pronounced in the U(VI) complex, particularly amongst the σ-bonding orbitals (and their weakly occupied, correlating, antibonding orbitals) but is almost entirely absent once the oxidation state is lowered to +4. The degree of multiconfigurational character in the cerium complexes is negligible. The absence of multiconfigurational character in both the U and Ce complexes with oxidation states +4 and lower may provide another source for the relative overestimation of bond lengths derived from the wavefunction-based approach: in these systems, the lack of static correlation means that the RASSCF/CASSCF calculations reduce to little more than Hartree-Fock (HF) calculations: HF theory is known to underbind molecular systems [26], although the perturbational treatment included in the optimisation partially corrects for this.

Table 3. RASSCF and CASSCF natural orbital occupancies of bonding (antibonding) orbitals in metal hexachlorides. All values are averaged of three equivalent orbitals except where f-electron occupation breaks this equivalency. [a] average taken over two weakly occupied orbitals; [b] no weakly unoccupied orbitals present.

Complex	σ		δ	
	CASSCF	RASSCF	CASSCF	RASSCF
$[U(VI)Cl_6]$	1.938 (0.061)	1.997 (0.009)	1.973 (0.033)	1.999 (0.007)
$[U(V)Cl_6]^-$	1.973 (0.030)	2.000 (0.007)	1.993 (0.012)	2.000 (0.007)
$[U(IV)Cl_6]^{2-}$	1.999 (0.008)	2.000 (0.007)	2.000 (0.006 [a])	2.000 (0.006)
$[U(III)Cl_6]^{3-}$	2.000 (0.091)	2.000 (0.007)	2.000 [b]	2.000 (0.999)
$[Ce(IV)Cl_6]^{2-}$	1.999 (0.004)	2.000 (0.007)	2.000 (0.004)	2.000 (0.007)
$[Ce(III)Cl_6]^{3-}$	2.000 (0.004 [a])	2.000 (0.007)	2.000 (0.003)	2.000 (0.007)

Visual inspection of the σ- and δ-bonding NOs (Figure 1) reveals increasing localisation on the chloride ligand set as the metal oxidation state lowers, which may correspond to a commensurate reduction in covalent character. U(VI) exhibits significant f-orbital contributions to both the σ- and δ-bonding NOs. This 5f character, though reduced, is still visible even for the σ-bonding NOs of the U(III) complex, whereas the corresponding δ-bonding orbitals rapidly localise on the ligands. This suggests more pronounced covalency in the σ-type interactions.

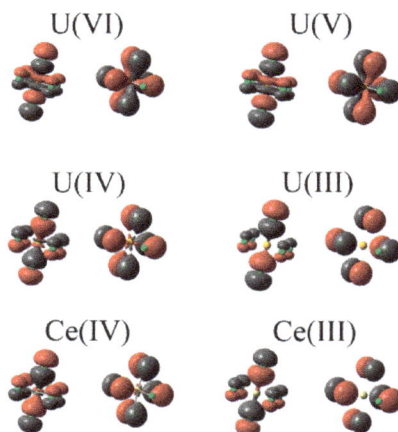

Figure 1. CASSCF-calculated σ- and δ-bonding natural orbitals for each system considered in this study. All orbitals rendered at an isosurface of 0.02 a.u.

Comparing the uranium and cerium complexes, Figure 1 reveals extremely similar orbital characteristics, suggesting that Ce covalency may be comparable to that of U in these systems. Both M(IV) complexes show significant f-character in the σ-type orbitals, whereas this character, whilst still present, is much less pronounced in the δ-type orbitals. f-character is almost completely absent from the corresponding M(III) orbitals.

2.3. Analysis of the Electron Density

2.3.1. QTAIM Derived Topological Properties

In order to further investigate the covalent character of M–Cl in these systems, the quantum theory of atoms in molecules (QTAIM) was employed. QTAIM is able to provide a robust and quantitative measure of the covalent contribution to bonding in the f-element complexes under consideration here. Table 4 summarises the key topological properties of the electron density at the M–Cl bond critical point (BCP). A key parameter is the magnitude of the electron density at the M–Cl BCP (ρ_{BCP}): this metric has been used extensively in the quantification of covalency, and provides a measure of the accumulation of electron density in the chemical bonding region. Broadly speaking, values of ρ_{BCP} greater than 0.2 are indicative of covalent interactions, while values lower than 0.1 indicate predominant ionic character. The variation of this metric can therefore provide information regarding the variation of covalency in the systems under consideration.

Table 4. Values of the electron density (ρ) and its Laplacian ($\nabla^2\rho$) at the M–Cl critical point of metal hexachlorides, evaluated at theoretically optimised geometries. All values are in a.u.

System	ρ_{BCP}				$\nabla^2\rho_{BCP}$			
	CASPT2	RASPT2	B3LYP	PBE	CASPT2	RASPT2	B3LYP	PBE
[UCl$_6$]	0.105	0.102	0.099	0.096	+0.148	+0.126	+0.156	+0.159
[UCl$_6$]$^-$	0.086	0.084	0.082	0.081	+0.148	+0.150	+0.145	+0.146
[UCl$_6$]$^{2-}$	0.064	0.059	0.063	0.065	+0.154	+0.144	+0.130	+0.140
[UCl$_6$]$^{3-}$	0.038	0.037	0.041	0.044	+0.121	+0.111	+0.119	+0.124
[CeCl$_6$]$^{2-}$	0.064	0.061	0.058	0.060	+0.144	+0.141	+0.122	+0.128
[CeCl$_6$]$^{3-}$	0.037	0.036	0.040	0.044	+0.105	+0.104	+0.101	+0.107

The low values of ρ_{BCP} reported in Table 4 demonstrate that, unsurprisingly, M–Cl bonds in all complexes should be considered primarily ionic in character. This assertion is supported by the values of $\nabla^2\rho_{BCP}$, which are expected to be positive for predominantly ionic interactions. However, Figure 2 demonstrates that, when considering the variation in ρ_{BCP} with respect to uranium oxidation state, all methodologies employed here display a clear and common trend, namely a reduction in magnitude from U(VI) to U(III). All methodologies are in broad agreement, with the correlated wavefunction methods demonstrating the greatest variation, commensurate with the greater variation in U–Cl bond lengths at this level of theory. The reduction in covalent character is partly due to the increasing M–Cl bond length as the oxidation state is lowered, but Figure 1 implies that this reduction may also be due to increased energetic mismatch between metal and ligand orbitals in lower oxidation state complexes.

Figure 3 compares the magnitude of ρ at the M–Cl BCP for uranium and cerium complexes in the same oxidation state. Remarkably, Ce values are almost identical to those of the U complexes, and with the same decrease when moving from the M(IV) to M(III) oxidation state. These data are a clear indicator of uranium-like levels of covalency in an analogous cerium complex. In previous QTAIM studies of Ce and U complexes [4,9,12] cerium covalency has always been noticeably lower, with ρ_{BCP} values approximately 80% of those obtained for analogous uranium complexes. For the systems under consideration here, Ce–Cl ρ_{BCP} values are, on average, 96.6% of those obtained for the analogous U complexes.

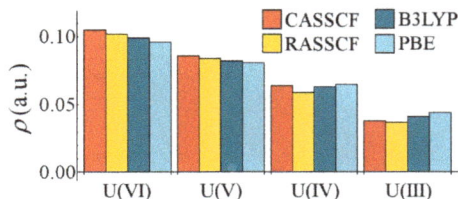

Figure 2. Variation of ρ at the U–Cl bond critical point as a function of oxidation state.

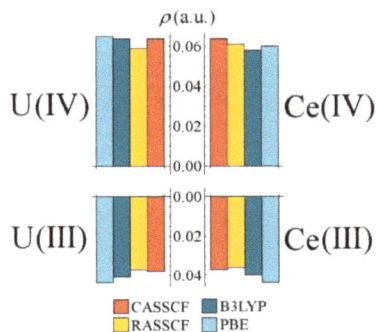

Figure 3. Comparison of ρ at the M–Cl bond critical point for tri- and tetravalent cerium and uranium complexes.

2.3.2. QTAIM Derived Integrated Properties

In addition to the topological properties of the electron density considered in the previous section, integrated properties can also be evaluated. Of interest here is the atomic electron population, a one-electron property obtained by integrating the electron density over a given atomic basin, and from which atomic charges can be derived. In addition to this, a pair of related two-electron properties, the localisation (λ) and delocalisation (δ) indices, can also be analysed in order to give detailed information regarding the nature and magnitude of bonding interactions: our previous research has identified a strong relationship between localisation index and oxidation state [4,12], while the delocalisation index, which quantifies the degree of electron sharing between a pair of atoms [27,28] can be considered as an alternative measure of covalency to ρ_{BCP} [4,12,15,27], providing data regarding electron delocalisation between atomic centres, which may occur independently of charge accumulation in the bonding region.

Table 5 summarises the QTAIM derived atomic charges and reveals that metal charges obtained from correlated wavefunction calculations are noticeably higher than those calculated using DFT. Although it is tempting to interpret this as being indicative of increased ionic character when employing wavefunction based methodologies, this is in contradiction to the topological data of Table 4, which demonstrates that all methodologies predict similar degrees of covalent character. This issue will be revisited in light of integrated two-electron data. There is a clear trend in all methodologies of increased electronic charge located on the Cl ion as the oxidation state is lowered, commensurate with the increased total electronic charge. A proportion of this charge, however, is found to be located on the uranium centre, whose charge decreases with decreasing oxidation state. Finally, it should be noted that, when considering equivalent oxidation states, cerium and uranium charges are very similar, again indicating similar electronic structures.

Table 5. QTAIM derived atomic charges of metal hexachlorides, evaluated at theoretically optimised geometries. All values are in a.u.

System	$q(M)$				$q(Cl)$			
	CASPT2	RASPT2	B3LYP	PBE	CASPT2	RASPT2	B3LYP	PBE
$[UCl_6]$	+3.01	+3.25	+2.55	+2.36	−0.50	−0.54	−0.42	−0.39
$[UCl_6]^-$	+3.02	+3.23	+2.52	+2.33	−0.67	−0.71	−0.59	−0.55
$[UCl_6]^{2-}$	+2.88	+2.95	+2.39	+2.24	−0.81	−0.82	−0.73	−0.71
$[UCl_6]^{3-}$	+2.38	+2.41	+2.11	+2.00	−0.90	−0.90	−0.85	−0.83
$[CeCl_6]^{2-}$	+2.81	+2.86	+2.26	+1.97	−0.80	−0.81	−0.71	−0.68
$[CeCl_6]^{3-}$	+2.39	+2.41	+2.10	+2.07	−0.90	−0.90	−0.85	−0.83

Localisation indices, $\lambda(M)$, are summarised in Table 6. Here, a trend reflecting the change in oxidation state is much more pronounced that of the atomic charges, with λ increasing significantly as the uranium oxidation state reduces from VI to III (see Figure 4). In a purely ionic system, it might be expected that $\lambda(U)$ would increase by unity for each change in oxidation state: in a previous study [12] we considered the variation in λ across the actinide series in the largely ionic $An(C_8H_8)_2$ (An = Th–Cm) and found changes close to unity. In the present work, variation in $\lambda(U)$ is less marked, implying greater covalency in the interactions.

Table 6. QTAIM derived localisation (λ) along with $Z(M)–\lambda(M)$, the number of electron donated to, or shared with, the ligand set by the metal centre evaluated at theoretically optimised geometries. Figures in parentheses are rounded to the nearest integer. All values are in a.u.

System	$\lambda(M)$				$Z(M)–\lambda(M)$			
	CASPT2	RASPT2	B3LYP	PBE	CASPT2	RASPT2	B3LYP	PBE
$[UCl_6]$	86.48	86.16	86.15	86.19	5.52 (6)	5.84 (6)	5.85 (6)	5.81 (6)
$[UCl_6]^-$	86.96	86.83	86.86	86.88	5.04 (5)	5.17 (5)	5.14 (5)	5.12 (5)
$[UCl_6]^{2-}$	87.68	87.73	87.68	87.91	4.32 (4)	4.27 (4)	4.32 (4)	4.09 (4)
$[UCl_6]^{3-}$	88.65	88.69	88.60	88.55	3.35 (3)	3.31 (3)	3.40 (3)	3.45 (3)
$[CeCl_6]^{2-}$	53.74	53.79	53.79	53.78	4.26 (4)	4.21 (4)	4.21 (4)	4.22 (4)
$[CeCl_6]^{3-}$	54.72	54.75	54.71	54.69	3.28 (3)	3.25 (3)	3.29 (3)	3.31 (3)

In order to make comparison between the cerium and uranium complexes under consideration, a more useful measure is $Z(M)–\lambda(M)$, which gives the number of electrons donated to, or shared with, the ligand set by the metal centre. We expect this measure to correlate with oxidation state, as we have previously reported [4,12]. Table 6 shows that this correlation is also present in systems considered here: rounding $Z(M)–\lambda(M)$ to the nearest integer returns the formal oxidation state for all systems at all levels of theory. The similarity between uranium and cerium complexes is again pronounced, with a maximum difference of just 0.15 a.u. and, on average, differences approximately half of this.

Finally we consider the delocalisation indices as an alternative measure of covalency. The high (O_h) symmetry of the systems (and therefore the atomic basins of the central ions: see Figure 5) under consideration here, along with the formal definition of the delocalisation index [27], allows us to decompose the total index into *gerade* and *ungerade* parity contributions. Since orbitals comprising the d-manifold have *gerade* parity and those comprising the f-manifold have *ungerade* parity, this decomposition provides a mechanism for assessing the d- and f-contributions to electron sharing. Total and decomposed delocalisation indices are given in Table 7 and data corresponding to uranium complexes is visualised in Figure 6.

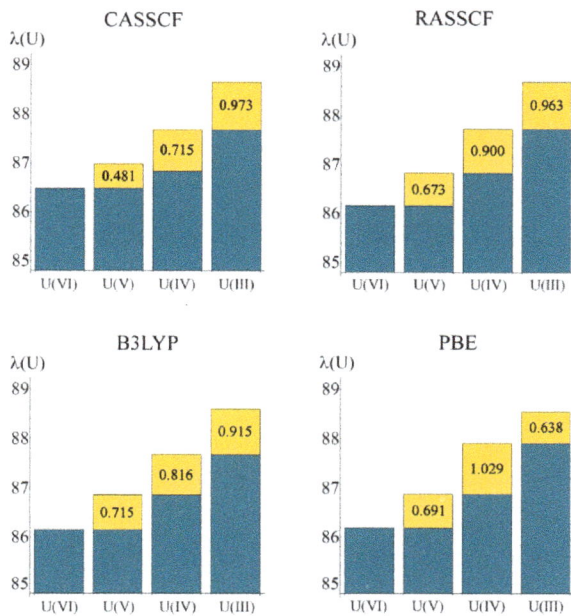

Figure 4. Localisation index as a function of oxidation state, calculated at all levels of theory. Values in yellow boxes indicate increase in electron localisation as oxidation state is lowered. All values are in atomic units.

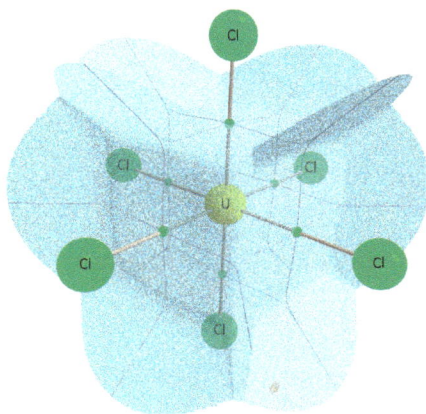

Figure 5. QTAIM calculated atomic basin of uranium in UCl_6, illustrating the O_h symmetry of the basin.

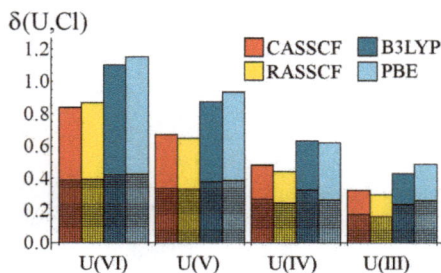

Figure 6. Variation of the U–Cl delocalisation index, δ, as a function of oxidation state. Cross-hatched regions correspond to the *gerade* contributions to δ. All values are given in a.u.

Table 7. Delocalisation indices, along with decomposed *gerade* (*g*) and *ungerade* (*u*) contributions, evaluated at theoretically optimised geometries. All values are in a.u.

System	$\delta(M,Cl)$											
	CASPT2			RASPT2			B3LYP			PBE		
	Total	*u*	*g*	Total	*u*	*g*	Total	*u*	*g*	Total	*u*	*g*
[UCl$_6$]	0.838	0.447	0.391	0.868	0.475	0.393	1.102	0.678	0.424	1.151	0.725	0.426
[UCl$_6$]$^-$	0.673	0.335	0.338	0.647	0.316	0.331	0.872	0.494	0.378	0.932	0.547	0.385
[UCl$_6$]$^{2-}$	0.482	0.210	0.272	0.442	0.197	0.245	0.633	0.307	0.326	0.619	0.353	0.266
[UCl$_6$]$^{3-}$	0.324	0.145	0.179	0.300	0.138	0.162	0.432	0.193	0.239	0.487	0.224	0.263
[CeCl$_6$]$^{2-}$	0.485	0.202	0.283	0.450	0.184	0.266	0.651	0.435	0.216	0.716	0.482	0.234
[CeCl$_6$]$^{3-}$	0.297	0.124	0.173	0.279	0.122	0.157	0.395	0.163	0.232	0.446	0.195	0.251

The delocalisation index data bears some resemblance to the ρ_{BCP} data of Figure 2, in that there is a clear reduction in the degree of electron sharing between the metal and chloride ions as the oxidation state is lowered. However, whereas quantitative agreement between correlated wavefunction and density functional methods was found with respect to ρ_{BCP}, Figure 6 reveals a significant disparity: density functional methods predict significantly larger M–Cl delocalisation indices than those obtained from correlated wavefunctions approaches. Decomposition of these indices shows that, for the U(VI) and U(V) oxidation states, this disparity is due to increased *ungerade*, e.g., f-electron, contributions, whereas for the lower U(IV) and U(III) the disparity has a growing *gerade* contribution. Kohn-Sham DFT suffers from the self-interaction problem, in which the interaction of a given electron with itself is not identically equal to zero (as is the case in correlated wavefunction approaches). This self-interaction error can lead to unexpected apparent electron delocalisation which vanishes when a correlated wavefunction approach is employed [29,30]. We therefore suggest that the origin of the enhanced delocalisation in DFT-calculated densities is due to self-interaction, a view supported by the fact that this effect is less pronounced in B3LYP-derived densities: B3LYP would be expected to be less prone to self-interaction error due to its incorporation of a component of exact exchange. We can deduce from our ρ_{BCP} data, however, that the enhanced delocalisation does not lead to an increase in electron density at the U–Cl bond-critical point.

Focusing on the correlated wavefunction data in Table 6, we find that the degree of total electron sharing in the U(III) complex is just 39% (35%) of that found in the U(VI) complex at the CASSCF (RASSCF) level. The *ungerade* contribution reduces to 32% (29%) and the *gerade* contribution reduces to 46% (41%) at the CASSCF (RASSCF) level. This shows that f-electron contributions to covalency, as quantified by this measure, drop more rapidly than those to d-electron contributions as the oxidation state is lowered.

Comparing uranium and cerium complexes, delocalisation indices largely mirror ρ_{BCP}, with all methodologies giving comparable values for the same oxidation state (see Figure 7).

Correlated wavefunction approaches, in particular, give almost identical results for the uranium and cerium complexes.

When considering *gerade* and *ungerade* contribution to δ(M,Cl), similar behaviour is again seen, with the exception being the Ce(IV) DFT-generated densities. Table 1 shows that both exchange-correlation functionals overestimate the Ce–Cl bond length in this system, which presumably accounts for the *gerade* contributions.

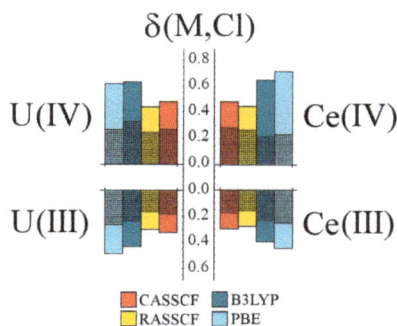

Figure 7. Comparison of M–Cl delocalisation index, δ, for tri- and tetravalent uranium and cerium hexahalides. Cross-hatched regions correspond to the *gerade* contributions to δ. All values are given in a.u.

3. Computational Details

Quantum chemical simulations were performed using version 7.6 of the MOLCAS software package [31,32] and employed ANO-RCC basis sets of polarised triple-zeta quality [30,33,34]. Scalar relativistic effects were included via use of the 2nd order Douglas-Kroll-Hess Hamiltonian [35,36]. Simulations were performed using both wavefunction- and density-based methodologies: complete/restricted active space self-consistent field (CASSCF/RASSCF) theory and density functional theory (DFT). In order to investigate the effects of including exact exchange, DFT calculations compared the GGA PBE [37] and hybrid-GGA B3LYP [38,39] exchange-correlation functionals: our group has successfully employed B3LYP in previous studies of actinide complexes [4,40–42], while the application of multiconfigurational approaches to the study of actinide complexes is well established: see [43] for a recent example.

RASSCF calculations were performed using a large active space which incorporated the chlorine 2p orbitals as well as the metal valence d- and f-orbitals. The RASSCF methodology requires the active space to be divided into three subspaces: in these calculations Cl 2p orbitals were placed in the RAS1 subspace, metal 4f/5f-orbitals in RAS2, and metal 5d/6d-orbitals in RAS3. Whilst, by definition, all configurations within the RAS2 subspace were allowed in the CI expansion of the wavefunction, only configurations involving double excitations from (to) RAS1 (RAS3) were included. Using notation defined previously [44,45] these calculations can be referred to as RASSCF(36 + n, 2, 2; 18, 7, 5) calculations, where n ranges from 0 to 3 depending on the oxidation state of the metal centre. A schematic of the orbitals included in the RASSCF and CASSCF calculations can be found in the electronic supporting information (See Figure S1 in ESI). Dynamical correlation effects were included perturbatively employing the RASPT2 approach. In these calculations the standard IPEA shift of 0.25 was applied, along with a 0.2 a.u. imaginary level shift of the energies to suppress the potential effect of weakly interacting intruder states.

In order to investigate the effects of truncation of the wavefunction implicit in the RASSCF methodology, CASSCF calculations were also performed. In these calculations it was intended that only the metal 4f/5f orbitals and those linear combinations of Cl 2p orbitals which could contribute to σ- or π-bonding interactions with them were included in the active space, however it was found that for lower oxidation state complexes it became impossible to stabilise such an active space, which was therefore enlarged to incorporate three additional orbitals which led to comparable active spaces between systems. This choice of active space resulted in a series of CASSCF $(18 + n, 16)$ calculations, where again where n ranges from 0 to 3 depending on the oxidation state of the metal centre. Dynamical correlation effects were included perturbatively.

The high (O_h) symmetry of the complexes studied allowed for numerical structural optimisation to be performed at all levels of theory: this was achieved via a series of single point energy calculations, varying the M–Cl bond length in steps of 0.01 Å. In this way, energetic minima were identified. Limitations of the code required simulations to be performed in the lower symmetry D_{2h} point group. For this reason, all irreducible representations (irreps) discussed here refer to the D_{2h} point group. All calculations employed a polarizable continuum solvent model [46,47] simulating bulk water in order to stabilise the anionic systems and provide an approximate description of the chemical environment experienced by the complexes.

In order to validate the numerical geometry optimisations performed here, version 6.4 of the TURBOMOLE quantum chemistry code [48] was used in order to perform analytical optimisations at the B3LYP level. These calculations employed Alrichs basis sets of polarised triple-zeta quality [49] (def-TZVP for U, def2-TZVP for all other atoms) along with associated relativistic effective core potentials in order to include scalar relativistic effects. We have previously shown [12,50] that the inclusion of the effects of spin-orbit coupling within the RASSI (restricted active space state interaction) framework has negligible effects on bond lengths and therefore the nature of the metal ligand interaction. Since this study is focused on the characterisation of M–Cl bonds, such effects have been omitted. Solvation was included via the COSMO continuum solvent model [51].

4. Summary and Conclusions

In this study we have optimised both Ce and U hexachlorides in a variety of oxidation states at the CASPT2, RASPT2 and density functional levels of theory, employing a continuum solvation model to represent environmental effects. These simulations were found to be in good agreement with experimentally determined structures.

The degree of multiconfigurational character exhibited by these systems was explored by examination of natural orbital occupancies (NOOs). Whilst there was very little deviation from integer occupation in σ- and δ-type bonding/antibonding orbitals using the RASSCF methodology, CASSCF electronic structures displayed a degree of multiconfigurational character for higher oxidation states (U(VI), U(V)), particularly with respect to the σ-bonds. This multiconfigurational character effectively vanishes for lower oxidation states (M(IV), M(III)). U(III) provides an interesting case, where multiconfigurational character manifests itself not in the M–Cl bonds, but within the f-manifold itself: f_δ orbitals have NOOs of 0.913 and f_σ orbitals have NOOs of 0.091. This is presumably due to a weakening of the ligand field, which results in near degeneracy with the f-manifold.

Visual inspection of the CASSCF natural orbitals revealed a trend in the localisation of the σ- and δ- bonding orbitals, revealing an increase in the localisation of the orbitals on the chloride ligands with decreased oxidation state, which may indicate a decrease in the covalency of these bonds. There is also a remarkable similarity between the U(IV) and Ce(IV), and U(III) and Ce(III) orbitals, indicating a comparable level of covalency between U and Ce complexes. Analysis based on molecular orbitals, however, should be treated with caution: unitary transformations of the orbital set can have profound effects on apparent delocalisation, whilst leaving the (experimentally observable) electron density unchanged. Employing density-based analysis methods, such as QTAIM, instead, allows for a robust, unambiguous and quantitative characterisation of bonding. In this case, the similarity between the U and Ce compounds is supported by the QTAIM data where the magnitude of the electron density at the bond critical point (ρ_{BCP}), an accepted measure of covalency, is 0.064 a.u. for both U(IV) and Ce(IV), reducing to 0.038 a.u. and 0.037 a.u. for U(III) and Ce(III) respectively at the CASSCF level:

similar agreement was found for other methodologies. The calculated ρ_{BCP} values also demonstrated an increase in covalency with increased oxidation state for all systems studied.

The delocalisation indices, $\delta(M,Cl)$, provide an alternative measure of covalency, one more closely associated with orbital mixing. Here, $\delta(M,Cl)$, gave a broadly similar picture to that obtained from ρ_{BCP}, namely greater covalency in higher oxidation state systems, and extremely similar covalency between complexes of U and Ce in the same oxidation state. However, there was a noticeable difference between correlated wavefunction- and DFT-derived densities: the latter were consistently higher than the former. To obtain further insight into the origin of this difference, advantage was taken of the high symmetry of the hexachlorides, which allowed decomposition of $\delta(M,Cl)$ into *gerade* and *ungerade* contributions, which can be broadly associated with the d- and f-orbital manifolds, respectively. This decomposition revealed that the increased magnitude of $\delta(M,Cl)$ obtained from DFT-derived densities has its origin in the *ungerade* contribution. We suggest that this is an overestimation, due to the self-interaction error present in DFT-based methodologies. However, this overestimation of electron sharing does not manifest itself as an increase in density at the QTAIM-derived M–Cl bond critical point.

Recent DFT-based comparisons of trivalent actinide and lanthanide bonding [52,53] have demonstrated greater f-orbital contributions to bonding in the former, with Ce nonetheless having enhanced covalent character when compared to other lanthanides. The findings of the present study show that, in agreement with experimental observations, there are chemical environments in which Ce can adopt even greater actinide-like bonding character.

In summary, recent XAS studies have identified unexpected evidence of U(IV) levels of covalency in Ce(IV) compounds, with this covalency having its origins in both 4f and 5d orbital contributions. All theoretical methods used in this study support this notion, without having to resort to orbital-based analysis techniques. Correlated wavefunction methods, combined with density-based analysis, provide the most robust description of electronic structure and bonding in these f-element hexahalides.

Supplementary Materials: Supplementary materials can be found at http://www.mdpi.com/2304-6740/3/4/0482/s1.

Acknowledgments: Andrew Kerridge thanks the Engineering and Physical Sciences Research Council (EPSRC) for the award of a career acceleration fellowship (grant EP/J002208/1) and Reece Beekmeyer thanks University College London (UCL) for the award of IMPACT studentship. We thank the National service for Computational Chemistry Software (NSCCS) for access to the "slater" High performance Computing (HPC) facility and Lancaster University for access to the High End Computing (HEC) HPC facility. We also thank Nik Kaltsoyannis and Michael Patzschke for helpful discussions.

Author Contributions: Andrew Kerridge conceived and designed the simulations; Reece Beekmeyer performed the simulations; Andrew Kerridge and Reece Beekmeyer analysed the data; Andrew Kerridge and Reece Beekmeyer wrote the manuscript.

Conflicts of Interest: The authors declare no conflict of interest.

References

1. Kaltsoyannis, N. Does covalency increase or decrease across the actinide series? Implications for minor actinide partitioning. *Inorg. Chem.* **2013**, *52*, 3407–3413. [CrossRef] [PubMed]

2. Dolg, M.; Fulde, P.; Küchle, W.; Neumann, C.S.; Stoll, H. Ground state calculations of di-π-cyclooctatetraene cerium. *J. Chem. Phys.* **1991**, *94*, 3011. [CrossRef]

3. Kerridge, A.; Coates, R.; Kaltsoyannis, N. Is cerocene really a Ce(III) compound? All-electron spin-orbit coupled CASPT2 calculations on $M(\eta^8-C_8H_8)_2$ (M = Th, Pa, Ce). *J. Phys. Chem. A* **2009**, *113*, 2896–2905. [CrossRef] [PubMed]

4. Kerridge, A. Oxidation state and covalency in f-element metallocenes (M = Ce, Th, Pu): A combined CASSCF and topological study. *Dalton Trans.* **2013**, *42*, 16428–16436. [CrossRef] [PubMed]

5. Mooßen, O.; Dolg, M. Two interpretations of the cerocene electronic ground state. *Chem. Phys. Lett.* **2014**, *594*, 47–50. [CrossRef]

6.   Zhurov, V.V.; Zhurova, E.A.; Stash, A.I.; Pinkerton, A.A. Characterization of bonding in cesium uranyl chloride: Topological analysis of the experimental charge density. *J. Phys. Chem. A* **2011**, *115*, 13016–13023. [CrossRef] [PubMed]

7.   Bader, R.F.W. *Atoms in Molecules: A Quantum Theory*; Oxford University Press: Oxford, UK, 1990.

8.   Tassell, M.J.; Kaltsoyannis, N. Covalency in AnCp$_4$ (An = Th–Cm): A comparison of molecular orbital, natural population and atoms-in-molecules analyses. *Dalton Trans.* **2010**, *39*, 6576–6588. [CrossRef] [PubMed]

9.   Arnold, P.L.; Turner, Z.R.; Kaltsoyannis, N.; Pelekanaki, P.; Bellabarba, R.M.; Tooze, R.P. Covalency in CeIV and UIV halide and N-heterocyclic carbene bonds. *Chem. Eur. J.* **2010**, *16*, 9623–9629. [CrossRef] [PubMed]

10.   Kirker, I.; Kaltsoyannis, N. Does covalency really increase across the 5f series? A comparison of molecular orbital, natural population, spin and electron density analyses of AnCp$_3$ (An = Th–Cm; Cp = η^5-C$_5$H$_5$). *Dalton Trans.* **2011**, *40*, 124–131. [CrossRef] [PubMed]

11.   Vallet, V.; Wahlgren, U.; Grenthe, I. Probing the nature of chemical bonding in uranyl(VI) complexes with quantum chemical methods. *J. Phys. Chem. A* **2012**, *116*, 12373–12380. [CrossRef] [PubMed]

12.   Kerridge, A. f-Orbital covalency in the actinocenes (An = Th–Cm): Multiconfigurational studies and topological analysis. *RSC Adv.* **2014**, *4*, 12078–12086. [CrossRef]

13.   Wu, Q.Y.; Wang, C.Z.; Lan, J.H.; Xiao, C.L.; Wang, X.K.; Zhao, Y.L.; Chai, Z.F.; Shi, W.Q. Theoretical investigation on multiple bonds in terminal actinide nitride complexes. *Inorg. Chem.* **2014**, *53*, 9607–9614. [CrossRef] [PubMed]

14.   Jones, M.B.; Gaunt, A.J.; Gordon, J.C.; Kaltsoyannis, N.; Neu, M.P.; Scott, B.L. Uncovering f-element bonding differences and electronic structure in a series of 1:3 and 1:4 complexes with a diselenophosphinate ligand. *Chem. Sci.* **2013**, *4*, 1189–1203. [CrossRef]

15.   Behrle, A.C.; Barnes, C.L.; Kaltsoyannis, N.; Walensky, J.R. Systematic investigation of thorium(IV)- and uranium(IV)-ligand bonding in dithiophosphonate, thioselenophosphinate, and diselenophosphonate complexes. *Inorg. Chem.* **2013**, *52*, 10623–10631. [CrossRef] [PubMed]

16.   Huang, Q.R.; Kingham, J.R.; Kaltsoyannis, N. The strength of actinide-element bonds from the quantum theory of atoms-in-molecules. *Dalton Trans.* **2015**, *44*, 2554–2566. [CrossRef] [PubMed]

17.   Kozimor, S.A.; Yang, P.; Batista, E.R.; Boland, K.S.; Burns, C.J.; Clark, D.L.; Conradson, S.D.; Martin, R.L.; Wilkerson, M.P.; Wolfsberg, L.E. Trends in covalency for d- and f-element metallocene dichlorides identified using chlorine K-edge X-ray absorption spectroscopy and time-dependent density functional theory. *J. Am. Chem. Soc.* **2009**, *131*, 12125–12136. [CrossRef] [PubMed]

18.   Minasian, S.G.; Keith, J.M.; Batista, E.R.; Boland, K.S.; Clark, D.L.; Conradson, S.D.; Kozimor, S.A.; Martin, R.L.; Schwarz, D.E.; Shuh, D.K.; *et al.* Determining relative f and d orbital contributions to M–Cl covalency in MCl$_6{}^{2-}$ (M = Ti, Zr, Hf, U) and UOCl$_5{}^-$ using Cl K-edge X-ray absorption spectroscopy and time-dependent density functional theory. *J. Am. Chem. Soc.* **2012**, *134*, 5586–5597. [CrossRef] [PubMed]

19.   Minasian, S.G.; Keith, J.M.; Batista, E.R.; Boland, K.S.; Clark, D.L.; Kozimor, S.A.; Martin, R.L.; Shuh, D.K.; Tyliszczak, T. New evidence for 5f covalency in actinocenes determined from carbon K-edge XAS and electronic structure theory. *Chem. Sci.* **2014**, *5*, 351–359. [CrossRef]

20.   Löble, M.W.; Keith, J.M.; Altman, A.B.; Stieber, S.C.E.; Batista, E.R.; Boland, K.S.; Conradson, S.D.; Clark, D.L.; Lezama Pacheco, J.; Kozimor, S.A.; *et al.* Covalency in lanthanides. An X-ray absorption spectroscopy and density functional theory study of LnCl$_6{}^{x-}$ (x = 3, 2). *J. Am. Chem. Soc.* **2015**, *137*, 2506–2523. [CrossRef] [PubMed]

21.   Bossé, E.; den Auwer, C.; Berthon, C.; Guilbaud, P.; Grigoriev, M.S.; Nikitenko, S.; le Naour, C.; Cannes, C.; Moisy, P. Solvation of UCl$_6{}^{2-}$ anionic complex by MeBu$_3$N$^+$, BuMe$_2$Im$^+$, and BuMeIm$^+$ cations. *Inorg. Chem.* **2008**, *47*, 5746–5755. [CrossRef] [PubMed]

22.   Taylor, J.C.; Wilson, P.W. Neutron and X-ray powder diffraction studies of the structure of uranium hexachloride. *Acta Crystallogr. Sect. B Struct. Crystallogr. Cryst. Chem.* **1974**, *30*, 1481–1484. [CrossRef]

23.   De Wet, J.F.; Caira, M.R.; Gellatly, B.J. The crystal structures of hexahalouranates. II. Triphenylbenzylphosphonium hexachlorouranate(V). *Acta Crystallogr. Sect. B Struct. Crystallogr. Cryst. Chem.* **1978**, *34*, 1121–1124. [CrossRef]

24.   Löwdin, P. Quantum theory of may-particle systems. I. Physical interpretations by means of density matrices, natural spin-orbitals, and convergence problems in the method of configurational interaction. *Phys. Rev.* **1955**, *376*, 1474–1489. [CrossRef]

25. Schmidt, M.W.; Gordon, M.S. The construction and interpretation of MCSCF wavefunctions. *Annu. Rev. Phys. Chem.* **1998**, *49*, 233–266. [CrossRef] [PubMed]

26. Becke, A.D. A new mixing of Hartree-Fock and local density-functional theories. *J. Chem. Phys.* **1993**, *98*, 1372. [CrossRef]

27. Fradera, X.; Austen, M.A.; Bader, R.F.W. The lewis model and beyond. *J. Phys. Chem. A* **1999**, *103*, 304–314. [CrossRef]

28. Bader, R.F.W.; Matta, C.F. Atoms in molecules as non-overlapping, bounded, space-filling open quantum systems. *Found. Chem.* **2013**, *15*, 253–276. [CrossRef]

29. Clavaguéra, C.; Dognon, J.P.; Pyykkö, P. Calculated lanthanide contractions for molecular trihalides and fully hydrated ions: The contributions from relativity and 4f-shell hybridization. *Chem. Phys. Lett.* **2006**, *429*, 8–12. [CrossRef]

30. Roos, B.O.; Lindh, R.; Malmqvist, P.A.; Veryazov, V.; Widmark, P.O.; Borin, A.C. New relativistic atomic natural orbital basis sets for lanthanide atoms with applications to the Ce diatom and LuF_3. *J. Phys. Chem. A* **2008**, *112*, 11431–11435. [CrossRef] [PubMed]

31. Karlström, G.; Lindh, R.; Malmqvist, P.Å.; Roos, B.O.; Ryde, U.; Veryazov, V.; Widmark, P.O.; Cossi, M.; Schimmelpfennig, B.; Neogrady, P.; *et al.* MOLCAS: A program package for computational chemistry. *Comput. Mater. Sci.* **2003**, *28*, 222–239. [CrossRef]

32. Aquilante, F.; Vico, L.D.E.; Ferré, N.; Ghigo, G.; Malmqvist, P.Å.; Neogrády, P.; Pedersen, T.B.; Nák, M.P.; Reiher, M.; Roos, B.O.; *et al.* Software news and update MOLCAS 7: The next generation. *J. Comput. Chem.* **2010**, *31*, 224–247. [CrossRef] [PubMed]

33. Roos, B.O.; Lindh, R.; Malmqvist, P.Å.; Veryazov, V.; Widmark, P.O. New relativistic ANO basis sets for actinide atoms. *Chem. Phys. Lett.* **2005**, *409*, 295–299. [CrossRef]

34. Roos, B.O.; Lindh, R.; Malmqvist, P.Å.; Veryazov, V.; Widmark, P.O. Main group atoms and dimers studied with a new relativistic ANO basis set. *J. Phys. Chem. A* **2004**, *108*, 2851–2858. [CrossRef]

35. Douglas, M.; Kroll, N. Quantum electrodynamical corrections to the fine structure of helium. *Ann. Phys.* **1974**, *155*, 89–155. [CrossRef]

36. Hess, B. Relativistic electronic-structure calculations employing a two-component no-pair formalism with external-field projection operators. *Phys. Rev. A* **1986**, *33*, 3742–3748. [CrossRef] [PubMed]

37. Perdew, J.; Burke, K.; Ernzerhof, M. Generalized gradient approximation made simple. *Phys. Rev. Lett.* **1996**, *77*, 3865–3868. [CrossRef] [PubMed]

38. Becke, A.D. Density-functional thermochemistry. III. The role of exact exchange. *J. Chem. Phys.* **1993**, *98*, 5648. [CrossRef]

39. Stephens, P.; Devlin, F.; Chabalowski, C.; Frisch, M. *Ab initio* calculation of vibrational absorption and circular dichroism spectra using density functional force fields. *J. Phys. Chem.* **1994**, *98*, 11623–11627. [CrossRef]

40. Coates, R.; Coreno, M.; DeSimone, M.; Green, J.C.; Kaltsoyannis, N.; Kerridge, A.; Narband, N.; Sella, A. A mystery solved? Photoelectron spectroscopic and quantum chemical studies of the ion states of $CeCp_3^+$. *Dalton Trans.* **2009**, 5943–5953. [CrossRef] [PubMed]

41. Kerridge, A.; Kaltsoyannis, N. All-electron CASPT2 study of $Ce(\eta^8-C_8H_6)_2$. *Comptes Rendus Chim.* **2010**, *13*, 853–859. [CrossRef]

42. Hashem, E.; Swinburne, A.N.; Schulzke, C.; Evans, R.C.; Platts, J.A.; Kerridge, A.; Natrajan, L.S.; Baker, R.J. Emission spectroscopy of uranium(IV) compounds: A combined synthetic, spectroscopic and computational study. *RSC Adv.* **2013**, *3*, 4350–4361. [CrossRef]

43. Le Guennic, B.; Autschbach, J. Magnetic properties and electronic structures of Ar_3U^{IV}–L complexes with Ar = $C_5(CH_3)_4H^-$ or $C_5H_5^-$ and L = CH_3, NO, and Cl. *Inorg. Chem.* **2014**, *53*, 13174–13187.

44. Sauri, V.; Serrano-Andrés, L.; Shahi, A.R.M.; Gagliardi, L.; Vancoillie, S.; Pierloot, K. Multiconfigurational second-order perturbation theory restricted active space (RASPT2) method for electronic excited states: A benchmark study. *J. Chem. Theory Comput.* **2011**, 153–168. [CrossRef]

45. Kerridge, A. A RASSCF study of free base, magnesium and zinc porphyrins: Accuracy *versus* efficiency. *Phys. Chem. Chem. Phys.* **2013**, *15*, 2197–2209. [CrossRef] [PubMed]

46. Barone, V.; Cossi, M. Quantum calculation of molecular energies and energy gradients in solution by a conductor solvent model. *J. Phys. Chem. A* **1998**, *102*, 1995–2001. [CrossRef]

47. Cossi, M.; Rega, N.; Scalmani, G.; Barone, V. Polarizable dielectric model of solvation with inclusion of charge penetration effects. *J. Chem. Phys.* **2001**, *114*, 5691–5701. [CrossRef]

48. Ahlrichs, R.; Bär, M.; Häser, M.; Horn, H.; Kölmel, C. Electronic structure calculations on workstation computers: The program system turbomole. *Chem. Phys. Lett.* **1989**, *162*, 165–169. [CrossRef]

49. Weigend, F.; Ahlrichs, R. Balanced basis sets of split valence, triple zeta valence and quadruple zeta valence quality for H to Rn: Design and assessment of accuracy. *Phys. Chem. Chem. Phys.* **2005**, *7*, 3297–3305. [CrossRef] [PubMed]

50. Kerridge, A.; Kaltsoyannis, N. Are the ground states of the later actinocenes multiconfigurational? All-electron spin-orbit coupled CASPT2 calculations on An(η^8-C_8H_8)$_2$ (An = Th, U, Pu, Cm). *J. Phys. Chem. A* **2009**, *113*, 8737–8745. [CrossRef] [PubMed]

51. Klamt, A.; Schüürmann, G. COSMO: A new approach to dielectric screening in solvents with explicit expressions for the screening energy and its gradient. *Perkins Trans.* **1993**, *2*, 799–805. [CrossRef]

52. Zaiter, A.; Amine, B.; Bouzidi, Y.; Belkhiri, L.; Boucekkine, A.; Ephritikhine, M. Selectivity of azine ligands toward lanthanide(III)/actinide(III) differentiation: A relativistic DFT based rationalization. *Inorg. Chem.* **2014**, *53*, 4687–4697. [CrossRef] [PubMed]

53. Herve, A.; Bouzidi, Y.; Berthet, J.; Belkhiri, L.; Thuery, P.; Boucekkine, A.; Ephritikhine, M. U–CN *versus* Ce–NC coordination in trivalent complexes derived from M[N(SiMe$_3$)$_2$]$_3$ (M = Ce, U). *Inorg. Chem.* **2014**, *53*, 6995–7013. [CrossRef] [PubMed]

![inorganics logo] *inorganics*

MDPI

Communication

Holmium(III) Supermesityl-Imide Complexes Bearing Methylaluminato/Gallato Ligands

Dorothea Schädle [1], Cäcilia Maichle-Mössmer [1], Karl W. Törnroos [2] and Reiner Anwander [1,*]

[1] Institut für Anorganische Chemie, Universität Tübingen, Auf der Morgenstelle 18, 72076 Tübingen, Germany; dorothea.schaedle@uni-tuebingen.de (D.S.); Caecilia.Maichle-Moessmer@uni-tuebingen.de (C.M.-M.)

[2] Department of Chemistry, University of Bergen, Allégaten 41, 5007 Bergen, Norway; karl.tornroos@uib.no

* Correspondence: reiner.anwander@uni-tuebingen.de; Tel.: +49-7071-29-72069; Fax: +49-7071-29-2436

Academic Editors: Stephen Mansell and Steve Liddle

Received: 22 September 2015; Accepted: 30 October 2015; Published: 10 November 2015

Abstract: Heterobimetallic μ_2-imide complexes [Ho(μ_2-Nmes*){Al(CH$_3$)$_4$}]$_2$ (**1**, supermesityl = mes* = C$_6$H$_2$tBu$_3$-2,4,6) and [Ho(μ_2-Nmes*){Ga(CH$_3$)$_4$}]$_2$ (**2**) have been synthesized from homoleptic complexes Ho[M(CH$_3$)$_4$]$_3$ (M = Al, Ga) via deprotonation of H$_2$Nmes* or with K[NH(mes*)] according to a salt metathesis-protonolysis tandem reaction. Single-crystal X-ray diffraction of isostructural complexes [Ho(μ_2-Nmes*){M(CH$_3$)$_4$}]$_2$ (M = Al, Ga) revealed asymmetric Ho$_2$N$_2$ metallacycles with very short Ho–N bond lengths and secondary Ho$\cdots$arene interactions.

Keywords: f-element; aluminum; gallium; imide

1. Introduction

The emerging field of rare-earth metal imide chemistry has revealed interesting structural motifs [1,2], but studies regarding their fundamental properties and reactivity are lagging behind. Bochkarev and Schumann were the first to report on a tetranuclear ytterbium(III) phenylimide complex obtainable via reduction of azobenzene by ytterbium naphthalenide [3]. Another synthesis strategy was developed by Evans *et al.*, adventitiously identifying [Nd(NPh)(AlMe$_2$)(AlMe$_4$)$_2$]$_2$ (Figure 1, **I**) via attempted alkylation of [Nd(NHPh)$_3$(KCl)$_3$] with excess AlMe$_3$ [4]. We and others succeeded in the isolation of a series of rare-earth metal complexes (Figure 1, **II–III**, secondary interactions are not shown), by adopting the strategy of organoaluminum-assisted imide formation [5–10]. Similarly, deprotonation of lanthanide anilide complexes was achieved via treatment with butyllithium, affording complexes of type **IV** [11]. A different synthesis approach has been employed in the reactions of alkyl complexes with amine-boranes to yield Lewis acid (LA)-stabilized imido entities [Ln(NR)(BH$_3$)] (R = H, teraryl) [12,13].

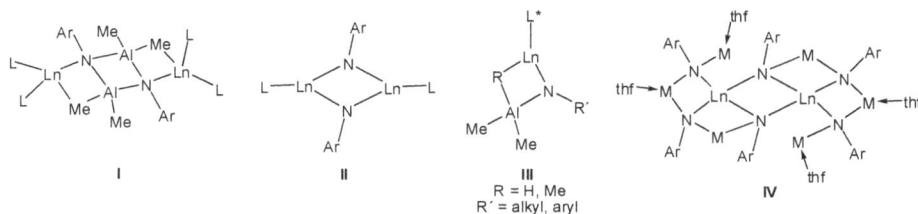

Figure 1. Structural motifs of heterobimetallic rare-earth metal imide complexes: **I** (L = AlMe$_4$$^-$, Ar = Ph), **II** (L = AlMe$_3$(NHAr)$^-$, Ar = C$_6$H$_3$iPr$_2$-2,6; L = AlMe$_4$$^-$, Ar = C$_6H_2tBu_3$-2,4,6), **III** (L* = monoanionic ancillary ligand), **IV** (M = Na, Li, Ar = C$_6$H$_3$iPr$_2$-2,6).

In general, complexes of type [(L)Ln(NR)] or [(L)Ln(NR)(AlMe$_3$)] have shown promising usability in a range of areas, including catalytic cyclotrimerization of benzonitrile [14], synthesis of substituted pyridines [15], preparation of Ln/M heterobimetallics [16], hydroelementation reactions [17], and polymerization of dienes [18]. Further, such complexes give access to many different types of new Ln(III) imide complexes comprising alkyl-imide, amide-imide, alkoxide-imide, and cyclopentadienyl-imide compounds [18]. Moreover, derivatization of the Ln=N(R) functionality with small molecules and organic substrates revealed interesting reaction patterns and bonding features [19,20].

To fully investigate the implications of the choice of Ln(III) alkyl precursor for any envisioned synthesis, we also employed rare-earth metal tetramethylgallate complexes. Although the structural parameters of the Ln/M heterobimetalic complexes Ln[M(CH$_3$)$_4$]$_3$ (M = Al, Ga) [21–23] and the derived half-sandwich [24], metallocene [25], and scorpionate complexes [10,26,27] are similar, the reactivity, in some cases, is dramatically different [10,26,27]. For example, previous studies from our laboratory revealed that [(Tp^{tBu,Me})Ln(CH$_3$)$_2$] (Tp^{tBu,Me} = hydrotris(3-*tert*-butyl-5-methylpyrazolyl)borate) and [(Tp^{tBu,Me})Ln(CH$_3$){Ga(CH$_3$)$_4$}] gave the Ln(III) anilide complexes [(Tp^{tBu,Me})Ln(CH$_3$)(NHR)] (R = alkyl, aryl), as opposed to the aluminum congener [(Tp^{tBu,Me})Ln(CH$_3$){Al(CH$_3$)$_4$}], which yields complexes of type **III** [9,10].

While a series of dimeric LA-stabilized rare-earth metal imides (type **II**) has been reported previously [7], we now report on additional aniline-derived rare-earth metal imide complexes employing methylaluminate and methylgallate complexes. The intention was to investigate fundamental differences of organogallium *versus* organoaluminum moieties since GaMe$_3$ should behave as a weaker LA towards nitrogen than AlMe$_3$.

2. Results and Discussion

The μ_2-imide complexes [Ln{Al(CH$_3$)$_4$}(μ_2-Nmes*)]$_x$ (Ln = Y, La, Nd, Lu; mes* = C$_6$H$_2$tBu$_3$-2,4,6) were previously obtained from homoleptic heterobimetallic complexes Ln[Al(CH$_3$)$_4$]$_3$ utilizing two distinct protocols: reaction with 2,4,6-tri-*tert*-butylaniline in *n*-hexane via methane elimination or with potassium (2,4,6-tri-*tert*-butylphenyl)amide in toluene according to a salt metathesis-protonolysis tandem reaction [7]. Similarly, the reaction of Ho[M(CH$_3$)$_4$]$_3$ (M = Al, Ga) with H$_2$Nmes* or K[NH(mes*)], respectively, led to dimeric complexes [Ho{M(CH$_3$)$_4$}(μ_2-Nmes*)]$_2$ (M = Al (**1**), Ga (**2**)) (Scheme 1). The identities of the holmium imide complexes **1** and **2** were confirmed by elemental analysis and single-crystal X-ray diffraction studies revealing isomorphous dimeric arrangements, featuring a Ho$_2$N$_2$ core on a crystallographic inversion center. The solid-state structures of **1** and **2** are depicted in Figure 2 with selected bond lengths and angles shown in Table 1. The geometry about the four-coordinate metal centers can best be described as distorted tetrahedral with two methyl groups and two imido nitrogen atoms at the four vertices. The rare-earth metal atoms are asymmetrically bridged by two μ_2-imido nitrogen atoms, displaying one short (Ho–N, **1**: 2.107(1) Å; **2**: 2.102(3) Å) and one long contact (Ho–N, **1**: 2.283(1) Å; **2**: 2.288(3) Å). The Ho–N bond lengths in **1** and **2** are comparable to the distances in yttrium imide complexes, considering the similar ionic radii (Table 2).

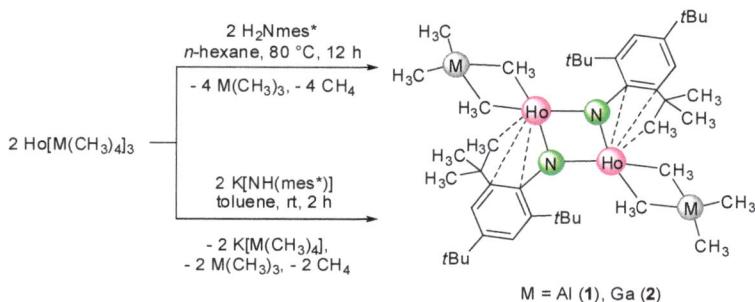

Scheme 1. Synthesis of holmium(III) imide complexes.

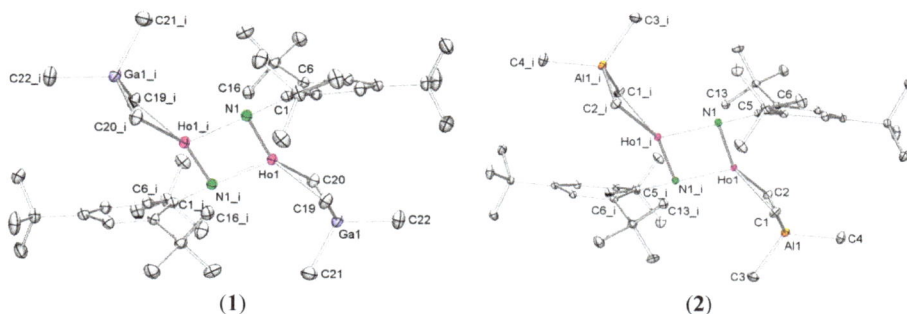

(1)   **(2)**

Figure 2. Solid-state structures of [Ho{Al(CH₃)₄}(μ₂-Nmes*)]₂ (**1**) and [Ho{Ga(CH₃)₄}(μ₂-Nmes*)]₂ (**2**) with 30% probability ellipsoids. Hydrogen atoms are omitted for clarity.

Table 1. Selected bond lengths [Å] and angles [°] of complexes **1** and **2**.

	1		2
Ho1–N1′	2.107(1)	Ho1–N1′	2.102(3)
Ho1–N1	2.283(1)	Ho1–N1	2.288(3)
Ho1′–N1	2.107(1)	Ho1′–N1	2.102(3)
Ho1–C1	2.512(2)	Ho1–C19	2.512(4)
Ho1–C2	2.598(2)	Ho1–C20	2.594(5)
Ho1–C5	2.626(1)	Ho1–C1	2.629(3)
Ho1–C6	2.715(1)	Ho1–C6	2.750(3)
Ho1–C13	2.833(1)	Ho1–C16	2.864(3)
Ho1···Al1	3.0838(5)	Ho1···Ga1	3.0573(4)
Al1–C1	2.0836(16)	Ga1–C19	2.115(4)
Al1–C2	2.0718(16)	Ga1–C20	2.092(4)
Al1–C3	1.9681(19)	Ga1–C21	1.978(4)
Al1–C4	1.9727(17)	Ga1–C22	1.974(4)
N1–C5	1.3811(18)	N1–C1	1.386(4)
N1′–Ho1–N1	84.63(4)	N1′–Ho1–N1	84.68(11)
Ho1′–N1–Ho1	95.37(4)	Ho1′–N1–Ho1	95.32(11)
Ho1′–N1–C5	175.67(10)	Ho1′–N1–C1	176.8(2)
Ho1–N1–C5	87.95(8)	Ho1–N1–C1	87.80(18)
C1–Ho1–C2	82.76(5)	C19–Ho1–C20	83.98(14)
C1–Al1–C2	108.80(6)	C19–Ga1–C20	108.63(15)
C3–Al1–C4	118.82(8)	C21–Ga1–C22	117.5(2)
C1–Ho1–C2–Al1	−11.34(5)	C19–Ho1–C20–Ga1	14.20(14)

Table 2. Selected Y–N(imido) and Ho–N(imido) bond lengths [Å].

Compounds	Ln–N	CN [a]	Reference
[(Tp^{tBu,Me})Y{NC$_6$H$_3$(CH$_3$)$_2$-2,6}(AlMe$_3$)] [c]	2.123(2)–2.128(3)	5	[10]
[(Tp^{tBu,Me})Y{NC$_6$H$_3$(CH$_3$)$_2$-2,6}(HAlMe$_2$)] [c]	2.133(2)	5	[8]
[(Tp^{tBu,Me})Y(NtBu)(AlMe$_3$)] [c]	2.081(3)–2.088(3)	5	[9]
[(Tp^{tBu,Me})Ho(NtBu)(AlMe$_3$)] [c]	2.083(2)–2.084(2)	5	[9]
[(Tp^{tBu,Me})Y(NAd)(AlMe$_3$)] [c,d]	2.092(2)–2.099(2)	5	[9]
[(Tp^{tBu,Me})Ho(NAd)(AlMe$_3$)] [c,d]	2.087(2)–2.090(2)	5	[9]
[(Tp^{tBu,Me})Y{NC$_6$H$_3$(CH$_3$)$_2$-2,6}(DMAP) [c,e]	2.024(4)	5	[10]
[(C$_5$Me$_4$SiMe$_3$)$_4$Y$_4$(μ$_3$-NCH$_2$CH$_3$)$_2$(μ$_2$-NCHPh)$_4$]	2.116(6)–2.418(6)	6/7	[14]
[L$_3$Y$_3$(μ$_2$-CH$_3$)$_3$(μ$_3$-CH$_3$)(μ$_3$-NR)] [f]	2.308(3)–2.435(7)	6	[28]
[Y{Al(CH$_3$)$_4$}(μ$_2$-Nmes*)]$_2$	2.1089(9)–2.2909(9)	6	[7]
[Ho{Al(CH$_3$)$_4$}(μ$_2$-Nmes*)]$_2$	2.107(1)–2.283(1)	6	b
[Ho{Ga(CH$_3$)$_4$}(μ$_2$-Nmes*)]$_2$	2.102(3)–2.288(3)	6	b

[a] CN = coordination number; [b] this work; [c] Tp^{tBu,Me} = hydrotris(3-*tert*-butyl-5-methylpyrazolyl)borate; [d] Ad = 1-adamantyl; [e] DMAP = 4-(dimethylamino)pyridine; [f] L = [PhC(NC$_6$H$_3$*i*Pr$_2$-2,6)$_2$]$^-$; R = alkyl, aryl.

The pronounced asymmetry of the Ho$_2$N$_2$ core most likely originates from secondary interactions between the holmium centers and the *ipso* and *ortho* carbons as well as one CH$_3$ *tert*-butyl group of the bridging μ$_2$-Nmes* ligands. One distinct difference between [Ho{Al(CH$_3$)$_4$}(μ$_2$-Nmes*)]$_2$ (**1**) and [Ho{Ga(CH$_3$)$_4$}(μ$_2$-Nmes*)]$_2$ (**2**) is the bending of the M(CH$_3$)$_4$ moiety being slightly more pronounced for the gallium derivative, since the softer Ga(III) center can achieve shorter Ln(III) ⋯ Ga(III) contacts.

A preliminary reactivity study of yttrium congener [Y{Al(CH$_3$)$_4$}(μ$_2$-Nmes*)]$_2$ [7] was performed in order to assess the feasibility of exchanging the AlMe$_4^-$ ligand by other ancillary ligands. The reaction of [Y{Al(CH$_3$)$_4$}(μ$_2$-Nmes*)]$_2$ with KCp* (Cp* = C$_5$Me$_5$) in toluene at 80 °C led to the formation of an off-white solid, which is insoluble in *n*-hexane and toluene. The ^{1}H NMR spectrum of the product indicated the formation of [Cp*Y(Nmes*)]$_n$. However, as shown previously, toluene-soluble imide complexes [(AlMe$_4$)Ln(NC$_6$H$_3$*i*Pr-2,6)(AlMe$_3$)$_x$]$_2$ readily undergo salt-metathesis reactions with a variety of alkaline metal salts [M^I(L)] (L = silylamide, cyclopentadienyl, aryloxide) to generate heteroleptic Ln(III) imide complexes [18].

3. Experimental Section

3.1. General Procedures

All operations were performed with rigorous exclusion of air and water, using standard Schlenk, high-vacuum, and glovebox techniques (MBraun MBLab; <1 ppm O$_2$, <1 ppm H$_2$O). Toluene and *n*-hexane were purified by using Grubbs columns (MBraun SPS, solvent purification system, MBraun, Garching, Germany) and stored in a glovebox; [D$_6$]benzene was obtained from Aldrich (St. Louis, MO, USA), degassed, dried over Na for 24 h, and filtered. Then 2,4,6-tri-*tert*-butylaniline was obtained from Aldrich and used as received. Potassium (2,4,6-tri-*tert*-butylphenyl)amide was synthesized according to literature procedures [7]. Homoleptic complexes [Ho{Al(CH$_3$)$_4$}$_3$] [22,29] and [Ho{Ga(CH$_3$)$_4$}$_3$] [30,31] (**2**) were prepared according to literature methods. DRIFT spectra were recorded on a NICOLET 6700 FTIR spectrometer (Thermo Scientific, Dreieich, Germany) using dried KBr and KBr window. Elemental analyses were performed on an Elementar Vario Micro Cube (Elementar, Hanau, Germany).

3.2. [Ho{M(CH$_3$)$_4$}(μ$_2$-Nmes*)]$_2$ (**1** and **2**)

3.2.1. Procedure A

A solution of Ho[M(CH$_3$)$_4$]$_3$ in toluene (3 mL) was added to a vigorously stirred suspension of potassium (2,4,6-tri-*tert*-butylphenyl)amide in toluene (2 mL). The reaction mixture was stirred for 2 h at ambient temperature and the toluene solution then separated by centrifugation, decanted, and filtrated. The solid residue (product and K[M(CH$_3$)$_4$] was extracted with additional toluene (5 × 2 mL). The extract was dried under vacuum and triturated with *n*-hexane (2 × 2 mL). After that the solid was washed with *n*-hexane (2 × 2 mL), followed by drying under reduced pressure. Compounds **1** and **2** were obtained as powder or by crystallization from the mother liquor at ambient temperature.

3.2.2. Procedure B

A solution of 2,4,6-tri-*tert*-butylaniline in *n*-hexane (3 mL) was added to a solution of Ho[M(CH$_3$)$_4$]$_3$ in *n*-hexane (2 mL). The reaction mixture was stirred for 8 h at 80 °C. The solution turned orange and a precipitate was formed. The mixture was chilled to ambient temperature, the solid product was separated by centrifugation and washed with *n*-hexane (2 × 2 mL). The procedure was repeated twice with the combined extracts. Compounds **1** and **2** were dried *in vacuo* and obtained as orange powder.

3.3. [Ho{Al(CH$_3$)$_4$}(μ$_2$-Nmes*)]$_2$ (**1**)

3.3.1. Procedure A

Following the procedure described above, Ho[Al(CH$_3$)$_4$]$_3$ (85.3 mg, 0.20 mmol) and potassium(2,4,6-tri-*tert*-butylphenyl)amide (59.9 mg, 0.20 mmol) yielded [Ho{Al(CH$_3$)$_4$}(μ$_2$-Nmes*)]$_2$ as orange crystals (102.3 mg, 0.10 mmol, ≥99%).

3.3.2. Procedure B

Following the procedure described above, Ho[Al(CH$_3$)$_4$]$_3$ (85.3 mg, 0.20 mmol) and 2,4,6-tri-*tert*-butylaniline (52.3 mg, 0.20 mmol) yielded [Ho{Al(CH$_3$)$_4$}(μ$_2$-Nmes*)]$_2$ as orange powder (38.4 mg, 0.04 mmol, 40%).

3.3.3. Physical Data of **1**

DRIFT IR (KBr): 3014 w, 2962 s, 2869 m, 2773 w, 1588 w, 1477 w, 1462 w, 1394 m, 1377 m, 1361 m, 1340 w, 1274 m, 1244 s, 1227 m, 1200 s, 1108 w, 916 w, 894 w, 874 w, 856 m, 783 w, 755 w, 716 s, 694 vs, 663 w, 638 w, 609 w, 578 m, 547 w, 504 m, 476 w, 458 w, 437 m·cm^{-1}; elemental analysis calcd (%) for C$_{44}$H$_{82}$Al$_2$N$_2$Ho$_2$ (1022.96 g·mol^{-1}): C 51.66, H 8.08, N 2.74; found: C 51.76, H 7.84, N 2.74.

3.4. [Ho(GaMe$_4$)(μ$_2$-Nmes*)]$_2$ (**2**)

3.4.1. Procedure A

Following the procedure described above, Ho[Ga(CH$_3$)$_4$]$_3$ (269.9 mg, 0.47 mmol) and potassium(2,4,6-tri-*tert*-butylphenyl)amide (139.9 mg, 0.47 mmol) yielded [Ho{Ga(CH$_3$)$_4$}(μ$_2$-Nmes*)]$_2$ as orange crystals (520.1 mg, 0.47 mmol, ≥99%).

3.4.2. Procedure B

Following the procedure described above, Ho[Ga(CH$_3$)$_4$]$_3$ (236.7 mg, 0.43 mmol) and 2,4,6-tri-*tert*-butylaniline (127.9 mg, 0.43 mmol) yielded [Ho{Ga(CH$_3$)$_4$}(μ$_2$-Nmes*)]$_2$ as orange powder (100.8 mg, 0.09 mmol, 42%).

3.4.3. Physical Data of **2**

DRIFT IR (KBr): 2962 vs, 2904 m, 2869 m, 2774 w, 1588 vw, 1477 w, 1460 w, 1394 s, 1378 m, 1360 m, 1274 m, 1246 vs, 1227 m, 1193 m, 1109 w, 915 vw, 875 w, 856 s, 783 w, 755 w, 716 w, 662 vw, 594 w, 556 w, 530 w, 505 m, 476 vw, 428 w·cm^{-1}; elemental analysis calcd for C$_{44}$H$_{82}$Ga$_2$N$_2$Ho$_2$ (1108.44 g/mol): C 47.68, H 7.46, N 2.53; found: C 47.91, H 7.42, N 2.52.

3.5. X-Ray Crystallography

Crystal data for compounds **1** and **2** are given in Table 3. Bond lengths and angles are listed in Table 2. Crystals of **1** and **2** were grown using standard techniques from saturated toluene solutions. Suitable single crystals for X-ray structure analyses were selected in a glovebox and coated with Parabar 10312 and fixed on a nylon loop/glass fiber.

Table 3. Crystallographic data for compounds **1** and **2**.

	1	2
Formula	$C_{44}H_{82}Al_2Ho_2N_2$	$C_{44}H_{82}Ga_2Ho_2N_2$
Color	Yellow	Yellow
M_r (g·mol^{-1})	1022.94	1108.41
Cryst system	Monoclinic	Triclinic
Space group	$C2/c$	$P\bar{1}$
a [Å]	24.0888(13)	10.1896(4)
b [Å]	11.7687(6)	11.5565(5)
c [Å]	20.4327(11)	11.5966(5)
α [°]	90	65.090(3)
β [°]	111.5590(10)	83.184(3)
γ [°]	90	84.962(4)
V [Å^3]	5387.3(5)	1228.69(9)
Z	4	1
F(000)	2080	556
T [K]	103(2)	173(2)
ρ_{calcd} (g cm^3)	1.261	1.498
μ(mm^{-1})	2.974	4.297
R_1 (obsd.) [a]	0.0172	0.0296
wR_2 (all) [b]	0.0466	0.0661
S [c]	1.042	1.082

[a] $R_1 = \Sigma(||F_0| - |F_c||)/\Sigma|F_0|$, $F_0 > 4\sigma(F_0)$; [b] $wR_2 = \{\Sigma[w(F_0^2 - F_c^2)^2/\Sigma[w(F_0^2)^2]\}^{1/2}$; [c] $S = [\Sigma w(F_0^2 - F_c^2)^2/(n_0 - n_p)]^{1/2}$.

Data for compound **2** were collected on a Stoe IPDS 2T instrument equipped with a fine focus sealed tube and graphite monochromator using MoK$_\alpha$ radiation (λ = 0.71073 Å) performing ω scans. Raw data were collected and integrated using Stoe's X-Area software package [32]. A numerical absorption correction based on crystal shape optimization was applied using Stoe's X-Red [33] and X-Shape [34]. X-ray data for compound **1** were collected on a Bruker AXS, TXS rotating anode instrument using a Pt^{135} CCD detector, and graphite monochromated using MoK$_\alpha$ radiation (λ = 0.71073 Å), employing ω-scans. Raw data were processed using APEX [35] and SAINT [36], corrections for absorption effects were applied using SADABS [37]. The structure was solved by direct methods and refined against all data by full-matrix least-squares methods on F^2 using SHELXTL [38] and ShelXle [39]. All graphics were produced employing ORTEP-3 [40] and POV-Ray [41]. Further details of the refinement and crystallographic data are listed in the CIF files. CCDC 1426090 (**1**) and 1426091 (**2**) contain all the supplementary crystallographic data for this paper. These data can be obtained free of charge from The Cambridge Crystallographic Data Centre via www.ccdc.cam.ac.uk/data_request/cif.

4. Conclusions

Like the homoleptic tetramethylaluminate Ho[Al(CH$_3$)$_4$]$_3$, the respective gallate complex can be converted into imido-bridged complexes [Ho{M(CH$_3$)$_4$}(μ_2-Nmes*)]$_2$ (M = Al (**1**), Ga (**2**)). Depending on the synthesis protocol, meaning either protonolysis or the salt metathesis-protonolysis tandem reaction, the imide complexes are formed in moderate to excellent yields, respectively. Preliminary experiments on the capability of [Ln{Al(CH$_3$)$_4$}(μ_2-Nmes*)]$_2$ to engage in ligand exchange reactions were hampered by solubility issues. Ongoing studies in our group address the feasibility of imide-tetramethylgallate complexes derived from other primary amines/anilines. We have recently shown that switching to the H_2Ndipp (dipp = $C_6H_3iPr_2$-2,6) proligand significantly enhanced the solubility of the isolated bimetallic rare-earth metal imide complexes $Ln_2(\mu_2$-Ndipp)(μ_3-Ndipp)[(μ_2-$CH_3)_2$Al(CH_3)][Al($CH_3)_4]_2$, thus allowing for facile derivatization [18].

Supplementary Materials: Supplementary materials can be found at http://www.mdpi.com/2304-6740/3/4/0500/s1.

Acknowledgments: We are grateful to the German Science Foundation for support (Grant: AN 238/15-1).

Author Contributions: All reactions and analyses described were planned and conducted by Dorothea Schädle. Analyses include DRIFT spectroscopy and elemental analysis. Publication writing was done by Dorothea Schädle and Reiner Anwander. The structural analyses by single crystal X-ray diffraction were performed by Cäcilia Maichle-Mössmer and Karl W. Törnroos.

Conflicts of Interest: The authors declare no conflict of interest.

References

1. Giesbrecht, G.R.; Gordon, J.C. Lanthanide alkylidene and imido complexes. *Dalton Trans.* **2004**, 2387–2393. [CrossRef] [PubMed]
2. Summerscales, O.T.; Gordon, J.C. Complexes containing multiple bonding interactions between lanthanoid elements and main-group fragments. *RSC Adv.* **2013**, *3*, 6682–6692. [CrossRef]
3. Trifonov, A.A.; Bochkarev, M.N.; Schumann, H.; Loebel, J. Reduction of azobenzene by naphthaleneytterbium: A tetranuclear ytterbium(III) complex combining 1,2-diphenylhydrazido(2−) and phenylimido ligands. *Angew. Chem. Int. Ed. Engl.* **1991**, *30*, 1149–1151. [CrossRef]
4. Evans, W.J.; Ansari, M.A.; Ziller, J.W.; Khan, S.I. Utility of arylamido ligands in yttrium and lanthanide chemistry. *Inorg. Chem.* **1996**, *35*, 5435–5444. [CrossRef] [PubMed]
5. Gordon, J.C.; Giesbrecht, G.R.; Clark, D.L.; Hay, P.J.; Keogh, D.W.; Poli, R.; Scott, B.L.; Watkin, J.G. The first example of a μ_2-imido functionality bound to a lanthanide metal center: X-ray crystal structure and DFT study of [(μ-ArN)Sm(μ-NHAr)(μ-Me)AlMe$_2$]$_2$ (Ar = 2,6-^{i}Pr$_2$C$_6$H$_3$). *Organometallics* **2002**, *21*, 4726–4734. [CrossRef]
6. Scott, J.; Basuli, F.; Fout, A.R.; Huffmann, J.C.; Mindiola, D.J. Evidence for the existence of a terminal imidoscandium compound: Intermolecular C–H activation and complexation reactions with the transient Sc–NAr species. *Angew. Chem. Int. Ed.* **2008**, *47*, 8502–8505. [CrossRef] [PubMed]
7. Schädle, D.; Schädle, C.; Törnroos, K.W.; Anwander, R. Organoaluminum-assisted formation of rare-earth metal imide complexes. *Organometallics* **2012**, *31*, 5101–5107. [CrossRef]
8. Schädle, C.; Schädle, D.; Eichele, K.; Anwander, R. Methylaluminum-supported rare-earth-metal dihydrides. *Angew. Chem. Int. Ed.* **2013**, *52*, 13238–13242. [CrossRef] [PubMed]
9. Schädle, D.; Maichle-Mössmer, C.; Schädle, C.; Anwander, R. Rare-earth-metal methyl, amide, and imide complexes supported by a superbulky scorpionate ligand. *Chem. Eur. J.* **2014**, *21*, 662–670. [CrossRef] [PubMed]
10. Schädle, D.; Meermann-Zimmermann, M.; Schädle, C.; Maichle-Mössmer, C.; Anwander, R. Rare-earth metal complexes with terminal imido ligands. *Eur. J. Inorg. Chem.* **2015**, 1334–1339. [CrossRef]
11. Chan, H.S.; Li, H.W.; Xie, Z. Synthesis and structural characterization of imido–lanthanide complexes with a metal–nitrogen multiple bond. *Chem. Commun.* **2002**, 652–653. [CrossRef]
12. Li, J.; Gao, D.; Hu, H.; Cui, C. Reaction of a bulky amine borane with lanthanide trialkyls. Formation of alkyl lanthanide imide complexes. *New J. Chem.* **2015**, *39*, 7567–7570. [CrossRef]
13. Rad'kov, V.; Dorcet, V.; Carpentier, J.F.; Trifonov, A.; Kirillov, E. Alkylyttrium complexes of amidine–amidopyridinate ligands. Intramolecular C(sp^3)–H activation and reactivity studies. *Organometallics* **2013**, *32*, 1517–1527. [CrossRef]
14. Cui, D.; Nishiura, M.; Hou, Z. Lanthanide-imido complexes and their reactions with benzonitrile. *Angew. Chem. Int. Ed.* **2005**, *44*, 959–962. [CrossRef] [PubMed]
15. Wicker, B.F.; Scott, J.; Fout, A.R.; Pink, M.; Mindiola, D.J. Atom-economical route to substituted pyridines via a scandium imide. *Organometallics* **2011**, *30*, 2453–2456. [CrossRef]
16. Lu, E.; Zhou, Q.; Li, Y.; Chu, J.; Chen, Y.; Leng, X.; Sun, J. Reactivity of scandium terminal imido complexes towards metal halides. *Chem. Commun.* **2012**, *48*, 3403–3405. [CrossRef] [PubMed]
17. Chu, J.; Lu, E.; Chen, Y.; Leng, X. Reversible addition of the Si–H bond of phenylsilane to the Sc=N bond of a scandium terminal imido complex. *Organometallics* **2012**, *32*, 1137–1140. [CrossRef]

18. Schädle, D.; Schädle, C.; Schneider, D.; Maichle-Mössmer, C.; Anwander, R. Versatile Ln$_2$(μ-NR)$_2$-imide platforms for ligand exchange and isoprene polymerization. *Organometallics* **2015**, *34*, 4994–5008. [CrossRef]
19. Chu, J.; Lu, E.; Liu, Z.; Chen, Y.; Leng, X.; Song, H. Reactivity of a scandium terminal imido complex towards unsaturated substrates. *Angew. Chem. Int. Ed.* **2011**, *50*, 7677–7680. [CrossRef] [PubMed]
20. Chu, J.; Han, X.; Kefalidis, C.E.; Zhou, J.; Maron, L.; Leng, X.; Chen, Y. Lewis acid triggered reactivity of a Lewis base stabilized scandium-terminal imido complex: C–H bond activation, cycloaddition, and dehydrofluorination. *J. Am. Chem. Soc.* **2014**, *136*, 10894–10897. [CrossRef] [PubMed]
21. Evans, W.J.; Anwander, R.; Doedens, R.J.; Ziller, J.W. The use of heterometallic bridging moieties to generate tractable lanthanide complexes of small ligands. *Angew. Chem. Int. Ed. Engl.* **1994**, *33*, 1641–1644. [CrossRef]
22. Zimmermann, M.; Frøystein, N.Å.; Fischbach, A.; Sirsch, P.; Dietrich, H.M.; Törnroos, K.W.; Herdtweck, E.; Anwander, R. Homoleptic rare-earth metal (III) tetramethylaluminates: Structural chemistry, reactivity, and performance in isoprene polymerization. *Chem. Eur. J.* **2007**, *13*, 8784–8800. [CrossRef] [PubMed]
23. Dietrich, H.M.; Raudaschl-Sieber, G.; Anwander, R. Trimethylyttrium and trimethyllutetium. *Angew. Chem. Int. Ed.* **2005**, *44*, 5303–5306. [CrossRef] [PubMed]
24. Dietrich, H.M.; Maichle-Mössmer, C.; Anwander, R. Donor-assisted tetramethylaluminate/gallate exchange in organolanthanide complexes: Pushing the limits of Pearson's HSAB concept. *Dalton Trans.* **2010**, *39*, 5783–5785. [CrossRef] [PubMed]
25. Dietrich, H.M.; Törnroos, K.W.; Herdtweck, E.; Anwander, R. Tetramethylaluminate and tetramethylgallate coordination in rare-earth metal half-sandwich and metallocene complexes. *Organometallics* **2009**, *28*, 6739–6749. [CrossRef]
26. Zimmermann, M.; Takats, J.; Kiel, G.; Törnroos, K.W.; Anwander, R. Ln(III) methyl and methylidene complexes stabilized by a bulky hydrotris(pyrazolyl)borate ligand. *Chem. Commun.* **2008**, 612–614. [CrossRef] [PubMed]
27. Zimmermann, M.; Litlabø, R.; Törnroos, K.W.; Anwander, R. "Metastable" Lu(GaMe$_4$)$_3$ reacts like masked [LuMe$_3$]: Synthesis of an unsolvated lanthanide dimethyl complex. *Organometallics* **2009**, *28*, 6646–6649. [CrossRef]
28. Hong, J.; Zhang, L.; Wang, K.; Zhang, Y.; Weng, L.; Zhou, X. Methylidene rare-earth-metal complex mediated transformations of C=N, N=N and N–H bonds: New routes to imido rare-earth-metal clusters. *Chem. Eur. J.* **2013**, *19*, 7865–7873. [CrossRef] [PubMed]
29. Evans, W.J.; Anwander, R.; Ziller, J.W. Inclusion of Al$_2$Me$_6$ in the crystalline lattice of the organometallic complexes LnAl$_3$Me$_{12}$. *Organometallics* **1995**, *14*, 1107–1109. [CrossRef]
30. Dietrich, H.M.; Meermann, C.; Törnroos, K.W.; Anwander, R. Sounding out the reactivity of trimethylyttrium. *Organometallics* **2006**, *25*, 4316–4321. [CrossRef]
31. Zimmermann, M.; Rauschmaier, D.; Eichele, K.; Törnroos, K.W.; Anwander, R. Amido-stabilized rare-earth metal mixed methyl methylidene complexes. *Chem. Commun.* **2010**, *46*, 5346–5348. [CrossRef] [PubMed]
32. *X-Area v. 1.55*; Stoe & Cie GmbH: Darmstadt, Germany, 2009.
33. *X-Red 32 v. 1.53*; Stoe & Cie GmbH: Darmstad, Germany, 2009.
34. *X-Shape v.2.12.2*; Stoe & Cie GmbH: Darmstadt, Germany, 2009.
35. *APEX v. 2012.10_0*; Bruker AXS Inc.: Madison, WI, USA, 2012.
36. *SAINT v. 7.99A*; Bruker AXS Inc.: Madison, WI, USA, 2012.
37. Sheldrick, G.M. *SADABS v. 2012/1*; Bruker AXS Inc.: Madison, WI, USA, 2012.
38. Sheldrick, G.M. Crystal structure refinement with *SHELXL*. *Acta Crystallogr. Sect. C* **2015**, *71*, 3–8. [CrossRef] [PubMed]
39. Hübschle, C.B.; Sheldrick, G.M.; Dittrich, B. *ShelXle*: A Qt graphical user interface for *SHELXL*. *J. Appl. Crystallogr.* **2011**, *44*, 1281–1284. [CrossRef] [PubMed]
40. Farrugia, L.J. *WinGX* and *ORTEP* for *Windows*: An update. *J. Appl. Crystallogr.* **2012**, *45*, 849–854. [CrossRef]
41. *POV-Ray v. 3.7*; Persistence of Vision Pty. Ltd.: Williamstown, Australia, 2004; Available online: http://www.povray.org/ (accessed on 18 March 2014).

![inorganics logo] *inorganics*

MDPI

Article

Gadolinium(III)-DOTA Complex Functionalized with BODIPY as a Potential Bimodal Contrast Agent for MRI and Optical Imaging

Matthias Ceulemans, Koen Nuyts, Wim M. De Borggraeve and Tatjana N. Parac-Vogt *

Department of Chemistry, KU Leuven, Celestijnenlaan 200F, Leuven 3001, Belgium;
Matthias.Ceulemans@chem.kuleuven.be (M.C.); Koen.Nuyts@chem.kuleuven.be (K.N.);
Wim.DeBorggraeve@chem.kuleuven.be (W.M.D.B.)
* Correspondence: Tatjana.Vogt@chem.kuleuven.be; Tel.: +32-16-327-612.

Academic Editors: Stephen Mansell and Steve Liddle
Received: 29 September 2015; Accepted: 17 November 2015; Published: 25 November 2015

Abstract: The synthesis and characterization of a novel gadolinium(III) DOTA complex functionalized with a boron-dipyrromethene derivative (BODIPY) is described. The assembly of the complex relies on azide diazotransfer chemistry in a copper tube flow reactor. The azide thus formed is coupled directly with an alkyne via click chemistry, resulting into a paramagnetic and luminescent gadolinium(III) complex. Luminescent data and relaxometric properties of the complex have been evaluated, suggesting the potential applicability of the complexes as a bimodal contrast agent for magnetic resonance and optical imaging. The complex displays a bright emission at 523 nm with an absorption maximum of 507 nm and high quantum yields of up to 83% in water. The proton relaxivity of the complex measured at 310 K and at frequencies of 20 and 60 MHz had the values of 3.9 and 3.6 $s^{-1} \cdot mM^{-1}$, respectively.

Keywords: Gadolinium(III); BODIPY; click chemistry; MRI; contrast agent

1. Introduction

Magnetic resonance imaging (MRI) is a diagnostic tool that has experienced large growth over the past years. Consequently, the search for highly-efficient, responsive, and tissue-specific markers has followed the same trend. Most of the clinically-used contrast agents are based on gadolinium(III) chelates [1–6]. Gadolinium(III), with its symmetric $^8S_{7/2}$ ground state and large magnetic moment (7.94 μ_B), is a superior ion to efficiently increase relaxation rates of water molecules [1]. Due to the toxicity of free gadolinium(III) (LD_{50} = 0.2 mmol·kg^{-1} in mice), and the fact that relative high concentrations of contrast agents are needed for MRI scan (up to 0.3 mmol per kg body weight) [7,8], strong chelating agents are used to ensure kinetic and thermostatic stability of the gadolinium(III) complex. The most widespread contrast agents used in modern molecular imaging techniques are the acyclic diethylenetriaminepentaacetic acid (Gd(III)-DTPA, Magnevist®, Bayer Healthcare, Berlin, Germany) and the cyclic 1,4,7,10-tetraazacyclododecane-1,4,7,10-tetraacetic acid (Gd(III)-DOTA, Dotarem®, Guerbet, Hongkong, China) [2]. The eight-fold coordination ensures high stability (log$K$ = 25.3 and 20.1 for DOTA and DOTA-propylamide respectively), and allows the binding of one water molecule directly to the metal center [9–12].

MRI excels in its special resolution and depth penetration, but it suffers from a low sensitivity. Combining it with a complimentary imaging technique would allow for obtaining a better diagnostic tool, as it was recently demonstrated by a successful combination of MRI and PET probes [13–15]. The combined contrast agent gives both insight into the morphology (MRI) and information about biomedical processes (PET) of the human body. Another promising imaging modality complementary to MRI is optical imaging, which has a very good sensitivity, but lacks the spatial resolution and depth penetration. The latter drawback can be circumvented by choosing the absorption and emission wavelength close to the biological window [16]. Different optical probes have already been coupled to DTPA or DOTA, such as luminescent polymers in lipophilic aggregates [17], luminescent metal complexes [18–27], dendrimers [28], and organic dyes such as quinolone [29], rhodamine [30,31], fluoresceine [32], and naphalimide [33,34].

A promising class of organic dyes are 4,4-Difluoro-4-bora-3a,4a-diaza-s-indacene (abbreviated as BODIPY) and its derivatives. This class of small organic molecules generally has very high extinction coefficients, fairly sharp fluorescence peaks, and high quantum yields. The BODIPY core is pretty stable in physiological conditions and relatively insensitive to the environment. Their emission wavelength is tunable by increasing the resonance within the BODIPY core [35]. The BODIPY dyes have been used in a variety of applications such as biological labels and probes [36–38], fluorescent probes [39], laser dyes [40,41], light emitting diodes [42], solar cells [43,44], and potential sensitizers in photodynamic therapy [45,46]. Although BODIPYs are versatile compounds with good optical properties, their use in developing bimodal contrast agents remain scarce [47–52]. ^{18}F labeled BODIPY derivatives have been recently suggested as bifunctional reporters for hybrid optical/positron emission tomography imaging [53,54]. An example of DOTA ligand functionalized with BODIPY compound was reported in 2010 by Bernhard *et al.* [47] with the idea to generate a bimodal imaging agent for optical and nuclear imaging (PET/CT) by introducing In(III), Ga(III) and Cu(II) into the DOTA moiety. In 2012 the same group expanded their compound to a DOTAGA derivative (GA = glutaric acid) which gave more stable complexes with the transition metals [48,51,52]. Considering the versatility and favorable optical properties of BODIPY organic dyes, there are surprisingly very few examples of its use as an optical probe in the development of bimodal contrast agents. The main disadvantage of BODIPY dyes is their hydrophobicity, which limits their solubility and therefore hinders their straightforward application as a probe for medical imaging. Although overcoming the solubility issue poses a challenge, it is also beneficial as it was demonstrated that hydrophilic dyes can improve cell permeability [50].

We have recently shown that BODIPY dyes can be functionalized in a copper tube flow reactor via chemistry that converts primary amines into azides using the catalyst generated *in situ* from the metallic copper [55]. In this paper we further expand this strategy for performing azide-alkyne cycloaddition in order to create a Gd-DOTA-BODIPY derivative. We report on the synthesis and characterization of this novel compound and evaluate its potential as a bimodal contrast agent for MRI and optical imaging.

2. Results and Discussion

2.1. Synthesis of Gd-DOTA-BODIPY Derivative

The main concept used in order to create a Gd-DOTA-BODIPY derivative is based on using click chemistry to couple a BODIPY dye to a DOTA moiety, resulting in a bimodal agent with both optical and paramagnetic entities (Figure 1). The novel BODIPY derivative developed in this work is 5,5-difluoro-1,3,7,9-tetramethyl-10-(2-(4-((2-gadolinium(III)-(4,7,10-tris(carboxylatomethyl)-1,4,7,10-tetraazacyclododecan-1-yl)acetamido)methyl)-1H-1,2,3-triazol-1-yl)ethyl)-5H-dipyrrolo[1,2-c:2′,1′-f]diazaborinin-4-ium-5-uide, presented in Figure 1. The synthesis of the BODIPY core starts with the protection of β-analine with phtalic acid in solvent free conditions. Subsequently reaction with thionyl chloride converts the acid group into an acid chloride, which can react further with two equivalent of 2,4-dimethylpyrrole to form a dipyrromethene derivative. Complexation of this product with borontrifluoride dietherate and a base forms BODIPY-$(CH_2)_2$-phtalimide. This product can easily be transformed into BODIPY-$(CH_2)_2$-amine by using hydrazine in ethanol, which was clearly demonstrated by the disappearance of the aromatic peaks at 7.77 and 7.88 ppm, and appearance of a new peak at 1.61 ppm in the ^{1}H NMR spectrum (Scheme 1). The final BODIPY product contains an amine function which can be transformed in a copper tube into an azide via a diazotranfer reagent (ISA·H_2SO_4) giving BODIPY-$(CH_2)_2$-azide shown in Figure 2a [56,57].

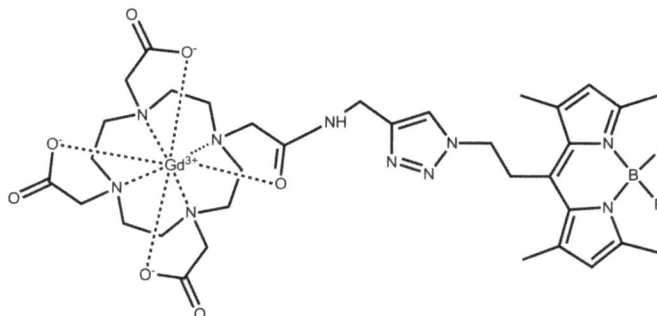

Figure 1. The Gd-DOTA-BODIPY derivative.

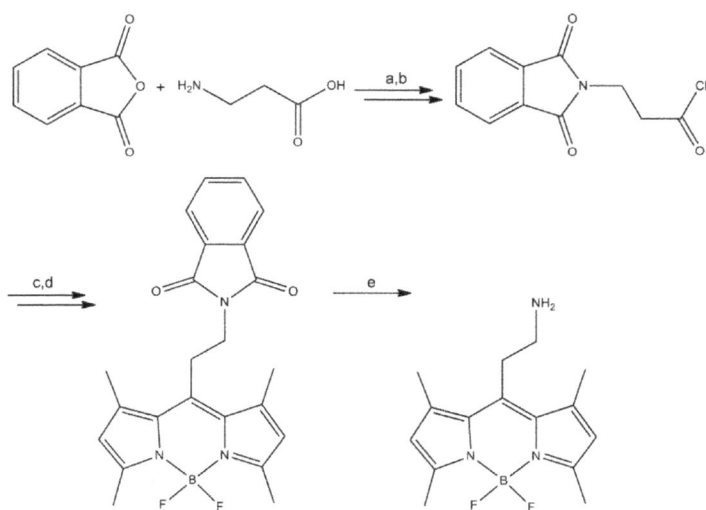

Scheme 1. Synthetic procedure for BODIPY-(CH$_2$)$_2$-amine: (**a**) 160 °C, 6 h; (**b**) SOCl$_2$, reflux, 3 h; (**c**) 2 equivalents 2,4-dimethylpyrrole, DCM, 0 °C reflux, 4 h; (**d**) 10 eq. DIPEA, 11 eq. BF$_3$·Et$_2$O, DCM, 0 °C R.T., overnight; and (**e**) N$_2$H$_4$·H$_2$O, EtOH, reflux, 3 h.

The propargylated Ln(III)-DOTA complex (Ln = La, Gd) that was used for coupling to BODIPY dye is shown in Figure 2b [58,59]. The coupling was performed by using a flow chemistry approach, which can mitigate some safety related issues regarding working with azides [60]. Organic azides can be synthesized from primary amines via a diazotransfer reaction. The potential short shelf life and highly explosive nature of some diazotransfer reagents requires their careful consideration of the safety issues for their handling [61]. Some safety issues can be circumvented by introduction of imidazole-1-sulfonyl azide (ISA) or its hydrogen sulfate salt (ISA·H$_2$SO$_4$) [56,62]. The flow reaction was performed by mixing solutions of BODIPY-(CH$_2$)$_2$-amine and ISA·H$_2$SO$_4$, which is graphically represented in Scheme 2. Upon leaving the reactor, the reaction mixture was directly added to a solution of propargylated Ln(III)-DOTA (Ln = La or Gd), resulting in the final product Ln(III)-DOTA-BODIPY.

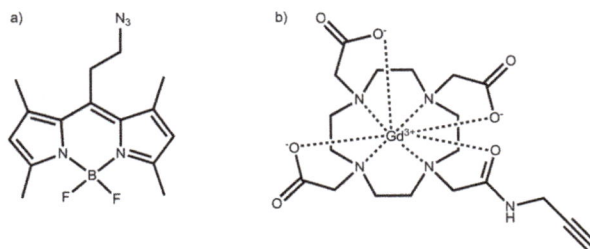

Figure 2. (**a**) Structure of BODIPY-(CH₂)₂-azide made in a copper tube reactor; (**b**) Structure of proparylated Gd(III)-DOTA complex used to couple to the azide.

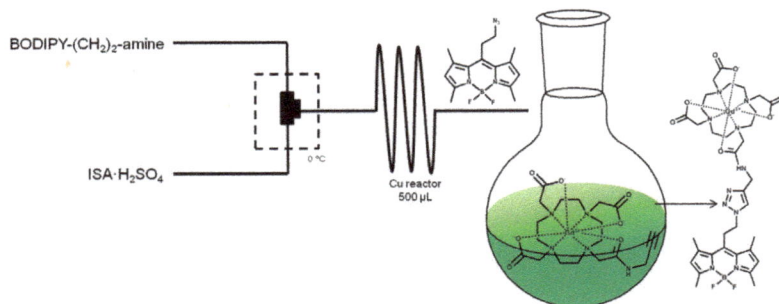

Scheme 2. Schematic representation of the flow reactor.

The Ln-DOTA-BODIPY complexes have been isolated and purified (see Figure S13) and the diamagnetic lanthanum(III) complex La-DOTA-BODIPY has been characterized by ^{1}H NMR spectroscopy in solution (Figure 3). The ^{1}H NMR spectrum shows a distinct peak of the triazole proton at 7.55 ppm and two protons on the BODIPY core at 6.00 ppm, indicating a successful linkage of the DOTA moiety with the BODIPY derivative. The ^{1}H NMR spectrum also shows broad peaks in the region from 2.20–3.82 ppm which are typical for the protons in the DOTA ring [63].

2.2. Photophysical Properties of the Gd-DOTA-BODIPY

The BODIPY derivatives are typically strongly red colored. Most BODIPY derivatives are apolar in nature and dissolve very well in apolar organic solvents, like chloroform or dichloromethane, and will emit a bright green fluorescence. Due to the coupling to hydrophilic Ln-DOTA complex the final Ln-DOTA-BODIPY adducts are water soluble and give bright green fluorescence in aqueous solution. Figure 4 shows a picture of an aqueous solution of the synthesized Gd-DOTA-BODIPY in the absence and in the presence of excitation light at 366 nm. The green fluorescence is clearly visible during excitation. Furthermore, the electronic spectra of Gd-DOTA-BODIPY are given in Figure 4 and depict the characteristic and rather narrow absorption and emission bands which are typical for BODIPY dyes. A main absorption band (blue dotted line) with a maximum λ_{abs}(max) at 503 nm is observed for Gd-DOTA-BODIPY in aqueous solution. These visible absorption bands can be assigned to the $S_0 \rightarrow S_1$ transition [64]. An additional, considerably weaker broad absorption band is observed in the UV-VIS region around 360 nm, and is attributed to the $S_0 \rightarrow S_2$ transition. It should be noted that the optical properties of the BODIPY dyes can be tuned by increasing conjugation through placing different substituents on the BODIPY core. Such substitutions increase the resonance of the whole structure and provide a red shift of the main emission band, which is favorable for biological applications. A tradeoff between tissue penetration and image resolution for *in vivo* imaging can be made in the optical imaging window (from 665 to 900 nm) [65]. In this region the extinction coefficients of the main sources of absorption such as hemoglobin, deoxyhemoglobin, and water, are at their minimum. However, it must be taken into account that placing additional substituents on the BODIPY usually renders the system more hydrophobic and decreases the solubility of the entire complex in water.

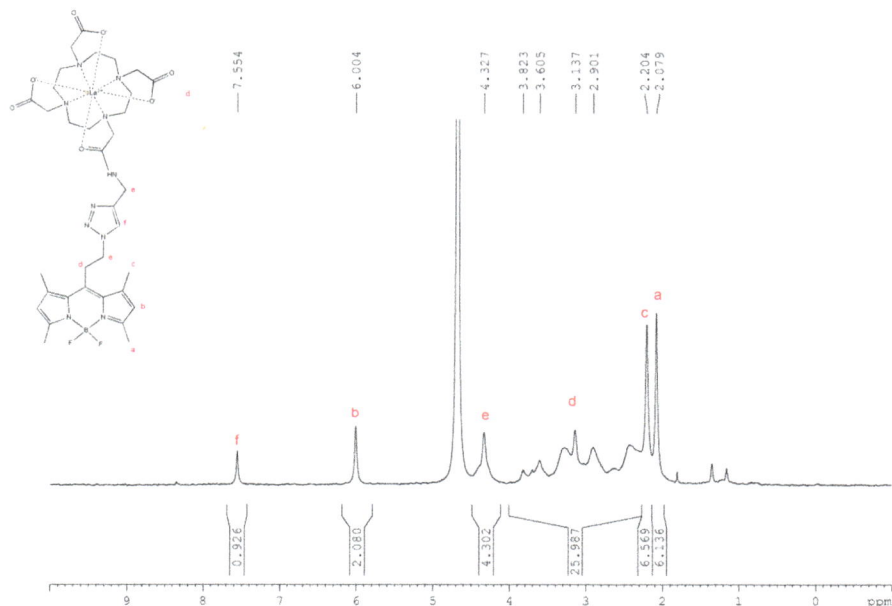

Figure 3. ^{1}H NMR spectrum of the final clicked La-DOTA-BODIPY derivative measured in deuterated water.

The emission maximum in water of Gd-DOTA-BODIPY (Figure 4) is observed at 523 nm and is in accordance with the previously reported BODIPY derivatives [66]. The excitation maximum (red dashed line) recorded at 560 nm shows a peak at 503 nm.

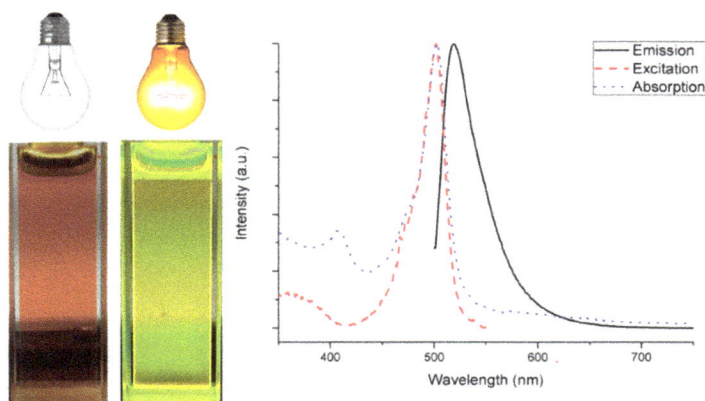

Figure 4. The emission (black line), excitation (red dashed line) and absorption (blue dotted line) spectrum of Gd-DOTA-BODIPY in water.

The fluorescent quantum yields Q_L of La-DOTA-BODIPY and Gd-DOTA-BODIPY were determined upon ligand excitation by a comparative method, using a solution of Rhodamine 6G (Q = 78%) [67] in water as the standard. The advantage of BODIPY dyes is that the quantum yields of their derivatives are generally high, even in water [39]. The quantum yield was determined according to the following equation:

$$Q_L = Q_S \times \frac{I_X}{I_S} \times \frac{A_S(\lambda_{exc})}{A_X(\lambda_{exc})} \times \frac{\eta_X^2}{\eta_S^2} \qquad (1)$$

In this equation the s and x refer to the standard and the unknown sample respectively, I represents the corrected total integrated emission intensity, A: the absorbance at the excitation wavelength, η: the refractive index of the solution (η_{water} = 1.33), and Q_s: the quantum yield of the standard (Q = 78%) [67]. The samples are diluted until the absorbance at the excitation wavelength is between 0.02–0.05. Quantum yield is recorded with ligand excitation wavelength of 490 nm and gave values of 80% and 83% for La-DOTA-BODIPY and Gd-DOTA-BODIPY, respectively.

2.3. Relaxivity of Gd-DOTA-BODIPY

The relaxivity r_1 is the efficiency of a 1 mM solution of gadolinium(III) agent to shorten the longitudinal relaxation time (T_1). According to the Solomon Bloembergen Morgan theory [68,69] several parameters can alter the relaxivity of a contrast agent. High relaxivity can be obtained with a higher amount of water molecules directly bound to the paramagnetic centre (q). Although an easy parameter to adjust, increasing the q value leads to less stable complexes. Other parameters are the distance between gadolinium(III) and water (r), the water residence time (τ_M), the rotational correlation time of the paramagnetic center (τ_R), the electronic relaxation time of gadolinium(III) at zero field (τ_{S0}) and the correlation time modulating the electronic relaxation (τ_v). The relaxivity of Gd-DOTA-BODIPY derivative measured at frequencies of 20 and 60 MHz at 310 K and was found to be 3.9 and 3.6 $s^{-1} \cdot mM^{-1}$ respectively. These values are in close comparison with previously reported values for Gd(III)-DOTA complex, which gave values of 3.5 and 3.1 $s^{-1} \cdot mM^{-1}$ at 20 and 60 MHz respectively [70]. A slight increase compared to the parent Gd-DOTA complex can be attributed to the increase in molecular weight of the Gd-DOTA-BODIPY complex, resulting in the increase of the rotational correlation time (τ_R) and in overall increase of the relaxivity. The limiting factors for the increase of Gd-DOTA-BODIPY relaxivity are probably the presence of an amide bond in the DOTA scaffold which may slow down the water exchange rate, or flexible linker used for attaching the BODIPY core to DOTA, which may result in high internal rotations within the Gd-DOTA-BODIPY.

3. Experimental Section

3.1. Materials, Reagents, and Solvents

β-alanine, boron trifluoride diethyl etherate, phosphoryl chloride, thionyl chloride was obtained from Sigma-Aldrich (Bornem, Belgium); Copper(II)sulphate, 2,4-dimethylpyrrole, dry dichloromethane, lanthanum(III) chloride heptahydrate, methanol, sodium ascorbate, triethylamine were purchased from Acros Organics (Geel, Belgium); acetonitrile, diethyl ether, ethyl acetate, petroleum ether, magnesium sulphate from VWR chemicals (Leuven, Belgium); dichloromethane, from hydrogen chloride Fisher Scientific (Loughborough, UK); Gadolinium(III) was obtained from Alfa Aesar (Ward Hill, Shrewsbury, MA, USA) and were used without further purification.

3.2. Characterization

NMR spectra were measured on a Bruker Avance 300 (Bruker, Karlsruhe, Germany) (operating at 300 MHz for ^{1}H NMR spectra, operating at 75 MHz for ^{13}C NMR), Bruker Avance 400 (operating at 400 MHz for ^{1}H NMR spectra, operating at 100 MHz for ^{13}C NMR spectra). Chemical shifts are reported in parts per million (ppm) and are referenced to the internal standard tetramethylsilane. For ^{13}C spectra, residual solvent signals are used as the internal standard. Spectra are taken at room temperature unless otherwise stated. Relaxation rates at 20 and 60 MHz were obtained on a Minispec mq-20, and a Minispec mq-60 (Bruker, Karlsruhe, Germany) respectively.

FT-IR spectra were measured by using a Bruker Vertex 70 FT-IR spectrometer (Bruker, Ettlingen, Germany). ESI-MS measurements were taken on a Thermo Electron LCQ Advantage apparatus (Thermo Scientific, Waltham, MA, USA) with Agilent 1100 pump (Agilent, Waldbronn, Germany) and injection system coupled to Xcalibur data analyzing software. Methanol is used as eluent. TXRF measurements were done on a Bruker S2 Picofox (Bruker, Berlin, Germany) by analyzing approximately 100 ppm gadolinium or europium solutions with respect to a Chem-Lab gallium standard solution (500 μg/mL, 2%–5% HNO_3).

The flow setup was constructed with copper GC tubing (0.065″ inner diameter, Restek, Middelburg, Netherlands) and the reagents are pumped using a Chemyx Fusion 200 syringe pump (Chemyx, Stafford, TX, USA). LC–MS was performed on an Alltech Prevail RP-C18 5 μm 150 mm × 2.1 mm column coupled to an Agilent 1100 degasser, quaternary pump, auto sampler, UV-DAD detector, and thermostated column module coupled to Agilent 6110 single-quadrupole MS. Use of Agilent LC/MSD Chemstation software. HPLC was performed on Waters Delta 600 analytical/preparative system equipped with a Waters 996 Photo Diode Array detector (Waters, Milford, MA, USA). Preparative column: Alltech C18 Prevail 5 μm 150 mm × 22 mm.

UV-VIS absorption spectra were measured on a Varian Cary 5000 spectrophotometer (Agilent, Santa Clara, CA, USA) on freshly prepared aqua solutions in quartz Suprasil® cells (115F-QS) with an optical path-length of 1.0 cm. Excitation and emission data were recorded on an Edinburgh Instruments FS900 steady state spectrofluorimeter (Edinburgh Instruments, Livingston, UK). This instrument is equipped with a 450 W xenon arc lamp and an extended red-sensitive photomultiplier (Hamamatsu R 2658P). Quantum yields were determined by a comparative method with an estimated experimental error of ± 10% using a solution of Rhodamine 6G in water (Q = 78%) as standard. The solutions are diluted to get an optical density lower than 0.05 at the excitation wavelength.

3.3. Synthetic Procedures

3.3.1. Synthesis of Product (**1**) (Figure 5)

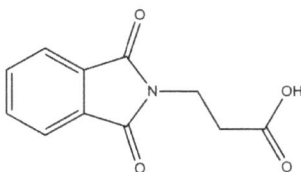

Figure 5. Chemical structure of Product **1**.

Phthalic anhydride (1 eq, 40 mmol, 5.92 g) and β-alanine (1 eq, 40 mmol, 3.56 g) were added in a round bottom flask and heated up to 160 °C for 6 h where it became a smelt. The mixture was cooled and the resulting solid was dissolved in DCM. The solution was washed with 0.1 M HCl (3 × 30 mL). The organic layer was dried with $MgSO_4$ and evaporated till dryness, yield: 5.68 g, 25.9 mmol, 65%. CI–MS (MeOH, *m/z*): calcd: 219.19 g/mol, found: 220.0 g/mol [M + H]⁺. ¹H NMR (300 MHz, CDCl₃, δ ppm): 2.80 (t, *J* = 7.4 Hz, 2H), 4.00 (t, *J* = 7.4 Hz, 2H), 7.74 (m, 2H), 7.84 (m, 2H); ¹³C NMR (75 MHz, CDCl₃, δ ppm): 32.8, 33.4, 123.4, 132.0, 134.1, 168.0, 176.5.

3.3.2. Synthesis of BODIPY-$(CH_2)_2$-Phtalimide (**2**) (Figure 6)

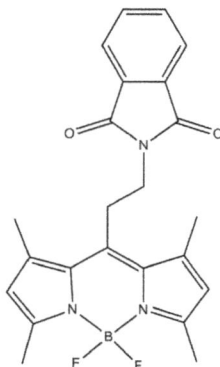

Figure 6. Chemical structure of Product **2**.

To a round bottom flask product **1** (1 eq, 5 mmol, 1.10 g) and thionyl chloride (5 eq, 25 mmol, 1.82 mL) were added and refluxed for 3 h. The thionyl chloride was removed *in vacuo* until dryness. The acid chloride was used further without any purification. The yellow residue was dissolved in 30 mL dry DCM and 2,4-dimethylpyrrole (2 eq, 10 mmol, 1.04 mL) was slowly added. The solution was stirred for 30 min at room temperature. The color changed from yellow to red-brown. The solution was subsequently refluxed for 4 h. The solution was then cooled to 0 °C and triethylamine (10 eq, 50 mmol, 6.97 mL) was added. After 5 min of stirring at 0 °C boron trifluoride diethyl etherate (11 eq, 55 mmol, 6.79 mL) was slowly added and the solution was stirred overnight at room temperature. Diethyl ether (100 mL) was added to the solution and the organic layer was washed with water (3 × 50 mL). The organic layer was dried with $MgSO_4$ and concentrated *in vacuo*. The resulting residue was purified using silica column (eluent: DCM/PET 50:50) yielding a dark orange solid, yield: 0.392 g, 0.9 mmol, 19%. CI–MS (MeOH, *m*/*z*): calcd: 421.25 g/mol, found: 421 g/mol [M]$^+$. ^{1}H NMR (300 MHz, CDCl$_3$, δ ppm): 2.54 (s, 6H), 2.67 (s, 6H), 3.42 (t, *J* = 8.60 Hz, 2H), 3.89 (t, *J* = 8.60 Hz, 2H), 6.11 (s, 2H), 7.77 (m, 2H), 7.88 (m, 2H); ^{13}C NMR (100 MHz, CDCl$_3$, δ ppm): 14.6, 16.4, 27.7, 39.0, 122.2, 123.5, 131.7, 131.9, 134.3, 139.8, 141.5, 154.9, 168.2.

3.3.3. Synthesis of BODIPY-$(CH_2)_2$-Amine (**3**) (Figure 7)

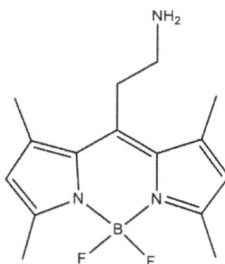

Figure 7. Chemical structure of Product 3.

BODIPY-$(CH_2)_2$-phtalimide (**2**) (1 eq, 0.37 mmol, 158 mg) and hydrazine hydrate (1 eq, 0.37 mmol, 18.8 mg) in 20 mL ethanol was refluxed for 3 h. After which the solution was cooled to room temperature and filtrated with a Millipore 0.45 μM filter and the solvent was concentrated *in vacuo* until an orange solid, yield: 0.106, 0.36 mmol, 97%. CI–MS (MeOH, *m/z*): calcd: 291.17 g/mol, found: 282 g/mol [M − F]⁺, and 291 [M]⁺. ¹H NMR (300 MHz, CDCl₃, δ ppm): 1.61, (broad, 2H), 2.45 (s, 6H), 2.51 (s, 6H), 2.99 (m, 2H), 3.14 (m, 2H), 6.05 (s, 2H); ¹³C NMR (100 MHz, CDCl₃, δ ppm): 10.5, 12.6, 28.2, 40.1, 117.9, 127.8, 136.6, 138.9, 150.2.

3.3.4. Synthesis of DO3A-tBu (**4**) (Figure 8)

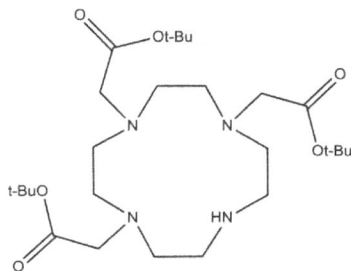

Figure 8. Chemical structure of Product **4**.

The synthesis uses an altered synthesis by Viguier *et al.* [58]. To a solution of tetraazacyclodedecaan (1 eq, 5.80 mmol, 1 g), sodium bicarbonate (3.5 eq, 20.3 mmol, 1.71 g) in 150 mL ACN under an argon atmosphere a solution of tert-butyl bromoacetate (3.5 eq, 20.3 mmol, 3.0 mL) in 50 mL ACN was added drop wise. The mixture was refluxed for 17 h. After removing the salts via filtration over Celite® the solvent was evaporated and the solid was recrystallized from toluene as a white powder, yield: 2.47 g, 4.8 mmol, 82%. ESI–MS: *m/z* calcd 516 [M + H]⁺, 538 [M + Na]⁺, found 516.0 [M + H]⁺, 537.5 [M + Na]⁺. ¹H NMR (CDCl₃, 300 MHz, δ ppm): 1.46 (s, 27H), 2.88 (t, 12H), 3.08 (t, 4H), 3.30 (s, 2H), 3.38 (s, 4H). ¹³C NMR (CDCl₃, 75 MHz, δ ppm): 28.2, 28.3, 47.5, 49.5, 51.3, 58.1, 81.6, 81.7, 169.8, 170.6.

3.3.5. Synthesis of {4,10-Bis-Tert-Butoxycarbonylmethyl-7-[(2-Propynylcarbamoyl)-Methyl]-1,4,7,10-Tetraaza-Cyclododec-1-yl}-Acetic Acid Tert-Butyl Ester (**5**) (Figure 9)

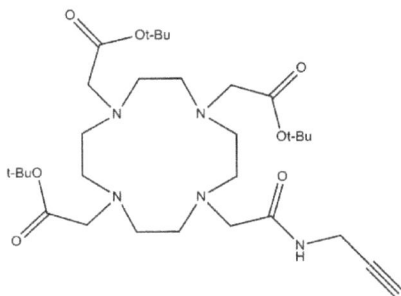

Figure 9. Chemical structure of Product **5**.

A solution of DO3A-tBu (**4**) (1 eq, 772 mg, 1.5 mmol), N-(2-propynyl)chloroacetamide (1.2 eq, 291 mg, 1.8 mmol) and K₂CO₃ (2 eq, 415 mg, 3.0 mmol) in 70 mL ACN was stirred under nitrogen at reflux temperature for 17 h. After filtration of over Celite®, the solvent was removed *in vacuo*, the residual mixture was purified by basic alumina (eluens CHCl₃:MeOH (98:2)) to give a colorless solid, yield: 1.45 g, 2.4 mmol, 94%. ESI–MS: *m/z* calcd 610.80 [M + H]⁺, found 632 [M + Na]⁺. ¹H NMR

(CDCl$_3$, 300 MHz, δ ppm): 1.45 (s, 18H), 1.46 (s, 9H), 2.15 (t, J = 2.3 Hz, 1H), 2.52 (broad, 4H), 2.70 (broad, 4H), 2.82 (broad, 4H), 2.91 (broad, 4H), 3.10 (s, 2H), 3.27 (s, 4H), 3.38 (s, 2H), 4.05 (dd, J = 5.5 Hz, J = 2.3 Hz, 2H), 9.27 (t, J = 5.5 Hz, 1H). ^{13}C NMR (CDCl$_3$, 75 MHz, δ ppm): 28.2, 28.3, 28.6, 29.3, 52.1, 52.5, 53.8, 55.2, 56.2, 57.6, 70.2, 72.9, 80.9, 170.8.

3.3.6. Synthesis of {4,10-Bis-Carboxymethyl-7-[(2-Propynylcarbamoyl)-Methyl]-1,4,7,10-Tetraaza-Cyclododec-1-yl}-Acetic Acid (**6**) (Figure 10)

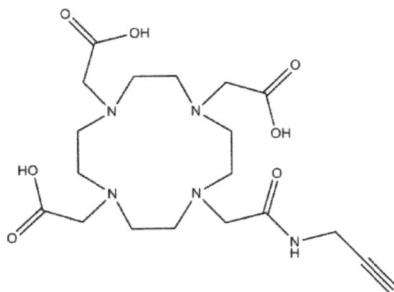

Figure 10. Chemical structure of Product **6**.

Product (**5**) (1 eq, 1.45 g, 2.4 mmol) was dissolved in 10 mL of a DCM/TFA (50:50) mixture and stirred under inert atmosphere overnight. After the reaction, 20 mL of DCM was added and evaporated (repeated two times), yield: 1.07 g, 2.4 mmol, quantitative. ESI-MS: m/z calcd 442.22 [M + H]$^+$, found 442 [M + H]$^+$, 464 [M + Na]$^+$. ^{1}H NMR (D$_2$O, 300 MHz, δ ppm): 2.42 (t, 1H), 2.48–4.01 (broad, 22H), 3.81 (broad, 2H). ^{13}C NMR (D$_2$O, 75 MHz, δ ppm): 29.1, 29.6, 49.7, 53.7, 58.6, 60.6, 60.8, 69.3, 72.7, 73.5, 78.7, 175.4, 179.4, 179.6.

3.3.7. Synthesis of Lanthanide(III) {4,10-Bis-Carboxymethyl-7-[(2-Propynylcarbamoyl)-Methyl]-1,4,7,10-Tetraaza-Cyclododec-1-yl}-Acetic Acid (Figure 11)

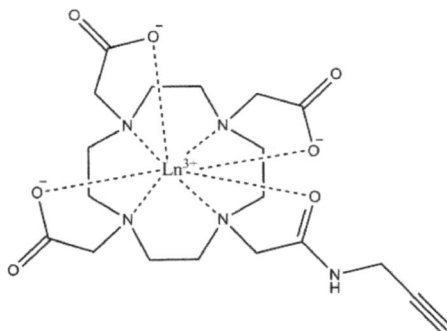

Figure 11. Chemical structure of Products **7** and **8**.

General procedure for complexation of propargylated DOTA: Product (**6**) (1 eq) was dissolved in 5 mL water and the appropriate lanthanide(III) chloride hydrate (1,1 eq) was added. The pH was adjusted to 5.5 with a 0.1 M KOH solution and stirred overnight at 60 °C. The solvent was evaporated. The residue was redissolved in 5 mL water and Chelex-100 was added and stirred for 2 h. This last step was repeated until no free lanthanide was found with an Arsenazo III indicator [71].

Propargylated Gd-DOTA (**7**): Yield: 0.446 g, 0.75 mmol, 67%. ESI-MS: *m/z* calcd 596.70 [M + H]$^+$, found 619.2 [M + Na]$^+$. IR (neat): 3416, 3245, 3098, 2873, 1675, 1609 cm^{-1}.

Propargylated La-DOTA (**8**): Yield: 0.445 g, 0.77 mmol, 65%. ESI-MS: *m/z* calcd 478.36 [M + H]$^+$, found 600.5 [M + Na]$^+$. ^{1}H NMR (CDCl$_3$, 300 MHz, δ ppm): 2.42 (t, 1H), 2.48–4.01 (broad, 22H), 3.81 (broad, 2H). ^{13}C NMR (CDCl$_3$, 75 MHz, δ ppm): 29.1, 29.6, 49.7, 53.7, 58.5, 60.6, 60.8, 69.2, 72.7, 73.5, 78.7, 175.4, 179.4, 179.6; IR (neat): 3429, 3258, 3103, 2857, 1673, 1606 cm^{-1}.

3.3.8. General Procedure Flow

Stock solution A: a 2 mL solution in a degassed mixture of 10:3:3 MeOH:DCM:H$_2$O of 1 eq BODIPY-amine and 11 equivalents of DIPEA.

Stock solution B: a 2 mL solution of three equivalents of ISA·H$_2$SO$_4$ in a degassed mixture of 10:3:3 MeOH:DCM:H$_2$O.

Both solutions in a separate syringe are combined via a T-mixer in an ice bath before introducing the mixture to the copper tube flow reactor (150 μL) at room temperature. Flow speed is adjusted so the retention time in the reactor is 300 s. The column is stabilized during three reactor volumes. Upon leaving the reactor, the reaction mixture was directly added to a solution of 1.5 equivalent of Ln-DOTA-propargyl and 1.5 eq sodium ascorbate. The mixture is left stirring overnight. The aqueous layer is lyophilized, redissolved in distilled water, purified via HPLC and lyophilized. The products are isolated as bright orange powders.

Product **9** (Gd) (Figure 12): Yield: 13 mg, 0.01 mmol, 6%. ESI–MS: *m/z* calcd 913.85 [M + H]$^+$, found 913.2 [M + H]$^+$. The concentration used for relaxivity was 1.45 mmol/L and was measured by TXRF against an internal gallium standard.

Product **10** (La) (Figure 12): Yield: 23 mg, 0.03 mmol, 9%. ESI-MS: *m/z* calcd 895.51 [M + H]$^+$, found 917.8 [M + Na]$^+$ (positive mode), 894.2 [M + e]$^-$ (negative mode). ^{1}H NMR (D$_2$O, 300 MHz, δ ppm): 2.08 (s, 6H), 2.20 (s, 6H), 2.20–3.82 (broad, 26H), 4.33 (broad, H), 6.00 (s, 2H), 7.55 (s, 1H). ^{11}B NMR (D$_2$O, 80 MHz, δ ppm): 2.16 ppm with respect to B(OMe)$_3$ as a reference.

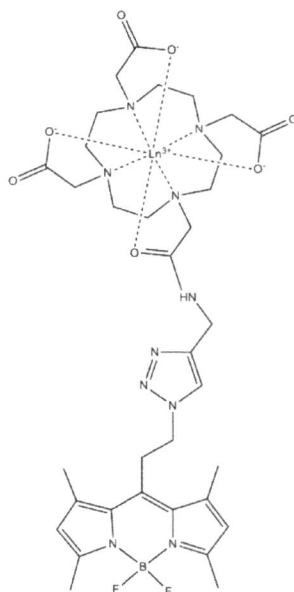

Figure 12. Chemical structure of Products **9** and **10**.

Inorganics **2015**, *3*, 133–147

After each reaction the copper tube flow reactor is cleansed in a sonication bath for 10 min while introducing a 1:1 mixture of MeOH:triethylamine at 4 mL/min, followed by 20 mL of pentane at 4 mL/min. The reactor is dried with argon and stored under argon.

4. Conclusions

In this paper a novel Gd-DOTA-BODIPY derivative is synthesized via a diazotransfer reaction under flow conditions in a copper tube. The complex is water-soluble and exhibits favorable fluorescent properties, thus offering the possibility for the use of BODIPY adducts for *in vivo* optical imaging. High quantum yield of 83% is obtained in water, and bright emission was observed at 523 nm. Shifting of the emission to the more optimal red-IR region of the spectrum will be attempted in future by increasing the electronic resonance of the BODIPY by adding appropriate substituents to the BODIPY core. The complex exhibits relaxivities which are comparable to the parent Gd-DOTA complex and, therefore, holds potential as a bimodal contrast agent for MRI and optical imaging.

Supplementary Materials: Supplementary materials can be found at http://www.mdpi.com/2304-6740/3/4/0516/s1.

Acknowledgments: Karel Duerinckx is acknowledged for his help with the NMR measurements and Michael Harris is acknowledged for his help with the relaxometric measurements.

Author Contributions: Matthias Ceulemans and Koen Nuyts have performed experimental work and spectroscopic measurements. Wim M. De Borggraeve and Tatjana N. Parac-Vogt have supervised the work and helped with interpretation of the results sand the writing of the final manuscript.

Conflicts of Interest: The authors declare no conflict of interest.

References

1. Hermann, P.; Kotek, J.; Kubícek, V.; Lukes, I. Gadolinium(III) complexes as MRI contrast agents: Ligand design and properties of the complexes. *Dalton Trans.* **2008**, *9226*, 3027–3047. [CrossRef] [PubMed]
2. Pierre, V.C.; Allen, M.J.; Caravan, P. Contrast agents for MRI: 30+ years and where are we going? *J. Biol. Inorg. Chem.* **2014**, *19*, 127–131. [CrossRef] [PubMed]
3. Waters, E.A.; Wickline, S.A. Contrast agents for MRI. *Basic Res. Cardiol.* **2008**, *103*, 114–121. [CrossRef] [PubMed]
4. Tóth, É.; Helm, L.; Merbach, A. Contrast Agents I. In *Topics in Current Chemistry*; Krause, W., Ed.; Springer Berlin Heidelberg: Berlin, Germany, 2002; Volume 221, pp. 61–101.
5. Debroye, E.; Parac-Vogt, T.N. Towards polymetallic lanthanide complexes as dual contrast agents for magnetic resonance and optical imaging. *Chem. Soc. Rev.* **2014**, *43*, 8178–8192. [CrossRef] [PubMed]
6. Cacheris, W.P.; Quay, S.C.; Rocklage, S.M. The relationship between thermodynamics and the toxicity of gadolinium complexes. *Magn. Reson. Imaging* **1990**, *8*, 467–481. [CrossRef]
7. Shellock, F.G.; Kanal, E. Safety of magnetic resonance imaging contrast agents. *J. Magn. Reson. Imaging* **1999**, *10*, 477–484. [CrossRef]
8. Bartolini, M.E.; Pekar, J.; Chettle, D.R.; McNeill, F.; Scott, A.; Sykes, J.; Prato, F.S.; Moran, G.R. An investigation of the toxicity of gadolinium based MRI contrast agents using neutron activation analysis. *Magn. Reson. Imaging* **2003**, *21*, 541–544. [CrossRef]
9. Accardo, A.; Tesauro, D.; Aloj, L.; Pedone, C.; Morelli, G. Supramolecular aggregates containing lipophilic Gd(III) complexes as contrast agents in MRI. *Coord. Chem. Rev.* **2009**, *253*, 2193–2213. [CrossRef]
10. Caravan, P.; Ellison, J.J.; McMurry, T.J.; Lauffer, R.B. Gadolinium(III) chelates as MRI contrast agents: Structure, dynamics, and applications. *Chem. Rev.* **1999**, *99*, 2293–2352. [CrossRef] [PubMed]
11. Villaraza, A.J.L.; Bumb, A.; Brechbiel, M.W. Macromolecules, dendrimers, and nanomaterials in magnetic resonance imaging: The interplay between size, function, and pharmacokinetics. *Chem. Rev.* **2010**, *110*, 2921–2959. [CrossRef] [PubMed]
12. Sherry, A.D.; Brown, R.D.; Geraldes, C.F.G.C.; Koenig, S.H.; Kuan, K.-T.; Spiller, M. Synthesis and characterization of the gadolinium(3+) complex of DOTA-propylamide: A model DOTA-protein conjugate. *Inorg. Chem.* **1989**, *28*, 620–622. [CrossRef]

13. Huang, W.-Y.; Davis, J.J. Multimodality and nanoparticles in medical imaging. *Dalton Trans.* **2011**, *40*, 6087–6103. [CrossRef] [PubMed]

14. Xie, J.; Liu, G.; Eden, H.S.; Ai, H.; Chen, X. Surface-engineered magnetic nanoparticle platforms for cancer imaging and therapy. *Acc. Chem. Res.* **2011**, *44*, 883–892. [CrossRef] [PubMed]

15. Louie, A. Multimodality imaging probes: Design and challenges. *Chem. Rev.* **2010**, *110*, 3146–3195. [CrossRef] [PubMed]

16. Joshi, B.P.; Wang, T.D. Exogenous molecular probes for targeted imaging in cancer: Focus on multi-modal imaging. *Cancers (Basel)* **2010**, *2*, 1251–1287. [CrossRef] [PubMed]

17. Soenen, S.J.; Vande Velde, G.; Ketkar-Atre, A.; Himmelreich, U.; de Cuyper, M. Magnetoliposomes as magnetic resonance imaging contrast agents. *Wiley Interdiscip. Rev. Nanomed. Nanobiotechnol.* **2002**, *3*, 197–211. [CrossRef] [PubMed]

18. Costa, J.; Ruloff, R.; Burai, L.; Helm, L.; Merbach, A.E. Rigid $M(II)L_2Gd_2(III)$ (M = Fe, Ru) complexes of a terpyridine-based heteroditopic chelate: A class of candidates for MRI contrast agents. *J. Am. Chem. Soc.* **2005**, *127*, 5147–5157. [CrossRef] [PubMed]

19. Livramento, J.B.; Sour, A.; Borel, A.; Merbach, A.E.; Tóth, É. A starburst-shaped heterometallic compound incorporating six densely packed Gd^{3+} Ions. *Chem. Eur. J.* **2006**, *12*, 989–1003. [CrossRef] [PubMed]

20. Livramento, J.B.; Weidensteiner, C.; Prata, M.I.M.; Allegrini, P.R.; Geraldes, C.F.G.C.; Helm, L.; Kneuer, R.; Merbach, A.E.; Santos, A.C.; Schmidt, P.; *et al.* First *in vivo* MRI assessment of a self-assembled metallostar compound endowed with a remarkable high field relaxivity. *Contrast Media Mol. Imaging* **2006**, *1*, 30–39. [CrossRef] [PubMed]

21. Parac-Vogt, T.N.; Vander Elst, L.; Kimpe, K.; Laurent, S.; Burtéa, C.; Chen, F.; van Deun, R.; Ni, Y.; Muller, R.N.; Binnemans, K. Pharmacokinetic and *in vivo* evaluation of a self-assembled gadolinium(III)-iron(II) contrast agent with high relaxivity. *Contrast Media Mol. Imaging* **2006**, *1*, 267–278. [CrossRef] [PubMed]

22. Paris, J.; Gameiro, C.; Humblet, V.; Mohapatra, P.K.; Jacques, V.; Desreux, J.F. Auto-assembling of ditopic macrocyclic lanthanide chelates with transition-metal ions. Rigid multimetallic high relaxivity contrast agents for magnetic resonance imaging. *Inorg. Chem.* **2006**, *45*, 5092–5102. [CrossRef] [PubMed]

23. Dehaen, G.; Verwilst, P.; Eliseeva, S.V.; Laurent, S.; Vander Elst, L.; Muller, R.N.; de Borggraeve, W.M.; Binnemans, K.; Parac-Vogt, T.N. A heterobimetallic ruthenium-gadolinium complex as a potential agent for bimodal imaging. *Inorg. Chem.* **2011**, *50*, 10005–10014. [CrossRef] [PubMed]

24. Dehaen, G.; Eliseeva, S.V.; Kimpe, K.; Laurent, S.; Vander Elst, L.; Muller, R.N.; Dehaen, W.; Binnemans, K.; Parac-Vogt, T.N.; Vanderelst, L.; *et al.* A self-assembled complex with a titanium(IV) catecholate core as a potential bimodal contrast agent. *Chem. Eur. J.* **2012**, *18*, 293–302. [CrossRef] [PubMed]

25. Verwilst, P.; Eliseeva, S.V.; Vander Elst, L.; Burtea, C.; Laurent, S.; Petoud, S.; Muller, R.N.; Parac-Vogt, T.N.; de Borggraeve, W.M. A tripodal ruthenium-gadolinium metallostar as a potential $\alpha_v\beta_3$ integrin specific bimodal imaging contrast agent. *Inorg. Chem.* **2012**, *51*, 6405–6411. [CrossRef] [PubMed]

26. Dehaen, G.; Eliseeva, S.V.; Verwilst, P.; Laurent, S.; Vander Elst, L.; Muller, R.N.; de Borggraeve, W.M.; Binnemans, K.; Parac-Vogt, T.N. Tetranuclear d-f metallostars: Synthesis, relaxometric, and luminescent properties. *Inorg. Chem.* **2012**, *51*, 8775–8783. [CrossRef] [PubMed]

27. Debroye, E.; Ceulemans, M.; Vander Elst, L.; Laurent, S.; Muller, R.N.; Parac-Vogt, T.N. Controlled synthesis of a novel heteropolymetallic complex with selectively incorporated lanthanide(III) ions. *Inorg. Chem.* **2014**, *53*, 1257–1259. [CrossRef] [PubMed]

28. Kuil, J.; Buckle, T.; Oldenburg, J.; Yuan, H.; Borowsky, A.D.; Josephson, L.; van Leeuwen, F.W.B. Hybrid peptide dendrimers for imaging of chemokine receptor 4 (CXCR4) expression. *Mol. Pharm.* **2011**, *8*, 2444–2453. [CrossRef] [PubMed]

29. Musonda, C.C.; Taylor, D.; Lehman, J.; Gut, J.; Rosenthal, P.J.; Chibale, K. Application of multi-component reactions to antimalarial drug discovery. Part 1: Parallel synthesis and antiplasmodial activity of new 4-aminoquinoline Ugi adducts. *Bioorganic Med. Chem. Lett.* **2004**, *14*, 3901–3905. [CrossRef] [PubMed]

30. Rivas, C.; Stasiuk, G.J.; Sae-Heng, M.; Long, N.J. Towards understanding the design of dual-modal MR/fluorescent probes to sense zinc ions. *Dalton Trans.* **2015**, *44*, 4976–4985. [CrossRef] [PubMed]

31. Hüber, M.M.; Staubli, A.B.; Kustedjo, K.; Gray, M.H.B.; Shih, J.; Fraser, S.E.; Jacobs, R.E.; Meade, T.J. Fluorescently detectable magnetic resonance imaging agents. *Bioconjug. Chem.* **1998**, *9*, 242–249. [CrossRef] [PubMed]

32. Mishra, A.A.K.; Pfeuffer, J.; Mishra, R.; Engelmann, J.; Mishra, A.A.K.; Ugurbil, K.; Logothetis, N.K. A new class of Gd-based DO3A-ethylamine-derived targeted contrast agents for MR and optical imaging. *Bioconjug. Chem.* **2006**, *17*, 773–780. [CrossRef] [PubMed]
33. Zhang, M.; Imm, S.; Bähn, S.; Neubert, L.; Neumann, H.; Beller, M. Efficient copper(II)-catalyzed transamidation of non-activated primary carboxamides and ureas with amines. *Angew. Chem. Int. Ed.* **2012**, *51*, 3905–3909. [CrossRef] [PubMed]
34. Jang, J.H.; Bhuniya, S.; Kang, J.; Yeom, A.; Hong, K.S.; Kim, J.S. Cu^{2+}-responsive bimodal (optical/MRI) contrast agent for cellular imaging. *Org. Lett.* **2013**, *15*, 4702–4705. [CrossRef] [PubMed]
35. Ulrich, G.; Ziessel, R.; Harriman, A. The chemistry of fluorescent bodipy dyes: Versatility unsurpassed. *Angew. Chem. Int. Ed.* **2008**, *47*, 1184–1201. [CrossRef] [PubMed]
36. Yee, M.-C.; Fas, S.C.; Stohlmeyer, M.M.; Wandless, T.J.; Cimprich, K.A. A cell-permeable, activity-based probe for protein and lipid kinases. *J. Biol. Chem.* **2005**, *280*, 29053–29059. [CrossRef] [PubMed]
37. West, R.; Panagabko, C.; Atkinson, J. Synthesis and characterization of BODIPY-α-tocopherol: A fluorescent form of vitamin E. *J. Org. Chem.* **2010**, *75*, 2883–2892. [CrossRef] [PubMed]
38. Kowada, T.; Maeda, H.; Kikuchi, K. BODIPY-based probes for the fluorescence imaging of biomolecules in living cells. *Chem. Soc. Rev.* **2015**, *44*, 4953–4972. [CrossRef] [PubMed]
39. Boens, N.; Leen, V.; Dehaen, W. Fluorescent indicators based on BODIPY. *Chem. Soc. Rev.* **2012**, *41*, 1130–1172. [CrossRef] [PubMed]
40. García-Moreno, I.; Amat-Guerri, F.; Liras, M.; Costela, A.; Infantes, L.; Sastre, R.; López Arbeloa, F.; Bañuelos Prieto, J.; López Arbeloa, Í. Structural changes in the BODIPY dye PM567 enhancing the laser action in liquid and solid media. *Adv. Funct. Mater.* **2007**, *17*, 3088–3098. [CrossRef]
41. Duran-Sampedro, G.; Agarrabeitia, A.R.; Garcia-Moreno, I.; Costela, A.; Bañuelos, J.; Arbeloa, T.; López Arbeloa, I.; Chiara, J.L.; Ortiz, M.J. Chlorinated BODIPYs: Surprisingly efficient and highly photostable laser dyes. *Eur. J. Org. Chem.* **2012**, *2012*, 6335–6350. [CrossRef]
42. Lakshmi, V.; Rajeswara Rao, M.; Ravikanth, M. Halogenated boron-dipyrromethenes: Synthesis, properties and applications. *Org. Biomol. Chem.* **2015**, *13*, 2501–2517. [CrossRef] [PubMed]
43. Bessette, A.; Hanan, G.S. Design, synthesis and photophysical studies of dipyrromethene-based materials: Insights into their applications in organic photovoltaic devices. *Chem. Soc. Rev.* **2014**, *43*, 3342. [CrossRef] [PubMed]
44. Singh, S.P.; Gayathri, T. Evolution of BODIPY dyes as potential sensitizers for dye-sensitized solar cells. *Eur. J. Org. Chem.* **2014**, *2014*, 4689–4707. [CrossRef]
45. González-Béjar, M.; Liras, M.; Francés-Soriano, L.; Voliani, V.; Herranz-Pérez, V.; Duran-Moreno, M.; Garcia-Verdugo, J.M.; Alarcon, E.I.; Scaiano, J.C.; Pérez-Prieto, J. NIR excitation of upconversion nanohybrids containing a surface grafted bodipy induces oxygen-mediated cancer cell death. *J. Mater. Chem. B* **2014**, *2*, 4554. [CrossRef]
46. Lim, S.H.; Thivierge, C.; Nowak-Sliwinska, P.; Han, J.; van den Bergh, H.; Wagnières, G.; Burgess, K.; Lee, H.B. *In vitro* and *in vivo* photocytotoxicity of boron dipyrromethene derivatives for photodynamic therapy. *J. Med. Chem.* **2010**, *53*, 2865–2874. [CrossRef] [PubMed]
47. Bernhard, C.; Goze, C.; Rousselin, Y.; Denat, F. First bodipy–DOTA derivatives as probes for bimodal imaging. *Chem. Commun.* **2010**, *46*, 8267. [CrossRef] [PubMed]
48. Bernhard, C.; Moreau, M.; Lhenry, D.; Goze, C.; Boschetti, F.; Rousselin, Y.; Brunotte, F.; Denat, F. DOTAGA-anhydride: A valuable building block for the preparation of DOTA-like chelating agents. *Chem. Eur. J.* **2012**, *18*, 7834–7841. [CrossRef] [PubMed]
49. Iwaki, S.; Hokamura, K.; Ogawa, M.; Takehara, Y.; Muramatsu, Y.; Yamane, T.; Hirabayashi, K.; Morimoto, Y.; Hagisawa, K.; Nakahara, K.; *et al.* A design strategy for small molecule-based targeted MRI contrast agents: Their application for detection of atherosclerotic plaques. *Org. Biomol. Chem.* **2014**, *12*, 8611–8618. [CrossRef] [PubMed]
50. Yamane, T.; Hanaoka, K.; Muramatsu, Y.; Tamura, K.; Adachi, Y.; Miyashita, Y.; Hirata, Y.; Nagano, T. Method for enhancing cell penetration of Gd^{3+}-based MRI contrast agents by conjugation with hydrophobic fluorescent dyes. *Bioconjug. Chem.* **2011**, *22*, 2227–2236. [CrossRef] [PubMed]
51. Duheron, V.; Moreau, M.; Collin, B.; Sali, W.; Bernhard, C.; Goze, C.; Gautier, T.; Pais de Barros, J.-P.; Deckert, V.; Brunotte, F.; *et al.* Dual labeling of lipopolysaccharides for SPECT-CT imaging and fluorescence microscopy. *ACS Chem. Biol.* **2014**, *9*, 656–662. [CrossRef] [PubMed]

52. Lhenry, D.; Larrouy, M.; Bernhard, C.; Goncalves, V.; Raguin, O.; Provent, P.; Moreau, M.; Collin, B.; Oudot, A.; Vrigneaud, J.-M.; *et al.* BODIPY: A highly versatile platform for the design of bimodal imaging probes. *Chemistry* **2015**, *21*, 13091–13099. [CrossRef] [PubMed]

53. Hendricks, J.A.; Keliher, E.J.; Wan, D.; Hilderbrand, S.A.; Weissleder, R.; Mazitschek, R. Synthesis of [^{18}F]BODIPY: Bifunctional reporter for hybrid optical/positron emission tomography imaging. *Angew. Chem. Int. Ed. Engl.* **2012**, *51*, 4603–4606. [CrossRef] [PubMed]

54. Liu, S.; Li, D.; Zhang, Z.; Surya Prakash, G.K.; Conti, P.S.; Li, Z. Efficient synthesis of fluorescent-PET probes based on [^{18}F]BODIPY dye. *Chem. Commun.* **2014**, *50*, 7371. [CrossRef] [PubMed]

55. Nuyts, K.; Ceulemans, M.; Parac-Vogt, T.N.; Bultynck, G.; de Borggraeve, W.M. Facile azide formation via diazotransfer reaction in a copper tube flow reactor. *Tetrahedron Lett.* **2015**, *56*, 1687–1690. [CrossRef]

56. Fischer, N.; Goddard-Borger, E.D.; Greiner, R.; Klapötke, T.M.; Skelton, B.W.; Stierstorfer, J. Sensitivities of some imidazole-1-sulfonyl azide salts. *J. Org. Chem.* **2012**, *77*, 1760–1764. [CrossRef] [PubMed]

57. Hansen, A.M.; Sewell, A.L.; Pedersen, R.H.; Long, D.-L.; Gadegaard, N.; Marquez, R. Tunable BODIPY derivatives amenable to "click" and peptide chemistry. *Tetrahedron* **2013**, *69*, 8527–8533. [CrossRef]

58. Viguier, R.F.H.; Hulme, A.N. A sensitized europium complex generated by micromolar concentrations of copper(I): Toward the detection of copper(I) in biology. *J. Am. Chem. Soc.* **2006**, *128*, 11370–11371. [CrossRef] [PubMed]

59. Verwilst, P.; Eliseeva, S.V.; Carron, S.; Vander Elst, L.; Burtea, C.; Dehaen, G.; Laurent, S.; Binnemans, K.; Muller, R.N.; Parac-Vogt, T.N.; *et al.* A modular approach towards the synthesis of target-specific MRI contrast agents. *Eur. J. Inorg. Chem.* **2011**, *2011*, 3577–3585. [CrossRef]

60. Hessel, V.; Kralisch, D.; Kockmann, N.; Noël, T.; Wang, Q. Novel process windows for enabling, accelerating, and uplifting flow chemistry. *ChemSusChem* **2013**, *6*, 746–789. [CrossRef] [PubMed]

61. Johansson, H.; Pedersen, D.S. Azide- and alkyne-derivatised α-amino acids. *Eur. J. Org. Chem.* **2012**, *2012*, 4267–4281. [CrossRef]

62. Goddard-Borger, E.D.; Stick, R.V. An efficient, inexpensive, and shelf-stable diazotransfer reagent: Imidazole-1-sulfonyl azide hydrochloride. *Org. Lett.* **2007**, *9*, 3797–3800. [CrossRef] [PubMed]

63. Duimstra, J.A.; Femia, F.J.; Meade, T.J. A gadolinium chelate for detection of β-glucuronidase: A self-immolative approach. *J. Am. Chem. Soc.* **2005**, *127*, 12847–12855. [CrossRef] [PubMed]

64. Wang, L.; Verbelen, B.; Tonnelé, C.; Beljonne, D.; Lazzaroni, R.; Leen, V.; Dehaen, W.; Boens, N. UV-VIS spectroscopy of the coupling products of the palladium-catalyzed C–H arylation of the BODIPY core. *Photochem. Photobiol. Sci.* **2013**, *12*, 835–847. [CrossRef] [PubMed]

65. Kobayashi, H.; Ogawa, M.; Alford, R.; Choyke, P.L.; Urano, Y. New strategies for fluorescent probe design in medical diagnostic imaging. *Chem. Rev.* **2010**, *110*, 2620–2640. [CrossRef] [PubMed]

66. Niu, S. Advanced Water Soluble BODIPY Dyes: Synthesis and Application. Ph.D. Thesis, Université de Strasbourg, Strasbourg, France, July 2011.

67. Olmsted, J. Calorimetric determinations of absolute fluorescence quantum yields. *J. Phys. Chem.* **1979**, *83*, 2581–2584. [CrossRef]

68. Solomon, I. Relaxation processes in a system of two spins. *Phys. Rev.* **1955**, *99*, 559–565. [CrossRef]

69. Bloembergen, N. Proton relaxation times in paramagnetic solutions. *J. Chem. Phys.* **1957**, *27*, 572. [CrossRef]

70. Laurent, S.; Elst, L.V.; Muller, R.N. Comparative study of the physicochemical properties of six clinical low molecular weight gadolinium contrast agents. *Contrast Media Mol. Imaging* **2006**, *1*, 128–137. [CrossRef] [PubMed]

71. Onishi, H. Spectrophotometric determination of zirconium, uranium, thorium and rare earths with arsenazo III after extractions with thenoyltrifluoroacetone and tri-*n*-octylamine. *Talanta* **1972**, *19*, 473–478. [CrossRef]

inorganics

MDPI

Article

Synthesis and Reactivity of a Cerium(III) Scorpionate Complex Containing a Redox Non-Innocent 2,2′-Bipyridine Ligand

Fabrizio Ortu, Hao Zhu, Marie-Emmanuelle Boulon and David P. Mills *

School of Chemistry, The University of Manchester, Oxford Road, Manchester, M13 9PL, UK;
fabrizio.ortu@manchester.ac.uk (F.O.); zhuhao@kans.cn (H.Z.);
marie.emanuelle.boulon@manchester.ac.uk (M.-E.B.)
* Correspondence: david.mills@manchester.ac.uk; Tel.: +44-161-275-4606 (ext. 45606); Fax: +44-161-275-4598

Academic Editors: Stephen Mansell and Steve Liddle
Received: 14 September 2015; Accepted: 17 November 2015; Published: 27 November 2015

Abstract: The Ce(III) hydrotris(3,5-dimethylpyrazolyl)borate complex [Ce(Tp^{Me2})$_2$(κ^2-dmpz)] (**1**) (Tp^{Me2} = {HB(dmpz)$_3$}$^-$; dmpz = 3,5-dimethylpyrazolide) was isolated in fair yield from the reaction of [Ce(I)$_3$(THF)$_4$] with two equivalents of [K(Tp^{Me2})] via the facile decomposition of Tp^{Me2}. [Ce(Tp^{Me2})$_2$(bipy)] (**2**) was synthesized in poor yield by the "one-pot" reaction of [Ce(I)$_3$(THF)$_4$], bipy (bipy = 2,2′-bipyridine), KC$_8$ and two equivalents of [K(Tp^{Me2})] in tetrahydrofuran (THF). The reaction of **2** with N-methylmorpholine-N-oxide produced the known decomposition product [Ce(Tp^{Me2})(μ-BOp^{Me2})]$_2$ (**3**) (BOp^{Me2} = {HBO(dmpz)$_2$}$^{2-}$) in poor yield, presumably by N–O and B–N bond cleavage of a reactive intermediate. The reaction of **2** with trimethylsilylazide gave [Ce(Tp^{Me2})$_2$(N$_3$)] (**4**) in poor yield; the fate of bipy and the trimethylsilyl group is unknown. Complexes **1**–**4** were characterized by single crystal XRD, NMR and FTIR spectroscopy and elemental analysis. Complex **2** was additionally probed by UV/Vis/NIR and Electron Paramagnetic Resonance (EPR) spectroscopies, Cyclic Voltammetry (CV) and magnetometry, which together indicate a formal 4f^1 Ce(III) center coordinated by a bipy·$^-$ radical anion in this system.

Keywords: lanthanide; cerium; scorpionate; tris(pyrazolyl)borate; radical; redox non-innocent

1. Introduction

Complexes that exhibit terminal unsupported multiple bonds between a transition metal and a p-block element are legion, and interest in these species has surged in tandem with their increasing applicability in synthetic processes [1]. In contrast, the corresponding lanthanide (Ln) chemistry is underdeveloped, as the large orbital energy mis-match between Ln valence orbitals and the 2p$_z$ orbital of C, N and O makes these bonds highly polarized, making them prone to decomposition and oligomerisation pathways [2,3]. As such the isolation of Ln=CR$_2$, Ln=NR and Ln=O bonds is a major synthetic challenge, with the first structurally characterized supported Ce(IV)=CR$_2$ [4], terminal unsupported Ln=NR (Ln = Y, Lu) [5] and terminal supported Ce(IV)=O [6] complexes reported very recently, whilst the corresponding terminal unsupported Sc=NR chemistry has flourished since the first example was reported in 2010 by Chen [7].

Synthetic routes to terminal unsupported actinide (An) An=CR$_2$, An=NR and An=O multiple bonds are far more developed than pathways to the corresponding Ln species [8,9]. Andersen [10], Bart [11–13] and Zi and Walter [14] have shown that U(IV) and Th(IV) oxo and imido complexes can be prepared by the respective addition of N-oxides or organoazides to An(II) synthons, which mimic the reactivity of An(II) complexes although the metal is formally in the +4 oxidation state. We set out to find if this methodology could be extended to cerium, which is unique amongst the Lns as it has a readily accessible +4 oxidation state [E$^{\ominus}$ Ce(IV) → Ce(III) = 1.74 V] [15].

The An(II) synthon utilized by Bart that is of most relevance here is [U(Tp^{Me2})$_2$(bipy)] (Tp^{Me2} = hydrotris(3,5-dimethylpyrazolyl)borate; bipy = 2,2′-bipyridine) [16], which formally contains a bipy dianion and a U(IV) center. Marques previously reported a Ln(II) synthon analogue, [La(Tp^{Me2})$_2$(bipy)] [17], which formally contains a monoreduced bipy·$^-$ radical anion and lanthanum in the +3 oxidation state. As Tp^{Me2} ligands have been shown to stabilize a wide variety of interesting motifs in Ln chemistry [18], we reasoned that they could be suitable ancillary ligands to support terminal Ce(IV)=O or Ce(IV)=NR multiple bonds.

We envisaged that an analogous Ce(II) synthon, [Ce(Tp^{Me2})$_2$(bipy)], could react with *N*-oxides or organoazides to give complexes that contain terminal unsupported Ce(IV)=O or Ce(IV)=NR multiple bonds, provided the Ce(III) → Ce(IV) oxidation potential in these complexes could be overcome by these reagents. [Ce(Tp^{Me2})$_2$(bipy)] should contain a formal Ce(III) 4f^1 center and the coupling of this unpaired spin with a ligand radical could lead to unusual physicochemical properties. It is noteworthy that interesting multiconfigurational behavior has previously been observed in [Yb(Cp*)$_2$(bipy)] [19,20], where the electronic ground state contains contributions from Yb(III)/bipy·$^-$ (4f^{13}) and Yb(II)/bipy (4f^{14}) configurations, but the donor properties of Cp* ligands are not comparable to those of Tp^{Me2} ligands. The synthesis and characterization of [Ce(Tp^{Me2})$_2$(bipy)] and its reactivity towards *N*-oxides and organoazides is described herein.

2. Results and Discussion

2.1. Preparation of a Ce(II) Synthon

2.1.1. Synthesis and Structural Characterization of [Ce(Tp^{Me2})$_2$(κ^2-dmpz)] (1)

In an effort to prepare the heteroleptic complex [Ce(Tp^{Me2})$_2$(I)], [Ce(I)$_3$(THF)$_4$] [21] was reacted with two equivalents of [K(Tp^{Me2})] [22] in THF by slight modifications of the procedures previously used to synthesize [Ln(Tp^{Me2})$_2$(X)] (Ln = La, X = Cl, I [23]; Ln = Nd, X = Cl [24]; Ln = Sm, X = Cl, Br [25]; Ln = Eu, X = Cl [25]). However, [Ce(Tp^{Me2})$_2$(I)] was not identified in the reaction mixture, and the only isolable product was [Ce(Tp^{Me2})$_2$(κ^2-dmpz)] (1), which was obtained in fair crystalline yield from a saturated toluene solution (Scheme 1). The relatively high yield of 1 showed that this route is not suitable for the preparation of [Ce(Tp^{Me2})$_2$(I)] under the conditions we employed but the one-pot method used in Section 2.1.2 circumvented the need to isolate [Ce(Tp^{Me2})$_2$(X)] (see below).

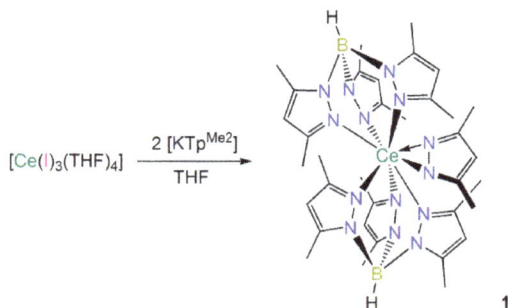

Scheme 1. Synthesis of **1**.

Complex **1** presumably forms by the decomposition of a Tp^{Me2} ligand via the mechanism previously postulated by Kunrath *et al.*, when [La(Tp^{Me2})$_2$(κ^2-dmpz)] was isolated during the synthesis of [La(Tp^{Me2})$_2$(X)] (X = Cl, I) [23]. Coordination of Tp^{Me2} to lanthanum was proposed to twist pyrazolyl rings and weaken B–N bonds, promoting bond cleavage to form (dmpz)$^-$ *in situ*, which reacts with [La(Tp^{Me2})$_2$(X)] to give [La(Tp^{Me2})$_2$(κ^2-dmpz)]. This mechanism has been corroborated by a control experiment, where [La(Tp^{Me2})$_2$(X)] was treated with dmpzH and triethylamine, and [La(Tp^{Me2})$_2$(κ^2-dmpz)] was detected in the ^{1}H NMR spectrum [23].

It is noteworthy that we were not able to identify any traces of [Ce(Tp^{Me2})$_2$(I)] in the reaction mixture during the synthesis of **1**, whereas [La(Tp^{Me2})$_2$(κ^2-dmpz)] was a minor product during the preparation of [La(Tp^{Me2})$_2$(I)] [23], and no dmpz-containing byproducts were reported from the syntheses of [Ln(Tp^{Me2})$_2$(Cl)] (Ln = Sm [25]; Nd or Eu [24]), and [Sm(Tp^{Me2})$_2$][I] [25]. However, the preparation of [Ce(Tp^{Me2})$_2$(Cl)] from CeCl$_3$ has previously been described as problematic by Sella [24], whereas the synthesis of [Ce(Tp^{Me2})$_2$(OTf)] from Ce(OTf)$_3$ is more straightforward due to OTf$^-$ being a better leaving group than Cl$^-$ [26]. Kunrath *et al.*, stated that it is difficult to coordinate two Tp^{Me2} ligands to lanthanides due to steric crowding [23], therefore we conclude that when [Ce(I)$_3$(THF)$_4$] is used as a precursor for the synthesis of [Ce(Tp^{Me2})$_2$(I)] intermediates with high coordination numbers form, which promote Tp^{Me2} decomposition by twisting and weakening B–N bonds.

The elemental analysis data for **1** correlated well with the predicted values, as with the other complexes (**2**–**4**, see below) reported in this paper. The ^{1}H NMR spectrum of **1** exhibited resonances over a wide chemical shift range (δ_H: −7.99 to +18.26 ppm) due to the 4f^1 formulation of Ce(III), and this precluded any interpretation of ^{13}C{^{1}H} NMR spectroscopic data. However, a single resonance was observed in the ^{11}B{^{1}H} NMR spectrum of **1** (δ_B: −0.66 ppm), which is downfield of that reported for [Ce(Tp^{Me2})(COT)] (δ_B: −21.0 ppm) [27]. The solution magnetic susceptibility of **1** was measured by the Evans method (μ_{eff} = 2.33 μ_B) [28], and this correlates well with the free ion approximation for a Ce(III) 4f^1, ^{2}F$_{5/2}$ ground term system (μ_{eff}: 2.54 μ_B [29]) considering weighing errors. The FTIR spectrum exhibited an absorption at 2547 cm^{-1} that is characteristic of the B–H stretching mode in tripodal Tp systems, and this was also seen in **2**–**4** (see below) [30,31].

The solid state structure of **1**·C$_7$H$_8$ was determined by a single crystal XRD study and is depicted in Figure 1, with selected bond lengths and angles compiled in Table 1. The presence of toluene in the lattice causes **1**·C$_7$H$_8$ to adopt a different space group to [La(Tp^{Me2})$_2$(κ^2-dmpz)] [23], which was crystallized from cyclohexane. Due to the similar 8-coordinate ionic radii of La(III) [1.160 Å] and Ce(III) [1.143 Å] [32], variable crystal packing effects are likely the cause of two minor discrepancies between the solid state structures of **1** and [La(Tp^{Me2})$_2$(κ^2-dmpz)]. Firstly, the dmpz ligand coordinates in a less symmetrical manner in **1** than in [La(Tp^{Me2})$_2$(κ^2-dmpz)], as evidenced by the respective Ln–N$_{dmpz}$ distances [2.385(5) and 2.505(5) Å *vs.* 2.468(4) and 2.514(3) Å] [23]. Secondly, the Ln–N$_{TpMe2}$ distances in **1** [2.546(4)–2.667(4) Å] are within a narrower range than those found in [La(Tp^{Me2})$_2$(κ^2-dmpz)] [2.601(3)–2.748(3) Å] [23]. All other metrical parameters for **1** are unremarkable.

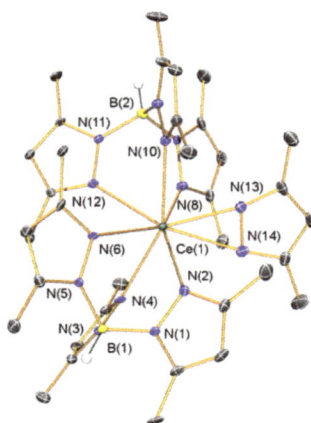

Figure 1. Molecular structure of **1**·C$_7$H$_8$ with selective atom labeling. Displacement ellipsoids set at 30% probability level and hydrogen atoms (except B–H) and lattice solvent omitted for clarity.

2.1.2. Synthesis and Structural Characterization of [Ce(Tp^{Me2})(bipy)] (2)

Marques and co-workers previously reported that the one-pot reaction of [La(Cl)$_3$(THF)$_{1.5}$] with two equivalents of [K(Tp^{Me2})] and one equivalent of bipy over a Na/Hg amalgam in THF gave [La(Tp^{Me2})$_2$(bipy)] in 90% yield [17]. We modified this procedure to prepare [Ce(Tp^{Me2})$_2$(bipy)] (2) using [Ce(I)$_3$(THF)$_4$] [21], two equivalents of [K(Tp^{Me2})] [22], one equivalent of bipy and KC$_8$ [33] in THF (Scheme 2), in an effort to avoid the use of mercury. A poor crystalline yield of 2 (22%) was reproducibly obtained using our method, likely a result of the difficulties encountered during the synthesis of 1 (see above), though no other products could be identified in the reaction mixtures. The intense dark red color of 2 indicated that the complex contained a bipy·$^-$ radical monoanion; therefore we collected a wide range of characterization data to probe this system.

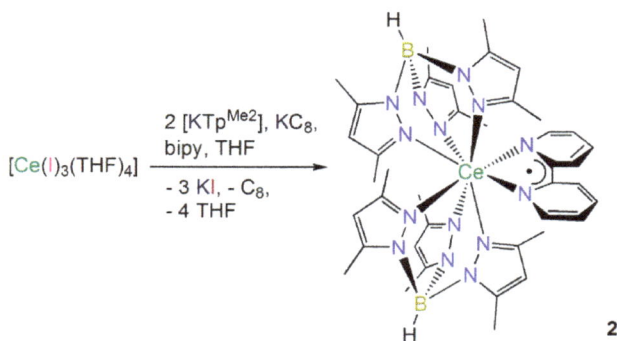

Scheme 2. Synthesis of 2.

The ^{1}H NMR spectrum of 2 exhibits a similar range of resonances to 1 [δ_H: −9.30 to 19.84 ppm], with a wide spectral window of +200 to −200 ppm evaluated as bipy resonances were observed at large negative chemical shifts for [Ce(Cp*)$_2$(bipy)] (Cp* = C$_5$Me$_5$) [34]. The bipy proton resonances in 2 could only be tentatively assigned, as neutral and monoanionic bipy·$^-$ resonances may be difficult to distinguish in a paramagnetic spectrum. In Marques' report of the ^{1}H NMR spectrum of the lanthanum analogue [La(Tp^{Me2})$_2$(bipy)] these signals were not observed [17]. One signal was seen in the ^{11}B{^{1}H} NMR spectrum of 2 (δ_B: 10.48 ppm); however the ^{13}C{^{1}H} NMR spectrum could not be interpreted. The FTIR spectrum of 2 exhibited characteristic absorptions for a bipy radical anion (v = 1541, 1494 and 941 cm^{-1}) with the lowest frequency peak correlating with that assigned for [La(Tp^{Me2})$_2$(bipy)] (v = 940 cm^{-1}) [17]. The solution susceptibility of 2 (μ_{eff} = 1.59 μ_B) was reproducibly much lower than the value obtained for 1, which could be attributed to either partial decomposition of 2 or antiferromagnetic coupling between Ce(III) and bipy·$^-$. To probe this result a variable temperature magnetic analysis was performed together with an EPR spectroscopy study (see below).

The solid state configuration of 2·(C$_4$H$_8$O)$_2$ was confirmed by single crystal XRD (depicted in Figure 2, with selected bond lengths and angles compiled in Table 1). The metrical parameters of structurally analogous [La(Tp^{Me2})$_2$(bipy)]·(C$_4$H$_8$O)$_2$ have been discussed previously [17], and as these are nearly identical to those in 2·(C$_4$H$_8$O)$_2$ considering the small difference in ionic radii between lanthanum and cerium [32] these will not be commented on here for brevity. The salient point of the discussion of the structure of [La(Tp^{Me2})$_2$(bipy)] is that the distances throughout the bipy scaffold were used as evidence of its radical monoanionic formulation (the NMR and FTIR spectroscopy and magnetic data of [La(Tp^{Me2})$_2$(bipy)] concurred with this assignment) [17], and statistically identical intra-ligand bond distances were observed in the coordinated bipy framework in 2.

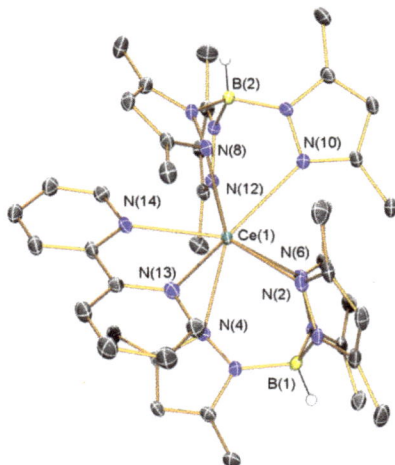

Figure 2. Molecular structure of **2**·$(C_4H_8O)_2$ with selective atom labeling. Displacement ellipsoids set at 30% probability level and hydrogen atoms (except B–H) and lattice solvent omitted for clarity.

2.1.3. Further Characterization of [Ce(Tp^Me2)(bipy)] (2)

The electronic structure of **2** was probed by additional techniques in an effort to unequivocally establish a formal Ce(III)/bipy·⁻ configuration. The UV/Vis/NIR spectrum of **2** exhibits a strong absorption at 387 nm ($\varepsilon_{max} \approx 4100$ M⁻¹·cm⁻¹) and a weaker, broad absorption at 812–972 ($\varepsilon_{max} \approx 500$ M⁻¹·cm⁻¹) nm (Figure 3). These absorptions are comparable to those observed for [Yb(Cp*)₂(bipy)], which were assigned as π→π* and π*→π* transitions that arise from the unpaired electron in the bipy radical π-system [19].

Bipy has neutral, monoanionic and dianionic forms that can be easily interchanged and these processes are readily observed by electrochemistry [35]; therefore the cyclic voltammogram of **2** was obtained (Figure 4). Only one Nernstian *quasi*-reversible process was observed within the range limited by solvent and electrolyte experimental conditions ($E = -2.34$ V *vs.* Fc/Fc⁺, $\Delta E = 95$ mV), which was assigned to a [Ce(Tp^Me2)₂(bipy)]⁺ / [Ce(Tp^Me2)₂(bipy·⁻)] process based on its similarity to the first reduction potential of bipy that we measured using identical conditions ($E = -2.72$ V *vs.* Fc/Fc⁺). The second reduction potential of bipy and the oxidation of Ce(III) to Ce(IV) could not be observed under the conditions employed, which indicates that strong oxidizing agents are required to form Ce(IV) complexes supported by Tp^Me2 ligands. This is unsurprising given both the considerable precedent of Tp ligands stabilizing metals in low oxidation states [18] and the dominance of electron-rich donor ligands in Ce(IV) chemistry [36].

The powder X-band EPR spectrum of **2** was collected in an effort to observe resonances arising from coupling of the Ce(III) and bipy·⁻ radical unpaired electrons. No signal was seen at room temperature but a highly anisotropic complex spectrum was obtained at 5 K. We performed a variable temperature study to track the thermal evolution of the spectrum (Figure 5). The bulk features at 5 K are almost fully maintained at 10 K, but at 15 K the spectrum is less distinct and at higher temperatures the fine structure cannot be discerned from the baseline. The 5 K spectrum could not be modeled and therefore we were not able to extract any parameters but this highlights the complex nature of the coupling between the bipy·⁻ radical electron and the $^2F_{5/2}$ ground state doublet deriving from Ce(III).

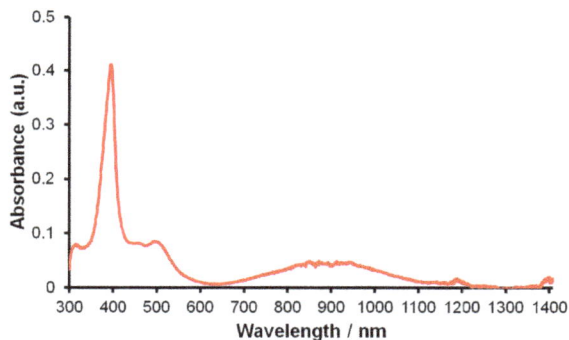

Figure 3. UV/Vis/NIR spectrum of **2** (0.1 mM in THF).

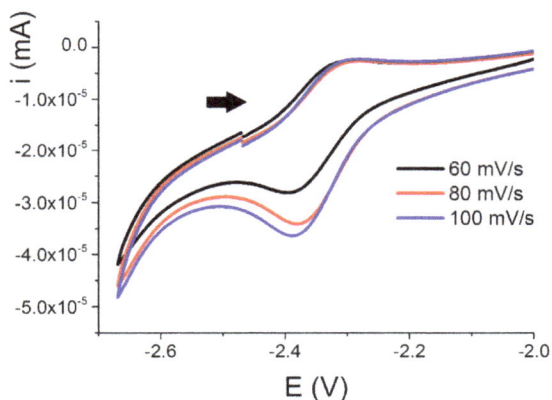

Figure 4. Cyclic voltammogram of **2** (0.5 mM in THF, 0.1 M [NBu$^n_4$][BF$_4$]). Black arrow shows the direction of applied potential from the origin.

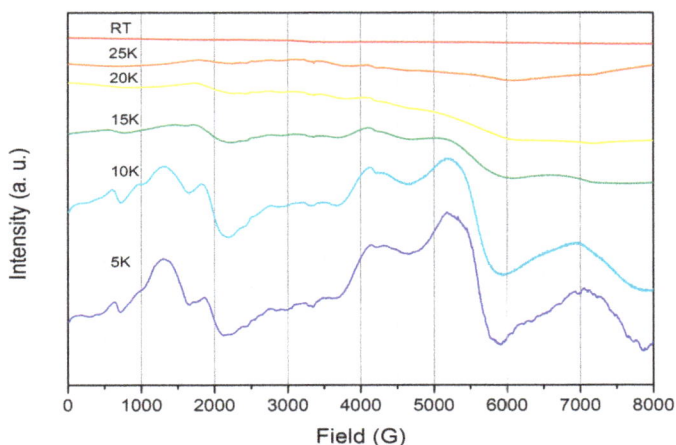

Figure 5. Variable temperature powder X-band EPR spectra of **2** (v = 9.385 GHz).

Magnetic measurements were performed on a solid sample of **2** suspended in eicosane and the magnetic susceptibility product ($\chi_M T$) was measured from 300 to 2 K in a 10,000 Oe applied dc field (Figure 6). At 300 K, $\chi_M T$ = 0.44 cm^3·Kmol^{-1} (1.21 μ_B), which is comparable to the moment measured in solution at 298 K (0.58 cm^3·Kmol^{-1}), taking into account weighing errors. These values are much lower than is predicted for a system containing one organic radical and one Ce(III) center (1.1 cm^3·Kmol^{-1}) and are closer to a Ce(III)-only system. The discrepancy between predicted and observed $\chi_M T$ values at 300 K can be attributed to the presence of diamagnetic impurities, or more complex magnetic behavior in **2**. The $\chi_M T$ value decreases to 0.15 cm^3·Kmol^{-1} (0.41 μ_B) at 2 K. We analyzed **2** further by measuring the magnetization against variable applied dc field at 5 K from 0–70,000 Oe, finding a near-linear correlation of *M vs. H*, with *M* = 0.39 μ_B at 70,000 Oe (Figure 7). In contrast at 2 K the data forms a curve, with a linear dependence of *M vs. H* from 0 to around 20,000 Oe (*M* = 0.27 μ_B), but at higher fields the effect of field on *M* is reduced. At both temperatures saturation was not reached at the maximum obtainable field, and this was observed reproducibility on separate samples, consistent with a system containing two unpaired electrons. There is also an essentially linear relationship of *M* and *H* from 35,000–70,000 Oe (*M* = 0.50 μ_B at 70,000 Oe) with $\partial M/\partial H$ for this range of fields lower than for the corresponding data at 5 K.

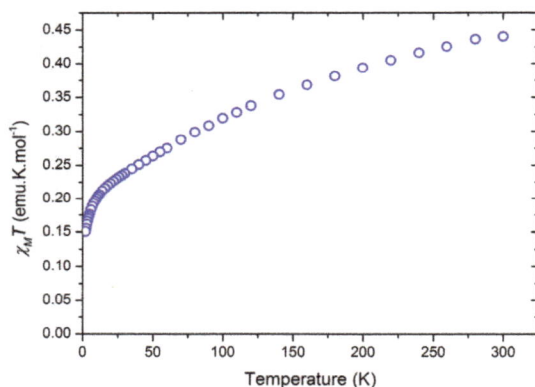

Figure 6. Variation of the product of the molar magnetic susceptibility product ($\chi_M T$, cm^3·Kmol^{-1}) with temperature (K) of **2**.

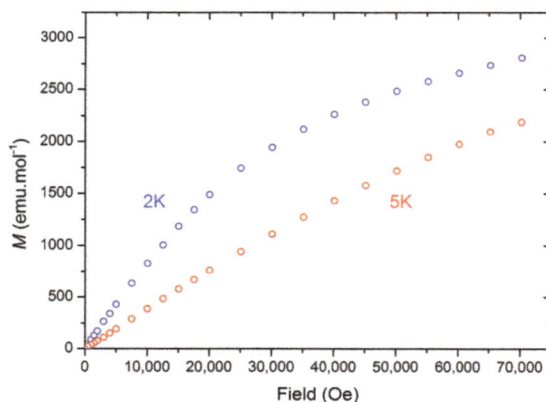

Figure 7. Magnetization (*M*, μ_B) *vs.* Field (Oe) of **2** at 2 K and 5 K.

2.2. Reactivity of a Ce(II) Synthon

2.2.1. Synthesis and Structural Characterization of [Ce(Tp^{Me2})(μ-BOp^{Me2})]$_2$ (3)

We were interested in finding out if **2** could react with 2e$^-$ oxidizing agents as a Ce(II) synthon, mimicking the reactivity profile of a Ce(II) complex, even though the measurements in Section 2.1.3 indicate that Ce is in the +3 oxidation state in **2**. The reactions of **2** with *N*-oxides were investigated as Bart previously showed that the treatment of [U(Tp^{Me2})(bipy)] with pyridine-*N*-oxide gave the U(IV) terminal oxo complex [U(Tp^{Me2})(O)] [11]. In contrast the reaction of **2** with *N*-methylmorpholine-*N*-oxide, pyridine-*N*-oxide or TEMPO (2,2,6,6-tetramethylpiperidine-1-oxyl) all gave the Ce(III) complex [Ce(Tp^{Me2})(μ-BOp^{Me2})]$_2$ (**3**) as the only isolable product in *ca.* 20% crystalline yield in each case (Scheme 3). Complex **3** was previously reported by Sella and co-workers to form in trace amounts via the partial hydrolysis of a range of heteroleptic complexes, [Ce(Tp^{Me2})$_2$(X)] (X = anionic ligand), and a mechanism for this decomposition was postulated whereby dmpzH was generated [37]. As no other products could be identified during the formation of **3** it is not certain if the oxygen atom derives from initial coordination of the *N*-oxide or adventitious O$_2$/H$_2$O; however the isolation of **3** was reproducible, performed with the strict exclusion of oxygen and water, and was formed in higher yields than previously reported.

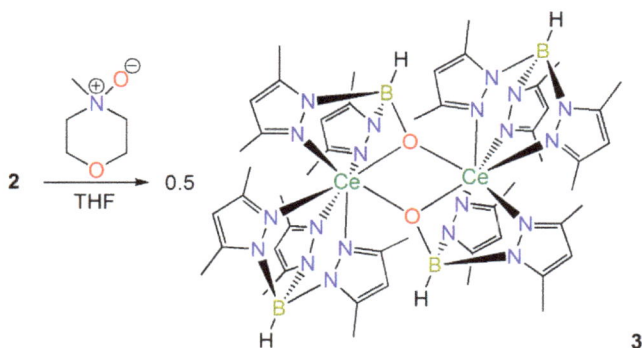

Scheme 3. Synthesis of **3**.

Whilst the solid state structure of **3** was obtained previously by Sella, no other characterization data were reported [37]; therefore we acquired elemental analysis and collected spectroscopic data for **3**. Unusually in the original report **3** was reported to exhibit a dark red color [37], whilst the samples we obtained were pale yellow. The ^{1}H NMR and FTIR spectra of **3** were unremarkable and as with **1** and **2** the ^{13}C{^{1}H} NMR spectrum could not be interpreted, though the ^{11}B{^{1}H} NMR spectrum of **3** contained one resonance (δ_B: 0.60 ppm). The Evans method susceptibility was also determined (μ_{eff}: 2.71 μ_B). Although the solid state structure of **3** has been reported previously [37], we include this data here for completeness as we obtained a polymorph (depicted in Figure 8, with selected bond lengths and angles compiled in Table 1).

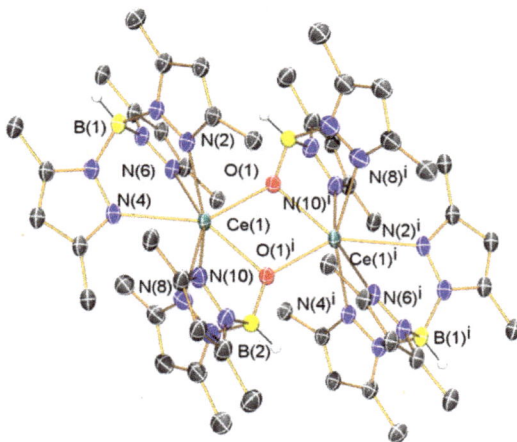

Figure 8. Molecular structure of **3**·(C$_4$H$_8$O)$_2$ with selective atom labeling. Displacement ellipsoids set at 30% probability level and hydrogen atoms (except B–H) and lattice solvent omitted for clarity. Symmetry operation to generate equivalent atoms: $i = 1 - x, 1 - y, 1 - z$.

Table 1. Selected bond lengths (Å) and angles (°) for **1–4**.

1·C$_7$H$_8$			
Ce(1)–N(2)	2.653(5)	Ce(1)–N(4)	2.582(4)
Ce(1)–N(6)	2.631(5)	Ce(1)–N(8)	2.624(5)
Ce(1)–N(10)	2.546(4)	Ce(1)–N(12)	2.667(4)
Ce(1)–N(13)	2.385(5)	Ce(1)–N(14)	2.505(5)
N(13)–N(14)	1.368(6)	Ce(1)–N(13)–N(14)	78.7(3)
Ce(1)–N(14)–N(13)	69.0(3)	N(13)–Ce(1)–N(14)	32.38(14)
2·(C$_4$H$_8$O)$_2$			
Ce(1)–N(2)	2.591(4)	Ce(1)–N(4)	2.650(3)
Ce(1)–N(6)	2.711(4)	Ce(1)–N(8)	2.648(4)
Ce(1)–N(10)	2.719(3)	Ce(1)–N(12)	2.577(4)
Ce(1)–N(13)	2.592(4)	Ce(1)–N(14)	2.612(4)
C(35)–C(36)	1.431(6)	C(35)–N(13)	1.373(6)
C(36)–N(14)	1.379(5)	N(13)–Ce(1)–N(14)	62.40(12)
Ce(1)–N(13)–C(35)	122.1(3)	Ce(1)–N(14)–C(36)	120.7(3)
N(13)–C(35)–C(36)	117.0(4)	N(14)–C(36)–C(35)	117.8(4)
3·(C$_4$H$_8$O)$_2$			
Ce(1)–N(2)	2.668(4)	Ce(1)–N(4)	2.584(4)
Ce(1)–N(6)	2.561(4)	Ce(1)–N(7)	2.591(4)
Ce(1)–N(9)	2.578(4)	Ce(1)–O(1)	2.292(3)
Ce(1)–O(1i)	2.378(3)	B(1)–O(1)	1.407(6)
Ce(1)···Ce(1i)	3.7987(5)	Ce(1)–O(1)–Ce(1i)	108.84(12)
O(1)–Ce(1)–O(1i)	71.16(12)		
4·C$_7$H$_8$			
Ce(1)–N(2)	2.656(5)	Ce(1)–N(4)	2.655(5)
Ce(1)–N(6)	2.565(4)	Ce(1)–N(8)	2.665(5)
Ce(1)–N(10)	2.539(5)	Ce(1)–N(12)	2.646(5)
Ce(1)–N(13)	2.340(6)	N(13)–N(14)	1.188(8)
N(14)–N(15)	1.165(8)	Ce(1)–N(13)–N(14)	165.5(5)
N(13)–N(14)–N(15)	178.4(8)		

2.2.2. Synthesis and Structural Characterization of [Ce(Tp^{Me2})$_2$(N$_3$)] (4)

Andersen [10], Bart [11] and Walter and Zi [13] have all shown that azides react with Th(II) synthons to yield Th(IV) imido complexes. Therefore we added trimethylsilyl azide to **2**, and when no reaction was observed at room temperature over three days the reaction mixture was refluxed for 16 h and the Ce(III) complex [Ce(Tp^{Me2})$_2$(N$_3$)] (**4**) was obtained as the only isolable product (Scheme 4). Whilst the mechanism for the formation of **4** and the fate of the silyl group is unclear, N–Si bond cleavage is relatively facile and Bart recently reported a range of reductive heterocouplings of the coordinated reduced bipy ligand in [U(Cp*)$_2$(bipy)] with ketones [38], though the electronics of the spectator ligands in this system are vastly different to **2**. Germane to this, no reaction was observed between **2** and mesityl azide, even after prolonged heating, showing that the preparation of Ce(IV) imido complexes by a formal two electron oxidation of **2** by silyl and aryl azides is not favored.

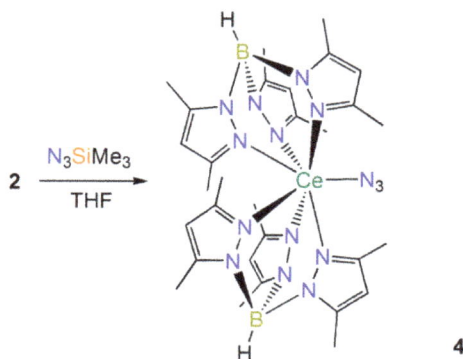

Scheme 4. Synthesis of **4**.

Gratifyingly, the ^{1}H and ^{13}C{^{1}H} NMR spectra of **4** could be fully assigned, with the ^{13}C signals all observed between 0 and 200 ppm. The Evans method susceptibility determined for **4** (μ_{eff} = 1.94 μ_B) was higher than that obtained for **2**. Only one signal was observed in the ^{11}B{^{1}H} NMR spectrum of **4** (δ_B: −25.29 ppm), albeit at a higher field than **1**–**3**, which we attribute to variable paramagnetic shift effects in the four complexes. The FTIR spectrum of **4** exhibited the diagnostic B–H stretching mode at 2555 cm^{-1}, together with characteristic absorptions for an azide group (v = 2083 and 2065 cm^{-1}) that are comparable to the absorption observed for [Ce(TRENDSAL)(N$_3$)] (TRENDSAL = [N{CH$_2$CH$_2$N=CH(C$_6$H$_2$Bu^t_2$-3,5-O-2)}$_3$]$^{3-}$) ($v$ = 2044 cm^{-1}) [39]. The composition of **4** was determined by single crystal XRD (depicted in Figure 9, with selected bond lengths and angles compiled in Table 1). The Ce(III)–N$_{azide}$ distance in **4** [2.340(6) Å] is shorter than the Ce(IV)–N$_{azide}$ distance observed in [Ce(TRENDSAL)(N$_3$)] [2.437(3) Å] [39] and the azide in **4** binds with a less bent geometry [Ce(1)–N(13)–N(14) 165.5(5)°] than the azide in [Ce(TRENDSAL)(N$_3$)] [Ce–N–N 133.4(2)°] [39], which is likely caused by the bulky Tp^{Me2} ligands in **4**. All other metrical parameters in **4** are unremarkable.

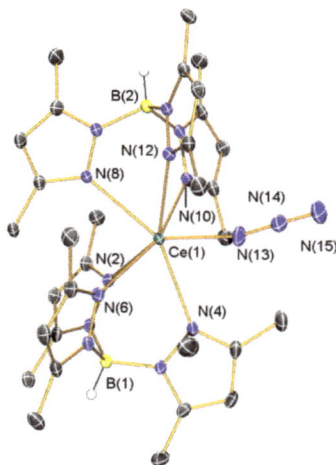

Figure 9. Molecular structure of **4**·C_7H_8 with selective atom labeling. Displacement ellipsoids set at 30% probability level and hydrogen atoms (except B–H) and lattice solvent omitted for clarity.

3. Experimental Section

3.1. General Procedures

All manipulations were carried out using standard Schlenk and glove box techniques under an atmosphere of dry argon. Solvents were dried by refluxing over potassium and degassed before use. All solvents were stored over potassium mirrors (with the exception of THF which was stored over activated 4 Å molecular sieves). Deuterated solvents were distilled from potassium, degassed by three freeze-pump-thaw cycles and stored under argon. $[Ce(I)_3(THF)_4]$ [21], $[K(Tp^{Me2})]$ [22] and KC_8 [33] were prepared according to published procedures. Most solid reagents were dried under vacuum for 4 h and most liquid reagents were dried over 4 Å molecular sieves and distilled before use. ^1H, $^{13}C\{^1H\}$ and $^{11}B\{^1H\}$ NMR spectra were recorded on a spectrometer operating at 400.2, 100.6 and 128.3 MHz, respectively; chemical shifts are quoted in ppm and are relative to TMS (^1H, ^{13}C) or external $BF_3 \cdot Et_2O$ (1.0 M in $CDCl_3$) (^{11}B). Magnetic susceptibility was determined by the Evans method [27]. FTIR spectra were recorded as Nujol mulls in KBr discs. Elemental microanalyses were carried out by Stephen Boyer at the Microanalysis Service, London Metropolitan University, UK. UV/Vis spectra were recorded in sealed 10 mm pathlength cuvettes. Redox potentials are referenced to the $[Fe(Cp)_2]^+/[Fe(Cp)_2]$ couple, which was used as an internal standard. Cyclic voltammetry was carried out using a sealed cell and a three-electrode arrangement, with a Pt wire working electrode, Pt flag secondary electrode and an AgCl/Ag wire pseudo-reference electrode. The susceptibility and magnetization curves were obtained using a SQUID magnetometer (Quantum Design, San Diego, CA, USA). The powdered sample was suspended in eicosane and sealed in a borosilicate glass NMR tube under vacuum, with diamagnetic contributions subtracted from the data. X-band EPR spectra were recorded using a MD5 dielectric resonator (Bruker, Billerica, MA, USA). The spectrometer is equipped with a CF935 cryostat connected to an Intelligent Temperature Controller (Oxford Instruments, Abingdon, UK).

3.2. Synthesis

$[Ce(Tp^{Me2})_2(\kappa^2\text{-dmpz})]$ (**1**). THF (30 mL) was added to a Schlenk containing $[Ce(I)_3(THF)_4]$ (1.62 g, 2 mmol) and $[K(Tp^{Me2})]$ (1.35 g, 4 mmol) at −78 °C. The yellow suspension was allowed to slowly warm to room temperature and was stirred for 72 h. Volatiles were removed *in vacuo* and the solid was extracted with toluene (13 mL), filtered and stored at −25 °C overnight to give yellow crystals of **1**. A second crop was obtained (1.06 g, 64%). Analysis Calculated (Anal. Calcd) for $C_{35}H_{51}B_2CeN_{14}$: C 50.65, H 6.19, N 23.62. Found: C 50.77, H 6.28, N, 23.55. ^1H NMR (d_6-benzene, 298 K): δ −7.60 (s, 6H, dmpz-CH_3), 3.61 (br, 36H, Tp^{Me2}-CH_3), 8.33 (br, 1H, dpmz-CH), 8.93 (br, 6H, Tp^{Me2}-CH), 18.49 (s, 3H, Tp^{Me2}-CH). $^{11}B\{^1H\}$ NMR (d_6-benzene, 298 K): δ −0.66 (s, BH). Evans method susceptibility (d_6-benzene, 298 K) μ_{eff}: 2.33 μ_B. FTIR v/cm^{-1} (Nujol): 2547 (w, B–H str), 1538 (m), 1438 (m), 1200 (m), 1067 (m), 1030 (m), 982 (m), 809 (m), 732 (w), 697 (m), 648 (m).

[Ce(Tp^{Me2})$_2$(bipy)] (**2**). [Ce(I)$_3$(THF)$_4$] (5.67 g, 7.0 mmol) and [K(Tp^{Me2})] (4.71 g, 14.0 mmol) were suspended in THF (40 mL) at −78 °C. The mixture was stirred for 20 min at −78 °C and bipy (1.09 g, 7.0 mmol) in THF (5 mL) was added dropwise. The mixture was stirred for 20 min at −78 °C and the reaction mixture was transferred into a Schlenk containing KC_8 (1.00 g, 7.4 mmol) and THF (10 mL) at −78 °C using a wide bore cannula. The dark brown reaction mixture was allowed to warm to room temperature and was stirred for 96 h. The reaction mixture was concentrated to *ca.* 30 mL, filtered and stored at −25 °C for 16 h to give dark red crystals of **2**. Several more crops were obtained (1.37 g, 22%). Anal. Calcd for $C_{40}H_{52}B_2CeN_{14}$: C, 53.92; H, 5.88; N, 22.00. Found: C, 53.86; H, 5.71; N, 22.25. ^1H NMR (d_6-benzene, 298 K): δ −9.30 (br, 18H, Tp^{Me2}-CH_3), −6.14 (br, 1H, bipy-CH), 3.16 (br, 2H, bipy-CH), 3.44 (br, 1H, bipy-CH), 5.05 (br, 18H, Tp^{Me2}-CH_3), 6.13 (br, 6H, Tp^{Me2}-CH), 12.86 (br, 3H, bipy-CH), 19.84 (br, 1H, bipy-CH). $^{11}B\{^1H\}$ NMR (d_6-benzene, 298 K): δ 10.48 (s, BH). Evans method susceptibility (d_6-benzene, 298 K) μ_{eff}: 1.59 μ_B. FTIR v/cm^{-1} (Nujol): 2513 (w, B–H str), 1541 (m, bipy radical anion), 1494 (m, bipy radical anion), 1410 (m), 1286 (m), 1265 (m), 1208 (m), 1168 (m), 1146 (m), 1115 (m), 1072 (m), 1028 (m), 992 (m), 977 (m), 941 (m, bipy radical anion), 829 (m), 773 (m), 641 (m).

[Ce(Tp^{Me2})(μ-BOp^{Me2})]$_2$ (**3**). THF (20 mL) was added to **2** (1.71 g, 1.9 mmol) and *N*-methylmorpholine-*N*-oxide (0.225 g, 2.0 mmol) at −78 °C. The orange reaction mixture was allowed to slowly warm to room temperature and was stirred for 16 h, forming a white precipitate. Volatiles were removed *in vacuo* and the solid was extracted with toluene (2 mL), filtered, and stored at −25 °C for 16 h to give pale yellow crystals of **3** (0.25 g, 21%). Anal. Calcd for $C_{50}H_{74}B_4Ce_2N_{20}O_2$: C 45.82, H 5.69, N 21.60. Found: C 45.63, H 5.43, N 21.36. ^1H NMR (d_6-benzene, 298 K): δ 3.55 (br, 36H, Tp^{Me2}-CH_3), 5.85 (s, 12H, BOp^{Me2}-CH_3), 8.15 (s, 6H, Tp^{Me2}-CH), 8.78 (s, 12H, BOp^{Me2}-CH_3), 18.51 (s, 4H, BOp^{Me2}-CH). $^{11}B\{^1H\}$ NMR (d_6-benzene, 298 K): δ 0.60 (s, BH). Evans method susceptibility (d_6-benzene, 298 K) μ_{eff}: 2.71 μ_B. FTIR v/cm^{-1} (Nujol): 2551 (w, B–H str), 1540 (m), 1415 (m), 1198 (m), 1076 (m), 1037 (m), 694 (m), 649 (m).

[Ce(Tp^{Me2})$_2$(N$_3$)] (**4**). Me$_3$SiN$_3$ (0.12 g, 1.0 mmol) in THF (10 mL) was added dropwise to a suspension of **2** (0.89 g, 1.0 mmol) in THF (10 mL). The reaction mixture was stirred for 3 days and refluxed at 80 °C for 16 h. Volatiles were removed *in vacuo* and the solid was extracted with toluene (10 mL), filtered, reduced in volume to *ca.* 2 mL and stored at −25 °C overnight to give yellow crystals of **4** (0.22 g, 29%). Anal. Calcd for $C_{30}H_{44}B_2CeN_{15}$: C, 46.38; H, 5.71; N, 27.04. Found: C, 46.49; H, 5.63; N, 26.86. ^1H NMR (d_6-benzene, 298 K): δ: −19.32 (s, 2H, Tp^{Me2}-CH), −19.09 (s, 6H, Tp^{Me2}-CH_3), −2.19 (s, 6H, Tp^{Me2}-CH_3), −0.07 (s, 6H, Tp^{Me2}-CH_3), 2.42 (s, 6H, Tp^{Me2}-CH_3), 5.52 (s, 2H, Tp^{Me2}-CH), 7.00 (s, 6H, Tp^{Me2}-CH_3), 10.02 (s, 2H, Tp^{Me2}-CH), 12.63 (s, 2H, Tp^{Me2}-CH), 18.93 (s, 6H, Tp^{Me2}-CH_3). $^{13}C\{^1H\}$ NMR (d_6-benzene, 298 K): δ: 11.08 (Tp^{Me2}-CH_3), 14.91 (Tp^{Me2}-CH_3), 101.76 (Tp^{Me2}-CH), 104.06 (Tp^{Me2}-CH), 120.78 (Tp^{Me2}-C), 123.24 (Tp^{Me2}-CH), 140.12 (Tp^{Me2}-C), 141.65 (Tp^{Me2}-C), 153.88 (Tp^{Me2}-C), 154.89 (Tp^{Me2}-C), 189.32 (Tp^{Me2}-C). $^{11}B\{^1H\}$ NMR (d_6-benzene, 298 K): δ −25.29 (s, BH). Evans method susceptibility (d_6-benzene, 298 K) μ_{eff}: 1.94 μ_B. FTIR v/cm^{-1} (Nujol): 2555 (w, B–H str), 2083 (s, N$_3$), 2065 (s, N$_3$), 1537 (m), 1201 (m), 1169 (m), 1072 (m), 1031 (m), 978 (w), 936 (w), 640 (w).

3.3. X-ray Crystallography

The crystal data for complexes **1**–**4** are compiled in Tables 2 and 3. Crystals were examined on CCD area detector diffractometers using graphite- or mirror-monochromated Mo *K*α (λ = 0.71073 Å) or Cu *K*α (λ = 1.54184 Å) radiation. Intensities were integrated from data recorded on 1° frames by ω (**1–3**) or φ and ω rotation (**4**). Cell parameters were refined from the observed positions of all strong reflections in each data set. A Gaussian grid face-indexed (**1–3**) or multi-scan (**4**) absorption correction with a beam profile correction was applied. The structures were solved variously by direct and heavy atom methods and were refined by full-matrix least-squares on all unique F^2 values, with anisotropic displacement parameters for all non-hydrogen atoms, and with constrained riding hydrogen geometries; $U_{iso}(H)$ was set at 1.2 (1.5 for methyl groups) times U_{eq} of the parent atom. The largest features in final difference syntheses were close to heavy atoms and were of no chemical significance. CrysAlisPro [40] was used for control and integration, SHELXTL [41] and OLEX2 [42] were employed for structure solution and refinement and POVRAY [43] was used for molecular graphics. Highly disordered lattice solvent in **3** could not be modelled and Platon SQUEEZE was used to resolve this component [44]. CCDC 1423263–1423266 (**1–4**) contain the supplementary crystal data for this article. These data can be obtained free of charge from the Cambridge Crystallographic Data Centre via www.ccdc.cam.ac.uk/data_request/cif.

Table 2. Crystallographic data for **1–2**.

Parameter	$1 \cdot C_7H_8$	$2 \cdot (C_4H_8O)_2$
Formula	$C_{42}H_{59}B_2CeN_{14}$	$C_{48}H_{68}B_2CeN_{14}O_2$
Fw	921.77	1034.90
cryst size, mm	$0.171 \times 0.152 \times 0.095$	$0.116 \times 0.110 \times 0.095$
cryst syst	Triclinic	Monoclinic
space group	$P-1$	$P21/c$
a, Å	11.1262(9)	14.2034(4)
b, Å	14.6920(14)	20.6250(7)
c, Å	14.7874(14)	17.1106(6)
α, °	70.037(9)	90
β, °	85.302(8)	95.066(3)
γ, °	81.302(7)	90
V, Å^3	2244.7(4)	4992.9(3)
Z	2	4
ρ_{calcd}, g cm^{-3}	1.364	1.377
μ, mm^{-1}	1.061	0.965
no. of reflections measd	15541	22174
no. of unique reflns, R_{int}	8201, 0.0954	9125, 0.0630
no. of reflns with $F^2 > 2\sigma(F^2)$	6502	7004
transmn coeff range	0.806–1.000	0.965–1.000
R, R_w [a] ($F^2 > 2\sigma(F^2)$)	0.0642, 0.1312	0.0523, 0.1058
R, R_w [a] (all data)	0.0845, 0.1414	0.0746, 0.1163
S [a]	1.046	1.029
Number of parameters	547	708
max., min. diff map, e Å^{-3}	3.59, −0.17	1.55, −1.61

[a] Conventional $R = \Sigma||F_o| - |F_c||/\Sigma|F_o|$; $R_w = [\Sigma w(F_o^2 - F_c^2)^2/\Sigma w(F_o^2)^2]^{1/2}$; $S = [\Sigma w(F_o^2 - F_c^2)^2/\text{no. data} - \text{no. params})]^{1/2}$ for all data.

Table 3. Crystallographic data for **3–4**.

Parameter	3·(C$_4$H$_8$O)$_2$	4·C$_7$H$_8$
Formula	C$_{62}$H$_{90}$B$_4$Ce$_2$N$_{20}$O$_2$	C$_{37}$H$_{52}$B$_2$CeN$_{15}$
Fw	1447.62	868.67
cryst size, mm	0.320 × 0.260 × 0.164	0.279 × 0.204 × 0.129
cryst syst	Triclinic	Triclinic
space group	P–1	P–1
a, Å	10.0579(4)	11.2609(3)
b, Å	13.3606(6)	12.5209(3)
c, Å	13.6905(7)	16.7464(4)
α, °	87.438(4)	105.035(2)
β, °	73.681(4)	96.314(2)
γ, °	71.869(4)	112.313(3)
V, Å^3	1676.10(14)	2051.38(10)
Z	1	2
ρ_{calcd}, g cm^{-3}	1.291	1.406
μ, mm^{-1}	10.730	1.157
no. of reflections measd	13,249	28,373
no. of unique reflns, R_{int}	5891, 0.0581	28,373, 0.060
no. of reflns with $F^2 > 2\sigma(F^2)$	5342	22,381
transmn coeff range	0.197–1.000	0.687–1.000
R, R_w [a] ($F^2 > 2\sigma(F^2)$)	0.0448, 0.1119	0.0515, 0.1456
R, R_w [a] (all data)	0.0496, 0.1153	0.0654, 0.0515
S [a]	1.015	1.097
Number of parameters	368	510
max., min. diff map, e Å^{-3}	1.53, −2.10	1.63, −1.08

[a] Conventional $R = \Sigma||F_o| - |F_c||/\Sigma|F_o|$; $R_w = [\Sigma w(F_o{}^2 - F_c{}^2)^2/\Sigma w(F_o{}^2)^2]^{1/2}$; $S = [\Sigma w(F_o{}^2 - F_c{}^2)^2/no.$ data—no. params)]$^{1/2}$ for all data.

4. Conclusions

We have investigated novel synthetic routes to prepare heteroleptic cerium tris(pyrazolyl)borate complexes, including a Ce(II) synthon, **2**, that formally contains a monoreduced bipy·$^-$ ligand and a Ce(III) center. We used a wide range of techniques to probe the electronic structure of **2** and conclude that the coordinated bipy exhibits features of a monoreduced radical. However, the EPR spectrum and magnetic measurements indicate that the coupling of this radical with the formal 4f^1 electron is complex, and future analysis of a more extensive library of Ce(II) synthons containing radical ligands would facilitate a deeper understanding of this coupling. A selected reactivity study of **2** with potential 2e$^-$ oxidants was performed and whilst bipy was eliminated in each isolated product cerium was not found in the +4 oxidation state. The formation of **3** and **4** suggests that the reaction of *N*-oxides and azides with Ce(II) synthons supported by bipy·$^-$ and Tp^{Me2} is not a favored route to Ce(IV) complexes exhibiting Ce=O or Ce=NR multiple bonds under the conditions we employed.

Supplementary Materials: Supplementary materials can be found at http://www.mdpi.com/2304-6740/3/4/0534/s1.

Acknowledgments: We thank the Engineering and Physical Sciences Research Council (Grant No EP/K039547/1, EP/L018470/1 and EP/L014416/1), including the national Electron Paramagnetic Resonance Facility, and The University of Manchester for funding this work. Additional research data supporting this publication are available from The University of Manchester eScholar repository at DOI:10.15127/1.280046.

Author Contributions: David P. Mills conceived and designed the experiments. Hao Zhu, Fabrizio Ortu and Marie-Emmanuelle Boulon performed the experiments. David P. Mills, Fabrizio Ortu, Marie-Emmanuelle Boulon and Hao Zhu analyzed the data. David P. Mills wrote the paper, with contributions from all co-authors.

Conflicts of Interest: The authors declare no conflict of interest.

References

1. Nugent, W.A.; Mayer, J.M. *Metal-Ligand Multiple Bonds*; Wiley-Interscience: New York, NY, USA, 1988.
2. Giesbrecht, G.R.; Gordon, J.C. Lanthanide alkylidene and imido complexes. *Dalton Trans.* **2004**, 2387–2393. [CrossRef] [PubMed]
3. Summerscales, O.T.; Gordon, J.C. Complexes containing multiple bonding interactions between lanthanoid elements and main-group fragments. *RSC Adv.* **2013**, *3*, 6682–6692. [CrossRef]
4. Gregson, M.; Lu, E.; McMaster, J.; Lewis, W.; Blake, A.J.; Liddle, S.T. A cerium(IV)–carbon multiple bond. *Angew. Chem. Int. Ed.* **2013**, *52*, 13016–13019. [CrossRef] [PubMed]
5. Schädle, D.; Meermann-Zimmerman, M.; Schädle, C.; Maichle-Mössmer, C.; Anwander, R. Rare-earth metal complexes with terminal imido ligands. *Eur. J. Inorg. Chem.* **2015**, *2015*, 1334–1339.
6. So, Y.-M.; Wang, G.-C.; Li, Y.; Sung, H.H.-Y.; Williams, I.D.; Lin, Z.; Leung, W.-H. A tetravalent cerium complex containing a Ce=O bond. *Angew. Chem. Int. Ed.* **2014**, *53*, 1626–1629. [CrossRef] [PubMed]
7. Lu, E.; Li, Y.; Chen, Y. A scandium terminal imido complex: Synthesis, structure and DFT studies. *Chem. Commun.* **2010**, *46*, 4469–4471. [CrossRef] [PubMed]
8. Hayton, T.W. Metal-ligand multiple bonding in uranium: Structure and reactivity. *Dalton Trans.* **2010**, *39*, 1129–1404. [CrossRef] [PubMed]
9. Hayton, T.W. Recent developments in actinide-ligand multiple bonding. *Chem. Commun.* **2013**, *49*, 2956–2973. [CrossRef] [PubMed]
10. Zi, G.; Jia, L.; Werkema, E.L.; Walter, M.D.; Gottfriedsen, J.P.; Andersen, R.A. Preparation and reactions of base-free bis(1,2,4-tri-*tert*-butylcyclopentadienyl)uranium oxide, Cp'₂UO. *Organometallics* **2005**, *24*, 4251–4264. [CrossRef]
11. Kraft, S.J.; Walensky, J.; Fanwick, P.E.; Hall, M.B.; Bart, S.C. Crystallographic evidence of a base-free uranium(IV) terminal oxo species. *Inorg. Chem.* **2010**, *49*, 7620–7622. [CrossRef] [PubMed]
12. Matson, E.M.; Kiemicki, J.J.; Anderson, N.H.; Fanwick, P.E.; Bart, S.C. Isolation of a uranium(III) benzophenone ketyl radical that displays redox-active ligand behavior. *Dalton Trans.* **2014**, *43*, 17885–17888. [CrossRef] [PubMed]
13. Anderson, N.H.; Odoh, S.O.; Yao, Y.; Williams, U.J.; Schaefer, B.A.; Kiernicki, J.J.; Lewis, A.J.; Goshert, M.D.; Fanwick, P.E.; Schelter, E.J.; et al. Harnessing redox activity for the formation of uranium tris(imido) compounds. *Nat. Chem.* **2014**, *6*, 919–926. [CrossRef] [PubMed]
14. Ren, W.; Zi, G.; Walter, M.D. Synthesis, structure and reactivity of a thorium metallocene containing a 2,2'-bipyridyl ligand. *Organometallics* **2012**, *31*, 672–679. [CrossRef]
15. Nugent, L.J.; Baybarz, R.D.; Burnett, J.L.; Ryan, J.L. Electron-transfer and f-d absorption bands of some lanthanide and actinide complexes and the standard (II–III) oxidation potential for each member of the lanthanide and actinide series. *J. Phys. Chem.* **1973**, *77*, 1528–1539. [CrossRef]
16. Kraft, S.J.; Fanwick, P.E.; Bart, S.C. Synthesis and characterization of a uranium(III) complex containing a redox-active 2,2'-bipyridine ligand. *Inorg. Chem.* **2010**, *49*, 1103–1110. [CrossRef] [PubMed]
17. Roitershtein, D.; Domingos, A.; Pereira, L.C.J.; Ascenso, J.R.; Marques, N. Coordination of 2,2'-bipyridyl and 1,10-phenanthroline to yttrium and lanthanum complexes based on a scorpionate ligand. *Inorg. Chem.* **2003**, *42*, 7666–7673. [CrossRef] [PubMed]
18. Marques, N.; Sella, A.; Takats, J. Chemistry of the lanthanides using pyrazolylborate ligands. *Chem. Rev.* **2002**, *102*, 2137–2159. [CrossRef] [PubMed]
19. Booth, C.H.; Walter, M.D.; Kazhdan, D.; Hu, Y.-J.; Lukens, W.W.; Bauer, E.D.; Maron, L.; Eisenstein, O.; Andersen, R.A. Decamethylytterbocene complexes of bipyridines and diazabutadienes: Multiconfigurational ground states and open-shell singlet formation. *J. Am. Chem. Soc.* **2009**, *131*, 6480–6491. [CrossRef] [PubMed]
20. Nocton, G.; Booth, C.H.; Maron, L.; Andersen, R.A. Influence of the torsion angle in 3,3'-dimethyl-2,2'-bipyridine on the intermediate valence of Yb in (C₅Me₅)₂Yb(3,3'-Me₂-bipy). *Organometallics* **2013**, *32*, 5305–5312. [CrossRef]
21. Izod, K.; Liddle, S.T.; Clegg, W. A convenient route to lanthanide triiodide THF solvates. Crystal structures of LnI₃(THF)₄ [Ln = Pr] and LnI₃(THF)₃.₅ [Ln = Nd, Gd, Y]. *Inorg. Chem.* **2004**, *43*, 214–218. [CrossRef] [PubMed]
22. Trofimenko, S. Boron-pyrazole chemistry. IV. Carbon- and boron-substituted poly(1-pyrazolyl)borates. *J. Am. Chem. Soc.* **1967**, *89*, 6288–6294. [CrossRef]

23. Kunrath, F.A.; Casagrande, O.L., Jr.; Toupet, L.; Carpentier, J.-F. Synthesis and reactivity in salt metathesis reactions of trivalent [La(Tp^{Me2})$_2$X] (X = Cl, I) complexes: Crystal structures of [La(Tp^{Me2})$_2$Cl] and [La(Tp^{Me2})$_2$(κ^2-pz^{Me2})]. *Polyhedron* **2004**, *23*, 2437–2445. [CrossRef]

24. Galler, J.L.; Goodchild, S.; Gould, J.; McDonald, R.; Sella, A. Tris-pyrazolylborate complexes of redox inactive lanthanides—The structures of [(Tp^{Me2})$_2$NdX] (X = Cl, NPh$_2$, dpm, H$_2$BEt$_2$). *Polyhedron* **2004**, *23*, 253–262. [CrossRef]

25. Hillier, A.C.; Zhang, X.; Maunder, G.H.; Liu, S.Y.; Eberspacher, T.A.; Metz, M.V.; McDonald, R.; Domingos, A.; Marques, N.; Day, V.W.; *et al.* Synthesis and structural comparison of a series of divalent Ln(Tp$^{R,R'}$)$_2$ (Ln = Sm, Eu, Yb) and trivalent Sm(Tp^{Me2})$_2$X (X = F, Cl, I, BPh$_4$) complexes. *Inorg. Chem.* **2001**, *40*, 5106–5116. [CrossRef] [PubMed]

26. Liu, S.-Y.; Maunder, G.H.; Sella, A.; Stevenson, M.; Tocher, D.A. Synthesis and molecular structures of hydrotris(dimethylpyrazolyl)borate complexes of the lanthanides. *Inorg. Chem.* **1996**, *35*, 76–81. [CrossRef] [PubMed]

27. Amberger, H.-D.; Edelmann, F.T.; Gottfriedsen, J.; Herbst-Irmer, R.; Jank, S.; Kilimann, M.; Reddmann, H.; Schäfer, M. Synthesis, molecular, and electronic structure of (η^8-C$_8$H$_8$)Ln(scorpionate) half-sandwich complexes: An experimental key to a better understanding of f-element-cyclooctatetraenyl bonding. *Inorg. Chem.* **2009**, *48*, 760–772. [CrossRef] [PubMed]

28. Evans, D.F. The determination of the paramagnetic susceptibility of substances in solution by nuclear magnetic resonance. *J. Chem. Soc.* **1959**, 2003–2005. [CrossRef]

29. Van Vleck, J.H. *Theory of Electric and Magnetic Susceptibilities*; Oxford University Press: Oxford, UK, 1932.

30. Akita, M.; Otha, K.; Takahashi, Y.; Hikichi, S.; Moro-oka, Y. Synthesis and structure determination of Rh–diene complexes with the hydridotris(3,5-diisopropylpyrazolyl)borate ligand, Tp^{iPr}Rh(diene) (diene = cod, nbd): Dependence of the ν(B–H) values on the hapticity of the Tp^{iPr} ligand (κ^2 *vs.* κ^3). *Organometallics* **1997**, *16*, 4121–4128. [CrossRef]

31. Lopes, I.; Lin, G.Y.; Domingos, A.; Marques, N.; Takats, J. Unprecedented transformation of a hydrotris(pyrazolyl)borate ligand at a metal center: Synthesis and rearrangement of the first mixed Tp/Cp lanthanide complex, Sm(Tp^{Me2})$_2$(Cp). *J. Am. Chem. Soc.* **1999**, *121*, 8110–8111. [CrossRef]

32. Shannon, R.D.; Prewitt, C.T. Revised values of effective ionic radii. *Acta Crystallogr. Sect. B* **1970**, *26*, 1046–1048. [CrossRef]

33. Bergbreiter, D.E.; Killough, J.M. Reactions of potassium-graphite. *J. Am. Chem. Soc.* **1978**, *100*, 2126–2134. [CrossRef]

34. Mehdoui, T.; Berthet, J.-C.; Thuéry, P.; Salmon, L.; Rivière, E.; Ephritikhine, M. Lanthanide(III)/actinide(III) differentiation in the cerium and uranium complexes [M(C$_5$Me$_5$)$_2$(L)]$^{0,+}$ (L = 2,2'-bipyridine, 2,2':6',2''-terpyridine): Structural magnetic and reactivity studies. *Chem. Eur. J.* **2005**, *11*, 6994–7006. [CrossRef] [PubMed]

35. Krishnan, C.V.; Creutz, C.; Schwarz, H.A.; Sutin, N. Reduction potentials for 2,2'-bipyridine and 1,10'-phenanthroline couples in aqueous solutions. *J. Am. Chem. Soc.* **1983**, *105*, 5617–5623. [CrossRef]

36. Schelter, E.J. Cerium under the lens. *Nat. Chem.* **2013**, *5*, 348. [CrossRef] [PubMed]

37. Domingos, A.; Elsegood, M.R.J.; Hillier, A.C.; Lin, G.; Liu, S.Y.; Marques, N.; Maunder, G.H.; McDonald, R.; Sella, A.; *et al.* Facile pyrazolylborate ligand degradation at lanthanide centers: X-ray crystal structures of pyrazolylborinate-bridged bimetallics. *Inorg. Chem.* **2002**, *41*, 6761–6768. [CrossRef] [PubMed]

38. Mohammad, A.; Cladis, D.P.; Forrest, W.P.; Fanwick, P.E.; Bart, S.C. Reductive heterocoupling mediated by Cp*$_2$U(2,2'-bpy). *Chem. Commun.* **2012**, *48*, 1671–1673. [CrossRef] [PubMed]

39. Dröse, P.; Gottfriedsen, J.; Hrib, C.G.; Jones, P.G.; Hilfert, L.; Edelmann, F.T. The first cationic complex of tetravalent cerium. *Z. Anorg. Allg. Chem.* **2011**, *637*, 369–373. [CrossRef]

40. *CrysAlis PRO*, version 37; Agilent Technologies: Yarnton, UK, 2010.

41. Sheldrick, G.M. Crystal structure refinement with SHELX. *Acta Cryst. Sect. C* **2015**, *71*, 3–8. [CrossRef] [PubMed]

42. Dolomanov, O.V.; Bourhis, L.J.; Gildea, R.J.; Howard, J.A.K.; Puschmann, H. Olex2: A complete structure solution, refinement and analysis program. *J. Appl. Crystallogr.* **2009**, *42*, 339–341. [CrossRef]

43. POVRAY, version 3.7.0. Persistence of Vision Pty. Ltd.: Williamstown, Australia, 2004. Available online: http://www.povray.org/ (Accessed on 16 August 2015).
44. Spek, A.L. Single-crystal structure validation with the program PLATON. *J. Appl. Cryst.* **2003**, *36*, 7–13. [CrossRef]

inorganics

MDPI

Article

Magnetic and Photo-Physical Properties of Lanthanide Dinuclear Complexes Involving the 4,5-Bis(2-Pyridyl-N-Oxidemethylthio)-4′,5′-Dicarboxylic Acid-Tetrathiafulvalene-, Dimethyl Ester Ligand

Fabrice Pointillart [1,*], Saskia Speed [1], Bertrand Lefeuvre [1], François Riobé [2], Stéphane Golhen [1], Boris Le Guennic [1], Olivier Cador [1], Olivier Maury [2] and Lahcène Ouahab [1]

[1] Institut des Sciences Chimiques de Rennes UMR 6226 CNRS-UR1, Université de Rennes 1, 35042 Rennes Cedex, France; saskia.speed@univ-rennes1.fr (S.S.); bertrand.lefeuvre@univ-rennes1.fr (B.L.); stephane.golhen@univ-rennes1.fr (S.G.); boris.leguennic@univ-rennes1.fr (B.L.G.); olivier.cador@univ-rennes1.fr (O.C.); lahcene.ouahab@univ-rennes1.fr (L.O.)

[2] Laboratoire de Chimie de l'ENS-LYON-UMR 5182, 46 Allée d'Italie, 69364 Lyon Cedex 07, France; francois.riobe@ens-lyon.fr (F.R.); olivier.maury@ens-lyon.fr (O.M.)

* Author to whom correspondence should be addressed; fabrice.pointillart@univ-rennes1.fr; Tel.: +33-2-23-23-57-62; Fax: +33-2-23-23-68-40.

Academic Editors: Stephen Mansell and Steve Liddle

Received: 15 October 2015; Accepted: 23 November 2015; Published: 3 December 2015

Abstract: The reaction between the 4,5-bis(2-pyridyl-N-oxidemethylthio)-4′,5′-dicarboxylic acid-tetrathiafulvalene-, dimethyl ester ligand (**L**) and the metallo-precursors $Ln(hfac)_3 \cdot 2H_2O$ leads to the formation of two dinuclear complexes of formula $[Ln_2(hfac)_6(\mathbf{L})] \cdot (CH_2Cl_2) \cdot (C_6H_{14})_{0.5}$ (Ln^{III} = Dy^{III} (**1**) and Yb^{III} (**2**)). The X-ray structure reveals a quite regular square anti-prism symmetry for the coordination sphere of the lanthanide ion. UV-visible absorption properties have been experimentally measured and rationalized by TD-DFT calculations. The functionalization of the tetrathiafulvalene (TTF) core by two methyl ester moieties induces the appearance of an additional absorption band in the lowest-energy region of the spectrum. The latter has been identified as a HOMO (Highest Occupied Molecular Orbital)→LUMO (Lowest Unoccupied Molecular Orbital) Intra-Ligand Charge Transfer (ILCT) transition in which the HOMO and LUMO are centred on the TTF and methyl ester groups, respectively. Irradiation at $22,222$ cm^{-1} of this ILCT band induces an efficient sensitization of the Yb^{III}-centred emission that can be correlated to the magnetic properties.

Keywords: lanthanides; tetrathiafulvalene; magnetism; photo-physics; TD-DFT calculations

1. Introduction

Since the discovery of the first mononuclear single molecule magnet (SMM) of lanthanide [1], f-elements have taken a preponderant place in the elaboration of coordination complexes in the field of molecular magnetism [2–11]. Lanthanide ions indeed have several advantages, such as strong single-ion anisotropy and large magnetic moment [12,13]. Recently, actinides have started to contribute to the development of new SMMs due to their strong magnetic anisotropy and larger exchange interactions coming from the more extended nature of the 5f orbitals [14–20]. On one hand, SMMs are molecular objects with potential applications in quantum computing, high-density memory data storage devices, and spintronics [21–29]. On the other hand, lanthanide ions are widely exploited for their specific luminescence which displays sharp f–f transitions ranging from the visible to the near infrared (NIR) region. Nevertheless, the very weak intensity of the f–f absorption bands (Laporte forbidden) [30] makes a direct excitation less straightforward, and consequently, lanthanide ions are usually coordinated to an organic ligand that strongly absorbs light in the UV-visible range. Thus, this organic ligand plays the role of organic chromophore for the sensitization of the lanthanide's luminescence through an antenna effect [31,32]. In addition to the interest of lanthanide ions in coordination chemistry, a main challenge that chemists face is to use organic ligands that induce various behaviours such as redox activity, emission, magnetism, chirality, *etc.*, and lead to multi-properties (functional) coordination complexes or SMMs [33–35]. In this context, a few years ago we started to combine lanthanide ions and tetrathiafulvalene (TTF)-based ligands which were used for their electron donating ability, redox activity and electronic conductivity [36–44]. We have already demonstrated that such ligands are suitable to obtain SMMs [45–51] and to sensitize lanthanide luminescence [52–56]. Previously, we associated the Ln(hfac)$_3$·2H$_2$O (hfac$^-$ = 1,1,1,5,5,5-hexafluoroacetyacetonate) metallo-precursors with the (4,5-bis(2-pyridyl-*N*-oxidemethylthio)-4′,5′)-ethylenedithiotetrathiafulvene, -methyldithiotetrathiafulvene [57], or -4′,5′-ethylendioxotetrathiafulvene ligands, leading to several dinuclear complexes [58].

In the present contribution we go one step forward, using the new 4,5-bis(2-pyridyl-*N*-oxidemethylthio)-4′,5′-dicarboxylic acid-tetrathiafulvalene-, dimethyl ester ligand (**L**). The coordination reactions of Ln(hfac)$_3$·2H$_2$O (Ln = Dy^{III} (**1**) and Yb^{III} (**2**)) with (**L**) are presented and the X-ray structure of [Dy$_2$(hfac)$_6$(**L**)]·(CH$_2$Cl$_2$)·(C$_6$H$_{14}$)$_{0.5}$ is described. The photo-physical and magnetic properties of **1** and **2** are presented and discussed.

2. Results and Discussion

2.1. Synthesis

The target ligand **L** (Scheme 1) was prepared by deprotection of the thiol function of the 4,5-dicarboxylic acid-4′,5′-bis(2-cyanoethyl)thio)-tetrathiafulvalene-, dimethyl ester [59] with sodium methoxide or cesium hydroxide and then grafting of the 2-methylpyridine-1-oxide coordinating groups.

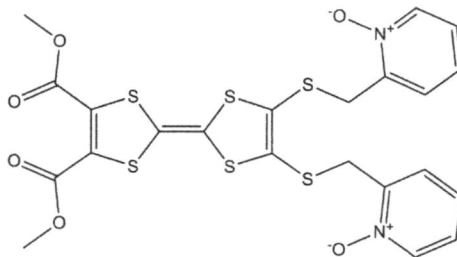

Scheme 1. Chemical structure of the new 4,5-bis(2-pyridyl-*N*-oxidemethylthio)-4′,5′-dicarboxylic acid-tetrathiafulvalene-, dimethyl ester ligand, **L**.

L can be described as a multiply functionalized ligand with two pyridine-*N*-oxide moieties that are suitable for the coordination of lanthanide ions, leading generally to the formation of dinuclear complexes on one side and two ester groups on the other side, whose electron-withdrawing character will monitor the ligand's electronic properties. These ester moieties could also be hydrolysed into carboxylate functions acting as additional coordinating sites [60].

2.2. Crystal Structure of [Dy$_2$(hfac)$_6$(L)]·(CH$_2$Cl$_2$)·(C$_6$H$_{14}$)$_{0.5}$ (1)

Compound **1** crystallizes in the Pcab (No.61) orthorhombic space group (Tables 1 and 2). The asymmetric unit is composed of two Dy(hfac)$_3$ moieties, two **L** ligands and one dichloromethane molecule of crystallization. An ORTEP (Oak Ridge Thermal Ellipsoid Plot) view is depicted in Figure 1.

Table 1. X-ray crystallographic data for **1**.

Compounds	$[Dy_2(hfac)_6(L)_2]\cdot(CH_2Cl_2)\cdot0.5C_6H_{14}$ (1)
Formula	$C_{78}H_{51}Cl_2Dy_2F_{36}N_4O_{24}S_{12}$
$M/g\cdot mol^{-1}$	2892.84
Crystal system	Orthorhombic
Space group	Pcab (No.61)
Cell parameters	$a = 17.3005(5)$ Å
	$b = 34.2288(10)$ Å
	$c = 37.6768(10)$ Å
Volume/Å^3	22,311.3(11)
Z	8
T/K	150 (2)
2θ range/°	$2.16 \leq 2\theta \leq 54.96$
ρcalc/g·cm^{-3}	1.722
μ/mm^{-1}	1.728
Number of reflections	103,297
Independent reflections	25,452
Rint	0.0815
$Fo^2 > 2\sigma(Fo)^2$	13,844
Number of variables	1396
R1, wR2	0.0805, 0.2175

Table 2. Selected bond lengths (Å) for compound **1**.

Compounds	1
Dy1–O9	2.263(8)
Dy1–O10	2.324(6)
Dy1–O11	2.376(7)
Dy1–O12	2.347(7)
Dy1–O13	2.359(7)
Dy1–O14	2.366(7)
Dy1–O15	2.381(7)
Dy1–O16	2.380(8)
Dy2–O17	2.344(7)
Dy2–O18	2.264(7)
Dy2–O19	2.359(7)
Dy2–O20	2.361(8)
Dy2–O21	2.352(8)
Dy2–O22	2.326(8)
Dy2–O23	2.382(7)
Dy2–O24	2.394(8)

The X-ray structure reveals that two Dy(hfac)$_3$ moieties are bridged by two **L** ligands. The two water molecules of the starting metallo-precursor have been substituted by two pyridine-N-oxide groups. The Dy1 and Dy2 ions are surrounded by eight oxygen atoms that belong to three hfac$^-$ ligands and the two pyridine-N-oxide coordinating acceptors. The average Dy–O distances are equal to 2.350(7) and 2.348(8) Å for Dy1 and Dy2, respectively. The arrangement of the ligands leads to a square antiprism (D$_{4d}$ symmetry) as coordination polyhedron for both dysprosium ions (CShM$_{SAPR-8}$ = 0.385 for Dy1 and CShM$_{SAPR-8}$ = 0.396 for Dy2). The distortion is visualized by continuous shape measures performed with SHAPE 2.1 [61]. The Dy coordination polyhedra in **1** are the most symmetrical of all the series of similar dinuclear complexes that we have already obtained [57,58].

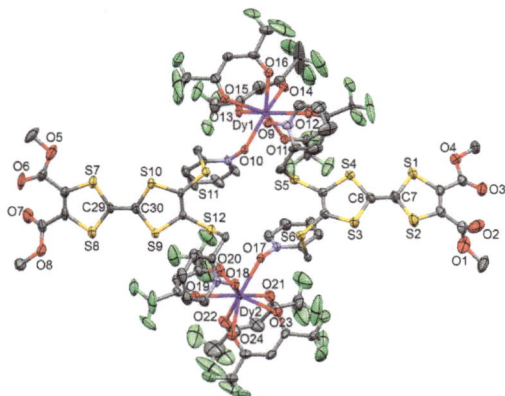

Figure 1. ORTEP view of **1** ([Dy$_2$(hfac)$_6$(**L**)$_2$]·(CH$_2$Cl$_2$)·0.5C$_6$H$_{14}$). Thermal ellipsoids are drawn at 30% probability. Hydrogen atoms and solvent molecules of crystallization are omitted for clarity.

The central C=C bonds of the TTF core are equal to 1.342(13) and 1.340(16) Å, which attests the neutral form of **L**. The ester functions play an important role in the cohesion of the crystal packing. While the ester groups involving O5–O8 are almost parallel, those involving O1–O4 are almost perpendicular, in order to optimize the intermolecular S···O short contacts.

The combination of the shortest S···S contacts (3.668 Å (S3···S3) and 3.889 Å (S3···S6)) and shortest S···O contacts (3.281 Å (S8···O1), 3.364 Å (S1···O5), and 3.493 Å (S3···O7)) leads to the formation of tetramers of donors (Figure 2). The shortest intra- and inter-molecular Dy–Dy distances are equal to 10.052 Å (Dy1···Dy2), 11.039 Å (Dy2···Dy2), and 11.340 Å (Dy1···Dy1).

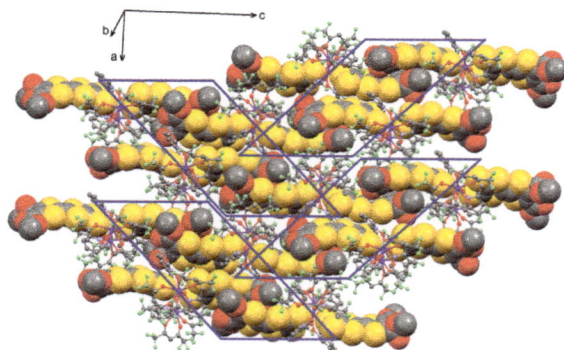

Figure 2. Crystal packing of **1** highlighting the formation of tetramers of **L** through short S···S and S···O contacts. Colour code: grey (C), red (O), orange (S), blue (D), green (F).

2.3. Electrochemical Properties

The redox properties of **L** and related complexes **1** and **2** ([Yb$_2$(hfac)$_6$(**L**)$_2$]·CH$_2$Cl$_2$·(C$_6$H$_{14}$)$_{0.5}$) are investigated by cyclic voltammetry (Figure S1) and the values of the oxidation potentials are listed in Table 3.

Table 3. Oxidation potentials (V *vs.* saturated calomel electrode, nBu$_4$NPF$_6$, 0.1 M in CH$_2$Cl$_2$ at 100 mV·s^{-1}) of the ligand **L** and complexes **1** and **2**.

	E$^1_{1/2}$/V		E$^2_{1/2}$/V		E$^3_{1/2}$/V	
	^{Ox}E$^1_{1/2}$	^{red}E$^1_{1/2}$	^{Ox}E$^2_{1/2}$	^{red}E$^2_{1/2}$	^{Ox}E$^3_{1/2}$	^{red}E$^3_{1/2}$
L	0.79	0.70	1.17	1.09	/	/
1	0.76	0.67	1.15	1.02	1.28	1.20
2	0.76	0.66	1.13	0.99	1.28	1.20

The cyclic voltammogram for **L** shows two mono-electronic oxidations at about 0.75 and 1.13 V, corresponding to the formation of a radical cation and a dication TTF fragment, respectively (Figure S1). These oxidation potentials are higher than those found for the functionalized TTF donor used in the synthesis of the other dinuclear complexes, due to the direct functionalization of the TTF core with two methyl ester groups [57,58]. The electrochemistry does not highlight significant effects on the oxidation potentials upon coordination of the electron attracting Ln(hfac)$_3$ fragments (Table 3). Nevertheless, an additional quasi-reversible oxidation wave is observed for the two complexes around 1.24 V. No clear explanation can be given for the origin of this additional redox activity at this point. The reversibility of the oxidation potentials is conserved and the electrochemical properties attest to the redox activity of **L** in the complexes.

2.4. Photo-Physical Properties

2.4.1. Absorption Properties

The UV-visible absorption properties of **L** and **2** have been studied in a CH$_2$Cl$_2$ solution and in a KBr pellet in solid-state (Figures 3a and 4a, Figures S2 and S3). Rationalization by TD-DFT calculations was performed on **L** and the Y(III) analogue of **1** and **2** (Figures 3b and 4b) following a computational strategy already used successfully on TTF-based systems [62]. The molecular orbital diagram is sketched in Figure 5.

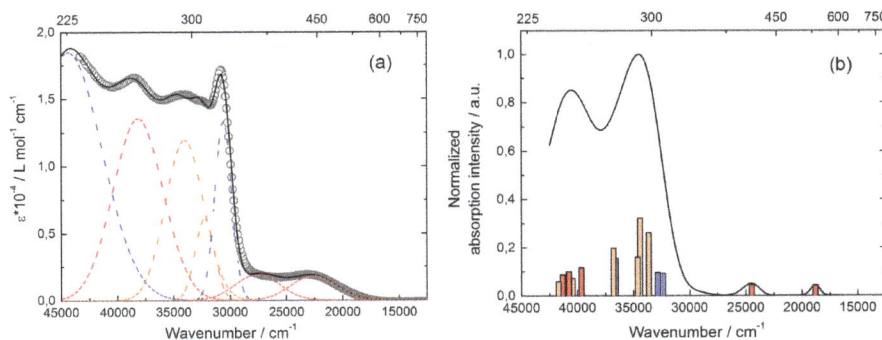

Figure 3. (a) Experimental UV-visible absorption spectra of **L** in CH$_2$Cl$_2$ solution ($C = 4 \times 10^{-5}$ mol·L^{-1}) (open grey circles). Respective Gaussian decompositions (red, orange, and blue dashed lines correspond to ILCT, intra-donor and intra-acceptor transitions, respectively) and best fit (full black line) ($R = 0.9994$) (**b**) Theoretical absorption spectra of **L** (black line). The bars represent the mean contributions of the absorption spectra.

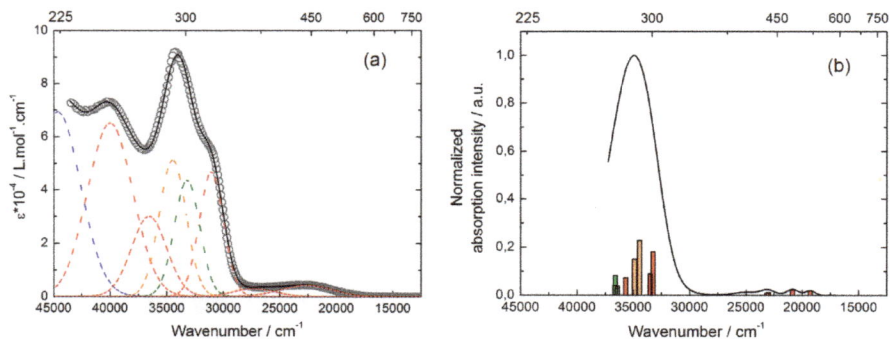

Figure 4. (**a**) Experimental UV-visible absorption spectra in CH$_2$Cl$_2$ solution of **2** ($C = 4 \times 10^{-5}$ mol·L^{-1}) (open grey circles). Respective Gaussian decompositions (green dashed line corresponds to intra-hfac transitions) and best fit (full black line) ($R = 0.9998$); (**b**) Theoretical absorption spectra of the Y^{III} analogue (black line). The bars represent the mean contributions of the absorption spectra.

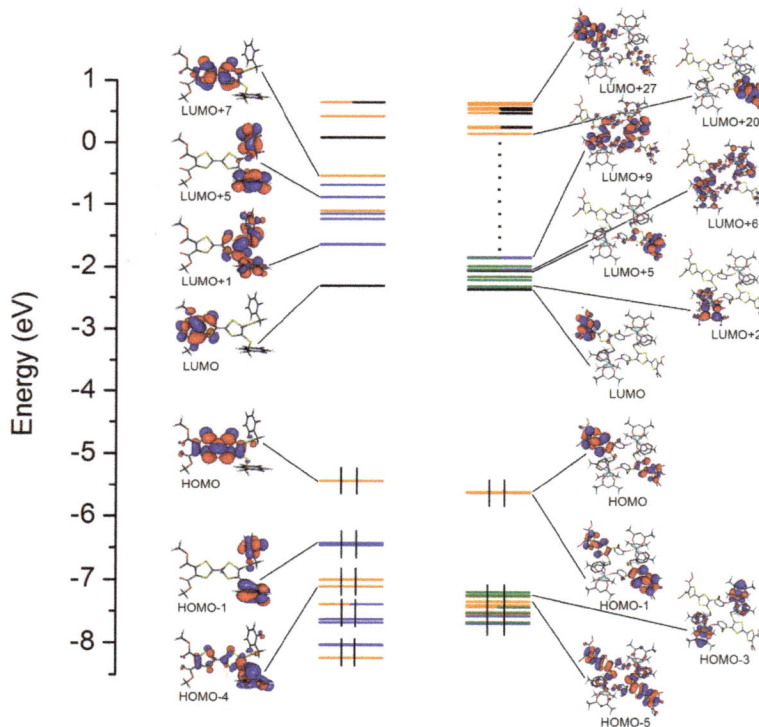

Figure 5. Molecular orbital diagram for **L** (**left**) and the Y^{III} analogue (**right**). The energy levels of the centred TTF, 2-methylpyridine-*N*-oxide, methyl ester, and hfac$^-$ orbitals are represented in orange, blue, black, and green, respectively.

The experimental absorption curve of **L** has been decomposed into seven bands (Figure 3a and Figure S2, Table 4). The calculated UV-visible absorption spectrum for **L** reproduces the experimental curve well (Figure 3b). The low energy bands (red Gaussian decomposition) were attributed to π–π*

HOMO→LUMO and HOMO→LUMO + 1 excitations. While the latter is identified as the classical TTF to Methyl-2-Py-*N*-oxide charge transfer (ILCT) (Table 4) [57,58], the lowest-energy excitation is identified as TTF to methyl ester charge transfer. Thus, the functionalization of the TTF core with methyl ester groups has successfully increased the intensity of the ILCT and induced the appearance of new ILCT at lower-energy (22,600 cm^{-1}, calculated at 18,810 cm^{-1}) than the TTF to Methyl-2-Py-*N*-oxide ILCT (27,600 cm^{-1}, calculated at 24,463 cm^{-1}). The remaining absorption bands were respectively identified as Intra-Acceptor (IA) (where the acceptor is the Py-*N*-oxide moiety), Intra-donor (ID), and ILCT excitations (Table 4).

Table 4. TD-DFT calculated excitations energies and main composition of low-lying electronic transitions for **L** and the Y(III) analogue of **1** and **2**. In addition, charge transfer and the pure intramolecular transitions are reported. ID, IA, Ihfac, H, and L represent respectively the intramolecular TTF (Donor), 2-pyridine-*N*-oxide (Py) (Acceptor) or hfac^{-} ligand, HOMO and LUMO; therefore ILCT stands for Intra-Ligand Charge Transfer (from TTF to 2-pyridine-*N*-oxide (PyNO) or methyl ester (ester)). The theoretical values are evaluated at the PCM(CH$_2$Cl$_2$)-PBE0/SVP level of approximation. Bold style is used to highlight the main contributions. Exp.: experimental, Th.: theoretical, Osc.: oscillator strength.

	Energy Exp. (cm^{-1})	Energy Th. (cm^{-1})	Osc.	Type	Assignment	Transition
L	22,600	18,810	0.03	ILCT	$\pi_{TTF} \rightarrow \pi^*_{Ester}$	H→L (99%)
	27,600	24,463	0.03	ILCT	$\pi_{TTF} \rightarrow \pi^*_{PyNO}$	H→L + 1 (88%)
	30,600	32,414 32,857	0.06 0.07	IA	$\pi_{PyNO} \rightarrow \pi^*_{PyNO}$	H-1→L + 1 (74%) H-2→L + 1 (70%)
	32,200	33,652 34,426 34,640	0.18 0.22 0.11	**ID** + ILCT	$\pi_{TTF} \rightarrow \pi^*_{TTF}$ + $\pi_{TTF} \rightarrow \pi^*_{PyNO}$	**H→L + 7 (60%)** H-3/-4→L + 1 (43/28%)
	34,000	36,605	0.11	**ID** + IA	$\pi_{TTF} \rightarrow \pi^*_{TTF}$ +	**H→L + 7/ + 9 (10/44%)**
		36,788	0.14		$\pi_{PyNO} \rightarrow \pi^*_{PyNO}$	H-1→L + 2/ + 5/ + 6 (9/27/10%)
	38,200	39,635 40,437 40,773 41,331 41,698	0.08 0.05 0.07 0.06 0.04	ID + **ILCT**	$\pi_{TTF} \rightarrow \pi^*_{TTF}$ + $\pi_{TTF} \rightarrow \pi^*_{PyNO}$	H-3/-4→L + 4 (38/40%) **H-5→L + 1 (38%)** H-3→L + 2 (25%) **H-10→L (41%)**
	44,400	/	/	IA	$\pi_{PyNO} \rightarrow \pi^*_{PyNO}$	/
Y	22,600	19,282 20,840	0.03 0.04	ILCT	$\pi_{TTF} \rightarrow \pi^*_{Ester}$	H/-1→L (76/22%) H/-1→L + 5 (22/72%)
	27,300	22,985	0.02	ILCT	$\pi_{TTF} \rightarrow \pi^*_{PyNO}$	H→L + 6/ + 8/ + 9 (20/19/21%)
		23,132	0.02			H-1→L + 6/ + 9 (17/33%)
	31,000	33,234 33,485	0.30 0.15	ID + **ILCT**	$\pi_{TTF} \rightarrow \pi^*_{TTF}$ + $\pi_{TTF} \rightarrow \pi^*_{Ester}$	H/-1→L + 20 (9/30%) H→L + 21 (15%) **H-4/-5→L (21/29%)**
	33,200	36,440 36,455 36,618 36,639	0.27 0.10	Ihfac	$\pi_{hfac} \rightarrow \pi^*_{hfac}$	H→L + 10 (53%) + H-3→L (65%)
	34,500	34,432 34,886	0.38	ID	$\pi_{TTF} \rightarrow \pi^*_{TTF}$	H/-1→L + 21 (12/41%) H/-1→L + 22 (9/30%)
	36,600	35,662	0.12	ILCT	$\pi_{TTF} \rightarrow \pi^*_{Ester}$	H-4/-5→L + 5 (13/7%)
	40,000	/	/	ILCT	$\pi_{TTF} \rightarrow \pi^*_{PyNO}$	/
	44,700	/	/	IA	$\pi_{L2} \rightarrow \pi^*_{PyImPy}$	/

For **2**, the absorption spectrum has been decomposed into eight bands for the solution and solid-state measurements.

The absorption spectra for the free ligand and the dinuclear complex are similar. Only an additional intense absorption band is observed around 33,000 cm^{-1} that corresponds to π–π^* intra-hfac$^-$ excitations [62]. Complexation induces a very weak red shift of the ILCT transitions involving the Py-*N*-oxide acceptor (300 cm^{-1}, calculated red shift 1400 cm^{-1}), due to the Lewis acid behaviour of the Ln(hfac)$_3$ moieties enforcing the electron withdrawing character of the 2-pyridine-*N*-oxide fragments. This effect is weak because of the poor electronic communication through the methylthio arms. Nevertheless, the red-shift of these absorption bands in coordination complexes compared to those in **L** is a first indication of the stability of the dinuclear complexes in such solvents.

2.4.2. Emission Properties

No significant emission for the free ligand was detected, as already observed in previous TTF-based ligands in which the TTF core is not conjugated with the coordinated moieties [57,58]. Emission properties of **2** were measured in solid state at 77 K (Figure 6). The characteristic luminescence profile of Yb^{III}, corresponding to the $^2F_{5/2} \to {}^2F_{7/2}$ transition, is observed upon irradiation at 22,222 cm^{-1} (450 nm). This low-energy irradiation is possible due to the ILCT involving the methyl ester groups. No residual emission centred on the ligand **L** was observed in the visible range after coordination. Based on the previous published TTF-based complexes of Yb^{III} [57,63,64], it is expected that in the antenna-effect sensitization process, such energy transfer takes place between the singlet CT excited state of the chromophore to the $^2F_{5/2}$ states of the Yb^{III} ion.

Four main emission maxima are clearly identified at the following energies: 9718, 9843, 9980, and 10,204 cm^{-1}. The total splitting is determined to be equal to 486 cm^{-1}. The values of this splitting for an Yb^{III} ion in a distorted and regular D$_3$ symmetry are 455 and 372 cm^{-1}, respectively [65], while a splitting of 528 cm^{-1} is found for an Yb complex in a lower symmetry, increasing up to 880 cm^{-1} for organometallic derivatives [66]. The value of 486 cm^{-1} thus seems to correspond to a quite low symmetry. Nevertheless, this value is lower than the one found in a dinuclear analogue (508 cm^{-1}) in which the distortion of the square anti-prism symmetry of the Yb^{III} coordination sphere is more pronounced (CShM$_{SAPR-8}$ = 0.484).

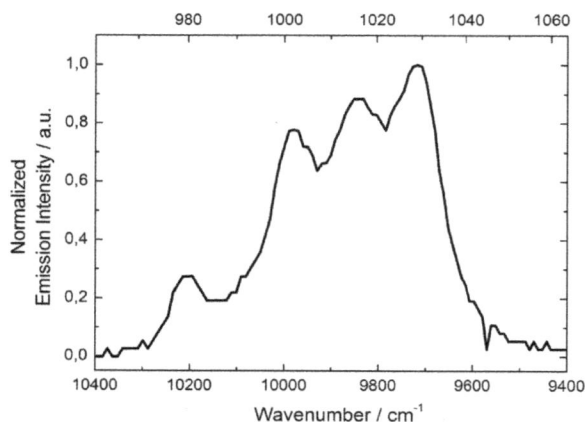

Figure 6. Luminescence spectrum of **2** in the Near Infrared (NIR) range for λ_{ex} = 22,222 cm^{-1} (450 nm) in solid-state at 77 K.

2.5. Magnetic Properties

The temperature dependence of $\chi_M T$, with χ_M being the molar magnetic susceptibility and T being the temperature in Kelvin, in powdered samples of compounds **1** and **2** are represented in Figure 7. The room temperature values are equal to 27.6 and 4.69 cm^3 K·mol^{-1} for **1** and **2**, respectively. These values are in good agreement with the two expected values for the two multiplet ground states: $^6H_{15/2}$ (14.17 × 2 cm^3 K·mol^{-1}) and $^2F_{7/2}$ (2.57 × 2 cm^3 K·mol^{-1}) for two Dy^{III} and two Yb^{III} ions, respectively [67]. $\chi_M T$ decreases on cooling due to the crystal field splitting of the multiplet ground states down to 20.2 cm^3·K·mol^{-1} for **1**, and 1.03 cm^3·K·mol^{-1} for **2**. At 2 K, the magnetization saturates around 10 Nβ for **1** and 2.8 Nβ for **2**. Neither of the two complexes show frequency dependence of ac susceptibility.

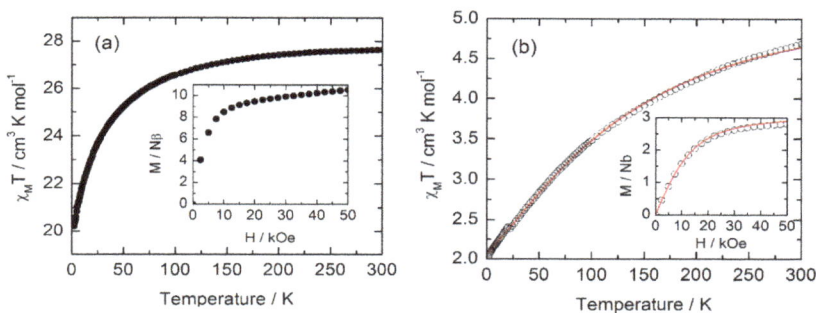

Figure 7. (**a**) Temperature dependence of $\chi_M T$ for **1**. Inset: field variation of the magnetization measured at 2 K; (**b**) Temperature dependence of $\chi_M T$ for **2**. Inset: field variation of the magnetization measured at 2 K. The red line corresponds to the best-fitted curve (see text). χ_M, molar magnetic susceptibility; T, temperature in Kelvin.

2.6. Correlation Magnetism-Luminescence

In the last five years, we have proven that magnetism and luminescence are intimately correlated [50,54,56,57,64]. In the case of compound **2**, the excitation $^2F_{5/2} \rightarrow {}^2F_{7/2}$ can be viewed as a picture of the crystal field splitting of the ground state multiplet, while temperature dependence of the magnetic response can be viewed as a thermal picture of the same ground state multiplet splitting. In the first approximation, the coordination polyhedron of Yb^{III} in **2** has a D$_{4d}$ symmetry (CShM$_{SAPR-8}$~0.4), with an environment close to an antiprismatic square. In this approximation, and in the formalism of Stevens, the ground state splitting can be accounted for by the following Hamiltonian for one metallic centre:

$$\hat{H} = B_2^0 \widehat{O_2^0} + B_4^0 \widehat{O_4^0} + B_6^0 \widehat{O_6^0}$$

Where $\widehat{O_k^q}$ are the operators equivalents expressed as polynomials of the total angular momentum ($\hat{J}^2$, $\hat{J}_z$, $\hat{J}_+$, and $\hat{J}_-$), and B_k^q are connected to the crystal field parameters [68,69]. The perturbation due to the application of an external magnetic field can be easily estimated and the magnetization calculated. In a D$_{4d}$ environment, only the terms with $q = 0$ and $k = 2$, 4, and 6 are non zero [70]. The temperature variation of $\chi_M T$ in **2** is fitted and the best curve is represented in Figure 7b (Yb1 and Yb2 are treated identically since they have a very similar environment). To ensure that the obtained set of parameters is correct, the M *vs.* H curve is also calculated at 2 K (inset of Figure 7b). In a D$_{4d}$ symmetry M_J is a good quantum number, so the eigenstates are pure M_J states. In this frame, the ground state corresponds to $M_J = \pm 5/2$, which is taken at the energy origin. The first excited state ($M_J = \pm 3/2$) is located at 215 cm^{-1}, the second at 468 cm^{-1}, and the last at 569 cm^{-1} above the ground state. The gap between the first two levels almost perfectly matches the gap between the two emission lines at 10,204 and 9980 cm^{-1}. The last two emission lines (9843 and 9780 cm^{-1}) are not so well reproduced, which might be due to the relative insensitivity of the magnetism to the high energy levels. Indeed, the last two levels are almost not thermally populated, even at room temperature.

3. Experimental Section

3.1. Synthesis

General Procedures and Materials: The precursors Ln(hfac)$_3$·2H$_2$O (Ln = Dy^{III} and Yb^{III}, hfac$^-$ = 1,1,1,5,5,5-hexafluoroacetylacetonate anion) [71] and the 4,5-dicarboxylic acid-4′,5′-bis(2-cyanoethyl)thio)-tetrathiafulvalene-, dimethyl ester [59] were synthesised following previously reported methods. All other reagents were purchased from Aldrich Co., Ltd. (Saint-Quentin Fallavier, France) and used without further purification.

3.2. Synthesis of 4,5-Bis(2-Pyridyl-N-Oxidemethylthio)-4′,5′-Dicarboxylic Acid-Tetrathiafulvalene-, Dimethyl Ester (L)

Method A. A solution of 0.5 M EtONa/EtOH (14 mL) was added to a suspension of 4,5-dicarboxylic acid-4′,5′-bis(2-cyanoethyl)thio)-tetrathiafulvalene-, dimethyl ester [72] (0.676 g, 1.38 mmol) in anhydrous degassed EtOH (41 mL) under argon. After being stirred at room temperature for 4 h, the mixture was reacted with a solution of 2-(chloromethyl)pyridine-1-oxide [73] (0.593 g, 4.13 mmol) in anhydrous degassed EtOH (27 mL), and then the mixture was stirred for 16 h. H$_2$O (41 mL) was added to quench the reaction, and the mixture was poured into CH$_2$Cl$_2$ (345 mL), washed with saturated NaHCO$_3$ (3 × 55 mL) and water (210 mL). The organic extract was concentrated in vacuum to give a red-brown oil, which was purified by chromatography on alumina gel, initially with CH$_2$Cl$_2$ (to remove the unreacted alkyl halide), and then with CH$_2$Cl$_2$/MeOH (10:1) to give the pure desired ligand. Yield: 437 mg (53%). Anal. Calcd (%) for C$_{22}$H$_{18}$N$_2$O$_6$S$_6$: C 44.15, H 3.01, N 4.68; found: C 44.03, H 3.05, N 4.74. ^{1}H-NMR (CDCl$_3$): 8.24–8.20 (m, 4H), 7.31–7.17 (m, 4H), 3.80 (s, 6H), 3.34 (s, 4H). I.R. (KBr): 2967 (w), 1738 (s), 1724 (s), 1584 (m), 1428 (m), 1415 (m), 1253 (s), 1092 (m), 1020 (m), 886 (w), and 769 (w) cm^{-1}.

Method B. CsOH·H$_2$O (335 mg, 2.00 mmol) dissolved in a minimum of MeOH was slowly added to 4,5-dicarboxylic acid-4′,5′-bis(2-cyanoethyl)thio)-tetrathiafulvalene-, dimethyl ester (0.490 g, 1.00 mmol) in 10 mL of DMF. After being stirred at room temperature for 30 min, the mixture was reacted with a solution of 2-(chloromethyl)pyridine-1-oxide [73] (0.430 g, 2.99 mmol) in anhydrous degassed MeOH (25 mL), and then the mixture was stirred overnight. Then the organic solvents were removed under vacuum and the residue extracted with CH$_2$Cl$_2$ and washed with H$_2$O. The organic phase was extracted and dried with MgSO$_4$. Pure ligand was obtained by chromatography on alumina gel, initially with CH$_2$Cl$_2$ (to remove the unreacted alkyl halide) and then with CH$_2$Cl$_2$/MeOH (10:1). Yield: 486 mg (59%).

3.3. Synthesis of Complexes 1 and 2

[Dy$_2$(hfac)$_6$(L)$_2$]·CH$_2$Cl$_2$·(C$_6$H$_{14}$)$_{0.5}$ (**1**). 32.8 mg of Dy(hfac)$_3$·2H$_2$O (0.04 mmol) were dissolved in 10 mL of CH$_2$Cl$_2$ and then added to a solution of 10 mL of CH$_2$Cl$_2$ containing 24.0 mg of **L** (0.04 mmol). After 15 min of stirring, 25 mL of *n*-hexane were layered at room temperature in the dark. Slow diffusion leads to dark red single crystals, which are suitable for X-ray studies. Yield: 100 mg (86%). Anal. Calcd (%) for C$_{75}$H$_{44}$Cl$_2$Dy$_2$F$_{36}$N$_4$O$_{24}$S$_{12}$: C 31.58, H 1.54, N 1.97; found: C 31.69, H 1.62 N, 2.01. I.R. (KBr): 2956 (w), 2928 (w), 2870 (w), 1734 (m), 1653 (s), 1556 (m), 1531 (m), 1508 (m), 1447 (w), 1256 (s), 1211 (s), 1144 (s), 1101 (w), 798 (w), 769 (w), 661 (m), and 587 (w) cm^{-1}.

[Yb$_2$(hfac)$_6$(L)$_2$]·CH$_2$Cl$_2$·(C$_6$H$_{14}$)$_{0.5}$ (**2**). This compound was obtained using the same protocol as for **1**, starting from 33.2 mg of Yb(hfac)$_3$·2H$_2$O (0.04 mmol) instead of Dy(hfac)$_3$·2H$_2$O. Yield: 82 mg (71%). Anal. Calcd (%) for C$_{75}$H$_{44}$Cl$_2$Yb$_2$F$_{36}$N$_4$O$_{24}$S$_{12}$: C 31.47, H 1.54, N 1.96; found: C 31.57, H 1.60,

N, 2.03. I.R. (KBr): 2954 (w), 2928 (w), 2871 (w), 1734 (m), 1657 (s), 1558 (m), 1533 (m), 1513 (m), 1448 (w), 1264 (s), 1216 (s), 1151 (s), 799 (w), 769 (w), 662 (m), and 588 (w) cm^{-1}.

3.4. Crystallography

A single crystal of **1** was mounted on a APEXII Bruker-AXS diffractometer for data collection (MoK$_\alpha$ radiation source, λ = 0.71073 Å), at the Centre de Diffractométrie (CDIFX), Université de Rennes 1, France. The structure was solved with a direct method using the SIR-97 program and refined with a full matrix least-squares method on F^2 using the SHELXL-97 program [74,75]. Crystallographic data are summarized in Table 1. Complete crystal structure results as a CIF file including bond lengths, angles, and atomic coordinates are deposited as Supplementary files.

3.5. Physical Measurements

The elementary analyses of the compounds were performed at the Centre Régional de Mesures Physiques de l'Ouest, Rennes. Cyclic voltammetry was carried out in CH$_2$Cl$_2$ solution, containing 0.1 M N(C$_4$H$_9$)$_4$PF$_6$ as supporting electrolyte. Voltammograms were recorded at 100 mV·s^{-1} with a platinum disk electrode. The potentials were measured *versus* a saturated calomel electrode (SCE). Absorption spectra were recorded on a Varian Cary 5000 UV-Visible-NIR spectrometer equipped with an integration sphere. The luminescence spectra were measured using a Horiba-Jobin Yvon Fluorolog-3® spectrofluorimeter, equipped with a three slit double grating excitation and emission monochromator with dispersions of 2.1 nm/mm (1200 grooves/mm). The steady-state luminescence was excited by unpolarized light from a 450 W xenon CW lamp. Quartz tubes containing the samples were immersed in liquid nitrogen and near infra-red spectra were recorded at right angle using a liquid nitrogen cooled, solid indium/gallium/arsenic detector (850–1600 nm). Spectra were reference corrected for both the excitation source light intensity variation (lamp and grating) and the emission spectral response (detector and grating). The dc magnetic susceptibility measurements were performed on a solid polycrystalline sample with a Quantum Design MPMS-XL SQUID magnetometer between 2 and 300 K in applied magnetic field of 0.2 T for temperatures of 2–20 K and 1 T for temperatures of 20–300 K. These measurements were all corrected for the diamagnetic contribution calculated with Pascal's constants.

3.6. Computational Details

DFT geometry optimizations and TD-DFT excitation energy calculations of the ligand **L** and Y^{III} analogue of the dinuclear complexes were carried out with the Gaussian 09 (revision A.02) package [76] employing the PBE0 hybrid functional [77,78]. The "Stuttgart/Dresden" basis sets and effective core potentials were used to describe the yttrium atom [79], whereas all other atoms were described with the SVP basis sets [80]. The first 50 monoelectronic excitations were calculated for ligand **L** while the first 100 monoelectronic excitations were calculated for the Y^{III} analogue of the dinuclear complexes. In all steps, a modelling of bulk solvent effects (solvent = dichloromethane) was included through the Polarisable Continuum Model (PCM) [81], using a linear-response non-equilibrium approach for the TD-DFT step [82,83]. Molecular orbitals were sketched using the Gabedit graphical interface [84].

4. Conclusions

Two dinuclear complexes of formula [Ln$_2$(hfac)$_6$(**L**)$_2$]·CH$_2$Cl$_2$·(C$_6$H$_{14}$)$_{0.5}$ (Ln = Dy (**1**) and Yb (**2**)) have been synthesised. The analysis of the polyhedron symmetry around the metallic centres reveals that the dinuclear complexes of the present work have the most symmetric coordination sphere (square anti-prism symmetry) compared to the previously published dinuclear compounds. Nevertheless, they do not display any out-of phase signals of their magnetic susceptibility, demonstrating that the symmetry of the coordination polyhedron is not the crucial parameter for displaying slow magnetic relaxation, but it might be the electronic distribution of the first neighbouring atoms. The functionalization of the TTF core with two methyl ester groups leads to a low-energy ILCT band, which is identified as the HOMO→LUMO excitation. The LUMO is localized on the methyl ester moieties. Irradiation of this absorption band leads to the sensitization of the Yb^{III}-centred luminescence. Static magnetic properties of the Yb^{III} analogue and its luminescence have been correlated. More work is in progress in the group in order to assemble this kind of dinuclear complexes through chemical modifications of the methyl ester groups, for example using carboxylic functions.

Supplementary Materials: Supplementary materials can be found at http://www.mdpi.com/2304-6740/3/4/0554/s1.

Acknowledgments: This work was supported by the CNRS, Rennes Métropole, Université de Rennes 1, Région Bretagne, FEDER and Agence Nationale de la Recherche (No. ANR-13-BS07-0022-01). Boris Le Guennic thanks the French GENCI-CINES center for high-performance computing resources (project x2015080649).

Author Contributions: Fabrice Pointillart, Saskia Speed and Bertrand Lefeuvre performed the syntheses and characterizations. Stéphane Golhen and Lahcène Ouahab performed single crystal X-ray diffraction analyses. François Riobé and Olivier Maury did the luminescence measurements and analyses. Boris Le Guennic performed the calculations and Olivier Cador the magnetic measurements.

Conflicts of Interest: The authors declare no conflict of interest.

References

1. Ishikawa, N.; Sugita, M.; Ishikawa, T.; Koshihara, S.; Kaizu, Y. Lanthanide double-decker complexes functioning as magnets at the single-molecular level. *J. Am. Chem. Soc.* **2003**, *125*, 8694–8695. [CrossRef] [PubMed]

2. Woodruff, D.N.; Winpenny, R.E.P.; Layfield, R.A. Lanthanide single-molecule magnets. *Chem. Rev.* **2013**, *113*, 5110–5148. [CrossRef] [PubMed]

3. Luzon, J.; Sessoli, R. Lanthanides in molecular magnetism: So fascinating so challenging. *Dalton Trans.* **2012**, *41*, 13556–13567. [CrossRef] [PubMed]

4. Sessoli, R.; Powell, A.K. Strategies towards single molecule magnets based on lanthanide ions. *Coord. Chem. Rev.* **2009**, *253*, 2328–2341. [CrossRef]

5. Layfield, R.A. Organometallic single-molecule magnets. *Organometallics* **2014**, *33*, 1084–1099. [CrossRef]

6. Feltham, H.L.C.; Brooker, S. Review of purely 4f and mixed-metal nd-4f single-molecule magnets containing only lanthanide ion. *Coord. Chem. Rev.* **2014**, *276*, 1–33. [CrossRef]

7. Benelli, C.; Gatteschi, D. Magnetism of lanthanides in molecular materials with transition-metal ions and organic radicals. *Chem. Rev.* **2002**, *102*, 2369–2387. [CrossRef] [PubMed]

8. Zhang, P.; Guo, Y.-N.; Tang, J. Recent advances in dysprosium-based single molecule magnets: Structural overview and synthetic strategies. *Coord. Chem. Rev.* **2013**, *257*, 1728–1763. [CrossRef]

9. Zhang, P.; Zhang, L.; Tang, J. Lanthanide single molecule magnets: Progress and perspective. *Dalton Trans.* **2015**, *44*, 3923–3929. [CrossRef] [PubMed]

10. Ungur, L.; Lin, S.-Y.; Tang, J.; Chibotaru, L.F. Single-molecule toroics in Ising-type lanthanide molecular clusters. *Chem. Soc. Rev.* **2014**, *43*, 6894–6905. [CrossRef] [PubMed]

11. Liddle, S.T.; van Slageren, J. Improving f-element single molecule magnets. *Chem. Soc. Rev.* **2015**, *44*, 6655–6669. [CrossRef] [PubMed]

12. Bünzli, J.C.G.; Piguet, C. Lanthanide-containing molecular and supramolecular polymetallic functional assemblies. *Chem. Rev.* **2002**, *102*, 1897–1928. [CrossRef] [PubMed]

13. Moro, F.; Mills, D.P.; Liddle, S.T.; van Slageren, J. The inherent single-molecule magnet character of trivalent uranium. *Angew. Chem. Int. Ed.* **2013**, *52*, 3430–3433. [CrossRef] [PubMed]

14. Coutinho, J.T.; Antunes, M.A.; Pereira, L.C.J.; Marçalo, J.; Almeida, M. Zero-field slow magnetic relaxation in a uranium(III) complex with a radical ligand. *Chem. Commun.* **2014**, *50*, 10262–10264. [CrossRef] [PubMed]

15. King, D.M.; Tuna, F.; McMaster, J.; Lewis, W.; Blake, A.J.; McInnes, E.J.L.; Liddle, S.T. Single-molecule magnetism in a single-ion triamidoamine uranium(V) terminal mono-oxo complex. *Angew. Chem. Int. Ed.* **2013**, *52*, 4921–4924. [CrossRef] [PubMed]

16. Mougel, V.; Chatelain, L.; Pécaut, J.; Caciuffo, R.; Colineau, E.; Griveau, J.-C.; Mazzanti, M. Uranium and manganese assembled in a wheel-shaped nanoscale single-molecule magnet with high spin-reversal barrier. *Nat. Chem.* **2012**, *4*, 1011–1017. [CrossRef] [PubMed]

17. Magnani, N.; Colineau, E.; Griveau, J.-C.; Apostolidis, C.; Walter, O.; Caciuffo, R. A plutonium-based single-molecule magnet. *Chem. Commun.* **2014**, *50*, 8171–8173. [CrossRef] [PubMed]

18. Meihaus, K.R.; Minasian, S.G.; Lukens, W.W., Jr.; Kozimor, S.A.; Shuh, D.K.; Tyliszczak, T.; Long, J.R. Influence of pyrazolate *vs.* N-heterocyclic carbene ligands on the slow magnetic relaxation of homoleptic trischelate lanthanide(III) and uranium(III) complexes. *J. Am. Chem. Soc.* **2014**, *136*, 6056–6068. [CrossRef] [PubMed]

19. Le Roy, J.J.; Gorelsky, S.I.; Korobkov, I.; Murugesu, M. Slow magnetic relaxation in uranium(III) and neodymium(III) cyclooctatetraenyl complexes. *Organometallics* **2015**, *34*, 1415–1418. [CrossRef]

20. Dei, A.; Gatteschi, D. Molecular (Nano) magnet as test grounds of quantum mechanics. *Angew. Chem. Int. Ed.* **2011**, *50*, 11852–11858. [CrossRef] [PubMed]

21. Leuenberger, M.N.; Loss, D. Quantum computing in molecular magnets. *Nature* **2001**, *410*, 789–793. [CrossRef] [PubMed]

22. Hill, S.; Edwards, R.S.; Aliaga-Alcalde, N.; Christou, G. Quantum coherence in an exchange-coupled dimer of single-molecule magnets. *Science* **2003**, *302*, 1015–1018. [CrossRef] [PubMed]

23. Hosseini, M.W.; Rebic, S.; Sparkes, B.M.; Twamley, J.; Buchler, B.C.; Lam, P.K. Memory-enhanced noiseless cross-phase modulation. *Light Sci. Appl.* **2012**, *1*, e40. [CrossRef]

24. Sessoli, R.; Tsai, H.L.; Schake, A.R.; Wang, S.; Vincent, J.B.; Folting, K.; Gatteschi, D.; Christou, G.; Hendrickson, D.N. High-spin molecules: [Mn$_{12}$O$_{12}$(O$_2$CR)$_{16}$(H$_2$O)$_4$]. *J. Am. Chem. Soc.* **1993**, *115*, 1804–1816. [CrossRef]

25. Mannini, M.; Pineider, F.; Sainctavit, P.; Danieli, C.; Otero, E.; Sciancalepore, C.; Talarico, A.M.; Arrio, M.-A.; Cornia, A.; Gatteschi, D.; *et al.* Magnetic memory of a single-molecule quantum magnet wired to a gold surface. *Nat. Mater.* **2009**, *8*, 194–197. [CrossRef] [PubMed]

26. Gu, M.; Li, X.; Cao, Y. Optical storage arrays: A perspective for future big data storage. *Light Sci. Appl.* **2014**, *3*, e177. [CrossRef]

27. Papasimakis, N.; Thongrattanasiri, S.; Zheludev, N.I.; de Abajo, F.J. The magnetic response of grapheme split-ring metamaterials. *Light Sci. Appl.* **2013**, *2*, e78. [CrossRef]

28. Sanvito, S. Molecular Spintronics. *Chem. Soc. Rev.* **2011**, *40*, 3336–3355. [CrossRef] [PubMed]

29. Bogani, L.; Wernsdorfer, W. Molecular spintronics using single-molecule magnets. *Nat. Mater.* **2008**, *7*, 179–186. [CrossRef] [PubMed]

30. De Bettancourt-Dias, A. *Luminescence of Lanthanide Ions in Coordination Compounds and Nanomaterials*; Wiley: Hoboken, NJ, USA, 2014.

31. Weissman, S.I. Intramolecular energy transfer, the fluorescence of complexes of europium. *J. Chem. Phys.* **1942**, *10*, 214–217. [CrossRef]

32. Crosby, G.A.; Kasha, M. Intramolecular energy transfer in ytterbium organic chelates. *Spectrochim. Acta* **1958**, *10*, 377–382. [CrossRef]

33. Lin, S.-Y.; Wang, C.; Zhao, L.; Wua, J.; Tang, J. Chiral mononuclear lanthanide complexes and field-induced single-ion magnet behaviour of Dy analogue. *Dalton Trans.* **2015**, *44*, 223–229. [CrossRef] [PubMed]

34. Long, J.; Rouquette, J.; Thibaud, J.-M.; Ferreira, R.A.S.; Carlos, L.D.; Donnadieu, B.; Vieru, V.; Chibotaru, L.F.; Konczewicz, L.; Haines, J.; *et al.* A high-temperature molecular ferroelectric Zn/Dy complex exhibiting single-ion-magnet behavior and lanthanide luminescence. *Angew. Chem. Int. Ed.* **2015**, *54*, 2236–2240. [CrossRef] [PubMed]

35. Li, X.-L.; Chen, C.-L.; Xiao, H.-P.; Wang, A.-L.; Liu, C.-M.; Zheng, X.; Gao, L.-J.; Yanga, X.-G.; Fang, S.-M. Luminescent, magnetic and ferroelectric properties of noncentrosymmetric chain-like complexes composed of nine-coordinate lanthanide ions. *Dalton Trans.* **2013**, *42*, 15317–15325. [CrossRef] [PubMed]

36. Tanaka, H.; Kobayashi, H.; Kobayashi, A.; Cassoux, P. Superconductivity, antiferromagnetism, and phase diagram of a series of organic conductors: λ-(BETS)$_2$Fe$_x$Ga$_{1-x}$Br$_y$Cl$_{4-y}$. *Adv. Mater.* **2000**, *12*, 1685–1689. [CrossRef]

37. Uji, S.; Shinagawa, H.; Terashima, T.; Terakura, C.; Yakabe, T.; Terai, Y.; Tokumoto, M.; Kobayashi, A.; Tanaka, H.; Kobayashi, H. Magnetic-field-induced superconductivity in a two-dimensional organic conductor. *Nature* **2001**, *410*, 908–910. [CrossRef] [PubMed]

38. Kobayashi, A.; Fujiwara, E.; Kobayashi, H. Single-component molecular metals with extended-TTF dithiolate ligands. *Chem. Rev.* **2004**, *104*, 5243–5264. [CrossRef] [PubMed]

39. Enoki, T.; Miyasaki, A. Magnetic TTF-based charge-transfer complexes. *Chem. Rev.* **2004**, *104*, 5449–5478. [CrossRef] [PubMed]

40. Coronado, E.; Day, P. Magnetic molecular conductors. *Chem. Rev.* **2004**, *104*, 5419–5448. [CrossRef] [PubMed]

41. Ouahab, L.; Enoki, T. Multiproperty molecular materials: TTF-based conductiong and magnetic molecular materials. *Eur. J. Inorg. Chem.* **2004**, *2004*, 933–941. [CrossRef]

42. Fujiwara, H.; Wada, K.; Hiraoka, T.; Hayashi, T.; Sugimoto, T.; Nakazumi, H.; Yokogawa, K.; Teramura, M.; Yasuzuka, S.; Murata, K.; *et al.* Stable metallic behavior and antiferromagnetic ordering of Fe(III) *d* spins in (EDO-TTFVO)$_2$·FeCl$_4$. *J. Am. Chem. Soc.* **2005**, *127*, 14166–14167. [CrossRef] [PubMed]

43. Lorcy, D.; Bellec, N.; Fourmigué, M.; Avarvari, N. Tetrathiafulvalene-based group XV ligands: Synthesis, coordination chemistry and radical cation salts. *Coord. Chem. Rev.* **2009**, *253*, 1398–1438. [CrossRef]

44. Pointillart, F.; Golhen, S.; Cador, O.; Ouahab, L. Paramagnetic 3D coordination complexes involving redox-active tetrathiafulvalene derivatives: An efficient approach to elaborate multi-properties materials. *Dalton Trans.* **2013**, *42*, 1949–1960. [CrossRef] [PubMed]

45. Cosquer, G.; Pointillart, F.; Golhen, S.; Cador, O.; Ouahab, L. Slow magnetic relaxation in condensed *versus* dispersed dysprosium(III) mononuclear complexes. *Chem. Eur. J.* **2013**, *19*, 7895–7903. [CrossRef] [PubMed]

46. Da Cunha, T.T.; Jung, J.; Boulon, M.-E.; Campo, G.; Pointillart, F.; Pereira, C.L.M.; Le Guennic, B.; Cador, O.; Bernot, K.; Pineider, F.; *et al.* Magnetic poles determinations and robustness of memory effect upon solubilization in a Dy-III-based single ion magnet. *J. Am. Chem. Soc.* **2013**, *135*, 16332–16335. [CrossRef] [PubMed]

47. Pointillart, F.; Golhen, S.; Cador, O.; Ouahab, L. Slow magnetic relaxation in a redox-active tetrathiafulvalene-based ferromagnetic dysprosium complex. *Eur. J. Inorg. Chem.* **2014**, *2014*, 4558–4563. [CrossRef]

48. Gao, F.; Zhang, X.-M.; Cui, L.; Deng, K.; Zeng, Q.-D.; Zuo, J.-L. Tetrathiafulvalene-supported triple-decker phthalocyaninato dysprosium(III) complex: Synthesis, properties and surface assembly. *Sci. Rep.* **2014**, *4*, 5928. [CrossRef] [PubMed]

49. Pointillart, F.; Bernot, K.; Golhen, S.; le Guennic, B.; Guizouarn, T.; Ouahab, L.; Cador, O. Magnetic memory in an isotopically enriched and magnetically isolated mononuclear dysprosium complex. *Angew. Chem. Int. Ed.* **2015**, *54*, 1504–1507. [CrossRef] [PubMed]

50. Pointillart, F.; Jung, J.; Berraud-Pache, R.; le Guennic, B.; Dorcet, V.; Golhen, S.; Cador, O.; Maury, O.; Guyot, Y.; Decurtins, S.; *et al.* Luminescence and single-molecule magnet behavior in lanthanide complexes involving a tetrathiafulvalene-fused dipyridophenazine ligand. *Inorg. Chem.* **2015**, *54*, 5384–5397. [CrossRef] [PubMed]

51. Feng, M.; Pointillart, F.; Lefeuvre, B.; Dorcet, V.; Golhen, S.; Cador, O.; Ouahab, L. Multiple single-molecule magnet behaviors in dysprosium dinuclear complexes involving a multiple functionalized tetrathiafulvalene-based ligand. *Inorg. Chem.* **2015**, *54*, 4021–4028. [CrossRef] [PubMed]

52. Faulkner, S.; Burton-Pye, B.P.; Khan, T.; Martin, L.R.; Wray, S.D.; Skabara, P.J. Interaction between tetrathiafulvalene carboxylic acid and ytterbium DO3A: Solution state self-assembly of a ternary complex which is luminescent in the near IR. *Chem. Commun.* **2002**, 1668–1669. [CrossRef]

53. Pope, S.J.A.; Burton-Pye, B.P.; Berridge, R.; Khan, T.; Skabara, P.; Faulkner, S. Self-assembly of luminescent ternary complexes between seven-coordinate lanthanide(III) complexes and chromophore bearing carboxylates and phosphonates. *Dalton Trans.* **2006**, 2907–2912. [CrossRef] [PubMed]

54. Pointillart, F.; le Guennic, B.; Golhen, S.; Cador, O.; Maury, O.; Ouahab, L. High nuclearity complexes of lanthanide involving tetrathiafulvalene ligands: Structural, magnetic, and photophysical properties. *Inorg. Chem.* **2013**, *52*, 1610–1620. [CrossRef] [PubMed]

55. Ran, Y.-F.; Steinmann, M.; Sigrist, M.; Liu, S.-X.; Hauser, J.; Decurtins, S. Tetrathiafulvalene-based lanthanide coordination complexes: Synthesis, crystal structure, optical and electrochemical characterization. *C. R. Chim.* **2012**, *15*, 838–844. [CrossRef]

56. Feng, M.; Pointillart, F.; le Guennic, B.; Lefeuvre, B.; Golhen, S.; Cador, O.; Maury, O.; Ouahab, L. Unprecedented sensitization of visible and near-infrared lanthanide luminescence by using a tetrathiafulvalene-based chromophore. *Chem. Asian J.* **2014**, *9*, 2814–2825. [CrossRef] [PubMed]

57. Pointillart, F.; le Guennic, B.; Cauchy, T.; Golhen, S.; Cador, O.; Maury, O.; Ouahab, L. A series of tetrathiafulvalene-based lanthanide complexes displaying either single molecule magnet or luminescence-direct magnetic and photo-physical correlations in the ytterbium analogue. *Inorg. Chem.* **2013**, *52*, 5978–5990. [CrossRef] [PubMed]

58. Soussi, K.; Jung, J.; Pointillart, F.; le Guennic, B.; Lefeuvre, B.; Golhen, S.; Cador, O.; Guyot, Y.; Maury, O.; Ouahab, L. Magnetic and photo-physical investigations into Dy^{III} and Yb^{III} complexes involving tetrathiafulvalene ligand. *Inorg. Chem. Front.* **2015**, *2*, 1105–1117. [CrossRef]

59. Simonsen, B.; Svenstrup, N.; Lau, J.; Simonsen, O.; Mork, P.; Kristensen, G.J.; Becher, J. Sequential functionalisation of bis-protected tetrathiafulvalene-dithiolates. *Synthesis* **1996**, *3*, 407–418. [CrossRef]

60. Lin, H.-H.; Yan, Z.-M.; Dai, J.; Zhang, D.-Q.; Zuo, J.-L.; Zhu, Q.-Y.; Jia, D.-X. A water-soluble derivative of tetrathiafulvalene exhibiting pH sensitive redox properties. *New J. Chem.* **2005**, *29*, 509–513. [CrossRef]

61. Llunell, M.; Casanova, D.; Cirera, J.; Bofill, J.M.; Alemany, P.; Alvarez, S. *SHAPE*; Version 2.1; Universitat de Barcelona: Barcelona, Spain, 2013.

62. Cosquer, G.; Pointillart, F.; le Guennic, B.; le Gal, Y.; Golhen, S.; Cador, O.; Ouahab, L. 3d4f heterobimetallic dinuclear and tetranuclear complexes involving tetrathiafulvalene as ligands: X-ray structures and magnetic and photophysical investigations. *Inorg. Chem.* **2012**, *51*, 8488–8501. [CrossRef] [PubMed]

63. Pointillart, F.; Cauchy, T.; Maury, O.; le Gal, Y.; Golhen, S.; Cador, O.; Ouahab, L. Tetrathiafulvalene-amido-2-pyridine-*N*-oxide as efficient charge-transfer antenna ligand for the sensitization of Yb^{III} luminescence in a series of lanthanide paramagnetic coordination complexes. *Chem. Eur. J.* **2010**, *16*, 11926–11941. [CrossRef] [PubMed]

64. Pointillart, F.; le Guennic, B.; Golhen, S.; Cador, O.; Maury, O.; Ouahab, L. A redox-active luminescent ytterbium based single molecule magnet. *Chem. Commun.* **2013**, *49*, 615–617. [CrossRef] [PubMed]

65. Gonçalves Silva, F.R.; Malta, O.L.; Reinhard, C.; Güdel, H.U.; Piguet, C.; Moser, J.; Bünzli, J.C.G. Visible and near-infrared luminescence of lanthnaide-containing dimetallic triple-stranded helicates: Energy transfer mechanisms in the Sm^{III} and Yb^{III} molecular edifices. *J. Phys. Chem. A* **2002**, *106*, 1670–1677. [CrossRef]

66. Lapadula, G.; Bourdolle, A.; Allouche, F.; Conley, M.; del Rosa, I.; Maron, L.; Lukens, W.W.; Guyot, Y.; Andraud, C.; Brasselet, S.; *et al.* Near-IR two photon microscopy imaging of silica nanoparticles functionalized with isolated sensitized Yb(III) centers. *Chem. Mater.* **2014**, *26*, 1062–1073. [CrossRef]

67. Kahn, O. *Molecular Magnetism*; VCH: Weinhem, Germany, 1993.

68. Orbach, R. Spin-lattice relaxation in rare-earth salts. *Proc. R. Soc. Lond. A* **1961**, *264*, 458–484. [CrossRef]

69. Rudowicz, C. Transformation relations for the conventional O_k^q and normalised $O'_k{}^q$ Stevens operator equivalents with $k = 1$ to 6 and $-k \leq q \leq k$. *J. Phys. C Solid State Phys.* **1985**, *18*, 1415–1430. [CrossRef]

70. Görlder-Walrand, C.; Binnemans, K. Rationalization of Crystal-Field Parametrization. In *Hanbook on the Physics and Chemistry of Rare Earths*; Elsevier: Philadelphia, PA, USA, 1996; Volume 23, pp. 121–283.

71. Richardson, M.F.; Wagner, W.F.; Sands, D.E. Rare-earth trishexafluoroacetylacetonates and related compounds. *J. Inorg. Nucl. Chem.* **1968**, *30*, 1275–1289. [CrossRef]

72. Binet, L.; Fabre, J.M.; Montginoul, C.; Simonsen, K.B.; Becher, J. Preparation and chemistry of new unsymmetrically substituted tetrachalcogenofulvalenes bearing $CN(CH_2)_2X$ and $HO(CH_2)_2X$ groups (X = S or Se). *J. Chem. Soc. Perkin Trans. 1* **1996**, 783–788. [CrossRef]

73. Polasek, M.; Sedinova, M.; Kotek, J.; Vander Elst, L.; Muller, N.R.; Hermann, P.; Lukes, I. Pyridine-*N*-oxide analogues of DOTA and their gadolinium(III) complexes endowed with a fast water exchange on the square-antiprismatic isomer. *Inorg. Chem.* **2009**, *48*, 455–465. [CrossRef] [PubMed]

74. Sheldrick, G.M. *SHELX97—Programs for Crystal Structure Analysis (Release 97-2)*; Tammanstrasse 4, D-3400; Institüt für Anorganische Chemie der Universität: Göttingen, Germany, 1998.

75. Altomare, A.; Burla, M.C.; Camalli, M.; Cascarano, G.L.; Giacovazzo, C.; Guagliardi, A.; Moliterni, A.G.G.; Polidori, G.; Spagna, R. *SIR97*: A new tool for crystal structure determination and refinement. *J. Appl. Cryst.* **1999**, *32*, 115–119. [CrossRef]

76. Frisch, M.J.; Trucks, G.W.; Schlegel, H.B.; Scuseria, G.E.; Robb, M.A.; Cheeseman, J.R.; Scalmani, G.; Barone, V.; Mennucci, B.; Petersson, G.A.; *et al. Gaussian 09, Revision A.02*; Gaussian Inc.: Wallingford, CT, USA, 2009.

77. Perdew, J.P.; Burke, K.; Ernzerhof, M. Generalized gradient approximation made simple. *Phys. Rev. Lett.* **1996**, *77*, 3865–3868. [CrossRef] [PubMed]

78. Adamo, C.; Barone, V. Toward reliable density functional methods without adjustable parameters: The PBE0 model. *J. Chem. Phys.* **1999**, *110*, 6158–6170. [CrossRef]

79. Dolg, M.; Stoll, H.; Preuss, H. A combination of quasirelativistic pseudopotential and ligand field calculations for lanthanoid compounds. *Theor. Chim. Acta* **1993**, *85*, 441–450. [CrossRef]

80. Weigend, F.; Ahlrichs, R. Balanced basis sets of split valence, triple zeta valence and quadruple zeta valence quality for H to Rn: Design and assessment of accuracy. *Phys. Chem. Chem. Phys.* **2005**, *7*, 3297–3305. [CrossRef] [PubMed]

81. Tomasi, J.; Mennucci, B.; Cammi, R. Quantum mechanical continuum solvation models. *Chem. Rev.* **2005**, *105*, 2999–3094. [CrossRef] [PubMed]

82. Cossi, M.; Barone, V. Time-dependent density functional theory for molecules in liquid solutions. *J. Chem. Phys.* **2001**, *115*, 4708–4717. [CrossRef]

83. Improta, R.; Barone, V.; Scalmani, G.; Frisch, M.J. A state-specific polarizable continuum model time dependent density functional theory method for excited state calculations in solution. *J. Chem. Phys.* **2006**, *125*, 054103. [CrossRef] [PubMed]

84. Allouche, A.-R. Gabedit—A graphical user interface for computational chemistry softwares. *J. Comput. Chem.* **2011**, *32*, 174–182. [CrossRef] [PubMed]

![inorganics logo] *inorganics*

MDPI

Article

On the Dehydrocoupling of Alkenylacetylenes Mediated by Various Samarocene Complexes: A Charming Story of Metal Cooperativity Revealing a Novel Dual Metal σ-Bond Metathesis Type of Mechanism (DM I σ-BM)

Christos E. Kefalidis * and Laurent Maron *

CNRS (Centre National de la Recherche Scientifique) & INSA (Institut National des Sciences Appliquées),
UPS (Université Paul Sabatier), LPCNO (Laboratoire de Physique et Chimie des Nano-objets),
Université de Toulouse, 135 Avenue de Rangueil, Toulouse F-31077, France
* Correspondence: christos.kefalidis@gmail.com (C.E.K.); laurent.maron@irsamc.ups-tlse.fr (L.M.);
  Tel.: +33-561-559-664 (C.E.K. & L.M.)

Academic Editors: Stephen Mansell and Steve Liddle
Received: 30 September 2015; Accepted: 26 November 2015; Published: 4 December 2015

Abstract: The prevailing reductive chemistry of Sm(II) has been accessed and explored mostly by the use of samarocene precursors. The highly reducing character of these congeners, along with their Lewis acidity and predominantly ionic bonding, allows for the relatively facile activation of C–H bonds, as well as peculiar transformations of unsaturated substrates (e.g., C–C couplings). Among other important C–C coupling reactions, the reaction of phenylacetylene with different mono- or bimetallic samarocene complexes affords trienediyl complexes of the type {[(C_5Me_5)$_2$Sm]$_2$(μ-η^2:η^2-PhC$_4$Ph)}. In contrast, when *t*-butylacetylene is used, uncoupled monomers of the type (C_5Me_5)$_2$Sm(C≡C–^{t}Bu) were obtained. Although this type of reactivity may appear to be simple, the mechanism underlying these transformations is complex. This conclusion is drawn from the density functional theory (DFT) mechanistic studies presented herein. The operating mechanistic paths consist of: (i) the oxidation of each samarium center and the concomitant double reduction of the alkyne to afford a binuclear intermediate; (ii) the C–H scission of the acetylinic bond that lies in between the two metals; (iii) a dual metal σ-bond metathesis (DM I σ-SBM) process that releases H_2; and eventually (iv) the C–C coupling of the two bridged μ-alkynides to give the final bimetallic trienediyl complexes. For the latter mechanistic route, the experimentally used phenylacetylene was considered first as well as the aliphatic hex-1-yne. More interestingly, we shed light into the formation of the mono(alkynide) complex, being the final experimental product of the reaction with *t*-butylacetylene.

Keywords: samarium; σ-bond metathesis; trienediyl; C–C coupling; mechanism; DFT calculations; bimetallic complexes; terminal alkynes

1. Introduction

The versatility of coupling reactions mediated by single-electron transfer (SET) from Sm(II) complexes is clearly illustrated by the chemistry of SmI_2 [1,2]. This coupling chemistry has been expanded by the use of (Cp*)$_2$Sm(THF)$_n$ (where Cp* = C_5Me_5; n = 0–2) [3–7]. Interestingly, both the Lewis acidity and the ionic bonding of these Sm(II) complexes offer ligation of substrates that can undergo facile activation of C–H bonds, or partial reduction of unsaturated compounds [8]. These two elementary steps can combine to achieve dehydrocoupling of alkynes, as was initially reported by Evans in the case of phenylacetylene. This led to the characterization of a new class of dinuclear Sm(III) trienediyl complexes, {[(Cp*)$_2$Sm]$_2$(μ-η^2:η^2-PhC$_4$Ph)} [9]. These trienediyl species were obtained by four different reaction procedures: (a) the reaction of (Cp*)$_2$Sm[CH(SiMe$_3$)]$_2$ with HC≡CPh; (b) thermolysis of (Cp*)$_2$Sm(C≡CPh)(THF) at 120–145 °C; (c) the reaction of [(Cp*)$_2$Sm(μ-H)]$_2$ with HC≡CPh; and (d) the reaction of (Cp*)$_2$Sm with HC≡CPh. This reactivity was also extended to include aliphatic terminal alkynes, HC≡CR (R = CH$_2$CH$_2$Ph, ^{i}Pent, and ^{i}Pr), leading to similar dinuclear trienediyl complexes [7]. However, when a more bulky acetylide is used (R = ^{t}Bu), the reaction halts at the formation of the alkynide monomers, e.g., the uncoupled dimer of the type [(Cp*)$_2$Sm(C≡C–^{t}Bu)]$_2$, which was prepared from the use of two different precursors; [(Cp*)$_2$Sm^{III}(μ-H)]$_2$, and (Cp*)$_2$Sm^{II}(THF) [7]. Alkyne dehydrocoupling has also been observed with copper salts [10] and titanium complexes [11].

Around the same period of time as the Evans report, two related contributions appeared in the literature. In particular, Teuben *et al.* [12] reported a study on early lanthanide carbyls, e.g., (Cp*)$_2$LnCH(SiMe$_3$)$_2$, that promote C–C coupling of terminal alkynes (HC≡CR) to give the corresponding trienediyl bimetallic species, {[(Cp*)$_2$Ln]$_2$(μ-η^2:η^2-RC$_4$R)} (Ln = Ce, R = Me or ^{t}Bu; Ln = La, R = Me). At the same time, Marks *et al.*, also reported the reaction of (Cp*)$_2$LaCH(SiMe$_3$) with HC≡CR (R = ^{t}Bu and Ph) to form the corresponding trienediyl dinuclear complexes [13]. The latter group concluded that the uncoupled dimer [(Cp*)$_2$La(C≡C–R)]$_2$ is the direct kinetic precursor of this reaction, and that neither a redox-active lanthanide ion nor a phenyl substituent on the alkyne is required for its completion. In a recent report from our group, we were able to shed light onto this type of reactivity by means of computational techniques [14]. Nevertheless, up until now, there has not been a clear picture of the alkyne coupling mechanism leading to dinuclear Sm(III) trienediyl complexes starting from low-valent samarocene (Figure 1).

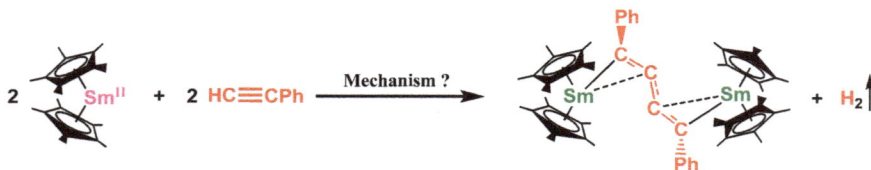

Figure 1. Reaction of samarocene with phenylacetylene to afford {[(C$_5$Me$_5$)$_2$Sm]$_2$(μ-η^2:η^2-PhC$_4$–Ph)} complex and H$_2$.

The bimetallic mechanism which was previously proposed in the case of N-anchored tris(aryloxide) uranium(III) can serve as a base [15], but how acetylenic C–H bonds break to release H$_2$ still needs to be answered. Hence, in this paper we attempt to understand the performance of samarocene in the formation of the trienediyl dinuclear complex, {[(Cp*)$_2$Sm]$_2$(μ-η^2:η^2-PhC$_4$–Ph)}, with plausible energetic reaction profiles being computed and suggested at the DFT (B3PW91) level of theory. The formation of the uncoupled complexes, instead of the trienediyl one where ^{t}BuC≡CH was used, is also discussed. Based on these studies, novel transition states which lead to unique reaction pathways in the domain of f-block chemistry are revealed.

2. Results and Discussion

2.1. Dehydrocoupling of PhC≡CH Using Cp*$_2$Sm

Based on chemical intuition and previous knowledge on related mechanisms [15], three different types of reaction sequence are envisioned for this particular reactivity (Figure 2).

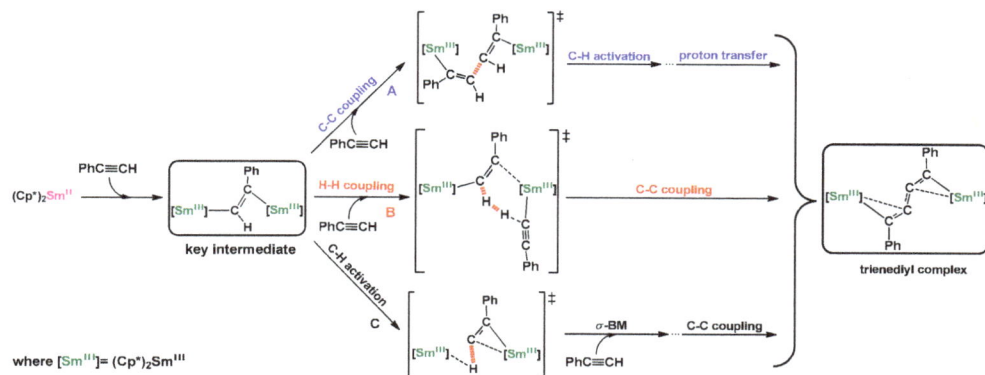

Figure 2. Possible reaction paths **A**, **B**, and **C** leading to the formation of the corresponding trienediyl complex using PhC≡CH as the substrate and (Cp*)$_2$Sm as a precursor.

All these sequences share a common initiation step, the double SET one, in which a doubly reduced phenylacetylene unit contained in a dinuclear samarium(III) complex is formed. From this "key intermediate" the reaction can proceed either through: (A) a direct insertion of PhC≡CH that leads to an alkynyl bis-Sm(III) complex; (B) a peculiar direct H–H coupling type of Transition State (TS); (C) a C–H activation leading to a μ-H:μ-C motif. Energetically the most favorable pathway relies on the C–H activation that corresponds to mechanism C and will be discussed in detail afterwards, together with the other routes. The whole energy profile for mechanism C is depicted in Figure 3.

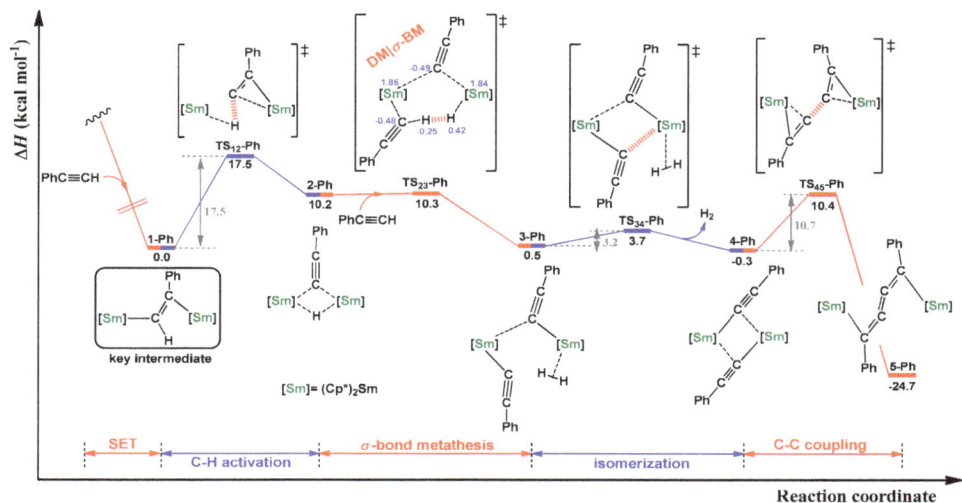

Figure 3. Enthalpy energy Δ*H* (kcal·mol^{-1}) profile for the formation of the corresponding trienediyl complex using PhC≡CH as substrate and (Cp*)$_2$Sm as a precursor (mechanism C). Cp* ligands are omitted for clarity. Numbers in blue correspond to natural charges.

As already reported by us in a recent theoretical work, the one electron reduction of phenylacetylene, induced by the coordination to the samarium center, is a highly exothermic process [16]. This is the outcome of high level *ab-initio* calculations (CAS-SCF), but it is also based on the newly introduced theoretical methodology that takes into account the "HOMO-LUMO gap" energy difference. As the overall redox process is known to be exothermic, it contributes to the total energy stabilization of the newly formed "key intermediate" [17]. Hence, the latter species will be the reference point for the whole mechanism. In the **1-Ph** intermediate, the α- and β-carbons of the alkylenic substrate bind to each distinct Sm center in an *E*-configuration, with the *Z*-isomer being less stable by 3.6 kcal·mol^{-1} in accordance with earlier structural interpretation [18]. Interestingly, in **1-Ph**, the C–C bond distance of the sandwiched alkyne is 1.36 Å which is 0.15 Å longer than in free phenylacetylene. This is a strong indication of the double reduction of the alkyne. In addition, the acetylenic C–H bond points towards one Sm center and is elongated by 0.06 Å with respect to free phenylacetylene, being readily available for a potential bond scission. At that point, the subsequent C–H bond activation (**TS$_{12}$-Ph**) yielding a μ-hydride-μ-alkynyl dinuclear complex (**2-Ph**) can take place with a moderate energy barrier of 17.5 kcal·mol^{-1} with respect to the "key intermediate". Hoffmann *et al.*, showed in a seminal work that extra stabilization is achieved by the mixing of the 1*s* orbital of the hydride with the available empty d_{z^2} and d_{yz} orbitals, of 1a_1 and 1b_2 symmetry respectively, of each Sm(Cp*)$_2$ fragment [19]. This can nicely explain the relative low energy barrier computed for this step. It is noteworthy that this bridged dimer shares features with geometrically close related yttrium complexes that were previously reported [20,21]. The reaction proceeds via **TS$_{23}$-Ph** that corresponds to the activation of the acetylenic proton of a second PhC≡CH molecule by the bridging hydride of complex **2-Ph**. This step is better described as a six-member σ-bond metathesis showing a negligible activation energy barrier. The alternation of the natural charge signs between the six atoms in **TS$_{23}$-Ph** (See Figure 3) and the respective shortening and lengthening of the bonds involved, confirms its σ-bond metathesis character. This is in line with previous studies [22–24]. To the best of our knowledge, this Dual Metal σ-Bond Metathesis transition state (DM I σ-BM) is the first report of such a type that implies two lanthanide centers working in concert. This type of transition state can be also found in other related lanthanide types of reactivity [25]. The latter step affords a μ-alkynyl-η^1-alkynyl dihydrogen complex (**3-Ph**) that is almost isoenergetic with the "key intermediate". In the subsequent step an isomerization takes place which induces the release of H$_2$, and leads to the formation of bis-μ-phenylacetylenyl complex, **4-Ph**. The formation of the latter requires a very low activation barrier ($\Delta_r H^\# = 3.2$ kcal·mol^{-1}). From **4-Ph**, the final trienediyl dinuclear complex **5-Ph** is obtained by the C–C homocoupling of the two terminal μ-alkynyl moieties via **TS$_{45}$-Ph**. Interestingly, similar transition states were postulated to proceed in analogous systems, as for instance the bis-μ-alkynyl titanocene complexes [26]. In addition, analogous structures were structurally isolated and characterized for other lanthanide centers [12,27–29]. As already reported from our group, the energy barrier for the C–C coupling step is surprisingly low, being only 10.7 kcal·mol^{-1} [14]. This is odd since two negatively charged moieties have to be homocoupled. One way to accomplish this is to reduce the electron density of the negatively charged α-carbons of the alkynyls. This is achieved by the nucleophilic assistance from the β-carbon of each triple bond which accumulates enough negative charge, resulting in a stronger interaction with the second samarium center. Finally, the rate determining step of the entire mechanism is the scission of the C–H acetylenic bond (17.5 kcal·mol^{-1}), with the overall exothermicity of the mechanism being 24.7 kcal·mol^{-1} with respect to **1-Ph** intermediate.

The other two pathways depicted in Figure 1 were considered starting from **1-Ph**. In the first case (mechanism A), and inspired by our recent mechanistic work [15], the possibility of a direct C–C coupling in 1,1-fashion was investigated. This led to the corresponding Sm(III) bis-vinyl complex **6-Ph** (Figure 4). Although this step is strongly exothermic (−31.0 kcal·mol^{-1}), the activation barrier ($\Delta_r H^\ddagger = 24.8$ kcal·mol^{-1}) is much higher than the highest barrier of mechanism C, by more than 7 kcal·mol^{-1}. The reaction involves an ionic transition state, **TS$_{16}$-Ph**, in which the incoming triple bond is polarized by one samarium center, allowing the formation of a lone-pair at the β-carbon atom and an empty *sp^2*-orbital at its α-carbon that overlaps with the filled *sp^2*-orbital of the α-carbon of the bridged reduced triple bond. This transition state closely resembles the one reported for the C–C coupling of the hex-1-yne using *N*-anchored tris-aryloxide complexes of U(III) [15]. Then the reaction proceeds through the activation of the C–H bond in close vicinity to the Sm center, surmounting an even higher activation barrier than the previous step ($\Delta_r H^{\ddagger}$ = 25.7 kcal·mol^{-1}). The geometry of the intermediate that results, **7-Ph**, corresponds to a weakly coordinated complex, in which two fragments develop an H–H interaction (Figure 4). The latter undergoes a proton transfer to give the trienediyl compound upon release of H$_2$, via a transition state of relatively low activation barrier ($\Delta_r H^{\ddagger}$ = 6.9 kcal· mol^{-1}). However, this mechanism can be ruled out from further consideration since it suffers kinetically from two relatively high activation barriers of *ca.* 24 kcal·mol^{-1}, and thermodynamically by the formation of the highly stable bis-vinyl complex **6-Ph**. The latter corresponds to the lowest energy point of the mechanism, being 6.3 kcal·mol^{-1} lower than the final trienediyl complex, as depicted in Figure 4.

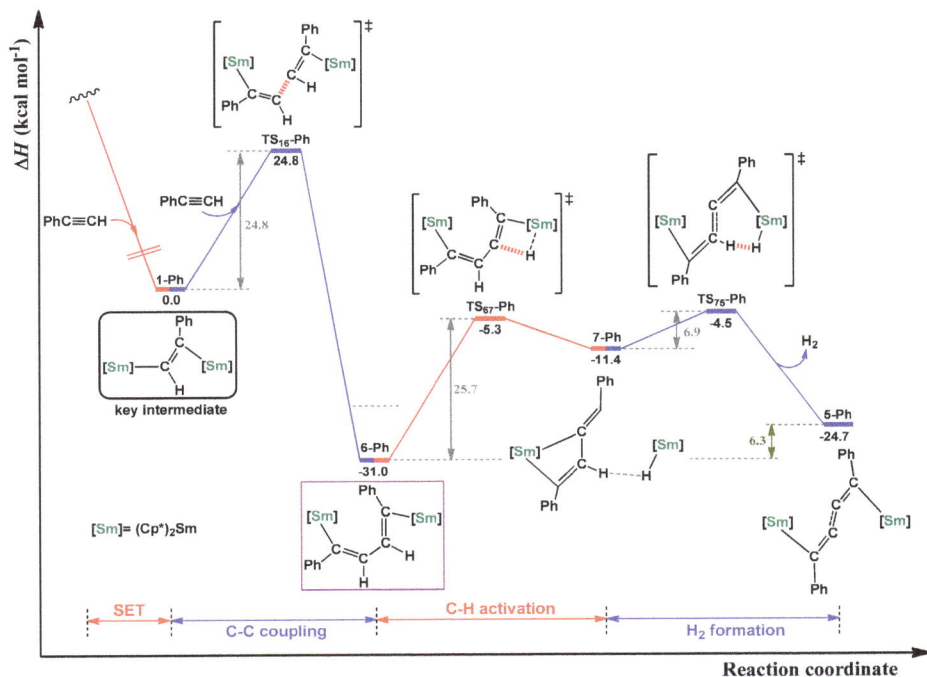

Figure 4. Enthalpy energy ΔH (kcal·mol^{-1}) profile for the formation of the corresponding trienediyl complex using PhC≡CH as substrate and (Cp*)$_2$Sm as a precursor (mechanism A). Cp* ligands are omitted for clarity.

Finally, from "key intermediate" **1-Ph**, a mechanism that eliminates H$_2$ and homocouples the two alkynyls in a concerted homolytic manner (mechanism B) was attempted. Despite our efforts, we were unable to locate any transition state of this type. Instead, we found the direct formation of H$_2$, but without any C–C coupling, leading to **5-Ph** after reorganization (Figure 5, mechanism B). By examining the nature of each hydrogen at the transition state, this can be regarded as an H–H coupling type of TS. Interestingly, the corresponding natural charges of the two hydrogens are −0.01 and 0.10 |e| (Figure 5). The first charge corresponds to the acetylenic bridged hydrogen, while the second belongs to the newly inserted alkyne. Nevertheless, the activation energy barrier for this step ($\Delta_r H^{\ddagger}$ = 27.2 kcal·mol^{-1}) allows us to discard this pathway from further consideration.

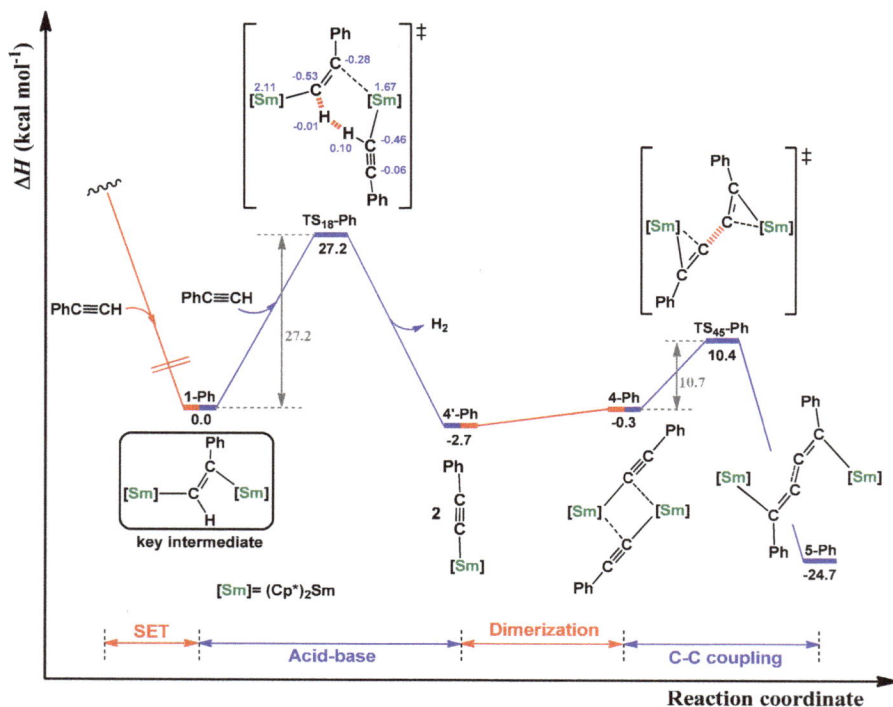

Figure 5. Enthalpy energy ΔH (kcal·mol⁻¹) profile for the formation of the corresponding trienediyl complex using PhC≡CH as substrate and (Cp*)₂Sm as a precursor (Mechanism B). Cp* ligands are omitted for clarity. Numbers in blue correspond to natural charges.

In contrast, when a monometallic mechanism is operating, the C–C coupling can occur in 1,1 or 1,2 fashion (Figure S1 in supplementary information). In particular, addition of a second alkyne molecule to the (Cp*)₂Smᴵᴵᴵ(HC≡CPh) intermediate can take place in two different ways; in 1,1 and 1,2 fashion, affording the corresponding bis-vinyl species. The activation barriers were found to be 9.8 and 5.9 kcal·mol⁻¹ respectively, with the exothermicities of each one being 31.1 and 27.7 kcal·mol⁻¹ with respect to the monoreduced intermediate, (C₅Me₅)₂Sm(HC≡CPh). The computed small energy difference between the two transition states, as well as the preference for 1,2 C–C coupling, cannot account for the observed experimental selectivity which is found to be the reverse. On top of that, another weak point of the monometallic mechanism is the fact that after the formation of bis-vinyl monomeric complex, an oxidation of a second Smᴵᴵ(Cp*)₂ has to take place. This will result in the formation of the same energetically buried bis-vinyl intermediate, **6-Ph**, as described previously in mechanism A. Since the formation of the latter is undesired, for the reasons described above, the existence of a monometallic pathway can also be ruled out.

2.2. Dehydrocoupling of ᵗBuC≡CH Using Cp*₂Sm

We then turned our attention to include the more bulky H–C≡C–ᵗBu alkyne that is found to give monomeric acetylide complexes rather than bimetallic trienediyl complexes [7], with the mechanism C applied to the new alkyne (Figure 6).

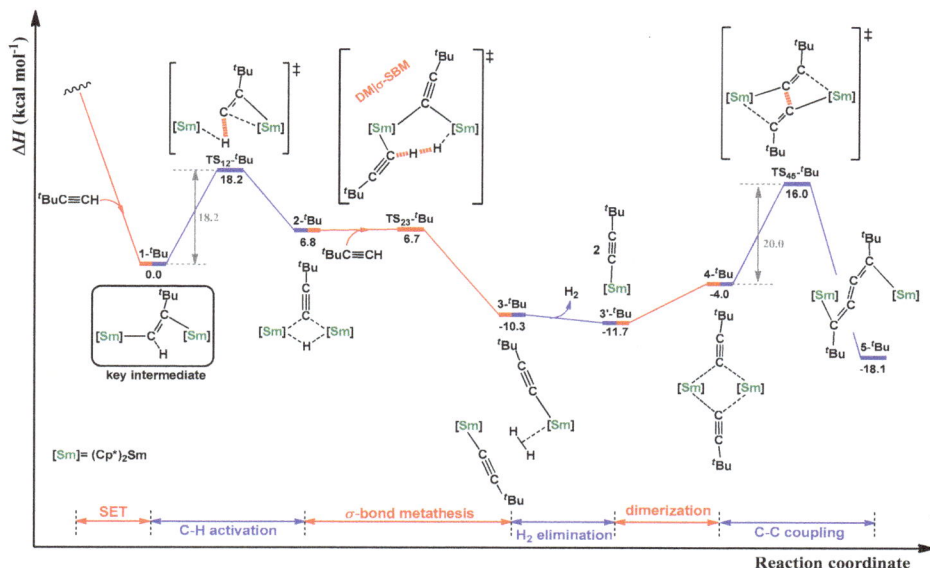

Figure 6. Enthalpy energy ΔH (kcal·mol^{-1}) profile for the formation of the uncoupled alkynide complex using ^{t}BuC≡CH as substrate and (Cp*)$_2$Sm as a precursor. Cp* ligands are omitted for clarity.

Unlike the previous case, the first step corresponds to SET but with a concomitant dissociation of a THF molecule; the latter molecule being coordinated to the precursor complex Sm(Cp*)$_2$. It is expected that the high energy gain observed in this step will be sufficient for the easy dissociation of the THF molecule. The latter is computed to cost around 10 kcal·mol^{-1} as it is depicted in Figure 7a. In addition, the isodesmic reaction of the net exchange of THF with the *t*-butylacetylene costs only 7.3 kcal·mol^{-1}, indicative of the low amount of energy needed for such a reaction (Figure 7b). It should be noted that all the attempts to locate a local minimum that corresponds to the "key intermediate" with at least one THF coordinated to a samarium center were not successful, due to steric reasons. Hence, most likely the THF dissociates to yield the more energetically stable bimetallic intermediate **1-^{t}Bu**.

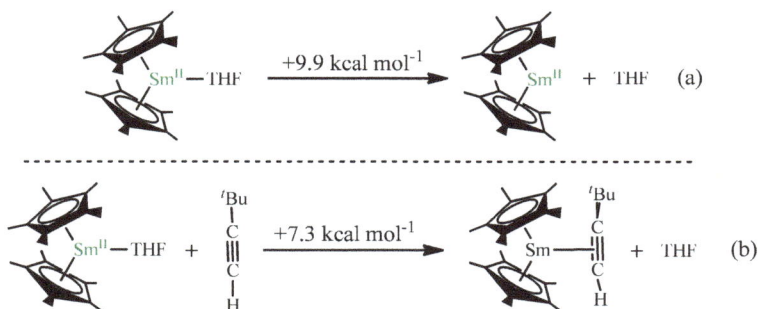

Figure 7. (**a**) Dissociation reaction of THF from the mono-solvated samarocene complex; (**b**) Exchange reaction of mono-solvated (THF) samarocene complex with ^{t}BuC≡CH.

Therefore, as in the phenylacetylene case, the "key intermediate" **1-^{t}Bu** is considered as the starting point of the whole enthalpy profile. It is noteworthy that the first two steps are found to possess similar energy variations with that of phenylacetylene. This was not expected initially since the bulkier ${}^{t}BuC\equiv CH$ will introduce more steric repulsions when it is sandwiched between the two samarocene moieties. This was especially anticipated in the C–H activation step, being in very close proximity to the two pentamethylcyclopentadienyl groups. Nevertheless, the inherent flexibility of the system probably allows the minimization of such repulsive forces by directing the methyl groups of the ${}^{t}Bu$ to lie in an optimal position among the four Cp* ligands. The only, but major differentiation with respect to the phenylacetylene case is the geometry of the product of the DM I σ-BM step. The outcome of this process, computationally, is the disruption of the bimetallic complex into two monomeric alkynyl complexes, with one bearing a weakly coordinated dihydrogen molecule. Hence, instead of the generation of a μ-alkynyl-η^{1}-alkynyl dihydrogen-like complex, as is found in the PhC≡CH case, e.g., **3-Ph**, the formation of the two monomers leads to a substantial energy stabilization (being 10.3 kcal·mol⁻¹ lower than the reference point). Then the system releases H_2 very easily to gain an extra energy stabilization of 1.4 kcal·mol⁻¹ and to yield consequently the well separated alkynyl complexes **3'-${}^{t}Bu$**. The latter were experimentally observed as the only product of such reactivity [7]. Following this, the reaction can potentially proceed into the subsequent C–C coupling requiring a considerable amount of energy ($\Delta_r H^{\ddagger} = 27.7$ kcal·mol⁻¹), kinetically hardly accessible [14]. Overall, the formation of the separated alkynyl species corresponds to the lowest point of the whole reaction route, being in perfect agreement with the experimental observations.

2.3. Dehydrocoupling of Hex-1-yne Using Cp*₂Sm

To understand the effect induced by an alkyl substituent on the alkyne, we considered computationally the aliphatic hex-1-yne in order to check how this differentiates from the bulky ${}^{t}BuC\equiv CH$, and the PhC≡CH in which there is an extended conjugation of the triple bond with the phenyl ring. Even though the gain in energy is large for the double reduction of the hex-1-yne, as it was computed in a previous work by our group [16], it is half of that obtained for the phenylacetylene, possibly due to the delocalization of the extra electron(s) into the phenyl ring. In Figure 8, the most favorable reaction pathway is presented, reverting again to mechanism C.

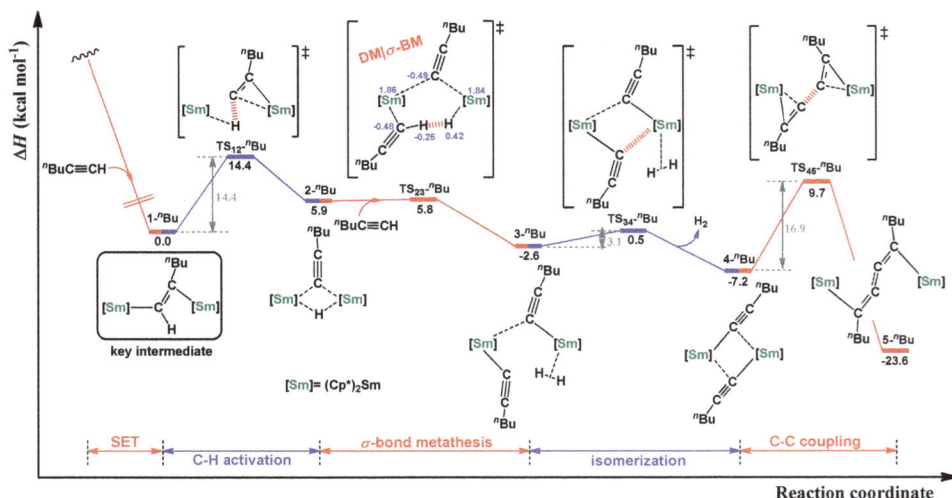

Figure 8. Enthalpy energy ΔH (kcal·mol⁻¹) profile for the formation of the trienediyl complex using ${}^{n}BuC\equiv CH$ as substrate and (Cp*)₂Sm as a precursor. Cp* ligands are omitted for clarity. Numbers in blue correspond to natural charges.

The geometry of the "key intermediate", **1-^{n}Bu**, is similar to that found in the two previously described cases with the α- and β-carbons of the doubly reduced triple bond adopting a zig-zag geometry. In the same way, the acetylenic C–H bond is elongated by 0.08 Å (with respect to the free hex-1-yne) being already prepared for the forthcoming intramolecular C–H bond scission. Next, the aforementioned C–H bond activation takes place yielding a μ-hydride-μ-alkynyl dinuclear intermediate with an enthalpy activation barrier of 14.4 kcal·mol^{-1}. This activation barrier is lower than the two previous by almost 4 kcal·mol^{-1}. This is probably attributed to the reduced steric hindrance that substituents as ^{n}Bu-, and Ph- to a smaller extent, are inducing compared to the ^{t}Bu-one. The reaction can further proceed with the addition of a second hex-1-yne equivalent to form the bis-μ-alkynyl complex, **4-^{n}Bu**, after release of H$_2$. This is achieved by passing through the **TS$_{23}$-^{n}Bu**, which corresponds to a barrierless process. The last step corresponds to the homo C–C coupling of the two μ-alkynyl moieties to afford the final trienediyl dinuclear complex. The activation barrier of this step ($\Delta_r H^{\ddagger}$ = 16.9 kcal·mol^{-1}), is the highest of the entire mechanism, being consequently the rate determining step, contrary to the previous cases. Finally, this process is exothermic by 16.4 kcal·mol^{-1}, while the overall reaction exothermicity is found to be 23.6 kcal·mol^{-1}. Therefore, we conclude that the tested alkyne can be possibly used experimentally to afford, upon reaction with the Cp*$_2$Sm complex, the corresponding trienediyl species. Concluding, apart from some small stabilization observed at the C–H bond activation transition state and the formation of the uncoupled dimeric species, **4-^{n}Bu**, with respect to the other two cases, the ^{n}Bu- substituent does not play any decisive role in the feasibility of such reactivity.

2.4. Dehydrocoupling of Various Terminal Alkynes Using [Cp*$_2$Sm(μ-H)$_2$]$_2$ Complex

As described in the introduction, two experiments were conducted by Evans *et al.*, for the formation of the potential trienediyl complexes using ^{t}BuC≡CH. The first was already investigated in detail in Section 2.2. In the second experiment, the bis-hydride dimer of Sm(III) was used, but as in the first case, it did not lead to the expected homocoupled product but instead to the formation of the uncoupled alkynide complexes as major product. It should also be noted that when PhC≡CH is considered using the same precursor, [Cp*$_2$Sm(μ-H)$_2$]$_2$, it yields the corresponding trienediyl complex. Therefore, a mechanistic investigation at the B3PW91 level of theory was carried out in order to discover the reason for this experimentally observed discrepancy and to propose a rational mechanism. As a reasonable mechanistic scenario we envisioned two consecutive (DM l σ-BM–H$_2$ release) steps (Figure 9a). Interestingly, for the phenylacetelyne case, the first DM l σ-BM type of transition state surmounts a relative low activation barrier ($\Delta_r H^{\ddagger}$ = 10.5 kcal·mol^{-1}). The intermediate **2-Ph** formed after the H$_2$ release (the isomerization step) is energetically stable enough, being 7.9 kcal·mol^{-1} lower than the starting material. The following part of the mechanism is essentially the same as proposed in mechanism C and consequently will not be discussed further. Finally, it is worth noting that a potential equilibrium between the dimer **9** and its mono-hydride structure was found computationally in favor of the dimer by 16.5 kcal·mol^{-1} in terms of enthalpy energy, in direct agreement with the experimentally observed thermal stability of **9** [18]. Hence, any mechanistic scenario starting from the monomeric complex was not considered in our studies.

The same energy reaction pattern was considered in the case of ^{t}BuC≡CH. Again, the first step corresponds to a low energy process, even lower than for the phenylacetylene, with sufficient energy stabilization upon H$_2$ release, being almost isoenergetic to that obtained for PhC≡CH. However, unlike PhC≡CH, the second DM l σ-BM step yields the uncoupled alkynyl complex, as was the case in the divalent samarocene, from which the C–C coupling is not kinetically easily accessible (Figure 9b). Hence, such type of two consecutive (DM l σ-BM–H$_2$ release) steps can serve as an explanation of the experimentally observed reactivity, shedding ample light into this peculiar reactivity in lanthanide chemistry.

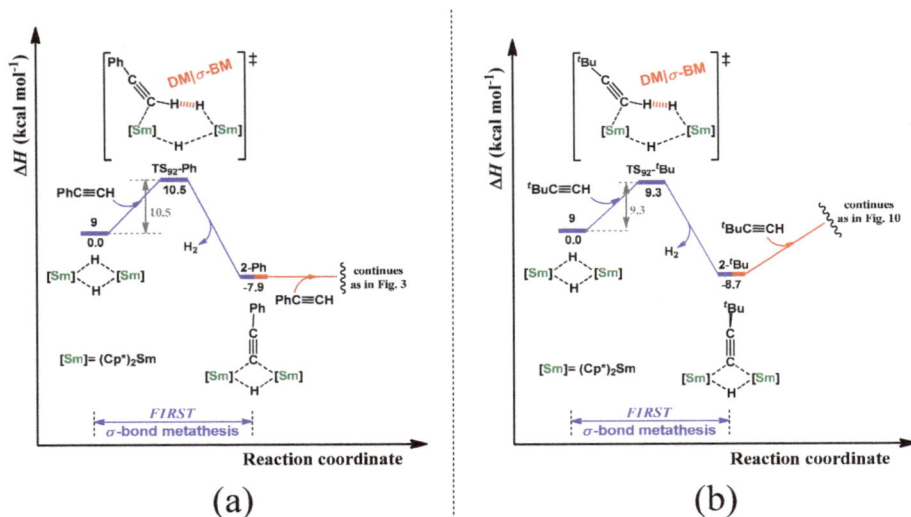

Figure 9. Enthalpy energy ΔH (kcal·mol^{-1}) profiles for the formation of the corresponding trienediyl complexes using (**a**) PhC≡CH, and (**b**) ^{t}BuC≡CH as substrates, and [(Cp*)$_2$Sm(μ-H)]$_2$ as a precursor. Cp* ligands are omitted for clarity.

3. Computational Section

All the quantum calculations, required for the delineation of the intermediate and transition state molecular structures, were performed at the density functional theory level using the B3PW91 [30,31] hybrid functional as implemented in the Gaussian program code [32]. The basis set used for the samarium atoms was the Stuttgart–Dresden–Koln (large-core) [33], augmented by an f polarization function in conjunction with the appropriate scalar relativistic pseudopotential [34]. For all the remaining atoms the all-electron double zeta basis set 6–31 G (d,p) was used [35,36]. Enthalpy energies were obtained at T = 298.15 K based on the harmonic approximation. Intrinsic Reaction Paths (IRPs) were traced from the various transition structures to verify the reactant to product linkage [37,38]. Natural population analysis (NPA) was performed using Weinhold's methodology [39,40].

4. Conclusions

Although the reaction of a low-valent samarocene with terminal alkynes affording a trienediyl binuclear complex and H$_2$ seems simple, the mechanism underlying this transformation is mechanistically complicated. In the present DFT mechanistic study, the cooperativity of the two lanthanides is highlighted, with a unique mechanistic scenario being proposed, involving novel type of transition states, such as the DM | σ-SBM, for example. In this particular saddle point of the potential energy surface, the two samarium centers are working synergistically to facilitate H$_2$ elimination, by forming a six-member transition state. It is worth noting that this type of transition state consists of a novel paradigm of σ-bond metathesis that can be added to the already well-established related types [41]. Among other mechanistic paths, the operating mechanism consists of (i) the oxidation of each samarium center and the concomitant double reduction of the alkyne to afford a binuclear intermediate; (ii) the C–H scission of the acetylinic bond that lies in between the two metals; (iii) a DM | σ-SBM process that releases the H$_2$ molecule; (iv) the C–C coupling of the two bridged μ-alkynides to give the final experimentally observed bimetallic trienediyl complex. For the latter mechanistic route the experimentally used substrate, phenylacetylene, was considered first, and afterwards the hex-1-yne in order to check the effect of the alkyl substituent. For the aliphatic alkyne, the applied mechanism leads to the same conclusions in terms of enthalpy of reaction, making this process feasible experimentally. On the other hand, the lack of obtaining the trienediyl analogs when the $^{t}BuC{\equiv}CH$ is used, as reported experimentally, is due to the thermodynamic preference for disruption of the bimetallic product of the DM I σ-SBM step over the homocoupling of the two bulky alkynyl moieties. We strongly believe that this computational contribution sheds light into the mechanism at work in this peculiar reactivity, and will help in the direction of additional understanding in the future of organolanthanide chemistry.

Supplementary Materials: Supplementary materials can be found at http://www.mdpi.com/2304-6740/3/4/0573/s1.

Acknowledgments: Laurent Maron is member of the Institute Universitaire de France. CINES (Centre Informatique National de l'Enseignement Supérieur) and CALMIP (Calcul en Midi-Pyrénées) are acknowledged for a generous grant of computing time. The Humboldt Foundation and The Chinese Academy of Science are also acknowledged. The authors are grateful to Lionel Perrin for fruitful discussions and comments on various aspects of the present work.

Author Contributions: Laurent Maron conceived the project; Christos E. Kefalidis performed the calculations; Laurent Maron and Christos E. Kefalidis analyzed the data and wrote the manuscript.

Conflicts of Interest: The authors declare no conflict of interest.

References

1. Krief, A.; Laval, A.-C. Coupling of organic halides with carbonyl compounds promoted by SmI$_2$, the Kagan reagent. *Chem. Rev.* **1999**, *99*, 745–777. [CrossRef] [PubMed]

2. Procter, D.J.; Flowers, R.A., II; Skrydstrup, T. *Organic Synthesis Using Samarium Diiodide: A Practical Guide*; Royal Society of Chemistry Publishing: Cambridge, UK, 2010.

3. Evans, W.J.; Chamberlain, L.R.; Ulibarri, T.A.; Ziller, J.W. Reactivity of trimethylaluminum with (C$_5$Me$_5$)$_2$Sm(THF)$_2$: Synthesis, structure, and reactivity of the samarium methyl complexes (C$_5$Me$_5$)$_2$Sm[(μ-Me)AlMe$_2$(μ-Me)]$_2$Sm(C$_5$Me$_5$)$_2$ and (C$_5$Me$_5$)$_2$SmMe(THF). *J. Am. Chem. Soc.* **1988**, *110*, 6423–6432. [CrossRef]

4. Nolan, S.P.; Stern, C.L.; Marks, T.J. Organo-*f*-element thermochemistry. Absolute metal-ligand bond disruption enthalpies in bis(pentamethylcyclopentadienyl)samarium hydrocarbyl, hydride, dialkylamide, alkoxide, halide, thiolate, and phosphide complexes. Implications for organolanthanide bonding and reactivity. *J. Am. Chem. Soc.* **1989**, *111*, 7844–7853.

5. Evans, W.J.; Ulibarri, T.A.; Ziller, J.W. Reactivity of (C$_5$Me$_5$)$_2$Sm with aryl-substituted alkenes: Synthesis and structure of a bimetallic styrene complex that contains an η^2-arene lanthanide interaction. *J. Am. Chem. Soc.* **1990**, *112*, 219–223. [CrossRef]

6. Evans, W.J.; Ulibarri, T.A.; Ziller, J.W. Reactivity of (C$_5$Me$_5$)$_2$Sm and related species with alkenes: Synthesis and structural characterization of a series of organosamarium allyl complexes. *J. Am. Chem. Soc.* **1990**, *112*, 2314–2324. [CrossRef]

7. Evans, W.J.; Keyer, R.A.; Ziller, J.W. Investigation of organolanthanide-based carbon–carbon bond formation: Synthesis, structure, and coupling reactivity of organolanthanide alkynide complexes, including the unusual structures of the trienediyl complex [(C$_5$Me$_5$)$_2$Sm]$_2$[μ-η^2:η^2-Ph(CH$_2$)$_2$C=C=C=C–(CH$_2$)$_2$Ph] and the unsolvated alkynide [(C$_5$Me$_5$)$_2$Sm(C{\equiv}CCMe$_3$)]$_2$. *Organometallics* **1993**, *12*, 2618–2633.

8. Evans, W.J.; Davis, B.L. Chemistry of tris(pentamethylcyclopentadienyl) *f*-element complexes, (C$_5$Me$_5$)$_3$M. *Chem. Rev.* **2002**, *102*, 2119–2136. [CrossRef] [PubMed]

9. Evans, W.J.; Keyer, R.A.; Ziller, J.W. Carbon–carbon bond formation by coupling of two phenylethynyl ligands in an organolanthanide system. *Organometallics* **1990**, *9*, 2628–2631. [CrossRef]

10. Wood, G.L.; Knobler, C.B.; Hawthorne, M.F. Synthesis of Cp$_2$Ti[C{\equiv}CSi(CH$_3$)$_3$]$_2$ and the synthesis and molecular structure of [Cp$_2$TiC{\equiv}CSi(CH$_3$)$_3$]$_2$. *Inorg. Chem.* **1989**, *28*, 382–384. [CrossRef]

11. Sekutowski, D.G.; Stucky, G.D. Oxidative coupling of the phenylethynyl group in μ-(1-3η:2-4η-*trans,trans*-1,4-diphenylbutadiene)-bis(bis(η^5-methylcyclopentadienyl)titanium) and the reaction of 1,4-diphenyl-1,3-butadiene with bis(cyclopentadienyl)titanium(II). *J. Am. Chem. Soc.* **1976**, *98*, 1376–1382. [CrossRef]

12. Heeres, H.J.; Nijhoff, J.; Teuben, J.H. Reversible carbon–carbon bond formation in organolanthanide systems. Preparation and properties of lanthanide acetylides [Cp*$_2$LnC≡CR]$_n$ and their rearrangement products [Cp*$_2$Ln]$_2$-μ-η^2:η^2-RC$_4$R) (Ln = La, Ce; R = alkyl). *Organometallics* **1993**, *12*, 2609–2617. [CrossRef]

13. Forsyth, C.M.; Nolan, S.P.; Stern, C.L.; Marks, T.J.; Rheingold, A.L. Alkyne coupling reactions mediated by organolanthanides. Probing the mechanism by metal and alkyne variation. *Organometallics* **1993**, *12*, 3618–3623. [CrossRef]

14. Kefalidis, C.E.; Perrin, L.; Maron, L. Computational insights into carbon–carbon homocoupling reactions mediated by organolanthanide(III) complexes. *Dalton Trans.* **2014**, *43*, 4520–4529. [CrossRef] [PubMed]

15. Kosog, B.; Kefalidis, C.E.; Heinemann, F.W.; Maron, L.; Meyer, K. Uranium(III)-mediated C–C-coupling of terminal alkynes: Formation of dinuclear uranium(IV) vinyl complexes. *J. Am. Chem. Soc.* **2012**, *134*, 12792–12797. [CrossRef] [PubMed]

16. Kefalidis, C.E.; Essafi, S.; Perrin, L.; Maron, L. Qualitative estimation of the single-electron transfer step energetics mediated by samarium(II) complexes: A "SOMO−LUMO gap" approach. *Inorg. Chem.* **2014**, *53*, 3427–3433. [CrossRef] [PubMed]

17. Labouille, S.; Nief, F.; Maron, L. Theoretical treatment of redox processes involving lanthanide(II) compounds: Reactivity of organosamarium(II) and organothulium(II) complexes with CO$_2$ and pyridine. *J. Phys. Chem. A* **2011**, *115*, 8295–8301. [CrossRef] [PubMed]

18. Evans, W.J.; Bloom, I.; Hunter, W.E.; Atwood, J.L. Organolanthanide hydride chemistry. 3. Reactivity of low-valent samarium with unsaturated hydrocarbons leading to a structurally characterized samarium hydride complex. *J. Am. Chem. Soc.* **1983**, *105*, 1401–1403. [CrossRef]

19. Ortiz, J.V.; Hoffmann, R. Hydride bridges between LnCp$_2$ centers. *Inorg. Chem.* **1985**, *24*, 2095–2104. [CrossRef]

20. Schaverien, C.J. Alkoxides as ancillary ligands in organolanthanide chemistry: Synthesis of, reactivity of, and olefin polymerization by the μ-hydride-μ-alkyl compounds [Y(C$_5$Me$_5$)(OC$_6$H$_3$^{t}Bu$_2$)]$_2$(μ-H)(μ-alkyl). *Organometallics* **1994**, *13*, 69–82. [CrossRef]

21. Duchateau, R.; van Wee, C.T.; Meetsma, A.; van Duijnen, P.T.; Teuben, J.H. Insertion and C–H bond activation of unsaturated substrates by bis(benzamidinato)yttrium alkyl, [PhC(NSiMe$_3$)$_2$]$_2$YR (R = CH$_2$Ph·THF, CH(SiMe$_3$)$_2$), and hydrido, {[PhC(NSiMe$_3$)$_2$]$_2$Y(μ-H)}$_2$, compounds. *Organometallics* **1996**, *15*, 2291–2302. [CrossRef]

22. Steigerwald, M.L.; Goddard, W.A. The 2$_s$ + 2$_s$ reactions at transition metals. 1. The reactions of D$_2$ with Cl$_2$TiH$^+$, Cl$_2$TiH, and Cl$_2$ScH. *J. Am. Chem. Soc.* **1984**, *106*, 308–311. [CrossRef]

23. Ziegler, T.; Folga, E.; Berces, A. A density functional study on the activation of hydrogen–hydrogen and hydrogen–carbon bonds by Cp$_2$Sc–H and Cp$_2$Sc–CH$_3$. *J. Am. Chem. Soc.* **1993**, *115*, 636–646. [CrossRef]

24. Maron, L.; Eisenstein, O. DFT study of H−H activation by Cp$_2$LnH d^0 complexes. *J. Am. Chem. Soc.* **2001**, *123*, 1036–1039. [CrossRef] [PubMed]

25. Ren, J.; Hu, J.; Lin, Y.; Xing, Y.; Shen, Q. The reactivity of lanthanide alkyl compounds with phenylacetylene: Synthesis and structure of [(Bu^{t}Cp)$_2$LnC≡CPh]$_2$ (Ln= Nd, Gd). *Polyhedron* **1996**, *15*, 2165–2169. [CrossRef]

26. Cuenca, T.; Gómez, R.; Gómez-Sal, P.; Rodriguez, G.M.; Royo, P. Reactions of titanium- and zirconium(III) complexes with unsaturated organic systems. X-ray structure of {[(η^5-C$_5$H$_5$)Zr(CH$_3$)]$_2$[μ-η^1-η^2-CN(Me$_2$ C$_6$H$_3$)](μ-η^5-η^5-C$_{10}$H$_8$)}. *Organometallics* **1992**, *11*, 1229–1234. [CrossRef]

27. Atwood, J.L.; Hunter, W.E.; Evans, W.J. Synthesis and crystallographic characterization of a dimeric alkynide-bridged organolanthanide: [(C$_5$H$_5$)$_2$ErC≡CC(CH$_3$)$_3$]$_2$. *Inorg. Chem.* **1981**, *20*, 4115–4119. [CrossRef]

28. Evans, W.J.; Bloom, I.; Hunter, W.E.; Atwood, J.L. Synthesis of organosamarium complexes containing Sm–C and Sm–P bonds. Crystallographic characterization of [(MeC$_5$H$_4$)$_2$SmC≡CCMe$_3$]$_2$. *Organometallics* **1983**, *2*, 709–714. [CrossRef]

29. Shen, Q.; Zheng, D.; Lin, L.; Lin, Y. Synthesis of (*t*-C$_4$H$_9$C$_5$H$_4$)$_2$Sm(DME) and its reactivity with phenylacetylene: Synthesis and structure of ((*t*-C$_4$H$_9$C$_5$H$_4$)$_2$SmC≡CPh)$_2$. *J. Organomet. Chem.* **1990**, *391*, 307–312. [CrossRef]

30. Becke, A.D. Density-functional thermochemistry. III. The role of exact exchange. *J. Chem. Phys.* **1993**, *98*, 5648–5652. [CrossRef]

31. Perdew, J.P.; Wang, Y. Accurate and simple analytic representation of the electron-gas correlation energy. *Phys. Rev. B* **1992**, *45*, 13244–13249. [CrossRef]

32. Gaussian 09, Revision A.02. Gaussian Inc.: Wallingford, CT, USA, 2009.

33. Dolg, M.; Stoll, H.; Savin, A.; Preuss, H. Energy-adjusted pseudopotentials for the rare earth elements. *Theor. Chim. Acta* **1989**, *75*, 173–194. [CrossRef]

34. Dolg, M.; Stoll, H.; Preuss, H. A combination of quasirelativistic pseudopotential and ligand field calculations for lanthanoid compounds. *Theor. Chim. Acta* **1993**, *85*, 441–450. [CrossRef]

35. Hehre, W.J.; Ditchfield, R.; Pople, J.A. Self-consistent molecular orbital methods. XII. Further extensions of Gaussian-type basis sets for use in molecular orbital studies of organic molecules. *J. Chem. Phys.* **1972**, *56*, 2257–2261. [CrossRef]

36. Hariharan, P.C.; Pople, J.A. The influence of polarization functions on molecular orbital hydrogenation energies. *Theor. Chim. Acta* **1973**, *28*, 213–223. [CrossRef]

37. Gonzalez, C.; Schlegel, H.B. An improved algorithm for reaction path following. *J. Chem. Phys.* **1989**, *90*, 2154–2161. [CrossRef]

38. Gonzalez, C.; Schlegel, H.B. Reaction path following in mass-weighted internal coordinates. *J. Phys. Chem.* **1990**, *94*, 5523–5527. [CrossRef]

39. Reed, A.E.; Curtiss, L.A.; Weinhold, F. Intermolecular interactions from a natural bond orbital, donor–acceptor viewpoint. *Chem. Rev.* **1988**, *88*, 899–926. [CrossRef]

40. Weinhold, F. Natural bond orbital methods. In *Encyclopedia of Computational Chemistry*; Schleyer, P.V.R., Allinger, N.L., Clark, T., Eds.; John Wiley & Sons: Chichester, UK, 1998; Volume 3, pp. 1792–1811.

41. Waterman, R. σ-Bond metathesis: A 30-year retrospective. *Organometallics* **2013**, *32*, 7249–7263. [CrossRef]

![inorganics logo] *inorganics*

MDPI

Article

Synthesis and Characterization of Cerium(IV) Metallocenes

Andrew D. Sutton [1,*], David L. Clark [2,*], Brian L. Scott [3] and John C. Gordon [1,*]

[1] Chemistry Division—Inorganic, Isotope and Actinide Chemistry (C-IIAC), Los Alamos National Laboratory, Los Alamos, NM 87545, USA

[2] National Security Education Centre (NSEC), Los Alamos National Laboratory, Los Alamos, NM 87545, USA

[3] Materials Synthesis and Integrated Devices (MPA-11), Los Alamos National Laboratory, Los Alamos, NM 87545, USA; bscott@lanl.gov

* Correspondence: adsutton@lanl.gov (A.D.S.); dlclark@lanl.gov (D.L.C.); jgordon@lanl.gov (J.C.G.); Tel.: +1-505-665-2931 (A.D.S.); +1-505-665-0983 (D.L.C.); +1-505-665-6962 (J.C.G.).

Academic Editors: Stephen Mansell and Steve Liddle

Received: 24 October 2015; Accepted: 4 December 2015; Published: 11 December 2015

Abstract: By applying a salt metathesis approach between $Ce(O^tBu_3)_2(NO_3)_2(THF)_2$ and the potassium salts of mono- and ditrimethylsilyl substituted cyclopentadienes, we were able to isolate two new Ce(IV) metallocenes, including to the best of our knowledge, the first structurally characterized bis-cyclopentadiene Ce(IV) compound.

Keywords: cerium; organometallic; cyclopentadiene; crystal structure

1. Introduction

Cerium(IV) is a strong one-electron oxidant and its use in areas such as organic synthesis [1–3], bioinorganic chemistry [4,5], materials science [6] and industrial catalysts such as automotive catalytic converters [7–13] is widespread. Soluble cerium(IV) complexes are also important precursors for the preparation of ceria nanoparticles [14] and this has increased the demand for well-characterized precursors for the development of this area. Ce(IV) alkoxides are well suited for this purpose and reports of Ce(IV) alkoxides exist where these have exhibited utility as precursors in CVD applications [15] and in ring-opening polymerization of lactide [16]. The synthesis of Ce(IV) organometallic complexes, however, is limited by the lack of available starting materials and this has constrained the growth and development of traditional organometallic chemistry and reactivity using this high-oxidation state lanthanide [17]. Two main approaches exist as entry points into Ce(IV) chemistry; the oxidation of Ce(III) compounds and the use of materials such as $Ce(OPr^i)_4$, $Ce(OTf)_4$, cerium ammonium nitrate (CAN) derived $[Ce(O^tBu)_2(NO_3)_2(THF)_2]$ [18] and most recently, the amide complex, $Ce[(N^iPr_2)_4]$ [19]. The oxidation of Ce(III) is challenging and the results are often unpredictable due to ligand redistribution reactions. For example, $CeX[N(SiMe_3)_2]_3$ can be prepared from $Ce[N(SiMe_3)_2]_3$ using $TeCl_4$ or PBr_2Ph_3 as oxidants for X = Cl, Br respectively in low yields (Cl: 24%, Br: 30%) [20,21] but these are unstable is solvents other than THF. In contrast, $TeCl_4$ was ineffective in the oxidation of $Ce[(OCMe_2CH_2(1-C\{NCHCHNPr^i\}))]$, while benzoquinone readily oxidized it to the tetravalent cerium carbene $Ce[(OCMe_2CH_2(1-C\{NCHCHNPr^i\}))]_4$ [22]. Benzoquinone was also used effectively to oxidize $[Ce\{N(SiMe_3)_2\}_3]$ to form the new cerium(IV) silylamide complex $[Ce\{N(SiMe_3)_2\}_3(bda)_{0.5}]_2$ (bda = 1,4-benzenediolato) [23]. While cyclopentadienyl Ce(IV) complexes have been characterized spectroscopically [24,25] or even claimed [26] and refuted [27], to the best of our knowledge and with the aid of the Cambridge Crystallographic Database (CCDC), we are only aware of two reports of structurally characterized Ce(IV) metallocene complexes. The first was reported by Evans et al. in 1989 and was prepared from $Ce(O^tBu_3)_n(NO_3)_m(THF)_2$ (m = 2 or 3; n + m = 4) and mNaCp to give $Cp_2Ce(O^tBu_3)_2$ and $Cp_3Ce(O^tBu_3)$, the latter being crystallographically characterized [28]. These complexes utilized a soluble Ce(IV) alkoxide nitrate species which is readily prepared from cerium ammonium nitrate (CAN) [18]. Following this in 2010, Anwander and Edelmann reported the oxidation of CeCp$_3$ with PhICl$_2$ to form Cp$_3$CeCl [29]. Herein we present further expansion of known cyclopentadiene Ce(IV) complexes, with the first structurally characterized biscyclopentadiene Ce(IV) compound, and we will briefly discuss attempts to perform subsequent transformations of these relatively rare compounds.

2. Results and Discussion

2.1. Preparation of Cerium(IV) Metallocenes

Our initial goal was to access a system containing sterically demanding ligands that would direct chemistry to the ancillary ligand sites. To this end we decided to utilize –SiMe$_3$-substituted cyclopentadienyl ligands. We added two equivalents of KCp′ (Cp′ = C$_5$H$_4$SiMe$_3$) to a solution of Ce(O^{t}Bu$_3$)$_2$(NO$_3$)$_2$(THF)$_2$ in THF and observed an immediate color change from orange to deep red. The solution was allowed to stir overnight and the solvent was removed. The solid was extracted with HMDSO, filtered and a microcrystalline purple solid was obtained after storage at −20 °C for several days. This was analyzed by ^{1}H- and ^{13}C{^{1}H}-NMR spectroscopy which permitted the characterization of **1** as (Cp′)$_2$Ce(OCMe$_3$)$_2$, produced in reasonable yield (Scheme 1). Despite numerous attempts using various solvents and solvent combinations, single crystals for X-ray crystallography could not be obtained.

The same methodology was applied to the 1,3-substituted cyclopentadiene, KCp″, (Cp″ = C$_5$H$_4$(SiMe$_3$)$_2$-1,3) (Scheme 1) and when extracted with hexamethyldisiloxane (HMDSO) this afforded red single crystals which enabled us to crystallographically characterize **2** as (Cp″)$_2$Ce(O^{t}Bu$_3$)$_2$ (Figure 1). The structure of **2** was solved in monoclinic P2$_1$/c and exhibits a pseudo-tetrahedral geometry with two coordinated Cp″ ligands and two coordinated O^{t}Bu ligands. Selected bond lengths and angles are provided in Table 1. Due to the steric bulk associated with the TMS groups, the angle between the rings is more obtuse causing a deviation from true T_d geometry with respect to the O–Ce–O angle between the two O^{t}Bu groups (103.9°). The Ce-centroid distances in **2** (2.511 Å average) are comparable to those in Cp$_3$Ce(O^{t}Bu$_3$) (2.509 Å average) [28] but significantly longer than those reported for Cp$_3$CeCl (2.460 Å average) [29] presumably due to the added steric congestion closer to the metal center resulting from the O^{t}Bu groups. The Ce–O bond lengths in **2** (2.077(5) and 2.097(6) Å) are slightly elongated compared to that in Cp$_3$Ce(O^{t}Bu$_3$) (2.045(6) Å) [28] which is most likely a steric effect rather than an electronic effect due to the comparable Ce-centroid distance between substituted and unsubstituted ring systems.

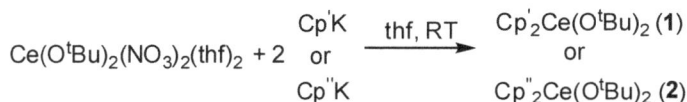

$$\text{Ce(O}^t\text{Bu)}_2\text{(NO}_3\text{)}_2\text{(thf)}_2 + 2 \quad \begin{array}{c} \text{Cp}'\text{K} \\ \text{or} \\ \text{Cp}''\text{K} \end{array} \quad \xrightarrow{\text{thf, RT}} \quad \begin{array}{c} \text{Cp}'_2\text{Ce(O}^t\text{Bu)}_2 \textbf{ (1)} \\ \text{or} \\ \text{Cp}''_2\text{Ce(O}^t\text{Bu)}_2 \textbf{ (2)} \end{array}$$

Scheme 1. Reaction scheme for the preparation of **1** and **2**.

2.2. Attempted Reactivity of **1** and **2**

In order to probe the reactivity of **1** and **2**, and synthesize additional cyclopentadienyl compounds of Ce(IV), each was reacted in tetrahydrofuran or toluene with one and two equivalents of KOTf, TMS-Cl, Ph$_3$SiOH, MeNH$_2$, Et$_3$N, MeMgBr, Bu$_2$Mg with no reaction observed by NMR spectroscopy. Reaction with dilute HCl as a solution in Et$_2$O led to reduction of both **1** and **2** as evidence by paramagnetic NMR spectra although no tractable product could be isolated and characterized.

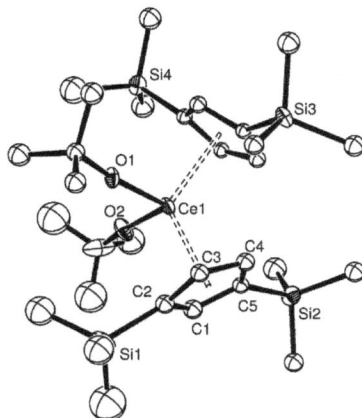

Figure 1. Thermal ellipsoid plot of the structure of **2** (hydrogen atoms omitted ellipsoids at 30% probability).

Table 1. Selected bond lengths (Å) and angles (°) of **2**.

Ce(1)–O(1)	2.077(5)	Ce(1)–Cn(1)	2.508(7)
Ce(1)–O(2)	2.097(6)	Ce(1)–C(12)	2.798(8)
Ce(1)–C(1)	2.799(9)	Ce(1)–C(13)	2.791(8)
Ce(1)–C(2)	2.811(9)	Ce(1)–C(14)	2.764(8)
Ce(1)–C(3)	2.743(9)	Ce(1)–C(15)	2.774(8)
Ce(1)–C(4)	2.733(9)	Ce(1)–C(16)	2.812(9)
Ce(1)–C(5)	2.810(8)	Ce(1)–Cn(2)	2.514(7)
O(1)–Ce(1)–Cn(1)	108.7(4)	O(1)–Ce(1)–Cn(2)	109.9(8)
O(1)–Ce(1)–Cn(2)	105.8(2)	O(1)–Ce(1)–O(2)	103.9(2)
O(2)–Ce(1)–Cn(1)	105.6(4)	Cn(1)–Ce(1)–Cn(2)	121.7(2)

3. Experimental Section

Unless otherwise noted, all reactions were performed using standard Schlenk-line techniques or in a Vacuum Atmosphere glove box. All glassware and Celite were stored in an oven at *ca.* 425 K. Hexamethyldisiloxane was distilled and degassed before use. Hexane, toluene, and THF were freshly distilled from sodium and degassed with nitrogen prior to use. C_6D_6 was vacuum transferred from sodium/benzophenone. NMR spectra were recorded at ambient temperature on Bruker AV-400. ^{1}H and $^{13}C\{^1H\}$ chemical shifts are given relative to residual protic solvent peaks and coupling constants (*J*) are given in Hz. Infra-red samples were prepared as Nujol mulls and taken between KBr plates. Melting points were determined using sealed capillaries prepared under nitrogen and are uncorrected. Unless otherwise noted, all reagents were acquired from commercial sources and used as received.

A crystal of **2** was mounted in a nylon cryoloop from Paratone-N oil. The data were collected on a Bruker D8 diffractometer, with an APEX II charge-coupled-device (CCD) detector, and a Bruker Kryoflex liquid nitrogen low temperature device (140 K). The instrument was equipped with a graphite monochromatized MoKα X-ray source (λ = 0.71073 Å), and a 0.5 mm monocapillary. A hemisphere of data was collected using ω scans, with 10-s frame exposures and 0.5° frame widths. Data collection and initial indexing and cell refinement were handled using APEX II [30] software. Frame integration, including Lorentz-polarization corrections, and final cell parameter calculations were carried out using SAINT+ [31] software. The data were corrected for absorption using redundant reflections and the SADABS [32] program. Decay of reflection intensity was not observed as monitored via analysis of redundant frames. The structure was solved using Direct methods and difference Fourier techniques. One t-butyl group was disordered, and the methyl groups were refined in two positions (C28/C28′, C29/C29′, and C30/C30′). Site occupancy factors for disordered pairs were tied to 1.0. A total of 31 bond distance and temperature factor restraints were used to force convergence of the disordered group. All hydrogen atom positions were idealized, and rode on the atom they were attached to. The final refinement included anisotropic temperature factors on all non-hydrogen atoms. Structure solution, refinement, graphics, and creation of publication materials were performed using SHELXTL [33] software. Additional details can be found in crystallographic information file (CCDC No. 1433004) and in the Supplementary Information.

3.1. Preparation of (Cp′)$_2$Ce(O^{t}Bu)$_2$ (1)

KCp′ (0.352 g, 2.0 mmol) was dissolved in THF (2.5 mL) and added to a suspension of Ce(O^{t}Bu)$_2$(NO$_3$)$_2$(THF)$_2$ (0.555 g, 1.0 mmol) in THF (2.5 mL), resulting in an immediate color change to deep red. The reaction was allowed to stir overnight and the solvent removed *in vacuo*. The resultant solid was extracted with HMDSO (10 mL) and stored at −20 °C to afford a microcrystalline purple solid (0.305 g, 54%). ^{1}H-NMR (400 MHz, C$_6$D$_6$): δ 0.48 (s, 18 H, Si(CH$_3$)$_3$); 1.32 (s, 18 H, C(CH$_3$)$_3$); 6.26 (m, 4 H); 6.65 (m, 4 H). ^{13}C{^{1}H}-NMR (100.6 MHz, C$_6$D$_6$): 0.73, 33.56, 84.56, 123.95, 127.55, 127.79. IR (cm^{-1}): 1622 (m); 1377 (s); 1314 (w); 1287 (w); 1268 (m); 1195 (w); 1099 (m); 1085 (w); 1042 (w); 1015 (m); 980 (w); 930 (s); 846 (w); 814 (w); 800 (w); 746 (w); 665 (w). m.p. > 260 °C (dec.). CHO for C$_{24}$H$_{44}$CeO$_2$Si$_2$ Calc. % (found) C 51.39 (51.61), H 7.91 (7.86), O 5.70 (5.44).

3.2. Preparation of (Cp″)$_2$Ce(O^{t}Bu)$_2$ (2)

KCp″ (0.497 g, 2.0 mmol) was dissolved in THF (2.5 mL) and added to a suspension of Ce(O^{t}Bu)$_2$(NO$_3$)$_2$(THF)$_2$ (0.555 g, 1.0 mmol) in THF (2.5 mL) resulting in an immediate color change to deep red. The reaction was allowed to stir overnight and the solvent was removed *in vacuo*. The resultant solid was extracted with HMDSO (20 mL) and stored at −20 °C for two days to afford X-ray quality crystals (0.474 g, 67%). ^{1}H-NMR (300 MHz, C$_6$D$_6$): δ 0.45 (s, 36 H, Si(CH$_3$)$_3$); 1.39 (s, 18 H, C(CH$_3$)$_3$); 6.20 (m, 4 H); 6.59 (m, 4 H). ^{13}C{^{1}H}-NMR (100.6 MHz, C$_6$D$_6$): 0.77, 34.87, 85.77, 125.23, 128.11, 129.02. IR (cm^{-1}): 1620 (m); 1381 (s); 1373 (s); 1322 (w); 1292 (w); 1258 (m); 1204 (w); 1123 (w); 1103 (m); 1074 (w); 1057 (w); 1035 (w); 1012 (m); 976 (w); 924 (s); 849 (w); 817 (w); 796 (w); 777 (m); 743 (w); 722 (w); 690 (w); 639 (w). m.p. > 260 °C (dec.). CHO for C$_{30}$H$_{60}$CeO$_2$Si$_4$ Calc. % (found) C 51.09 (51.18), H 8.58 (8.41), O 4.54 (4.38).

4. Conclusions

By applying a salt metathesis approach between Ce(O^{t}Bu$_3$)$_2$(NO$_3$)$_2$(THF)$_2$ and the potassium salts of mono- and ditrimethylsilyl substituted cyclopentadienes, we were able to isolate two new Ce(IV) metallocenes, including to the best of our knowledge, the first structurally characterized dicyclopentadiene Ce(IV) compound. With these two compounds in hand, we attempted to exploit the reactivity of these, but to no avail. The Ce–O bonds were inert to all our attempts to substitute the O^{t}Bu ligands, highlighting the oxophilic nature of the (higher valent) lanthanides. Due to this lack of reactivity, we believe that Ce(O^{t}Bu$_3$)$_2$(NO$_3$)$_2$(THF)$_2$ is not well suited as a starting material due to the resilience of the resultant Ce–O bonds to further substitution. It is our opinion that future efforts should be focused on more amenable starting compounds such as Ce(OTf)$_4$ and Ce(N^{i}Pr$_2$)$_4$ [19] and using oxidation strategies from Ce(III) to Ce(IV) [20–22,29].

Supplementary Materials: Supplementary materials can be found at http://www.mdpi.com/2304-6740/3/4/0589/s1.

Acknowledgments: Andrew D. Sutton acknowledges the Seaborg Institute at Los Alamos National Laboratory for funding a Postdoctoral Fellowship.

Author Contributions: Andrew D. Sutton designed and performed the experimental work and wrote the manuscript, David L. Clark designed the experiments, Brian L. Scott performed the X-ray crystallography and structural determination and John C. Gordon, designed the experiments and wrote the manuscript.

Conflicts of Interest: The authors declare no conflict of interest.

References

1. Das, A.K. Kinetic and mechanistic aspects of metal ion catalysis in cerium(IV) oxidation. *Coord. Chem. Rev.* **2001**, *213*, 307–325. [CrossRef]
2. Dziegiec, J.; Domagala, S. The oxidation activity of cerium(IV) ions toward some of the organic compounds. *Trends Inorg. Chem.* **2005**, *8*, 43–64. [CrossRef]
3. Nair, V.; Balagopal, L.; Rajan, R.; Mathew, J. Recent advances in synthetic transformations mediated by cerium(IV) ammonium nitrate. *Acc. Chem. Res.* **2004**, *37*, 21–30. [CrossRef] [PubMed]
4. Komiyama, M. Sequence-selective scission of DNA and RNA by lanthanide ions and their complexes. *Met. Ions Biol. Syst.* **2003**, *40*, 463–475. [PubMed]
5. Yamamoto, Y.; Komiyama, M. Development of new biotechnology by cerium(IV)-based artificial restriction enzyme. *Mater. Integr.* **2005**, *19*, 55–59.
6. Jian, H.-O.; Zhou, X.-R.; Zhao, D.-F. Recent progress of synthetic methods and applications of cerium β-diketones. *Huaxue Shiji* **2006**, *28*, 279–282.
7. Duprez, D.; Descorme, C. Oxygen storage/redox capacity and related phenomena on ceria-based catalysts. *Catal. Sci. Ser.* **2002**, *2*, 243–280.
8. Imamura, S. Ceria-based wet-oxidation catalysts. *Catal. Sci. Ser.* **2002**, *2*, 431–452.
9. Kaspar, J.; Fornasiero, P. Nanostructured materials for advanced automotive de-pollution catalysts. *J. Solid State Chem.* **2003**, *171*, 19–29. [CrossRef]
10. Kaspar, J.; Fornasiero, P.; Graziani, M. Use of CeO_2-based oxides in the three-way catalysis. *Catal. Today* **1999**, *50*, 285–298. [CrossRef]
11. Primet, M.; Garbowski, E. Fundamentals and applications of ceria in combustion reactions. *Catal. Sci. Ser.* **2002**, *2*, 407–429.
12. Shelef, M.; Graham, G.W.; McCabe, R.W. Ceria and other oxygen storage components in automotive catalysts. *Catal. Sci. Ser.* **2002**, *2*, 343–375.
13. Trovarelli, A.; de Leitenburg, C.; Boaro, M.; Dolcetti, G. The utilization of ceria in industrial catalysis. *Catal. Today* **1999**, *50*, 353–367. [CrossRef]
14. Droese, P.; Gottfriedsen, J.; Hrib, C.G.; Jones, P.G.; Hilfert, L.; Edelmann, F.T. The first cationic complex of tetravalent cerium. *Z. Anorg. Allg. Chem.* **2011**, *637*, 369–373. [CrossRef]
15. Aspinall, H.C.; Bacsa, J.; Jones, A.C.; Wrench, J.S.; Black, K.; Chalker, P.R.; King, P.J.; Marshall, P.; Werner, M.; Davies, H.O.; *et al.* Ce(IV) complexes with donor-functionalized alkoxide ligands: Improved precursors for chemical vapor deposition of CeO_2. *Inorg. Chem.* **2011**, *50*, 11644–11652. [CrossRef] [PubMed]
16. Broderick, E.M.; Diaconescu, P.L. Cerium(IV) catalysts for the ring-opening polymerization of lactide. *Inorg. Chem.* **2009**, *48*, 4701–4706. [CrossRef] [PubMed]
17. Arnold, P.L.; Casely, I.J.; Zlatogorsky, S.; Wilson, C. Organometallic cerium complexes from tetravalent coordination complexes. *Helv. Chim. Acta* **2009**, *92*, 2291–2303. [CrossRef]
18. Evans, W.J.; Deming, T.J.; Olofson, J.M.; Ziller, J.W. Synthetic and structural studies of a series of soluble cerium(IV) alkoxide and alkoxide nitrate complexes. *Inorg. Chem.* **1989**, *28*, 4027–4034. [CrossRef]
19. Schneider, D.; Spallek, T.; Maichle-Moessmer, C.; Toernroos, K.W.; Anwander, R. Cerium tetrakis(diisopropylamide)—A useful precursor for cerium(IV) chemistry. *Chem. Commun.* **2014**, *50*, 14763–14766. [CrossRef] [PubMed]
20. Eisenstein, O.; Hitchcock, P.B.; Hulkes, A.G.; Lappert, M.F.; Maron, L. Cerium masquerading as a group 4 element: Synthesis, structure and computational characterisation of [CeCl{N(SiMe$_2$)$_3$}]. *Chem. Commun.* **2001**, 1560–1561. [CrossRef]
21. Hitchcock, P.B.; Hulkes, A.G.; Lappert, M.F. Oxidation in nonclassical organolanthanide chemistry: Synthesis, characterization, and X-ray crystal structures of cerium(III) and -(IV) amides. *Inorg. Chem.* **2004**, *43*, 1031–1038. [CrossRef] [PubMed]
22. Casely, I.J.; Liddle, S.T.; Blake, A.J.; Wilson, C.; Arnold, P.L. Tetravalent cerium carbene complexes. *Chem. Commun.* **2007**, 5037–5039. [CrossRef] [PubMed]
23. Werner, D.; Deacon, G.B.; Junk, P.C.; Anwander, R. Cerium(III/IV) formamidinate chemistry, and a stable cerium(IV) diolate. *Chemistry* **2014**, *20*, 4426–4438. [CrossRef] [PubMed]

24. Greco, A.; Cesca, S.; Bertolini, W. New 7r-cyclooctate′I′raenyl and iT-cyclopentadienyl complexes of cerium. *J. Organomet. Chem.* **1976**, *113*, 321–330. [CrossRef]
25. Gulino, A.; Casarin, M.; Conticello, V.P.; Gaudiello, J.G.; Mauermann, H.; Fragala, I.; Marks, T.J. Efficient synthesis, redox characteristics, and electronic structure of a tetravalent tris(cyclopentadienyl)cerium alkoxide complex. *Organometallics* **1988**, *7*, 2360–2364. [CrossRef]
26. Kalsotra, B.L.; Multani, R.K.; Jain, B.D. Preparation and properties of tricyclopentadienyl cerium(IV) chloride and bisindenyl cerium(IV) dichloride. *Isr. J. Chem.* **1971**, *9*, 569–572. [CrossRef]
27. Deacon, G.B.; Tuong, T.D.; Vince, D.G. Refutation of the synthesis of tetrakis(cyclopentadienyl)cerium(IV). *Polyhedron* **1983**, *2*, 969–970. [CrossRef]
28. Evans, W.J.; Deming, T.J.; Ziller, J.W. The utility of ceric ammonium nitrate-derived alkoxide complexes in the synthesis of organometallic cerium(IV) complexes. Synthesis and first X-ray crystallographic detrmination of a tetravalent cerium cyclopentadienide complex, $(C_5H_5)_3Ce(OCMe_3)$. *Organometallics* **1989**, *8*, 1581–1583. [CrossRef]
29. Dröse, P.; Crozier, A.R.; Lashkari, S.; Gottfriedsen, J.; Blaurock, S.; Hrib, C.G.; Maichle-Mössmer, C.; Schädle, C.; Anwander, R.; Edelmann, F.T. Facile access to tetravalent cerium compounds: One-electron oxidation using iodine(III) reagents. *J. Am. Chem. Soc.* **2010**, *132*, 14046–14047. [CrossRef] [PubMed]
30. *APEX II 1.08*, Bruker AXS, Inc.: Madison, WI, USA, 2004.
31. *SAINT+ 7.06*, Bruker AXS, Inc.: Madison, WI, USA, 2003.
32. Sheldrick, G. *SADABS 2.03*, University of Göttingen: Göttingen, Germany, 2001.
33. *SHELXTL 5.10*, Bruker AXS, Inc.: Madison, WI, USA, 1997.

![inorganics logo] *inorganics*

MDPI

Article

Expanding the Chemistry of Actinide Metallocene Bromides. Synthesis, Properties and Molecular Structures of the Tetravalent and Trivalent Uranium Bromide Complexes: $(C_5Me_4R)_2UBr_2$, $(C_5Me_4R)_2U(O\text{-}2,6\text{-}^iPr_2C_6H_3)(Br)$, and $[K(THF)][(C_5Me_4R)_2UBr_2]$ (R = Me, Et)

Alejandro G. Lichtscheidl [1], Justin K. Pagano [1], Brian L. Scott [2], Andrew T. Nelson [3,*] and Jaqueline L. Kiplinger [1,*]

[1] Chemistry Division, Los Alamos National Laboratory, Mail Stop J-514, Los Alamos, NM 87545, USA; agl@lanl.gov (A.G.L.); pagano@lanl.gov (J.K.P.)

[2] Materials Physics & Applications Division, Los Alamos National Laboratory, Mail Stop J-514, Los Alamos, NM 87545, USA; bscott@lanl.gov

[3] Materials Science & Technology Division, Los Alamos National Laboratory, Mail Stop E-549, Los Alamos, NM 87545, USA; atnelson@lanl.gov

* Correspondence: atnelson@lanl.gov (A.T.N.) ; kiplinger@lanl.gov (J.L.K.); Tel.: +1-505-665-9553 (J.L.K.); Fax: +1-505-667-9905 (J.L.K.)

Academic Editors: Stephen Mansell and Steve Liddle

Received: 23 November 2015; Accepted: 11 December 2015; Published: 6 January 2016

Abstract: The organometallic uranium species $(C_5Me_4R)_2UBr_2$ (R = Me, Et) were obtained by treating their chloride analogues $(C_5Me_4R)_2UCl_2$ (R = Me, Et) with Me_3SiBr. Treatment of $(C_5Me_4R)_2UCl_2$ and $(C_5Me_4R)_2UBr_2$ (R = Me, Et) with $K(O\text{-}2,6\text{-}^iPr_2C_6H_3)$ afforded the halide aryloxide mixed-ligand complexes $(C_5Me_4R)_2U(O\text{-}2,6\text{-}^iPr_2C_6H_3)(X)$ (R = Me, Et; X = Cl, Br). Complexes $(C_5Me_4R)_2U(O\text{-}2,6\text{-}^iPr_2C_6H_3)(Br)$ (R = Me, Et) can also be synthesized by treating $(C_5Me_4R)_2U(O\text{-}2,6\text{-}^iPr_2C_6H_3)(Cl)$ (R = Me, Et) with Me_3SiBr, respectively. Reduction of $(C_5Me_4R)_2UCl_2$ and $(C_5Me_4R)_2UBr_2$ (R = Me, Et) with KC_8 led to isolation of uranium(III) "ate" species $[K(THF)][(C_5Me_5)_2UX_2]$ (X = Cl, Br) and $[K(THF)_{0.5}][(C_5Me_4Et)_2UX_2]$ (X = Cl, Br), which can be converted to the neutral complexes $(C_5Me_4R)_2U[N(SiMe_3)_2]$ (R = Me, Et). Analyses by nuclear magnetic resonance spectroscopy, X-ray crystallography, and elemental analysis are also presented.

Keywords: trivalent and tetravalent uranium; bromide; organometallic

1. Introduction

The bis(cyclopentadienyl) complexes of uranium $(C_5Me_4R)_2UX_2$ (R = Me or Et; X = Cl or I) [1–3], $(1,3\text{-}R'_2C_5H_3)_2UX_2$ and $(1,2,4\text{-}R'_3C_5H_3)_2UX_2$ (R' = ^tBu, $SiMe_3$; X = Cl, I) [4–7] have been known for years, owing to the ease of preparation of reliable chloride and iodide starting materials. With the vast amount of literature attention devoted to $(C_5Me_4R)UX_2$ compounds, it is curious that the corresponding bis(cyclopentadienyl) uranium bromide systems have not been investigated to a similar extent. Considering that the structure and reactivity of these uranium compounds are strongly influenced by the nature of the halide [8], it would be useful to have the $(C_5Me_4R)UBr_2$ congeners available for synthetic actinide chemistry.

Uranium bromide complexes are rare compared to their chloride and iodide counterparts. This is succinctly illustrated in Figure 1, which presents the organometallic tetravalent and trivalent uranium complexes that have been reported to date. The vast majority of uranium(IV) bromide complexes have been prepared by salt metathesis with either UBr$_4$ or UBr$_4$(NCCH$_3$)$_4$. These compounds include the dicarbollide complex [Li(THF)$_4$]$_2$[(C$_2$H$_{11}$B$_9$)$_2$UBr$_2$] (**1**) [9]; the cyclopentadienyl complexes (C$_5$H$_5$)$_3$UBr (**2**), (C$_5$H$_5$)UBr$_3$(L)$_2$ (L = Me$_2$NC(O)^{t}Bu (**3**), OPPh$_3$ (**4**), THF (**5**)), (C$_5$H$_5$)$_2$UBr$_2$(L) (L = Me$_2$NC(O)^{t}Bu (**6**), OPPh$_3$ (**7**)), {(C$_5$H$_5$)UBr$_2$X[OP(Ph)$_2$C$_2$H$_4$P(O)(Ph$_2$)]}$_2$ (X = C$_5$H$_5$ (**8**), Br (**9**)); and the indenyl complexes (C$_9$H$_7$)$_3$UBr (**10**), (C$_9$H$_7$)UBr$_3$(OPPh$_3$)(THF) (**11**), (C$_9$H$_7$)UBr$_3$(THF)$_2$ (**12**), (C$_9$H$_7$)UBr$_3$(OPPh$_3$) (**13**), [(C$_9$H$_7$)UBr$_2$(NCCH$_3$)$_4$][UBr$_6$] (**14**) and {[(C$_9$H$_7$)UBr(NCCH$_3$)$_4$]$_2$O}[UBr$_6$]$_2$ (**15**) [10–16]. An alternative method that has been employed to prepare uranium(IV) bromide complexes is halide exchange. For example, (1,2,4-^{t}Bu$_3$C$_5$H$_2$)$_2$UBr$_2$ (**16**) [5] and (1,3-R$_2$C$_5$H$_3$)$_2$UBr$_2$ (R = SiMe$_3$ (**17**), ^{t}Bu (**18**)) [4] were prepared by reacting the chloride analogues with excess Me$_3$SiBr, and (**17**) from its chloride analogue and BBr$_3$ [7].

In recent years, oxidation chemistry has been developed to access a variety of uranium(IV) bromide complexes such as (C$_5$Me$_5$)$_2$U(X)(Br) (X = N(SiMe$_3$)$_2$ (**19**) [17], (^{i}PrN)$_2$C(Me) (**20**) [18], MeNC(Me)NAd (**21**) [19]), which were synthesized by treating (C$_5$Me$_5$)$_2$U(X) with CuBr or AgBr. In a similar fashion, (C$_5$Me$_5$)$_3$U was shown to react with benzyl bromide to afford (C$_5$Me$_5$)$_3$UBr (**22**) [20]. The η^2-suldenamido derivative (C$_5$Me$_5$)$_2$U(η^2-^{t}BuNSPh)(Br) (**23**) was obtained after long reaction times of chloride complex with MeMgBr [21]. Finally, protonolysis of (η^3-C$_3$H$_5$)$_4$U with HBr provided (η^3-C$_3$H$_5$)$_3$UBr (**24**) [22].

Reduction chemistry has been the primary method for accessing the handful of known bromide complexes. For example, treatment of [Li(THF)$_4$]$_2$[(C$_2$H$_{11}$B$_9$)$_2$UBr$_2$] (**1**) [9] and (1,3-R$_2$C$_5$H$_3$)$_2$UBr$_2$ (R = SiMe$_3$ (**17**), ^{t}Bu (**18**)) [4] with *tert*-butyl lithium or sodium amalgam gave [Li(THF)$_4$]$_2$[(C$_2$H$_{11}$B$_9$)$_2$UBr(THF)] (**25**) and [(1,3-R$_2$C$_5$H$_3$)$_2$UBr]$_2$ (R = SiMe$_3$ (**26**), ^{t}Bu (**27**)), respectively [7,23–25]. Complex (C$_5$Me$_5$)$_3$UBr (**22**) is thermally unstable and leads to isolation of (C$_5$Me$_5$)$_2$UBr (**28**), from which (C$_5$Me$_5$)$_2$UBr(THF) (**29**) can be formed [20]. Finally, [1,3-(SiMe3)2C5H3]$_2$UBr(CN^{t}Bu)$_2$ (**30**) was obtained by treating complex **26** with excess CN^{t}Bu [26].

Tetravalent Uranium Complexes:

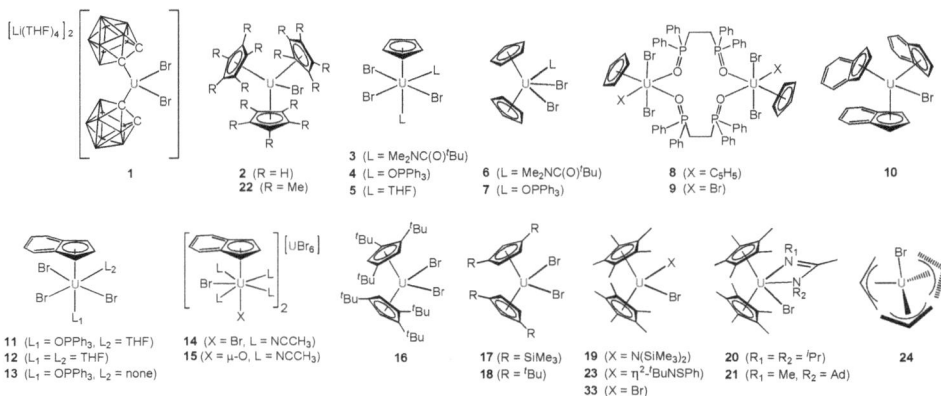

Trivalent Uranium Complexes:

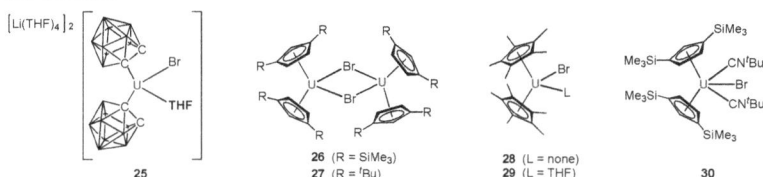

Figure 1. Structures of known organometallic uranium(IV) and uranium(III) bromide compounds.

Herein, we report the preparation of the bromide complexes $(C_5Me_4R)_2UBr_2$ (R = Me (**33**), Et (**34**)) and show that they and $(C_5Me_4R)_2UCl_2$ (R = Me (**31**), Et (**32**)) can be used as precursors to mixed-ligand species $(C_5Me_4R)_2U(O-2,6-^iPr_2C_6H_3)(X)$ (R = Me, Et; X = Cl and Br) and uranium(III) bromide and chloride "ate" species $[K(THF)_n][(C_5Me_4R)_2UX_2]$ (R = Me, Et; X = Cl, Br; n = 0.5 or 1). In addition, we show that $(C_5Me_4R)_2U[N(SiMe_3)_2]$ can be synthesized from such "ate" species, demonstrating that the $[K(THF)_n][(C_5Me_4R)_2UX_2]$ complexes are useful precursors for the synthesis of neutral uranium(III) compounds. The syntheses of all complexes are discussed and the characterization by ^{1}H NMR spectroscopy, melting point, elemental analysis and single crystal X-ray diffraction is provided.

2. Results and Discussion

Synthesis: Our study takes advantage of the halide exchange reagent, trimethylsilyl bromide, Me_3SiBr. The compounds $(C_5Me_4R)_2UBr_2$ (R = Me (**33**), Et (**34**)) were obtained by treating $(C_5Me_4R)_2UCl_2$ (R = Me (**31**), Et (**32**)) with excess Me_3SiBr (Scheme 1) similar to the method reported by Andersen for the preparation of the bent cyclopentadienyl species $(1,3-R'_2C_5H_3)_2UBr_2$ [4] and $(1,2,4-R'_3C_5H_2)_2UBr_2$ [5] (R' = ^tBu, $SiMe_3$). Complexes **33** and **34** can be obtained in 96% and 90% yield, respectively, using this procedure, but it is important to add fresh Me_3SiBr three times to the reaction mixture and to let the mixture react for at least 12 h each time. Lower time intervals and/or lesser number of treatments lead to incomplete reactions. The uranium(IV) bromide complex $(C_5Me_5)_2UBr_2$ (**33**) has been mentioned before, but no experimental data for this compound were provided [20,27]. To the best of our knowledge, this is the first time that compounds **33** and **34** have been fully characterized. The [1]H Nuclear Magnetic Resonance (NMR) spectrum of **33** in benzene-d_6 shows a singlet at 15.70 ppm, which is intermediate in value compared to $(C_5Me_5)_2UCl_2$ (13.5 ppm) [3] and $(C_5Me_5)_2UI_2$ (17.9 ppm) [28–30]; while that of complex **34** shows four singlets corresponding to the three inequivalent methyl groups (25.1, 15.6 and 13.1 ppm) and the methylene group (0.31 ppm) of the C_5Me_4Et ligand. The ^{1}H NMR shifts of **34** are also downfield compared to the analogous $(C_5Me_4Et)_2UCl_2$ (22.9, 13.8 and −2.8 ppm, respectively) and $(C_5Me_4Et)_2U(CH_3)_2$ (14.1, 5.6, 5.1 and −4.3 ppm, respectively) [31].

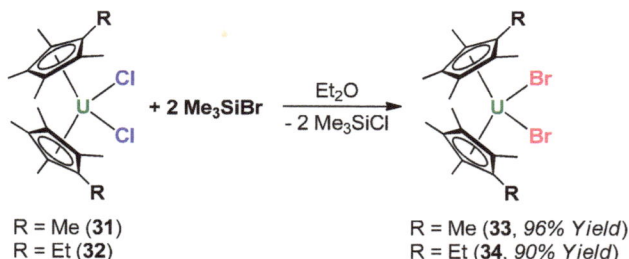

Scheme 1. Synthesis of the uranium(IV) bromide complexes $(C_5Me_4R)_2UBr_2$ (R = Me (**33**), Et (**34**)) by halide exchange with trimethylsilyl bromide.

The X-ray structure of complexes **33** and **34** are shown in Figure 2. The U(1)–Br(1) bond lengths of **33** (2.7578(5) Å) and **34** (2.7607(4) and 2.7609(4) Å) fall within the range of known neutral U(IV) terminal bromides (2.734(1)–2.831(7) Å) [6,7,9,16,32–45] and the Br(1)–U(1)–Br(2) bond angles of **33** (97.64(3)°) and **34** (96.224(14)°) are similar to those of $(C_5Me_5)_2UX_2$ (X = Cl (97.9(4)°), I (95.96(2)°)) [1,2] and $[1,3-(SiMe3)2C5H3]_2UBr_2$ (94.60(4)°) [4], but larger than most non-cyclopentadienyl-based complexes containing *cis*-bromide ligands (75.48(8)–103.83(12)°) [6,7,9,16,32–45]. The U(1)–C_{Cent} distances for **33** (2.438(4) Å) and **34** (2.444(2) Å) fall within the typical distances for C_5Me_5-based complexes (2.393–2.562 Å) [1,2,31,46–50]. Similarly, the $C_{Cent(1)}$–U(1)–$C_{Cent(2)}$ bond angles of **33** (137.57(15)°) and **34** (138.50(8)°) fall within the known range for bis(pentamethylcyclopentadienyl)-based complexes (126.2–142.6°) [1,2,31,46–50].

As outlined in Scheme 2, the halide complexes **31–34** react with bulky aryloxides such as K[O-2,6-^{i}Pr$_2$C$_6$H$_3$] to yield the corresponding halide aryloxide mixed-ligand complexes in excellent yields, (C$_5$Me$_4$R)$_2$U(O-2,6-^{i}Pr$_2$C$_6$H$_3$)(Cl) (**35**, R = Me, 92%; **36**, R = Et, 90%) and (C$_5$Me$_4$R)$_2$U(O-2,6-^{i}Pr$_2$C$_6$H$_3$)(Br) (**37**, R = Me, 99%; **38**, R = Et, 96%). Complex **35** has been previously synthesized by oxidation of (C$_5$Me$_4$R)$_2$U(O-2,6-^{i}Pr$_2$C$_6$H$_3$)(THF) with CuCl [8]. Synthesis of the halide aryloxide mixed-ligand complexes **35–38** by salt metathesis not only complements the known oxidative synthetic route to **35**, but also helps to introduce complexes **36–38** here for the first time. Finally, similar to the synthesis of complexes **33** and **34**, treatment of **35** and **36** with excess Me$_3$SiBr in THF also yields **37** and **38** in 93% and 91% yields, respectively (Scheme 3).

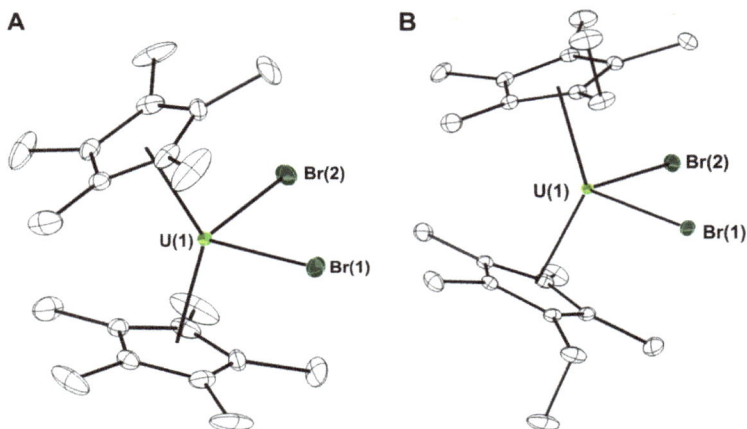

Figure 2. Molecular structure of (C$_5$Me$_5$)$_2$UBr$_2$ (**33**) (**A**) and (C$_5$Me$_4$Et)$_2$UBr$_2$ (**34**) (**B**) with thermal ellipsoids projected at the 50% probability level. Hydrogen atoms have been omitted for clarity. Selected bond distances (Å) and angles (°) for **33**: U(1)–Br(1), 2.7578(5); U(1)–C$_{Cent}$, 2.438(4); Br(1)–U(1)–Br(2), 97.64(3); C$_{Cent(1)}$–U(1)–C$_{Cent(2)}$, 137.57(15). For **34**: U(1)–Br(1), 2.7609(4); U(1)–Br(2), 2.7607(4); U(1)–C$_{Cent}$, 2.444(2); Br(1)–U(1)–Br(2), 96.224(14); C$_{Cent(1)}$–U(1)–C$_{Cent(2)}$, 138.50(8).

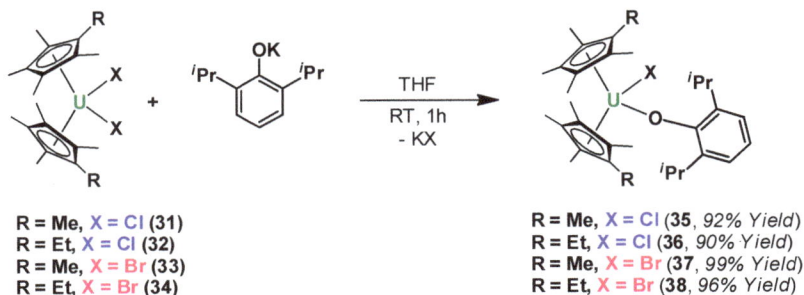

R = Me, X = Cl (**31**)
R = Et, X = Cl (**32**)
R = Me, X = Br (**33**)
R = Et, X = Br (**34**)

R = Me, X = Cl (**35**, *92% Yield*)
R = Et, X = Cl (**36**, *90% Yield*)
R = Me, X = Br (**37**, *99% Yield*)
R = Et, X = Br (**38**, *96% Yield*)

Scheme 2. Synthesis of the halide aryloxide mixed-ligand uranium(IV) complexes (C$_5$Me$_4$R)$_2$U(O-2,6-^{i}Pr$_2$C$_6$H$_3$)(Cl) (R = Me (**35**), Et (**36**)) and (C$_5$Me$_4$R)$_2$U(O-2,6-^{i}Pr$_2$C$_6$H$_3$)(Br) (R = Me (**37**), Et (**38**)) by salt metathesis.

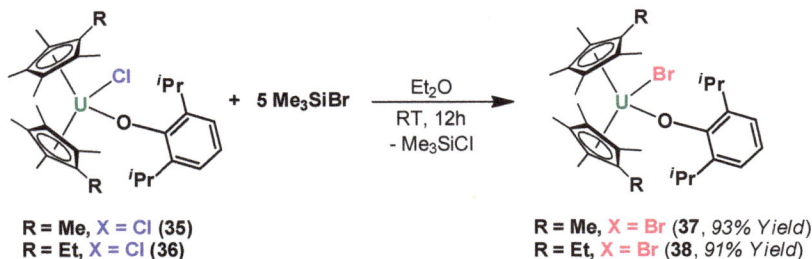

Scheme 3. Synthesis of the bromide aryloxide mixed-ligand uranium(IV) complexes $(C_5Me_4R)_2U(O\text{-}2,6\text{-}{}^iPr_2C_6H_3)(Br)$ (R = Me (**37**), Et (**38**)) by halide exchange with trimethylsilyl bromide.

It is worth mentioning that in the family of the halide aryloxide series of compounds $(C_5Me_5)_2U(O\text{-}2,6\text{-}{}^iPr_2C_6H_3)(X)$ (X = F, Cl, Br, I), the bromide complex is the only member that has not been previously reported [2,8,48]. Comparison of the methyl proton resonance of C_5Me_5 by ^1H NMR spectroscopy reveals an upfield shift in the order I (9.85 ppm) < Br (8.76 ppm) < Cl (7.85 ppm) << F (3.19 ppm). This trend has been observed before in other cyclopentadienyl complexes and it is directly correlated with the π-donating ability of the halide ligand [51]. Similarly, comparison of the methyl peaks of the C_5Me_4Et ligand of **36** (11.9, 8.5, 7.8, 7.8 and 6.6 ppm) and **38** (11.9, 9.11, 9.07, 8.3 and 7.8 ppm) shows the same pattern, where the peak positions of $(C_5Me_4Et)_2U(O\text{-}2,6\text{-}{}^iPr_2C_6H_3)(Br)$ appear downfield from those of $(C_5Me_4Et)_2U(O\text{-}2,6\text{-}{}^iPr_2C_6H_3)(Cl)$.

Compound **37** crystallizes with two asymmetric units in the crystal lattice and the structures are shown in Figure 3. The U–Br bond distances (2.7951(12) and 2.8023(12) Å) are somewhat longer than those of **33** (2.7578(5) Å) and **34** (U(1)–Br(1) = 2.7609(4) and U(1)–Br(2) = 2.7607(4) Å), but still fall within the range of the known neutral U–Br complexes (2.734(1)–2.831(7) Å) [6,7,9,16,33–45]. The U–O bond distances of **37** (2.110(7) and 2.134(7) Å) are similar to those of the known halide and pseudohalide derivatives $(C_5Me_5)_2U(O\text{-}2,6\text{-}{}^iPr_2C_6H_3)(X)$ (X = F (2.124(6) Å), Cl (2.110(5) Å), I (2.114(6) Å), N$_3$ (2.117(5) Å) and CH$_3$ (2.126(4) Å)) [2,48,50]. Likewise, the U–O–C$_{ipso}$ and O–U–Br bond angles of **37** (U–O–C$_{ipso}$ = 166.9(6) and 61.7(6)°; O–U–Br = 104.64(18)° and 104.04(17)°) compare favorably with the known complexes $(C_5Me_5)_2U(O\text{-}2,6\text{-}{}^iPr_2C_6H_3)(X)$ (X = F: U–O–C$_{ipso}$ = 165.4(6)°, O–U–F = 104.1(2)°; X = Cl: U–O–C$_{ipso}$ = 164.0(4)°, O–U–Cl = 102.06(14)°; X = I: U–O–C$_{ipso}$ = 166.6(6)°, O–U–I = 104.87(15)°; X = N$_3$: U–O–C$_{ipso}$ = 165.2(4)°, O–U–N = 100.7(2)° and X = CH$_3$: U–O–C$_{ipso}$ = 163.2(4)°, O–U–CH$_3$ = 98.80(19)°). Finally the U–C$_{Cent}$ bond distances and the C$_{Cent}$–U–C$_{Cent}$ bond angles of **37** (U–C$_{Cent}$ = 2.47(1), 2.46(1) and 2.45(1) Å; C$_{Cent}$–U–C$_{Cent}$ = 134.7(3)° and 133.2(2)°) are comparable with those of other bis(pentamethylcyclopentadienyl) complexes (U–C$_{Cent}$ = 2.420–2.562 Å; C$_{Cent}$–U–C$_{Cent}$ = 126.2–142.6°) [1,2,31,46–50] and to the other $(C_5Me_5)_2U(O\text{-}2,6\text{-}{}^iPr_2C_6H_3)(X)$ derivatives (X = F: U–C$_{Cent}$ = 2.444(2)–2.451(1) Å, C$_{Cent}$–U–C$_{Cent}$ = 135.8–135.9°; X = Cl: U–C$_{Cent}$ = 2.444(3)–2.457(3) Å; C$_{Cent}$–U–C$_{Cent}$ = 133.1–134.5°; X = I: U–C$_{Cent}$ = 2.447(3)–2.456(4) Å; C$_{Cent}$–U–C$_{Cent}$ = 133.5–133.6°; X = N$_3$: U–C$_{Cent}$ = 2.505(3)–2.532(3) Å; C$_{Cent}$–U–C$_{Cent}$ = 133.2°; X = CH$_3$: U–C$_{Cent}$ = 2.466(2)–2.471(3) Å; C$_{Cent}$–U–C$_{Cent}$ = 134.0°) [2,48,50].

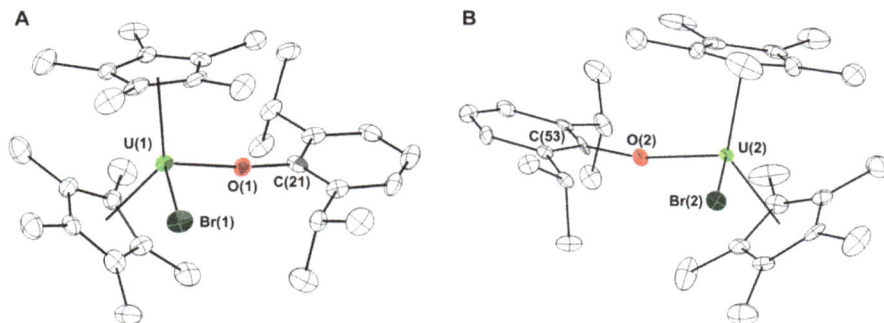

Figure 3. Molecular structures of the two crystallographically-independent $(C_5Me_5)_2U(O\text{-}2,6\text{-}{}^iPr_2C_6H_3)(Br)$ (**37**) complexes (**A** and **B**) with thermal ellipsoids projected at the 30% probability level. Hydrogen atoms have been omitted for clarity. Selected bond distances (Å) and angles (°): U(1)–O(1), 2.110(7); U(2)–O(2), 2.134(7); U(1)–Br(1), 2.7951(12); U(2)–Br(2), 2.8023(12); U(1)–$C_{Cent(1)}$, 2.47(1); U(1)–$C_{Cent(2)}$, 2.46(1); U(2)–$C_{Cent(3)}$, 2.45(1); U(2)–$C_{Cent(4)}$, 2.46(1); Br(1)–U(1)–O(1), 104.64(18); Br(2)–U(2)–O(2), 104.04(17); U(1)–O(1)–C(21), 166.9(6); U(2)–O(2)–C(53), 161.7(6); $C_{Cent(1)}$–U(1)–$C_{Cent(2)}$, 134.7(3); $C_{Cent(3)}$–U(2)–$C_{Cent(4)}$, 133.2(2).

Treatment of the uranium(IV) dihalide complexes **31–34** with one equivalent of KC_8 in toluene at room temperature affords the corresponding uranium(III) chloride and bromide "ate" species $[K(THF)_n][(C_5Me_4R)_2UX_2]$ (**39**, X = Cl, R = Me, n = 1; **40**, X = Cl, R = Et, n = 0.5; **41**, X = Br, R = Me, n = 1; **42**, X = Br, R = Et, n = 0.5) as illustrated in Scheme 4. During the course of the reaction, the vivid red-maroon color of complexes **31–34** changes to an insoluble green solid. After workup, all complexes were isolated in good to excellent yields (**39** (79%), **40** (93%), **41** (90%) and **42** (93%)).

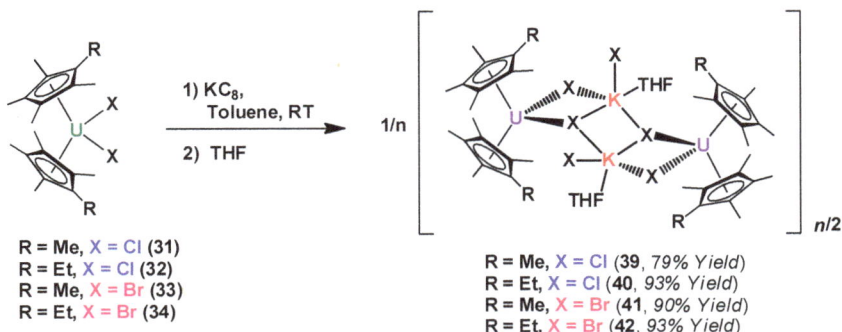

R = Me, X = Cl (**31**)
R = Et, X = Cl (**32**)
R = Me, X = Br (**33**)
R = Et, X = Br (**34**)

R = Me, X = Cl (**39**, *79% Yield*)
R = Et, X = Cl (**40**, *93% Yield*)
R = Me, X = Br (**41**, *90% Yield*)
R = Et, X = Br (**42**, *93% Yield*)

Scheme 4. Synthesis of uranium(III) complexes $[K(THF)][(C_5Me_5)_2UX_2]$ (X = Cl (**39**), Br (**41**)) and $[K(THF)_{0.5}][(C_5Me_4Et)_2UX_2]$ (X = Cl (**40**), Br (**42**)) by KC_8 reduction of the corresponding uranium(IV) dihalides.

Complexes **41** and **42** are rare examples of trivalent uranium bromides. A truncated form of the solid-state structure of compound **41** is shown in Figure 4. The complex is best described as a polymer with potassium ions linking the uranium centers with potassium–bromide bridges. The geometry at each uranium center is pseudo tetrahedral, while that at each potassium is distorted trigonal pyramidal. The K–Br bond lengths of **41** (3.1684(14), 3.1671(14), 3.2577(15) and 3.2477(15) Å) are comparable to those reported for other compounds with K–Br interactions (2.988(11)–3.844(3) Å) [52–66]. The U–Br bond lengths of **41** (U–Br = 2.9126(6) and 2.8953(6) Å) are comparable to the few known U(III) bromides in the literature: $\{[1,3\text{-}(SiMe3)2C5H3]_2UBr\}_2$ (**26**, U–Br = 2.93(1) and 2.94(1) Å) [24], $[U(Br)_2(H_2O)_5(NCMe)_2][Br]$ (U–Br = 3.074(4) Å) [37], $[Li(THF)_4]_2[(C_2H_{20}B_9)_2U(Br)(THF)]$ (**25**, U–Br = 2.883(2) Å) [25], $[1,3\text{-}(SiMe3)2C5H3]_2U(Br)(CN^tBu)_2$ (**30**, U–Br = 2.8761(10) Å) [26] and $[UBr_3(DME)_2]_2$ (U–Br = 2.898(2), 2.887(2), 3.016(2) and 3.098(2) Å) [40] and are longer than those of **33** (U–Br = 2.7578(5) Å) and **34** (U–Br = 2.7609(4) and 2.7607(4) Å).

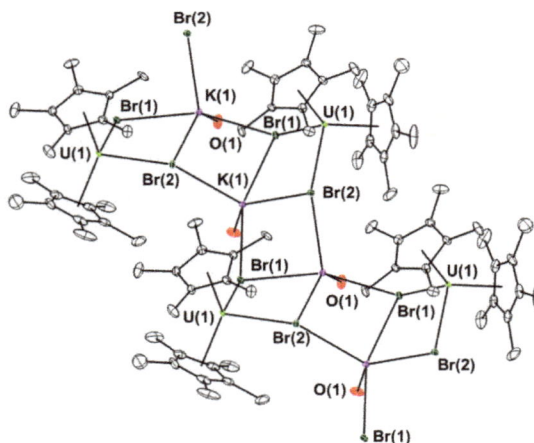

Figure 4. Solid state structure (30% probability ellipsoids) of a tetrameric unit of $[K(THF)][(C_5Me_5)_2UBr_2]$ (**41**) Hydrogen atoms and the carbon atoms of THF molecules are omitted for clarity. Selected bond distances (Å) and bond angles (°): U(1)–Br(1), 2.9126(6); U(1)–Br(2), 2.8953(6); K(1)–Br(1), 3.1681(14), 3.2577(15); K(1)–Br(2), 3.1671(14), 3.2477(15); K(1)–O(1), 2.641(6); U(1)–C$_{Cent(1)}$, 2.495(7); U(1)–C$_{Cent(2)}$, 2.485(7); Br(1)–U(1)–Br(2), 88.841(19); Br(1)–K(1)–Br(2), 114.32(15), 87.50(4), 87.35(3), 77.35(3); Br(1)–K(1)–O(1), 85.26(14), 93.41(16); Br(2)–K(1)–O(1), 100.91(18), 132.49(19); C$_{Cent(1)}$–U(1)–C$_{Cent(2)}$, 136.6(3).

In addition, the U–C$_{Cent}$ bond distances of **41** (2.495(7) and 2.485(7) Å) are comparable to those of **33** (2.438(4) Å), **34** (2.444(2) Å), and **37** (2.45(1), 2.46(1) and 2.47(1) Å). Finally, the C$_{Cent}$–U–C$_{Cent}$ bond angle of **41** (136.6(3)°) is similar to those of compounds **33** (137.57(15)°), **34** (138.50(8)°) and **37** (137.7(3)°), as well as other $(C_5Me_5)_2U(X)_2$ complexes (121.1–146.1°) [29,67–81].

Few other similar types of complexes have been reported in the past. For instance, Marks and co-workers reported that Na/Hg reduction of $(C_5Me_5)_2UCl_2$ formed $[Na(THF)_2][(C_5Me_5)_2UCl_2]$ [72], but no structure was presented and two other examples reported by Lappert and co-workers show monomeric motifs for $[M(THF)_2][\{1,3\text{-}(Me_3Si)_2C_5H_3\}_2UCl_2]$ (M = Li or Na) [7]. The polymeric structure of **41** can be explained by the presence of the considerably larger potassium ions, *versus* lithium and sodium adducts, which yield monomeric complexes.

As shown in Scheme 5, the synthetic utility of the "ate" complexes $[K(THF)][(C_5Me_4R)_2UX_2]$ (**39–42**) was demonstrated by their reaction with one equivalent of $Na[N(SiMe_3)_2]$ in toluene to afford $(C_5Me_4R)_2U[N(SiMe_3)_2]$ (R = Me, (**43**), Et (**44**)) in good yields (76%–86%).

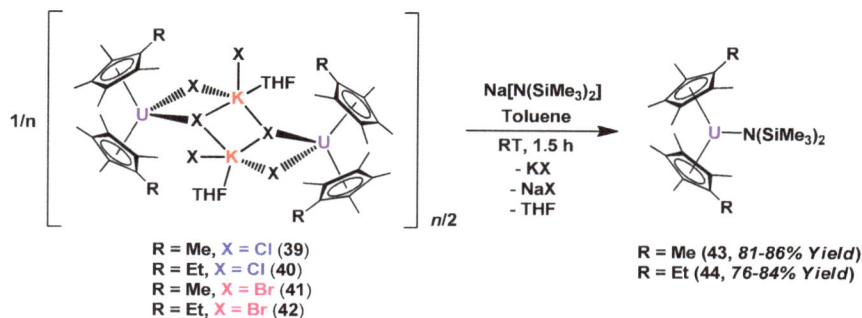

R = Me, X = Cl (39)
R = Et, X = Cl (40)
R = Me, X = Br (41)
R = Et, X = Br (42)

R = Me (43, *81-86% Yield*)
R = Et (44, *76-84% Yield*)

Scheme 5. Preparation of neutral uranium(III) complexes $(C_5Me_4R)_2U[N(SiMe_3)_2]$ (R = Me, (**43**), Et (**44**)).

3. Experimental Section

3.1. General Synthetic Considerations

Unless otherwise noted, all reactions and manipulations were performed at ambient temperatures in a recirculating Vacuum Atmospheres NEXUS model inert atmosphere (N_2) drybox equipped with a 40 CFM Dual Purifier NI–Train. Glassware was dried overnight at 150 °C before use. All NMR spectra were obtained using a Bruker Avance 400 MHz spectrometer at room temperature. Chemical shifts for ^1H NMR spectra were referenced to solvent impurities [82]. Melting points were determined with a Mel-Temp II capillary melting point apparatus equipped with a Fluke 50 S K/J thermocouple using capillary tubes flame-sealed under N_2; values are uncorrected. Elemental Analyses were performed by ALS Environmental (Tucson, AZ, USA) or Atlantic Microlab, Inc. (Norcross, GA, USA).

3.2. Materials

Unless otherwise noted, reagents were purchased from commercial suppliers and used without further purification. Celite (Aldrich, Saint Louis, MO, USA), neutral alumina (Aldrich), and 3 Å molecular sieves (Aldrich) were dried under dynamic vacuum at 220 °C for 48 h prior to use. All solvents (Aldrich), benzene-d_6 and THF-d_8 (Cambridge Isotopes, Tewksbury, MA, USA) were purchased anhydrous and dried over KH for 48 h, passed through a column of activated alumina, and stored over activated 3 Å molecular sieves prior to use. HO-2,6-$^iPr_2C_6H_3$ (Aldrich) was dried with activated 3 Å molecular sieves prior to use. Potassium metal (Aldrich) was rinsed with hexane, dried and used immediately. Me_3SiBr (Aldrich), graphite (Aldrich) and $Na[N(SiMe_3)_2]$ (Aldrich) were used as received. $(C_5Me_5)_2UCl_2$ (**31**) [3], $(C_5Me_4Et)_2UCl_2$ (**32**) [31], and $K[O-2,6-^iPr_2C_6H_3]$ [30] were prepared according to literature procedures.

3.3. Caution

Depleted uranium (primary isotope ^{238}U) is a weak α-emitter (4.197 MeV) with a half-life of 4.47×10^9 years; manipulations and reactions should be carried out in monitored fume-hoods or in an inert atmosphere drybox in a radiation laboratory equipped with α- and β-counting equipment.

3.4. Synthetic Procedures

$(C_5Me_5)_2UBr_2$ (33): A 20 mL scintillation vial was charged with a stir bar, $(C_5Me_5)_2UCl_2$ (**31**) (0.301 g, 0.518 mmol) and Et_2O (5 mL). To this solution, Me_3SiBr (0.397 g, 2.59 mmol) was added dropwise. The reaction mixture was stirred at ambient temperature for 12 h and the volatiles were removed under reduced pressure. Et_2O (5 mL) and Me_3SiBr (0.397 g, 2.59 mmol) were added again and the solution was left stirring at ambient temperature for another 12 h and the volatiles were removed under reduced pressure. Et_2O (5 mL) and Me_3SiBr (0.397 g, 2.59 mmol) were added for the third and final time and the reaction mixture was stirred at ambient temperature for 12 h and the volatiles were removed under reduced pressure. The resulting residue was dissolved in toluene (15 mL) and filtered through a Celite-padded coarse-porosity fritted filter. The red colored filtrate was collected and the volatiles were removed under reduced pressure to give $(C_5Me_5)_2UBr_2$ (**33**) as a red solid (0.333 g, 0.498 mmol, 96%). Single crystals suitable for X-ray diffraction were obtained by slow evaporation of a toluene solution at ambient temperature. ^1H NMR (benzene-d_6, 298 K): δ 15.70 (s, 30H, $v_{1/2}$ = 111.8 Hz, C$_5Me_5$). m.p. 276–278 °C. Anal. calcd. for $C_{20}H_{30}Br_2U$ (mol. wt. 668.29): C, 35.94; H, 4.52. Found: C, 35.20, H, 4.42.

(C$_5$Me$_4$Et)$_2$UBr$_2$ (34): A 20 mL scintillation vial was charged with a stir bar, $(C_5Me_4Et)_2UCl_2$ (**32**) (0.500 g, 0.823 mmol) and Et$_2$O (5 mL). To this solution Me$_3$SiBr (0.630 g, 4.12 mmol) was added dropwise. The reaction mixture was stirred at ambient temperature for 12 h and the volatiles were removed under reduced pressure. Et$_2$O (5 mL) and Me$_3$SiBr (0.630 g, 4.12 mmol) were added again and the solution was left stirring at ambient temperature for another 12 h and the volatiles were removed under reduced pressure. Et$_2$O (5 mL) and Me$_3$SiBr (0.630 g, 4.12 mmol) were added for the third and final time and the reaction mixture was stirred at ambient temperature for 12 h and the volatiles were removed under reduced pressure. The resulting residue was dissolved in toluene (15 mL) and filtered through a Celite-padded coarse-porosity fritted filter. The red colored filtrate was collected and the volatiles were removed under reduced pressure to give $(C_5Me_4Et)_2UBr_2$ (**34**) as a red solid (0.515 g, 0.740 mmol, 90%). Single crystals suitable for X-ray diffraction were obtained by slow evaporation of a toluene solution at ambient temperature. ^1H NMR (benzene-d_6, 298 K): δ 25.06 (s, 6H, $v_{1/2}$ = 43.3 Hz, CH$_2$CH$_3$), 15.57 (s, 12H, $v_{1/2}$ = 95.7 Hz, CH$_3$), 13.06 (s, 12H, $v_{1/2}$ = 93.8 Hz, CH$_3$), 0.31 (s, 4H, $v_{1/2}$ = 112.5 Hz CH$_2$). m.p. 160–162 °C. Anal. calcd. for $C_{22}H_{34}Br_2U$: C, 37.95; H, 4.92. Found: C, 37.59; H, 5.05.

(C$_5$Me$_5$)$_2$U(O-2,6-^{i}Pr$_2$C$_6$H$_3$)(Cl) (35): A 20 mL scintillation vial was charged with a stir bar, $(C_5Me_5)_2UCl_2$ (**29**) (0.149 g, 0.257 mmol), K[O-2,6-^{i}Pr$_2$C$_6$H$_3$] (0.056 g, 0.257 mmol) and THF (5 mL). The reaction mixture was stirred at ambient temperature for 35 min and the volatiles were removed under reduced pressure. The resulting residue was dissolved in toluene (10 mL) and filtered through a Celite-padded coarse-porosity fritted filter. The red filtrate was collected and the volatiles were removed under reduced pressure to give $(C_5Me_5)_2U(O$-2,6-^{i}Pr$_2C_6H_3)(Cl)$ (**35**) as a red solid (0.170 g, 0.236 mmol, 92%). The ^1H NMR spectrum collected in benzene-d_6 was consistent with data previously reported for complex **35** [8]. ^1H NMR (C$_6$D$_6$, 298 K): δ 8.37 (d, 1H, m-Ar-H), 8.35 (d, 1H, m-Ar-H), 7.85 (s, 30H, C$_5$(CH$_3$)$_5$), 7.06 (t, 1H, p-Ar-H), −6.01 (s, 6H, CH(CH$_3$)$_2$), −12.37 (s, 6H, CH(CH$_3$)$_2$), −35.74 (m, 1H, CH(CH$_3$)$_2$), −44.65 (m, 1H, CH(CH$_3$)$_2$).

(C$_5$Me$_4$Et)$_2$U(O-2,6-^{i}Pr$_2$C$_6$H$_3$)(Cl) (36): A 20 mL scintillation vial was charged with a stir bar, $(C_5Me_4Et)_2UCl_2$ (**32**) (0.300 g, 0.494 mmol), K[O-2,6-^{i}Pr$_2$C$_6$H$_3$] (0.107 g, 0.494 mmol) and THF (5 mL). The reaction mixture was stirred at ambient temperature for 35 min and the volatiles were removed under reduced pressure. The resulting residue was dissolved in toluene (5 mL) and filtered through a Celite-padded coarse-porosity fritted filter. The red colored filtrate was collected and the volatiles were removed under reduced pressure. The resulting red oil was dissolved in hexane (5 mL) and passed through a Celite-padded coarse-porosity fritted filter. The red filtrate was collected, the volatiles were removed under reduced pressure and the oily product triturated three times with acetonitrile (5 mL) to give $(C_5Me_4Et)_2U(O$-2,6-^{i}Pr$_2C_6H_3)(Cl)$ (**36**) as a red solid (0.332 g, 0.443 mmol, 90%). ^1H NMR (benzene-d_6, 298 K): δ 11.87 (s, 6H, $v_{1/2}$ = 13.6 Hz, CH$_3$), 8.49 (s, 6H, $v_{1/2}$ = 5.9 Hz, CH$_3$), 8.26 (d, 1H, J_{HH} = 8.5 Hz, m-Ar-CH), 8.15 (d, 1H, J_{HH} = 8.5 Hz, m-Ar-CH), 7.82 (s, 6H, $v_{1/2}$ = 13.6 Hz, CH$_3$), 7.78 (s, 6H, $v_{1/2}$ = 13.6 Hz, CH$_3$), 6.91 (t, 1H, J_{HH} = 8.5 Hz, p-Ar-H), 6.58 (s, 6H, $v_{1/2}$ = 4.7 Hz, CH$_3$), −0.55 (s, 2H, $v_{1/2}$ = 29 Hz, CH$_2$), −1.37 (s, 2H, $v_{1/2}$ = 29 Hz, CH$_2$), −5.85 (d, 6H, J_{HH} = 3.6 Hz, CH(CH$_3$)$_2$), −12.75 (s, 6H, $v_{1/2}$ = 6.9 Hz, CH(CH$_3$)$_2$), −35.67 (m, 1H, CH(CH$_3$)$_2$), −46.81 (s, 1H, $v_{1/2}$ = 28.6 Hz, CH(CH$_3$)$_2$). m.p. 92–95 °C. Anal. calcd. for $C_{34}H_{51}ClOU$ (mol. wt. 749.25): C, 54.50; H, 6.86. Found: C, 54.08; H, 7.16.

(C$_5$Me$_5$)$_2$U(O-2,6-^{i}Pr$_2$C$_6$H$_3$)(Br) (37): Method A: From (C$_5$Me$_5$)$_2$UBr$_2$. A 20 mL scintillation vial was charged with a stir bar, (C$_5$Me$_5$)$_2$UBr$_2$ (**33**) (0.100 g, 0.150 mmol), K[O-2,6-^{i}Pr$_2$C$_6$H$_3$] (0.0324 g, 0.150 mmol) and THF (5 mL). The reaction mixture was stirred at ambient temperature for 1 h and the volatiles were removed under reduced pressure. The resulting residue was dissolved in toluene (5 mL) and filtered through a Celite-padded coarse-porosity fritted filter. The red filtrate was collected and the volatiles were removed under reduced pressure to give (C$_5$Me$_5$)$_2$U(O-2,6-^{i}Pr$_2$C$_6$H$_3$)(Br) (**37**) as a red solid (0.113 g, 0.148 mmol, 99%).

Method B: From (C$_5$Me$_5$)$_2$U(O-2,6-^{i}Pr$_2$C$_6$H$_3$)(Cl). A 20 mL scintillation vial was charged with a stir bar, (C$_5$Me$_5$)$_2$U(O-2,6-^{i}Pr$_2$C$_6$H$_3$)(Cl) (**35**) (0.102 g, 0.141 mmol), Et$_2$O (5 mL) and Me$_3$SiBr (0.108 g, 0.707 mmol). The reaction mixture was stirred at ambient temperature for 12 h and the volatiles were removed under reduced pressure. The resulting residue was dissolved in toluene (5 mL) and filtered through a Celite-padded coarse-porosity fritted filter. The red filtrate was collected and the volatiles were removed under reduced pressure to give (C$_5$Me$_5$)$_2$U(O-2,6-^{i}Pr$_2$C$_6$H$_3$)(Br) (**37**) as a red solid (0.100 g, 0.131 mmol, 93%). Single crystals suitable for X-ray diffraction were obtained from a 50 mM toluene solution at −35 °C. ^{1}H NMR (benzene-d_6, 298 K): δ 9.41 (d, 1H, J_{HH} = 8.6 Hz, *m*-Ar-*H*), 9.07 (d, 1H, J_{HH} = 8.6 Hz, *m*-Ar-*H*), 8.76 (s, 30H, C$_5$(C*H$_3$*)$_5$), 7.40 (t, 1H, J_{HH} = 8.7 Hz, *p*-Ar-*H*), −5.89 (d, 6H, J_{HH} = 3.8 Hz, CH(C*H$_3$*)$_2$), −11.68 (s, 6H, CH(C*H$_3$*)$_2$), −35.67 (m, 1H, C*H*(CH$_3$)$_2$), −42.99 (m, 1H, C*H*(CH$_3$)$_2$). m.p. 214–216 °C. Anal. calcd. for C$_{32}$H$_{47}$BrOU (mol. wt. 765.65): C, 50.20; H, 6.19. Found: C, 49.46, H, 5.98.

(C$_5$Me$_4$Et)$_2$U(O-2,6-^{i}Pr$_2$C$_6$H$_3$)(Br) (38): Method A: From (C$_5$Me$_4$Et)$_2$UBr$_2$. A 20 mL scintillation vial was charged with a stir bar, (C$_5$Me$_4$Et)$_2$UBr$_2$ (**34**) (0.100 g, 0.144 mmol), K[O-2,6-^{i}Pr$_2$C$_6$H$_3$] (0.0311 g, 0.144 mmol) and THF (5 mL). The reaction mixture was stirred at ambient temperature for 1 h and the volatiles were removed under reduced pressure. The resulting residue was dissolved in toluene (5 mL) and filtered through a Celite-padded coarse-porosity fritted filter. The red filtrate was collected and the volatiles were removed under reduced pressure. The resulting red oil was dissolved in hexane (5 mL) and passed through a Celite-padded coarse-porosity fritted filter. The red filtrate was collected, the volatiles removed under reduced pressure and the oily product triturated three times with acetonitrile (5 mL) to give (C$_5$Me$_4$Et)$_2$U(O-2,6-^{i}Pr$_2$C$_6$H$_3$)(Br) (**38**) as a red solid (0.110 g, 0.139 mmol, 96%).

Method B: From (C$_5$Me$_4$Et)$_2$U(O-2,6-^{i}Pr$_2$C$_6$H$_3$)(Cl). A 20 mL scintillation vial was charged with a stir bar, (C$_5$Me$_4$Et)$_2$U(O-2,6-^{i}Pr$_2$C$_6$H$_3$)(Cl) (**36**) (0.100 g, 0.133 mmol), Et$_2$O (5 mL) and Me$_3$SiBr (0.102 g, 0.667 mmol). The reaction mixture was stirred at ambient temperature for 12 h and the volatiles were removed under reduced pressure. The resulting residue was dissolved in toluene (5 mL) and filtered through a Celite-padded coarse-porosity fritted filter. The red filtrate was collected and the volatiles were removed under reduced pressure. The resulting red oil was triturated with acetonitrile (3 mL) to give (C$_5$Me$_4$Et)$_2$U(O-2,6-^{i}Pr$_2$C$_6$H$_3$)(Br) (**38**) as a red solid (0.096 g, 0.121 mmol, 91%). ^{1}H NMR (benzene-d_6, 298 K): δ 11.92 (s, 6H, $v_{1/2}$ = 14.0 Hz, C*H$_3$*), 9.11 (s, 6H, $v_{1/2}$ = 5.3 Hz, C*H$_3$*), 9.07 (s, 6H, $v_{1/2}$ = 5.3 Hz, C*H$_3$*), 8.96 (d, 1H, J_{HH} = 8.3 Hz, *m*-Ar-C*H*), 8.33 (s, 6H, $v_{1/2}$ = 13.6 Hz, C*H$_3$*), 7.77 (s, 6H, $v_{1/2}$ = 5.4 Hz, C*H$_3$*), 7.23 (t, 1H, J_{HH} = 8.7 Hz, *p*-Ar-*H*), 0.35 (s, 2H, $v_{1/2}$ = 27 Hz, C*H$_2$*), −0.52 (s, 2H, $v_{1/2}$ = 31 Hz, C*H$_2$*), −5.72 (d, 6H, J_{HH} = 3.2 Hz, CH(C*H$_3$*)$_2$), −12.15 (s, 6H, $v_{1/2}$ = 7.0 Hz, CH(C*H$_3$*)$_2$), −35.62 (m, 1H, C*H*(CH$_3$)$_2$), −45.56 (m, 1H, C*H*(CH$_3$)$_2$). m.p. 69–72 °C. Anal. calcd. for C$_{34}$H$_{51}$BrOU (mol. wt. 793.70): C, 51.45; H, 6.48. Found: C, 51.51; H, 6.65.

KC$_8$: This is a modification of literature procedures [83,84]: A 20 mL scintillation vial was charged with potassium metal (0.040 g, 1.03 mmol) and graphite (0.099 g, 8.26 mmol) in an nitrogen atmosphere glove box. The reaction mixture was stirred with a spatula at 100 °C until the solid was completely bronze in color (~5 min), yielding quantitative solid KC$_8$ (0.140 g, 1.03 mmol, 100%).

[K(THF)][(C$_5$Me$_5$)$_2$UCl$_2$] (39): A 20 mL scintillation vial was charged with a stir bar, (C$_5$Me$_5$)$_2$UCl$_2$ (**31**) (0.600 g, 1.03 mmol), KC$_8$ (0.140 g, 1.03 mmol), and toluene (10 mL). The reaction mixture was stirred at ambient temperature for 4 h, yielding a green precipitate. The reaction mixture was filtered through a Celite-padded coarse-porosity fritted filter. The toluene-insoluble green solid, which was collected on the Celite-padded fritted filter was dissolved by adding THF

(15 mL). The green filtrate was collected and the volatiles were removed under reduced pressure to give [K(THF)][(C$_5$Me$_5$)$_2$UCl$_2$] (**39**) as a green solid (0.560 g, 0.405 mmol, 79%). Single crystals suitable for X-ray diffraction were grown from a saturated THF solution at −35 °C. ^{1}H-NMR (THF-d_8, 298 K): δ 3.62 (8H, m, THF-CH_2), 1.78 (8H, m, THF-CH_2), −5.49 (60H, $v_{1/2}$ = 52 Hz, C$_5$(CH_3)$_5$). m.p. 230.0 °C (Decomp). Anal. calcd. for C$_{48}$H$_{76}$Cl$_4$K$_2$O$_2$U$_2$ (mol. wt. 1381.18): C, 41.74; H, 5.55. Found: C, 41.41; H, 5.30.

[K(THF)$_{0.5}$][(C$_5$Me$_4$Et)$_2$UCl$_2$] (40): A 20 mL scintillation vial was charged with a stir bar, (C$_5$Me$_4$Et)$_2$UCl$_2$ (**32**) (0.500 g, 0.823 mmol), KC$_8$ (0.111 g, 0.823 mmol), and toluene (10 mL). The reaction mixture was stirred at ambient temperature for 2 h, yielding a green precipitate. The reaction mixture was filtered through a Celite-padded coarse-porosity fritted filter. The insoluble green solid on the Celite-padded coarse-porosity fritted filter. The toluene-insoluble green solid, which was collected on the Celite-padded frit was dissolved by adding THF (15 mL). The green filtrate was collected and the volatiles were removed under reduced pressure to give [K(THF)$_{0.5}$][(C$_5$Me$_4$Et)$_2$UCl$_2$] (**40**) as a green solid (0.549 g, 0.764 mmol, 93%). ^{1}H NMR (THF-d_8, 298 K): d 11.39 (s, 6H, $v_{1/2}$ = 29.0 Hz, CH_2CH$_3$), 3.62 (m, 4H, THF-CH_2), 1.79 (m, 4H, THF-CH_2), −4.61 (s, 12H, $v_{1/2}$ = 58.6 Hz, CH_3), −5.79 (s, 4H, $v_{1/2}$ = 87.3 Hz, CH_2), −6.43 (s, 12H, $v_{1/2}$ = 60.0 Hz, CH_3). m.p. 189–192 °C (Decomp). Anal. calcd. for C$_{48}$H$_{76}$Cl$_4$K$_2$OU$_2$ (mol. wt. 1365.18): C, 42.43; H, 5.61. Found: C, 42.54; H, 6.04.

[K(THF)][(C$_5$Me$_5$)$_2$UBr$_2$] (41): A 20 mL scintillation vial was charged with a stir bar, (C$_5$Me$_5$)$_2$UBr$_2$ (**33**) (0.263 g, 0.393 mmol), KC$_8$ (0.0530 g, 0.393 mmol), and toluene (10 mL). The reaction mixture was stirred at ambient temperature for 3 h, yielding a green precipitate. The reaction mixture was filtered through a Celite-padded coarse-porosity fritted filter. The toluene-insoluble green solid, which was collected on the Celite-padded fritted filter was dissolved by adding THF (15 mL). The green filtrate was collected and the volatiles were removed under reduced pressure to give [K(THF)][(C$_5$Me$_5$)$_2$UBr$_2$] (**41**) as a green solid (0.275 g, 0.353 mmol, 90%). Single crystals suitable for X-ray diffraction were obtained by slow evaporation of a THF solution at ambient temperature. ^{1}H NMR (THF-d_8, 298 K): δ 3.62 (m, 4H, THF-CH_2), 1.78 (m, 4H, THF-CH_2), −3.42 (s, 30H, $v_{1/2}$ = 84.5 Hz, C$_5$(CH_3)$_5$). m.p. 190–192 °C (Decomp). Anal. calcd. for C$_{48}$H$_{76}$Br$_4$K$_2$O$_2$U$_2$ (mol. wt. 1558.99): C, 36.98; H, 4.91. Found: C, 37.00; H, 4.98.

[K(THF)$_{0.5}$][(C$_5$Me$_4$Et)$_2$UBr$_2$] (42): A 20 mL scintillation vial was charged with a stir bar, (C$_5$Me$_4$Et)$_2$UBr$_2$ (**34**) (0.300 g, 0.431 mmol), KC$_8$ (0.058 g, 0.431 mmol), and toluene (10 mL). The reaction mixture was stirred at ambient temperature for 2 h, yielding a green precipitate. The reaction mixture was filtered through a Celite-padded coarse-porosity fritted filter. The toluene-insoluble green solid, which was collected on the Celite-padded fritted filter was dissolved by adding in THF (10 mL). The green colored filtrate was collected and the volatiles were removed under reduced pressure. The resulting solid was triturated twice with hexane (3 mL) to give [K(THF)$_{0.5}$][(C$_5$Me$_4$Et)$_2$UBr$_2$] (**42**) as a green solid (0.323 g, 0.400 mmol, 93%). ^{1}H NMR (THF-d_8, 298 K): δ 13.61 (s, 6H, $v_{1/2}$ = 76.5 Hz, CH_2CH$_3$), 3.62 (m, 4H, THF-CH_2), 1.79 (m, 4H, THF-CH_2), −3.10 (s, 12H, $v_{1/2}$ = 131.9 Hz, CH_3), −4.23 (s, 12H, $v_{1/2}$ = 135.8 Hz, CH_3), −4.47 (s, 4H, CH_2). m.p. 189–191 °C (Decomp). Anal. calcd. for C$_{48}$H$_{76}$Br$_4$K$_2$OU$_2$ (mol. wt. 1542.99): C, 37.36; H, 4.96. Found: C, 37.61; H, 5.23.

(C$_5$Me$_5$)$_2$U[N(SiMe$_3$)$_2$] (43): Method A: From [K(THF)][(C$_5$Me$_5$)$_2$UCl$_2$]. A 20 mL scintillation vial was charged with a stir bar, [K(THF)][(C$_5$Me$_5$)$_2$UCl$_2$] (**39**) (0.100 g, 0.148 mmol), Na[N(SiMe$_3$)$_2$] (0.027 g, 0.148 mmol), and toluene (5 mL). The reaction mixture was stirred at ambient temperature for 1 h, and the volatiles were removed under reduced pressure. The resulting residue was dissolved in hexane (5 mL) and filtered through a Celite-padded coarse-porosity fritted filter. The gray filtrate was collected and the volatiles were removed under reduced pressure to give (C$_5$Me$_5$)$_2$U[N(SiMe$_3$)$_2$] (**43**) as a gray solid (0.085 g, 0.127 mmoles, 86%).

Method B: From [K(THF)][(C$_5$Me$_5$)$_2$UBr$_2$]. A 20 mL scintillation vial was charged with a stir bar, [K(THF)][(C$_5$Me$_5$)$_2$UBr$_2$] (**41**) (0.100 g, 0.128 mmol), Na[N(SiMe$_3$)$_2$] (0.024 g, 0.128 mmol), and toluene (5 mL). The reaction mixture was stirred at ambient temperature for 1 h and the volatiles were removed under reduced pressure. The resulting residue was dissolved in hexane (5 mL) and filtered through a Celite-padded coarse-porosity fritted filter. The gray filtrate was collected and the volatiles were removed under reduced pressure to give $(C_5Me_5)_2U[N(SiMe_3)_2]$ (**43**) as a gray solid (0.065 g, 0.097 mmol, 76%). The ^1H NMR spectrum collected in benzene-d_6 was consistent with data previously reported for complex **43** [85]. ^1H NMR (benzene-d_6, 298 K): δ −5.69 (30H, s, $C_5(CH_3)_5$), −25.57 (18H, s, $Si(CH_3)_3$).

$(C_5Me_4Et)_2U[N(SiMe_3)_2]$ (**44**): Method A: From $[K(THF)_{0.5}][(C_5Me_4Et)_2UCl_2]$. A 20 mL scintillation vial was charged with a stir bar, $[K(THF)_{0.5}][(C_5Me_4Et)_2UCl_2]$ (**40**) (0.100 g, 0.139 mmol), $Na[N(SiMe_3)_2]$ (0.026 g, 0.139 mmol), and toluene (5 mL). The reaction mixture was stirred at ambient temperature for 1 h, and the volatiles were removed under reduced pressure. The resulting residue was dissolved in hexane (5 mL) and filtered through a Celite-padded coarse-porosity fritted filter. The gray filtrate was collected and the volatiles were removed under reduced pressure to give $(C_5Me_4Et)_2U[N(SiMe_3)_2]$ (**44**) as a gray solid (0.079 g, 0.113 mmol, 81%).

Method B: From $[K(THF)_{0.5}][(C_5Me_4Et)_2UBr_2]$. A 20 mL scintillation vial was charged with a stir bar, $[K(THF)_{0.5}][(C_5Me_4Et)_2UBr_2]$ (**42**) (0.100 g, 0.124 mmol), $Na[N(SiMe_3)_2]$ (0.023 g, 0.124 mmol), and toluene (5 mL). The reaction mixture was stirred at ambient temperature for 1 h, and the volatiles were removed under reduced pressure. The resulting residue was dissolved in hexane (5 mL) and filtered through a Celite-padded coarse-porosity fritted filter. The gray filtrate was collected and the volatiles were removed under reduced pressure to give $(C_5Me_4Et)_2U[N(SiMe_3)_2]$ (**44**) as a gray solid (0.073 g, 0.104 mmol, 84%). The ^1H NMR spectrum collected in benzene-d_6 was consistent with data previously reported for complex **44** [85]. ^1H NMR (benzene-d_6, 298 K): δ 10.28 (6H, s, CH_2CH_3), 4.03 (4H, s, CH_2CH_3), −5.62 (12H, s, CH_3), −6.52 (12H, s, CH_3), −25.48 (18H, s, $Si(CH_3)_3$).

X-ray Crystallography: Data for **33**, **34**, and **41** were collected on a Bruker D8 Quest diffractometer, with a CMOS detector in shutterless mode. The crystals were cooled to 100 K employing an Oxford Cryostream liquid nitrogen cryostat. Data for **37** were collected on a Bruker D8 diffractometer, with an APEX II CCD detector. The crystal was cooled to 140 K using a Bruker Kryoflex liquid nitrogen cryostat. Both data collections employed graphite monochromatized MoKα (λ = 0.71073 Å) radiation. Cell indexing, data collection, integration, structure solution, and refinement were performed using Bruker and SHELXTL software [86–89]. CIF files representing the X-ray crystal structures of **33**, **34**, **37** and **41** (Supplementary files) have been submitted to the Cambridge Crystallographic Database as submission numbers CCDC 1428938–1428941.

4. Conclusions

In summary, we have expanded the family of organouranium bromides with the preparation of the tetravalent and trivalent uranium bromide complexes $(C_5Me_4R)_2UBr_2$, $(C_5Me_4R)_2U(O-2,6-^iPr_2C_6H_3)(Br)$, and $[K(THF)][(C_5Me_4R)_2UBr_2]$ (R = Me, Et). The uranium(IV) compounds were easily accessed by treating the corresponding chloride analogues with excess Me_3SiBr. The uranium(IV) chloride complexes were all easily obtained from UCl_4 and their halide substitution chemistry with Me_3SiBr constitutes a practical and efficient route to new bromide derivatives, even in the presence of alkoxide ligands.

The reduction of $(C_5Me_4R)_2UX_2$ complexes with one equivalent of KC_8 allowed for the isolation of the uranium(III) species $[K(THF)][(C_5Me_5)_2UX_2]$ and $[K(THF)_{0.5}][(C_5Me_4Et)_2UX_2]$, which can be converted to the neutral uranium(III) complexes $(C_5Me_4R)_2U[N(SiMe_3)_2]$ upon treatment with $Na[N(SiMe_3)_2]$. Halide and anion compatibility can be very important in the synthesis of organouranium complexes, and this work provides new options for the actinide synthetic chemistry toolbox.

Supplementary Materials: Supplementary materials can be accessed at www.mdpi.com/2304-6740/4/1/1/s1.

Acknowledgments: For financial support of this work, we acknowledge the U.S. Department of Energy through the Los Alamos National Laboratory (LANL) Laboratory Directed Research and Development (LDRD) Program and the Los Alamos National Laboratory Glenn Theodore Seaborg Institute for Transactinium Science (PD Fellowship to Alejandro G. Lichtscheidl), the Advanced Fuels Campaign Fuel Cycle Research & Development Program

(Alejandro G. Lichtscheidl and Andrew T. Nelson), the Office of Workforce Development for Teachers and Scientists, Office of Science Graduate Student Research (SCGSR) program (Fellowship to Justin K. Pagano), and the Office of Basic Energy Sciences, Heavy Element Chemistry program (Jaqueline L. Kiplinger, Brian L. Scott, materials and supplies). The SCGSR program is administered by the Oak Ridge Institute for Science and Education for the DOE (contract DE-AC05-06OR23100). Los Alamos National Laboratory is operated by Los Alamos National Security, Limited Liability Corporation, for the National Nuclear Security Administration of U.S. Department of Energy (contract DE-AC52-06NA25396).

Author Contributions: Alejandro G. Lichtscheidl performed the experiments and drafted the manuscript. Justin K. Pagano solved the crystal structure of compound 37. Brian L. Scott solved the crystal structures of compounds 33, 34 and 41 and drafted the X-ray crystallography experimental section of the manuscript. Andrew T. Nelson and Jaqueline L. Kiplinger conceived the project and supervised the work. All authors discussed the results, interpreted the data, and contributed to the preparation of the final manuscript.

Conflicts of Interest: The authors declare no conflict of interest.

References

1. Spirlet, M.R.; Rebizant, J.; Apostolidis, C.; Kanellakopulos, B. Bis(cyclopentadienyl) actinide(IV) compounds. I. The structure of dichlorobis(pentamethyl-η^5-cyclopentadienyl)uranium(IV) and dichlorobis(pentamethyl-η^5-cyclopentadienyl)thorium(IV). *Acta Crystallogr. Sect. C Cryst. Commun.* **1992**, *48*, 2135–2137. [CrossRef]

2. Graves, C.R.; Schelter, E.J.; Cantat, T.; Scott, B.L.; Kiplinger, J.L. A mild protocol to generate uranium(IV) mixed-ligand metallocene complexes using copper(I) iodide. *Organometallics* **2008**, *27*, 5371–5378. [CrossRef]

3. Fagan, P.J.; Manriquez, J.M.; Maatta, E.A.; Seyam, A.M.; Marks, T.J. Synthesis and properties of bis(pentamethylcyclopentadienyl) actinide hydrocarbyls and hydrides. A new class of highly reactive f-element organometallic compounds. *J. Am. Chem. Soc.* **1981**, *103*, 6650–6667. [CrossRef]

4. Lukens, W.W., Jr.; Beshouri, S.M.; Blosch, L.L.; Stuart, A.L.; Andersen, R.A. Preparation, solution behavior, and solid-state structures of (1,3-$R_2C_5H_3$)$_2UX_2$, where R is CMe$_3$ or SiMe$_3$ and X is a one-electron ligand. *Organometallics* **1999**, *18*, 1235–1246. [CrossRef]

5. Zi, G.; Jia, L.; Werkema, E.L.; Walter, M.D.; Gottfriedsen, J.P.; Andersen, R.A. Preparation and reactions of base-free bis(1,2,4-tri-*tert*-butylcyclopentadienyl)uranium oxide, Cp'$_2$UO. *Organometallics* **2005**, *24*, 4251–4264. [CrossRef]

6. Blake, P.C.; Lappert, M.F.; Taylor, R.G.; Atwood, J.L.; Hunter, W.E.; Zhang, H. Synthesis, spectroscopic properties and crystal structures of [ML$_2$Cl$_2$] [M = Th or U; L = η-C$_5$H$_3$(SiMe$_3$)$_2$-1,3] and [UL$_2$X$_2$] (X = Br, I or BH$_4$). *Dalton Trans.* **1995**, 3335–3341. [CrossRef]

7. Blake, P.C.; Lappert, M.F.; Taylor, R.G.; Atwood, J.L.; Zhang, H. Some aspects of the coordination and organometallic chemistry of thorium and uranium (M^{III}, M^{IV}, U^V) in +3 and +4 oxidation states. *Inorg. Chim. Acta* **1987**, *139*, 13–20. [CrossRef]

8. Thomson, R.K.; Graves, C.R.; Scott, B.L.; Kiplinger, J.L. Organometallic uranium(IV) fluoride complexes: Preparation using protonolysis chemistry and reactivity with trimethylsilyl reagents. *Dalton Trans.* **2010**, *39*, 6826–6831. [CrossRef] [PubMed]

9. Rabinovich, D.; Haswell, C.M.; Scott, B.L.; Miller, R.L.; Nielsen, J.B.; Abney, K.D. New dicarbollide complexes of uranium. *Inorg. Chem.* **1996**, *35*, 1425–1426. [CrossRef] [PubMed]

10. Spirlet, M.R.; Rebizant, J.; Apostolidis, C.; Andreetti, G.D.; Kanellakopulos, B. Structure of tris(cyclopentadienyl)uranium bromide. *Acta Crystallogr. C* **1989**, *45*, 739–741. [CrossRef]

11. Goffart, J.; Fuger, J.; Gilbert, B.; Hocks, L.; Duyckaerts, G. On the indenyl compounds of actinide elements Part I: Triindenylthorium bromide and triindenyluranium bromide. *Inorg. Nucl. Chem. Lett.* **1975**, *11*, 569–583. [CrossRef]

12. Goffart, J.; Piret-Meunier, J.; Duyckaerts, G. On the indenyl compounds of actinide elements: Part V: Some oxygen-donor complexes of indenyl actinide halides. *Inorg. Nucl. Chem. Lett.* **1980**, *16*, 233–244. [CrossRef]

13. Spirlet, M.R.; Rebizant, J.; Goffart, J. Structure of tris(1-3-η-Indenyl)uranium bromide. *Acta Crystallogr. C* **1987**, *43*, 354–355. [CrossRef]

14. Bagnall, K.W.; Edwards, J. Uranium(IV) substituted poly(pyrazol-1-yl) borate complexes. *J. Less Common Met.* **1976**, *48*, 159–165. [CrossRef]

15. Bagnall, K.W.; Edwards, J.; Tempest, A.C. Some oxygen-donor complexes of cyclopentadienyluranium(IV) halides. *Dalton Trans.* **1978**, 295–298. [CrossRef]

16. Beeckman, W.; Goffart, J.; Rebizant, J.; Spirlet, M.R. The first cationic indenyl-f-metal complexes with pentagonal bipyramidal geometry. Crystal structures of $[C_9H_7UBr_2(CH_3CN)_4]_2^+[UBr_6]^{2-}$ and $[\{C_9H_7UBr(CH_3CN)_4\}_2O]^{2+}[UBr_6]^{2-}$. *J. Organomet. Chem.* **1986**, *307*, 23–37. [CrossRef]

17. Thomson, R.K.; Scott, B.L.; Morris, D.E.; Kiplinger, J.L. Synthesis, structure, spectroscopy and redox energetics of a series of uranium(IV) mixed-ligand metallocene complexes. *C. R. Chim.* **2010**, *13*, 790–802. [CrossRef]

18. Evans, W.J.; Walensky, J.R.; Ziller, J.W. Reaction chemistry of the U^{3+} metallocene amidinate $(C_5Me_5)_2[^iPrNC(Me)N^iPr]U$ including the isolation of a uranium complex of a monodentate acetate. *Inorg. Chem.* **2010**, *49*, 1743–1749. [CrossRef] [PubMed]

19. Evans, W.J.; Walensky, J.R.; Ziller, J.W. Reactivity of methyl groups in actinide metallocene amidinate and triazenido complexes with silver and copper salts. *Organometallics* **2010**, *29*, 101–107. [CrossRef]

20. Evans, W.J.; Nyce, G.W.; Johnston, M.A.; Ziller, J.W. How much steric crowding is possible in tris(η^5-pentamethylcyclopentadienyl) complexes? Synthesis and structure of $(C_5Me_5)_3UCl$ and $(C_5Me_5)_3UF$. *J. Am. Chem. Soc.* **2000**, *122*, 12019–12020. [CrossRef]

21. Danopoulos, A.A.; Hankin, D.M.; Cafferkey, S.M.; Hursthouse, M.B. η^2-Sulfenamido complexes of uranium. *Dalton Trans.* **2000**, 1613–1615. [CrossRef]

22. Lugli, G.; Mazzei, A.; Poggio, S. High 1,4-*cis*-polybutadiene by uranium catalysts, 1. Tris(π-allyl)uranium halide catalysts. *Makromol. Chem.* **1974**, *175*, 2021–2027. [CrossRef]

23. Lukens, W.W., Jr.; Beshouri, S.M.; Stuart, A.L.; Andersen, R.A. Solution structure and behavior of dimeric uranium(III) metallocene halides. *Organometallics* **1999**, *18*, 1247–1252. [CrossRef]

24. Blake, P.C.; Lappert, M.F.; Taylor, R.G.; Atwood, J.L.; Hunter, W.E.; Zhang, H. A complete series of uranocene(III) halides $[(UCp''_2X)_n]$ [X = F, Cl, Br, or I; $Cp'' = \eta^5$-$C_5H_3(SiMe_3)_2$]; Single-crystal X-ray structure determinations of the chloride and bromide (n = 2, X^- = μ-Cl, μ-Br). *Chem. Commun.* **1986**, 1394–1395. [CrossRef]

25. De Rege, F.M.; Smith, W.H.; Scott, B.L.; Nielsen, J.B.; Abney, K.D. Electrochemical properties of bis(dicarbollide)uranium dibromide. *Inorg. Chem.* **1998**, *37*, 3664–3666. [CrossRef] [PubMed]

26. Beshouri, S.M.; Zalkin, A. Bis[η^5-bis(trimethylsilyl)cyclopentadienyl]bromouranium(III) bis(tert-butyl isocyanide). *Acta Crystallogr. C* **1989**, *C45*, 1221–1222. [CrossRef]

27. Finke, R.G.; Hirose, Y.; Gaughan, G. $(C_5Me_5)_2UCl \cdot$ tetrahydrofuran. Oxidative-addition and related reactions. *Chem. Commun.* **1981**, 232–234. [CrossRef]

28. Maynadié, J.; Berthet, J.-C.; Thuéry, P.; Ephritikhine, M. An unprecedented type of linear metallocene with an f-element. *J. Am. Chem. Soc.* **2006**, *128*, 1082–1083. [CrossRef] [PubMed]

29. Maynadie, J.; Berthet, J.-C.; Thuery, P.; Ephritikhine, M. From bent to linear uranium metallocenes: Influence of counterion, solvent, and metal ion oxidation state. *Organometallics* **2006**, *25*, 5603–5611. [CrossRef]

30. Monreal, M.J.; Thomson, R.K.; Cantat, T.; Travia, N.E.; Scott, B.L.; Kiplinger, J.L. UI_4(1,4-dioxane)$_2$, $[UCl_4$(1,4-dioxane)]$_2$, and UI_3(1,4-dioxane)$_{1.5}$: Stable and versatile starting materials for low- and high-valent uranium chemistry. *Organometallics* **2011**, *30*, 2031–2038. [CrossRef]

31. Schelter, E.J.; Veauthier, J.M.; Graves, C.R.; John, K.D.; Scott, B.L.; Thompson, J.D.; Pool-Davis-Tournear, J.A.; Morris, D.E.; Kiplinger, J.L. 1,4-Dicyanobenzene as a scaffold for the preparation of bimetallic actinide complexes exhibiting metal–metal communication. *Chem. Eur. J.* **2008**, *14*, 7782–7790. [CrossRef] [PubMed]

32. LeMarechal, J.F.; Villiers, C.; Charpin, P.; Nierlich, M.; Lance, M.; Vigner, J.; Ephritikhine, M. Anionic tricyclopentadienyluranium(III) complexes. Crystal structure of $[Cp_3UClUCp_3][Na(18$-Crown-6)(THF)$_2]$ ($Cp = \eta$-C_5H_5). *J. Organomet. Chem.* **1989**, *379*, 259–269. [CrossRef]

33. Du Preez, J.G.H.; Gellatly, B.J.; Jackson, G.; Nassimbeni, L.R.; Rodgers, A.L. The chemistry of uranium. Part 23. Isomorphous tetrachlorobis(hexamethylphosphoramide)uranium(IV) and tetrabromobis (hexamethylphosphoramide)uranium(IV). *Inorg. Chim. Acta* **1978**, *27*, 181–184. [CrossRef]

34. Du Preez, J.G.H.; Zeelie, B.; Casellato, U.; Graziani, R. The chemistry of uranium. Part 34. Structural and thermal properties of UX_4L_2 complexes (X = Cl and Br; L = *N,N,N',N'*-tetramethylurea). *Inorg. Chim. Acta* **1986**, *122*, 119–126. [CrossRef]

35. Du Preez, J.G.H.; Rohwer, H.E.; van Brecht, B.J.A.M.; Zeelie, B.; Casellato, U.; Graziani, R. The chemistry of uranium. Part 41. Complexes of uranium tetrahalide, with emphasis on iodide, and triphenylphosphine oxide, tris(dimethylamino)phosphine oxide or tris(pyrrolidinyl)phosphine oxide. Crystal structure of tetrabromobis[tris(pyrrolidinyl)phosphine oxide]uranium(IV). *Inorg. Chim. Acta* **1991**, *189*, 67–75.

36. Du Preez, J.G.H.; Gouws, L.; Rohwer, H.; van Brecht, B.J.A.M.; Zeelie, B.; Casellato, U.; Graziani, R. The chemistry of uranium. Part 42. Cationic uranium(IV) complexes $UX_2L_4Y_2$ (X = Cl, Br or I; Y = ClO$_4$ or BPh$_4$; L = bulky strong neutral *O*-donor ligand) and [U(ClO$_4$)$_4$(OAsPh$_3$)$_4$]: Crystal structures of [UX$_2$L$_4$][BPh$_4$]$_2$ (L = tris(pyrrolidine-1-yl)phosphine oxide, X = Br or I). *Dalton Trans.* **1991**, 2585–2593. [CrossRef]
37. Zych, E.; Starynowicz, P.; Lis, T.; Drozdzynski, J. Crystal structure and spectroscopic properties of ammonium bis(acetonitrile)pentaaquadibromouranium dibromide. *Polyhedron* **1993**, *12*, 1661–1667. [CrossRef]
38. Berthet, J.-C.; Siffredi, G.; Thuéry, P.; Ephritikhine, M. Controlled chemical reduction of uranyl salts into $UX_4(MeCN)_4$ (X = Cl, Br, I) with Me$_3$SiX reagents. *Eur. J. Inorg. Chem.* **2007**, *2007*, 4017–4020. [CrossRef]
39. Caira, M.R.; de Wet, J.F.; Du Preez, J.G.H.; Gellatly, B.J. The crystal structures of hexahalouranates. I. Bis(triphenylethylphosphonium) hexachlorouranate(IV) and bis(triphenylethylphosphonium) hexabromouranate(IV). *Acta Crystallogr. B* **1978**, *34*, 1116–1120. [CrossRef]
40. Schnaars, D.D.; Wu, G.; Hayton, T.W. Reactivity of UH$_3$ with mild oxidants. *Dalton Trans.* **2008**, 6121–6126. [CrossRef] [PubMed]
41. Bombieri, G.; Benetollo, F.; Bagnall, K.W.; Plews, M.J.; Brown, D. Triphenylphosphine oxide complexes of actinide tetrahalides. Crystal and molecular structure of *trans*-tetrabromobis(triphenylphosphine oxide)uranium(IV). *Dalton Trans.* **1983**, 343–348. [CrossRef]
42. De Wet, J.F.; Caira, M.R. Structure and bonding in octahedral uranium(IV) complexes of the type $UX_4 \cdot 2L$ (X = halogen, L = unidentate, neutral oxygen donor). Part 1. The crystal structures of tetrabromobis(*NN'*-dimethyl-*NN'*-diphenylurea)uranium(IV) and α- and β-tetrachlorobis (*NN'*-dimethyl-*NN'*-diphenylurea)uranium(IV). *Dalton Trans.* **1986**, 2035–2041. [CrossRef]
43. De Wet, J.F.; Caira, M.R. Structure and bonding in octahedral uranium(IV) complexes of the type $UX_4 \cdot 2L$ (X = halogen, L = unidentate, neutral oxygen donor). Part 2. The crystal structures of tetrachlorobis[tris(pyrrolidinyl)phosphine oxide]-, tetrachlorobis (di-isobutyl sulphoxide)-, and tetrabromobis(triphenylarsine oxide)-uranium(IV). *Dalton Trans.* **1986**, 2043–2048. [CrossRef]
44. Kepert, D.; Patrick, J.; White, A. Structure and stereochemistry in "f-block" complexes of high coordination number. VI. Crystallographic characterizations of bis[(1,1-dimethyl-3-oxobutyl)triphenyl-phosphonium] hexabromouranate(IV) and tetrabromobis[ethane-1,2-diylbis(diphenylarsine) oxide]uranium(IV). *Aust. J. Chem.* **1983**, *36*, 469–476.
45. Jantunen, K.C.; Batchelor, R.J.; Leznoff, D.B. Synthesis, characterization, and organometallic derivatives of diamidosilyl ether thorium(IV) and uranium(IV) halide complexes. *Organometallics* **2004**, *23*, 2186–2193. [CrossRef]
46. Evans, W.J.; Miller, K.A.; Ziller, J.W.; Greaves, J. Analysis of uranium azide and nitride complexes by atmospheric pressure chemical ionization mass spectrometry. *Inorg. Chem.* **2007**, *46*, 8008–8018. [CrossRef] [PubMed]
47. Dietrich, H.M.; Ziller, J.W.; Anwander, R.; Evans, W.J. Reactivity of (C$_5$Me$_5$)$_2$UMe$_2$ and (C$_5$Me$_5$)$_2$UMeCl toward group 13 alkyls. *Organometallics* **2009**, *28*, 1173–1179. [CrossRef]
48. Thomson, R.K.; Graves, C.R.; Scott, B.L.; Kiplinger, J.L. Noble reactions for the actinides: Safe gold-based access to organouranium and azido complexes. *Eur. J. Inorg. Chem.* **2009**, *2009*, 1451–1455. [CrossRef]
49. Thomson, R.K.; Graves, C.R.; Scott, B.L.; Kiplinger, J.L. Straightforward and efficient oxidation of tris(aryloxide) and tris(amide) uranium(III) complexes using copper(I) halide reagents. *Inorg. Chem. Commun.* **2011**, *14*, 1742–1744. [CrossRef]
50. Thomson, R.K.; Graves, C.R.; Scott, B.L.; Kiplinger, J.L. Uncovering alternate reaction pathways to access uranium(IV) mixed-ligand aryloxide-chloride and alkoxide-chloride metallocene complexes: Synthesis and molecular structures of (C$_5$Me$_5$)$_2$U(O-2,6-^{i}Pr$_2$C$_6$H$_3$)(Cl) and (C$_5$Me$_5$)$_2$U(O-^{t}Bu)(Cl). *Inorg. Chim. Acta* **2011**, *369*, 270–273. [CrossRef]
51. Graves, C.R.; Yang, P.; Kozimor, S.A.; Vaughn, A.E.; Clark, D.L.; Conradson, S.D.; Schelter, E.J.; Scott, B.L.; Thompson, J.D.; Hay, P.J.; *et al.* Organometallic uranium(V)-imido halide complexes: From synthesis to electronic structure and bonding. *J. Am. Chem. Soc.* **2008**, *130*, 5272–5285. [CrossRef] [PubMed]
52. Gowda, B.T.; Foro, S.; Shakuntala, K. Potassium *N*-bromo-4-chlorobenzenesulfonamidate monohydrate. *Acta Crystallogr. E* **2011**, *67*, m962. [CrossRef] [PubMed]
53. Tonshoff, C.; Merz, K.; Bucher, G. Azidocryptands—Synthesis, structure, and complexation properties. *Org. Biomol. Chem.* **2005**, *3*, 303–308. [CrossRef] [PubMed]

54. Mahoney, J.M.; Nawaratna, G.U.; Beatty, A.M.; Duggan, P.J.; Smith, B.D. Transport of alkali halides through a liquid organic membrane containing a ditopic salt-binding receptor. *Inorg. Chem.* **2004**, *43*, 5902–5907. [CrossRef] [PubMed]

55. Jaroschik, F.; Nief, F.; le Goff, X.-F.; Ricard, L. Isolation of stable organodysprosium(II) complexes by chemical reduction of dysprosium(III) precursors. *Organometallics* **2007**, *26*, 1123–1125. [CrossRef]

56. Fender, N.S.; Fronczek, F.R.; John, V.; Kahwa, I.A.; McPherson, G.L. Unusual luminescence spectra and decay dynamics in crystalline supramolecular [(A18C6)$_4$MBr$_4$][TlBr$_4$]$_2$ (A = Rb, K; M = 3d element) complexes. *Inorg. Chem.* **1997**, *36*, 5539–5547. [CrossRef]

57. Danis, J.A.; Lin, M.R.; Scott, B.L.; Eichhorn, B.W.; Runde, W.H. Coordination trends in alkali metal crown ether uranyl halide complexes: The series [A(Crown)]$_2$[UO$_2$X$_4$] where A = Li, Na, K and X = Cl, Br. *Inorg. Chem.* **2001**, *40*, 3389–3394. [CrossRef] [PubMed]

58. Jia, Y.-Y.; Liu, B.; Liu, X.-M.; Yang, J.-H. Syntheses, structures and magnetic properties of two heterometallic carbonates: K$_2$Li[Cu(H$_2$O)$_2$Ru$_2$(CO$_3$)$_4$X$_2$]·5H$_2$O (X = Cl, Br). *CrystEngComm* **2013**, *15*, 7936–7942. [CrossRef]

59. Linti, G.; Li, G.; Pritzkow, H. On The chemistry of gallium: Part 19. Synthesis and crystal structure analysis of novel complexes containing Ga–FeCp(CO)$_2$-fragments. *J. Organomet. Chem.* **2001**, *626*, 82–91. [CrossRef]

60. Wu, F.; Tong, H.; Li, Z.; Lei, W.; Liu, L.; Wong, W.-Y.; Wong, W.-K.; Zhu, X. A white phosphorescent coordination polymer with Cu$_2$I$_2$ alternating units linked by benzo-18-crown-6. *Dalton Trans.* **2014**, *43*, 12463–12466. [CrossRef] [PubMed]

61. Gibney, B.R.; Wang, H.; Kampf, J.W.; Pecoraro, V.L. Structural evaluation and solution integrity of alkali metal salt complexes of the manganese 12-metallacrown-4 (12-MC-4) structural type. *Inorg. Chem.* **1996**, *35*, 6184–6193. [CrossRef]

62. Gao, X.; Zhai, Q.-G.; Li, S.-N.; Xia, R.; Xiang, H.-J.; Jiang, Y.-C.; Hu, M.-C. Two anionic [Cu^{I6}X$_7$]$_n^{n-}$ (X = Br and I) chain-based organic–inorganic hybrid solids with *N*-substituted benzotriazole ligands. *J. Solid State Chem.* **2010**, *183*, 1150–1158. [CrossRef]

63. Amo-Ochoa, P.; Jiménez-Aparicio, R.; Perles, J.; Torres, M.R.; Gennari, M.; Zamora, F. Structural diversity in paddlewheel dirhodium(II) compounds through ionic interactions: Electronic and redox properties. *Cryst. Growth Des.* **2013**, *13*, 4977–4985. [CrossRef]

64. Yan, H.; Jang, H.B.; Lee, J.-W.; Kim, H.K.; Lee, S.W.; Yang, J.W.; Song, C.E. A chiral-anion generator: Application to catalytic desilylative kinetic resolution of silyl-protected secondary alcohols. *Angew. Chem. Int.* **2010**, *49*, 8915–8917. [CrossRef] [PubMed]

65. Molcanov, K.; Kojic-Prodic, B.; Babic, D.; Zilic, D.; Rakvin, B. Stabilisation of tetrabromo- and tetrachlorosemiquinone (bromanil and chloranil) anion radicals in crystals. *CrystEngComm* **2011**, *13*, 5170–5178. [CrossRef]

66. Neumüller, B.; Gahlmann, F. Mesityltrifluorogallate. Die kristallstrukturen von Cs[MesGaF$_3$] und K[MesInBr$_3$]. *Z. Anorg. Allg. Chem.* **1993**, *619*, 1897–1904. [CrossRef]

67. Schelter, E.J.; Wu, R.; Scott, B.L.; Thompson, J.D.; Morris, D.E.; Kiplinger, J.L. Mixed valency in a uranium multimetallic complex. *Angew. Chem. Int.* **2008**, *47*, 2993–2996. [CrossRef] [PubMed]

68. Evans, W.J.; Kozimor, S.A.; Hillman, W.R.; Ziller, J.W. Synthesis and structure of the bis(tetramethylcyclopentadienyl)uranium metallocenes (C$_5$Me$_4$H)$_2$UMe$_2$, (C$_5$Me$_4$H)$_2$UMeCl, [(C$_5$Me$_4$H)$_2$U][(μ-η^6:η^1-Ph)(μ-η^1:η^1-Ph)BPh$_2$], and [(C$_5$Me$_4$)SiMe$_2$(CH$_2$CH=CH$_2$)]$_2$UI(THF). *Organometallics* **2005**, *24*, 4676–4683. [CrossRef]

69. Mehdoui, T.; Berthet, J.-C.; Thuery, P.; Salmon, L.; Riviere, E.; Ephritikhine, M. Lanthanide(III)/actinide(III) differentiation in the cerium and uranium complexes [M(C$_5$Me$_5$)$_2$(L)]$^{0,+}$ (L = 2,2'-bipyridine, 2,2':6',2''-terpyridine): Structural, magnetic, and reactivity studies. *Chem. Eur. J.* **2005**, *11*, 6994–7006. [CrossRef] [PubMed]

70. Mehdoui, T.; Berthet, J.-C.; Thuery, P.; Ephritikhine, M. The distinct affinity of cyclopentadienyl ligands towards trivalent uranium over lanthanide ions. evidence for cooperative ligation and back-bonding in the actinide complexes. *Dalton Trans.* **2005**, 1263–1272. [CrossRef] [PubMed]

71. Mehdoui, T.; Berthet, J.-C.; Thuery, P.; Ephritikhine, M. The remarkable efficiency of *N*-heterocyclic carbenes in lanthanide(III)/actinide(III) differentiation. *Chem. Commun.* **2005**, 2860–2862. [CrossRef] [PubMed]

72. Fagan, P.J.; Manriquez, J.M.; Marks, T.J.; Day, C.S.; Vollmer, S.H.; Day, V.W. Synthesis and properties of a new class of highly reactive trivalent actinide organometallic compounds. derivatives of bis(pentamethylcyclopentadienyl)uranium(III). *Organometallics* **1982**, *1*, 170–180. [CrossRef]

73. Cendrowski-Guillaume, S.M.; Lance, M.; Nierlich, M.; Vigner, J.; Ephritikhine, M. New actinide hydrogen transition metal compounds. Synthesis of [K(C$_{12}$H$_{24}$O$_6$)][(η-C$_5$Me$_5$)$_2$(Cl)UH$_6$Re(PPh$_3$)$_2$] and the crystal structure of its benzene solvate. *Chem. Commun.* **1994**, 1655–1656. [CrossRef]

74. Evans, W.J.; Kozimor, S.A.; Ziller, J.W. Methyl displacements from cyclopentadienyl ring planes in sterically crowded (C$_5$Me$_5$)$_3$M complexes. *Inorg. Chem.* **2005**, *44*, 7960–7969. [CrossRef] [PubMed]

75. Evans, W.J.; Kozimor, S.A.; Ziller, J.W. Bis(pentamethylcyclopentadienyl) U(III) oxide and U(IV) oxide carbene complexes. *Polyhedron* **2004**, *23*, 2689–2694. [CrossRef]

76. Evans, W.J.; Kozimor, S.A.; Ziller, J.W.; Kaltsoyannis, N. Structure, reactivity, and density functional theory analysis of the six-electron reductant, [(C$_5$Me$_5$)$_2$U]$_2$(μ-η^6:η^6-C$_6$H$_6$), synthesized via a new mode of (C$_5$Me$_5$)$_3$M reactivity. *J. Am. Chem. Soc.* **2004**, *126*, 14533–14547. [CrossRef] [PubMed]

77. Evans, W.J.; Nyce, G.W.; Forrestal, K.J.; Ziller, J.W. Multiple syntheses of (C$_5$Me$_5$)$_3$U. *Organometallics* **2002**, *21*, 1050–1055. [CrossRef]

78. Roger, M.; Belkhiri, L.; Thuéry, P.; Arliguie, T.; Fourmigué, M.; Boucekkine, A.; Ephritikhine, M. Lanthanide(III)/actinide(III) differentiation in mixed cyclopentadienyl/dithiolene compounds from X-ray diffraction and density functional theory analysis. *Organometallics* **2005**, *24*, 4940–4952. [CrossRef]

79. Arliguie, T.; Lescop, C.; Ventelon, L.; Leverd, P.C.; Thuéry, P.; Nierlich, M.; Ephritikhine, M. C–H and C–S bond cleavage in uranium(III) thiolato complexes. *Organometallics* **2001**, *20*, 3698–3703. [CrossRef]

80. Boisson, C.; Berthet, J.C.; Ephritikhine, M.; Lance, M.; Nierlich, M. Synthesis and crystal structure of [U(η-C$_5$Me$_5$)$_2$(OC$_4$H$_8$)$_2$][BPh$_4$], the first cationic cyclopentadienyl compound of uranium(III). *J. Organomet. Chem.* **1997**, *533*, 7–11. [CrossRef]

81. Manriquez, J.M.; Fagan, P.J.; Marks, T.J.; Vollmer, S.H.; Day, C.S.; Day, V.W. Pentamethylcyclopentadienyl organoactinides. trivalent uranium organometallic chemistry and the unusual structure of bis(pentamethylcyclopentadienyl)uranium monochloride. *J. Am. Chem. Soc.* **1979**, *101*, 5075–5078. [CrossRef]

82. Fulmer, G.R.; Miller, A.J.M.; Sherden, N.H.; Gottlieb, H.E.; Nudelman, A.; Stoltz, B.M.; Bercaw, J.E.; Goldberg, K.I. NMR chemical shifts of trace impurities: Common laboratory solvents, organics, and gases in deuterated solvents relevant to the organometallic chemist. *Organometallics* **2010**, *29*, 2176–2179. [CrossRef]

83. Lalancette, J.M.; Rollin, G.; Dumas, P. Metals intercalated in graphite. I. Reduction and oxidation. *Can. J. Chem.* **1972**, *50*, 3058–3062. [CrossRef]

84. Bergbreiter, D.E.; Killough, J.M. Reactions of potassium-graphite. *J. Am. Chem. Soc.* **1978**, *100*, 2126–2134. [CrossRef]

85. Thomson, R.K.; Cantat, T.; Scott, B.L.; Morris, D.E.; Batista, E.R.; Kiplinger, J.L. Uranium azide photolysis results in C–H bond activation and provides evidence for a terminal uranium nitride. *Nat. Chem.* **2010**, *2*, 723–729. [CrossRef] [PubMed]

86. APEX II 1.08, Bruker AXS, Inc. Madison, WI, USA, 2004. 53719.

87. SAINT+ 7.06, Bruker AXS, Inc. Madison, WI, USA, 2003. 53719.

88. SADABAS 2.03, University of Göttingen, Göttingen, Germany, 2001.

89. SHELXTL 5.10, Bruker AXS, Inc. Madison, WI, USA, 1997. 53719.

inorganics

MDPI

Article

Tuning of Hula-Hoop Coordination Geometry in a Dy Dimer

Yan Peng [1,2], Valeriu Mereacre [1], Christopher E. Anson [1] and Annie K. Powell [1,2,*]

[1] Institute of Inorganic Chemistry, Karlsruhe Institute of Technology, Engesserstrasse 15, 76131 Karlsruhe, Germany; Py1688@yahoo.com (Y.P.); valeriu.mereacre@kit.edu (V.M.); christopher.anson@kit.edu (C.E.A.)

[2] Institute of Nanotechnology, Karlsruhe Institute of Technology, Postfach 3640, 76021 Karlsruhe, Germany

* Correspondence: annie.powell@kit.edu; Tel.: +49-721-6084-2135; Fax: +49-721-6084-8142

Academic Editors: Stephen Mansell and Steve Liddle

Received: 18 September 2015; Accepted: 31 December 2015; Published: 8 January 2016

Abstract: The reaction of $DyCl_3$ with hydrazone Schiff base ligands and sodium acetate in the presence of triethylamine (Et_3N) as base affords two dysprosium dimers: $[Dy_2(HL1)_2(OAc)_2(EtOH)(MeOH)]$ (**1**) and $[Dy_2(L2)_2(OAc)_2(H_2O)_2] \cdot 2MeOH$ (**2**). The Dy^{III} ions in complexes **1** and **2** are linked by alkoxo bridges, and display "hula hoop" coordination geometries. Consequently, these two compounds show distinct magnetic properties. Complex **1** behaves as a field-induced single molecule magnet (SMM), while typical SMM behavior was observed for complex **2**. In addition, comparison of the structural parameters among similar Dy_2 SMMs with hula hoop-like geometry reveals the significant role played by coordination geometry and magnetic interaction in modulating the relaxation dynamics of SMMs.

Keywords: hydrazone Schiff base; Dy dimer; hula-hoop; single molecule magnets

1. Introduction

Recently, considerable attention has been paid to the design of new single molecule magnets (SMMs) involving 4f metal ions because of their significant magnetic anisotropy arising from the large unquenched orbital angular momentum [1,2]. Several reviews have already been published on SMMs which highlight strategies towards their synthesis and optimisation as well as understanding the factors determining the relaxation dynamics in such molecules [1,3–9]. However, it is still a challenge to understand the structure–property relationships. We know that the interplay between the ligand field effect, the geometry, and the strength of the magnetic interaction between the lanthanide sites will govern the SMM behavior of lanthanide-based SMMs [10]. The design of certain core motifs with different coordination environments (organic ligands, bridged ligands and co-ligands) provide an opportunity to probe the relaxation dynamics of polynuclear complexes, thus enriching the correlation between structure and magnetic properties in a family of dysprosium complexes. Furthermore, we can better understand the structure–property relationship to design Dy^{III}-based compounds with specific magnetic properties. Therefore, dinuclear lanthanide complexes, the simplest molecular units, have become a research hot topic in the field of molecule magnetism, due to their advantages compared with Single-Ion Magnets (SIMs) and the simple structural motif which can be easily controlled.

Hydrazone-based Schiff base ligands provide one of the most successful ligand types for assembling Dy dinuclear [11,12] or dimer [13–16] systems, exhibiting excellent performance in the construction of molecules with structurally distinct anisotropic centres. Minor changes can have significant implications on the physical characteristics and observed properties of the Dy_2 complexes. A suitable bridging ligand is crucial for assembling Dy_2 clusters with interesting magnetic properties, due to the difficulty in promoting magnetic interactions between the lanthanide ions through overlap of bridging ligand orbitals with the contracted 4f orbitals of the Ln ions [1].

As far as we know, no alkoxo-bridged Dy_2 complexes with hydrazone-containing Schiff base ligands have been reported to date. With this in mind, we designed two novel ligands, namely 3-hydroxy-naphthalene-2-carboxylic acid (6-hydroxymethyl-pyridin-2-ylmethylene)-hydrazide (H_3L1) and nicotinic acid (6-hydroxymethyl-pyridin-2-ylmethylene)-hydrazide (H_2L2), which are obtained by *in situ* reaction of 6-hydroxymethyl-pyridine-2-carbaldehyde with the corresponding hydrazine (Scheme 1). These ligands provide O, N, N, O-based chelating sites forming coordination pockets which are especially favorable for accommodating lanthanide ions. Here we report two Dy_2 compounds assembled from these two ligands showing so-called "hula hoop" coordination geometries. In this type of geometry, several coordinating atoms of the ligand(s) are arranged in a plane about the central lanthanide ion, which itself lies in this equatorial plane; two further ligands then occupy axial positions above and below the plane. The circle on which the equatorial ligands are situated thus defines a hula-hoop around the axis defined by the lanthanide and the two axial ligands. In contrast to previously published complexes of this type, in which the lanthanides are always bridged by phenoxo and/or hydrazone oxygens, the complexes reported here involve alkoxo-bridges. Furthermore, the relationship between structure and magnetic properties is discussed by comparison with the structural parameters of other Dy_2 SMMs in the literature with mono-hydrazone Schiff base ligands and "hula hoop" coordination geometries.

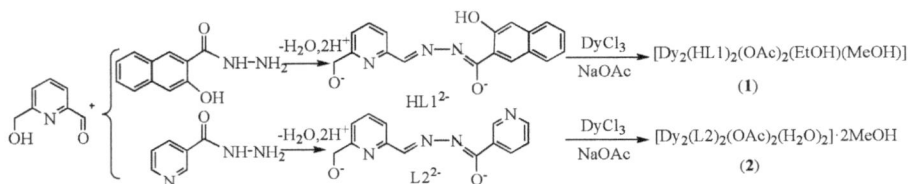

Scheme 1. Synthesis of complexes **1** and **2**.

2. Results and Discussion

2.1. Synthesis of compounds 1 and 2

The two compounds, $[Dy_2(HL1)_2(OAc)_2(EtOH)(MeOH)]$ (**1**) and $[Dy_2(L2)_2(OAc)_2(H_2O)_2] \cdot 2MeOH$ (**2**), were obtained in good yield from the reaction of 6-hydroxymethyl-pyridine-2-carbaldehyde with either 3-hydroxy-2-naphthoic acid hydrazide or 3-pyridinecarboxylic acid hydrazide, respectively, $DyCl_3 \cdot 6H_2O$, and $NaOAc \cdot 3H_2O$ in the presence of Et_3N as base in the molar ratio 3:3:3:10:10, in MeOH and EtOH or $CHCl_3$. Decreasing the amount of the ligands or increasing the amount of metal salt gave no crystals.

2.2. Molecular Structures of Compounds 1 and 2

The reaction of $DyCl_3 \cdot 6H_2O$ with either H_2L1 or HL2 in 2:1 MeOH and EtOH or $CHCl_3$ (*v/v*) in the presence of Et_3N and $NaOAc \cdot 3H_2O$ leads to the formation of pale yellow crystals of **1** or **2**. Crystal data and structure refinement details for **1** and **2** are summarized in Table S1. Their molecular structures are shown in Figure 1. Both compounds **1** and **2** crystallize in the triclinic $P\bar{1}$ space group with Z = 1, and phase purity was confirmed by powder XRD (Figure S1). These two compounds are dimeric Dy^{III} complexes consisting of eight-coordinated metal centers. The hydrazone Schiff base ligand provides an N_2O_2 donor set to each Dy^{III} with a fifth O atom provided by the bridging alkoxide function of the ligand from the second Dy^{III}. This circular environment forms the "hula-hoop" motif first identified in ($[I-(dibenzo[18]crown-6)La(\mu-OH)_2La(dibenzo[18]crown-6)I]]$) [17,18] for nine coordinated metal ion with six coordination sites in the plan ring and more recently extended to eight coordinated Dy ions with five coordination sites in the plane ring by Tang and co-workers [15]. The coordination sphere is completed by two O atoms from the bidentate OAc-*co*-ligand above the plane of the ring and an

O atom from the coordination solvent (EtOH/MeOH for **1** and H_2O for **2**) below the plane. In the asymmetric unit of **1** the coordination solvent is a disordered superposition of MeOH and EtOH; in the complete molecule there is one of each. The ligation corresponds to a $\eta^2{:}\eta^1{:}\eta^1{:}\eta^1{:}\eta^0{:}\mu_2$ coordination mode (Scheme 2), with similar Dy–O bond lengths (2.263(5) and 2.285(4) Å for **1**; 2.265(3) and 2.274(3) Å for **2**), similar Dy$\cdots$Dy distance (3.642(4) for **1** and 3.631(5) Å for **2**), as well as similar Dy–O–Dy angles (106.42(17) for **1** and 106.23(12)° for **2**). Additionally, there are intra-molecular hydrogen bonds in complex **1**, but no π–π interaction was found (Figure S3). Whereas for complex **2**, there are intra- and inter-molecular hydrogen bonds and short π–π (3.7972(4) Å) interactions were found (Figure S3). Furthermore, the closest intermolecular distances between Dy^{III} ions was found to be 10.65 Å in compound **1**, which is much longer than the distance found in compound **2** (7.09 Å).

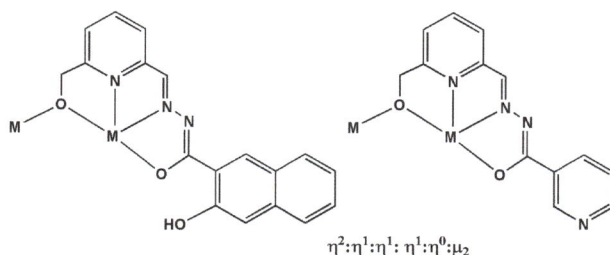

Scheme 2. The coordination modes of the H_3L1 and H_2L2 in compounds **1** and **2**.

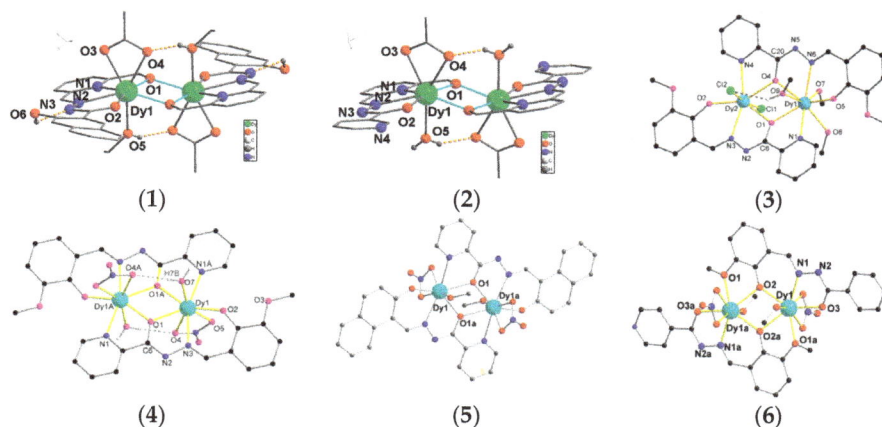

Figure 1. The molecular structures of compounds **1** and **2**, and the structures of Dy$_2$ compounds from hydrazine Schiff-base ligands (**3**, **4**, **5** and **6** [11,13–15]).

2.3. Magnetic Properties of Compounds 1 and 2

Direct-current (dc) magnetic susceptibilities of **1** and **2** were measured in an applied magnetic field of 1 kOe between 300 and 1.8 K. The two complexes show almost identical behavior (Figure 2) with observed χT values at 300 K of 28.47 $cm^3{\cdot}K{\cdot}mol^{-1}$ for **1** and 28.33 $cm^3{\cdot}K{\cdot}mol^{-1}$ for **2** in good agreement with what is expected for two uncoupled Dy^{III} ions ($S = 5/2$, $L = 5$, $^6H_{15/2}$, $g = 4/3$, 28.28 $cm^3{\cdot}K{\cdot}mol^{-1}$). The χT value gradually decreases on lowering the temperature to 50 K and then decreases rapidly to reach 4.84 $cm^3{\cdot}K{\cdot}mol^{-1}$ for **1** and 7.54 $cm^3{\cdot}K{\cdot}mol^{-1}$ for **2** at 1.8 K. The decline of χT is likely due to a combination of the progressive depopulation of Dy^{III} excited Stark sublevels [15,19,20] and possibly antiferromagnetic interaction within complexes **1** and **2**.

Magnetisation (M) data were collected in the 0–7 T field range at different temperatures. The lack of saturation of magnetisation (Figure S3) suggests the presence of a significant magnetic anisotropy and/or low-lying excited states. The magnetisation increases rapidly at low field and eventually reaches the value of 10.76 μ_B for **1** and 9.84 μ_B for **2** at 7 T without clear saturation. These values are lower than the expected saturation value of 20 μ_B (g = 4/3) for two non-interacting Dy^{III} ions, most likely due to the crystal-field effect [2].

In order to verify their potential SMM behavior, alternating current (ac) magnetic susceptibility studies were carried out on freshly filtered samples of **1** and **2**. In zero dc field, no out-of-phase signal (χ'') was observed in **1**, indicating the absence of SMM behaviour. Generally, this can be attributed to very fast quantum tunneling of the magnetization (QTM), which is commonly seen in pure lanthanide complexes [21–23]. The QTM may be shortcut by applying a static dc field and thus ac susceptibility measurements were obtained under a static dc field (Figure S4) (0–3 kOe). The optimal field is 2 kOe, therefore ac susceptibilities were carried out under this field. From the frequency dependencies of the ac susceptibility (Figure 3 and Figure S5) we can derive the magnetization time in the form of τ plotted as a function of 1/T between 4 and 8 K (Figure 4). At lower temperatures, the dynamics of **1** become temperature independent as expected in a pure quantum regime with a τ value of 1.22×10^{-3} s. Above 5 K, the data obey an Arrhenius law $\tau = \tau_0 exp(U_{eff}/k_B T)$ with an energy barrier of 35.4 K and pre-exponential factors (τ_0) of 3.15×10^{-7} s. For compound **2**, in zero dc field, the ac susceptibilities measured reveal the presence of slow relaxation of the magnetization, typical of SMM behaviour. The relaxation time shows two distinct regimes stemming from a temperature-independent quantum tunneling regime at low temperatures and a temperature-dependent thermally activated regime at temperatures above 6 K, following an Arrhenius law. The value of energy barrier (U_{eff}) and pre-exponential factor (τ_0) are 38.5 K and 1.04×10^{-6} s, comparable to those reported for similar Dy_2 SMMs [24].

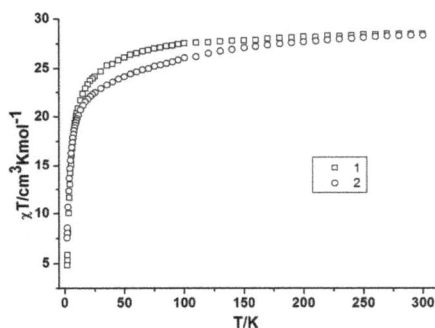

Figure 2. Temperature dependence of the χT products at 1000 Oe for **1** (square) and **2** (circle).

The Cole–Cole plots for compounds **1** and **2** from χ'' vs χ' at different temperatures are shown in Figure 5. For **1**, the shape is asymmetric. A reasonable fit to the generalized Debye model could only be obtained between 5.5 and 7.3 K with small α values (less than 0.1, Table S2), suggesting a small distribution of relaxation times in complex **1**. However, for **2**, a good fit was obtained according to the Debye model. The α values are below 0.33 (Table S2) which indicates a relatively narrow width of relaxation processes most likely due to a combination of QTM and thermally assisted relaxation pathways [25].

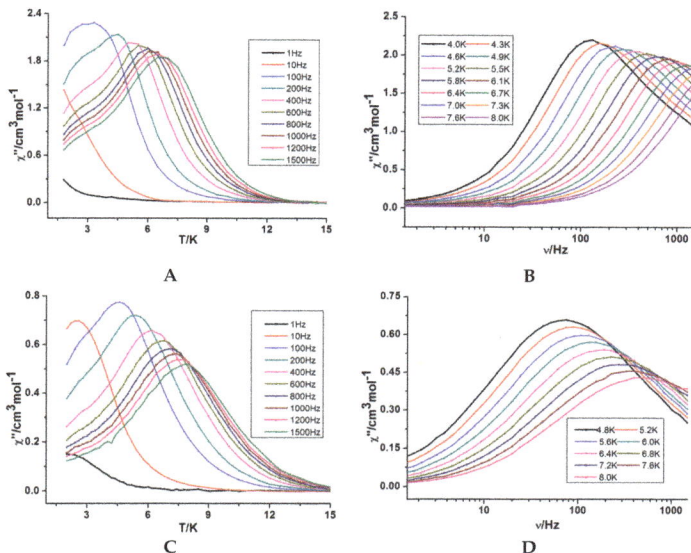

Figure 3. Temperature dependence (**A**) and frequency dependence (**B**) of the out-of phase ac susceptibility for **1** (**A**,**B**) under 2000 Oe dc field and Temperature dependence (**C**) and frequency dependence (**D**) of the out-of phase ac susceptibility for **2** (**C**,**D**) under zero dc field.

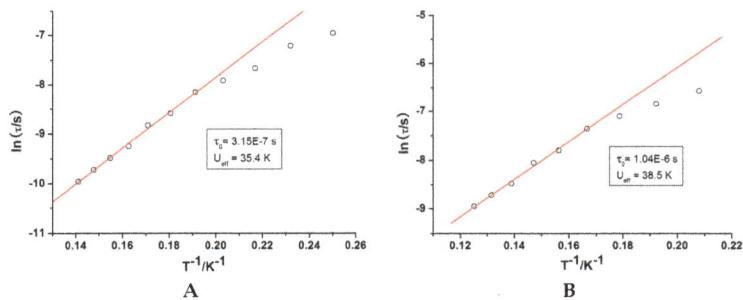

Figure 4. Magnetization relaxation time (τ) *vs.* $1/T$ plots for **1** (**A**) under 2000 Oe dc field and **2** (**B**) under a zero dc field.

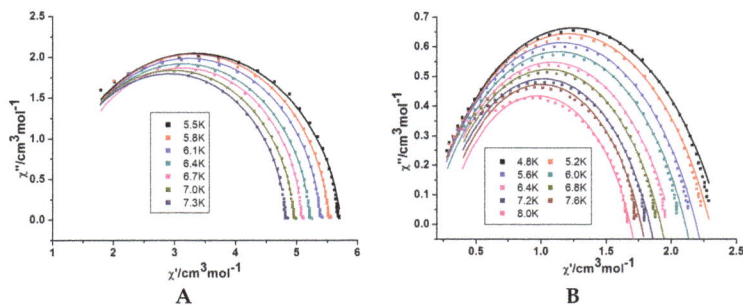

Figure 5. Plot of Cole–Cole for **1** (**A**) and **2** (**B**) at selected temperatures; Lines for fitting.

2.4. Structure–Property Relationship

To probe the structure–property relationship in mono-hydrazone Schiff-base ligand based Dy_2 SMM systems, some crucial parameters of structure of **1**, **2** along with those related compounds $[Dy_2(ovph)_2Cl_2(MeOH)_3] \cdot MeCN$ (where H_2ovph = pyridine-2-carboxylic acid [(2-hydroxy-3-methoxyphenyl)methylene] hydrazide) **3** [11], $[Dy_2(ovph)_2(NO_3)_2(H_2O)_2] \cdot 2H_2O$ **4** [15], $[Dy_2(L)_2(NO_3)_2(MeOH)_2] \cdot 4MeCN$ (where H_2L = N'-((2-hydroxy-1-naphthyl)methylene) picolinohydrazide) **5** [14] and $[Dy_2(hmi)_2(NO_3)_2(MeOH)_2]$ (where H_2hmi = 2-hydroxy-3-methoxyphenyl)methylene (isonicotino)hydrazine) **6** [13] are given in Table 1. In terms of other previously reported Dy_2 systems [26–28], it is clear that the mono-hydrazone Schiff-base provides a rigidly ligand with different kinds of coordination modes in terms of the available coordination pockets. The tautomeric nature of the arylhydrazone groups makes this system especially favourable for the isolation of such Dy_2 SMMs [15]. The "hula-hoop" geometry apparently provides a robust ligand field favouring slow magnetic relaxation of Dy^{III} ions. In compounds **1** and **2**, each Dy^{III} ion displays an eight-coordinate N_2O_6 coordination environment and links to the other Dy^{III} via alkoxide μ_2-O. As seen in Figure 2, both compounds have similar static magnetic behaviour and are antiferromagnetically coupled.

The dysprosium ions in compounds **3–5** are bridged by hydrazine–O, while in compound **6** they are bridged by phenolate. They show different magnetic behaviour and coupling. Thus it is not possible to predict the nature of the coupling between Dy ions from the Dy–O–Dy angle. Probably, the dominating factors governing the Dy–Dy coupling are a combination of Dy–O length and the local ligand field. For the ac susceptibilities, compound **3** has the highest energy barrier, because of without symmetry, high axiality and strong Ising exchange interaction, which efficiently suppresses quantum tunnelling of the magnetisation. The energy barriers of compound **1** and **2** are lower, which may be due to the coordinating anions which are different from those in **3** to **6**. A similar behaviour has been previously observed in a Dy dimer system [25]. Also, the axial solvent molecule contributes to this effect in terms of different energy barrier heights [29]. In order to compare the coordination geometries of the Dy ions, these were analysed using the *Shape* program [30–32]. The results are shown in Table 2, and indicate that the coordination geometries of the Dy ions in **1**, **2**, **4** and Dy_1 in **3** are close to triangular dodecahedral (TDD) with D_{2d} symmetry. In contrast, the geometries for Dy_2 in **3** and the Dy ions in **5**, are close to pentagonal bipyramidal (PBPY) with D_{5h} symmetry, and bicapped trigonal prismatic (BPTR) with C_{2v} symmetry, respectively. The differences of coordination geometry also contribute to the differences of the energy barrier of above compounds. Last but not least, it is noticeable that, by designing the ligands, suitable modulation of inter- or intramolecular hydrogen bonds or π–π interactions may have an unexpected effect on magnetic relaxation dynamics [11,33].

Table 1. Selected parameters of structure of **1**, **2** and reported Dy_2 compounds based on related ligands (**3–6**).

Compounds	1	2	3 *	4	5	6
Bridging atoms	Alkoxide (O)	Alkoxide (O)	Hydrazone (O)	Hydrazone (O)	Hydrazone (O)	Phenoxide (O)
Geometry (Dy_1)	hula hoop	hula hoop	pentagonal bipyramidal	hula hoop	hula hoop	hula hoop
Geometry ($Dy_{1'}/Dy_2$)	hula hoop	hula hoop	hula hoop	hula hoop	hula hoop	hula hoop
Donor atoms in cyclic ring	N_2O_3	N_2O_3	N_2O_3	N_2O_3	N_2O_3	NO_4
coordination anions	Ac^-	Ac^-	Cl^- or MeOH	NO_3^-	NO_3^-	NO_3^-
Coordination solvent	EtOH	H_2O	Cl^- or MeOH	H_2O	MeOH	MeOH
Coupling	antiferro	antiferro	ferro	ferro	antiferro	ferro
$d_{average}$ (Å)	2.384	2.384	2.370	2.355	2.387	
Dy–O–Dy (°)	106.40°	105.79°	111.67°	110.12°	114.88°	106.41°
Dy–Dy (Å)	3.643	3.631	3.769	3.8258	3.9225	
Field (Oe)	2000	0	0	0	0	0
Ueff (K)	35.36	38.46	198, 150	69	41.29	56

* Compound **3** is not centrosymmetric but an unsymmetrically coordinated Dy_2 dinuclear system.

223

Table 2. Analysis of lanthanide coordination geometry using *Shape*.

1	2	3 (Dy$_1$)	3 (Dy$_2$)	4	5
TDD-8 (D_{2d})	TDD-8 (D_{2d})	TDD-8 (D_{2d})	PBPY-7 (D_{5h})	TDD-8 (D_{2d})	BTPR-8 (C_{2v})
2.45	2.75	2.22	1.17	2.47	3.80

TDD: Triangular dodecahedron, PBPY: Pentagonal bipyramid, BTPR: Bicapped trigonal prism.

3. Experimental Section

3.1. General Information

All chemicals and solvents used for synthesis were obtained from commercial sources and used as received without further purification. All reactions were carried out under aerobic conditions. The elemental analyses (C, H, and N) were carried out using an Elementar Vario EL analyzer (Elementar Analysensysteme GmbH, Hanau, Germany). Fourier transform IR spectra (4000 to 400 cm^{-1}) were measured on a Perkin-Elmer Spectrum GX spectrometer (PerkinElmer LAS GmbH, Rodgau-Jügesheim, Germany) with samples prepared as KBr discs. Powder X-ray diffraction was carried out on a STOE STADI-P diffractometer (STOE & Cie. GmbH, Darmstadt, Germany), using Cu-Kα radiation with λ = 1.5406 Å.

3.2. The Preparation of [Dy$_2$(HL1)$_2$(OAc)$_2$(EtOH)$_2$] (1) and [Dy$_2$(L2)$_2$(OAc)$_2$(H$_2$O)$_2$]·2MeOH (2)

The ligands were prepared *in situ* from a solution of 6-hydroxymethyl-pyridine-2-carbaldehyde (21 mg, 0.15 mmol) and 3-hydroxy-2-naphthoic acid hydrazide (26.50 mg, 0.15 mmol) or 3-pyridinecarboxylic acid hydrazide (20 mg, 0.15 mmol) in 5 mL MeOH/EtOH and 10 mL CHCl$_3$ which was stirred for 30 min at room temperature. Then DyCl$_3$·6H$_2$O (56.50 mg, 0.15 mmol) was added under stirring. The resulting mixture was stirred for a further 30 min after NaOAc·3H$_2$O (68 mg, 0.50 mmol) and EtN$_3$ (0.5 mmol) was added. The solution was filtered and the yellow filtrate was left undisturbed to allow for the slow evaporation of the solvent. Yellow needle single crystals, suitable for X-ray diffraction analysis, were formed after one week in 55% yield (109.6 mg, based on Dy) for 1 [Dy$_2$(HL1)$_2$(OAc)$_2$(EtOH)$_2$] and 65% yield (111.6 mg, based on Dy) for 2 [Dy$_2$(L2)$_2$(OAc)$_2$(H$_2$O)$_2$]·2MeOH. Anal. Calcd. (Found) % for C$_{42}$H$_{40}$Dy$_2$N$_6$O$_{12}$ 1: C, 44.03 (44.01); H, 3.52 (3.69); N, 7.33 (7.25). Selected IR data (KBr, cm^{-1}) for 1: 3380 (w), 3053 (w), 1601 (s), 1562 (m), 1537 (s), 1508 (m), 1477 (s), 1455 (s), 1441 (m), 1425 (m), 1384 (m), 1341 (s), 1192 (m), 1150 (m), 1095 (w), 1052 (w), 1011 (w), 927(w), 802 (w), 751 (m), 652 (w), 574 (w), 481(w). Anal. Calcd. (Found) % for C$_{30}$H$_{30}$Dy$_2$N$_8$O$_{10}$ (minus two solvent molecules) 2: C, 36.48 (36.33); H, 3.06 (3.15); N, 11.35 (11.49). Selected IR data (KBr, cm^{-1}) for 2: 3402 (w), 3070 (w), 1637 (m), 1569 (m), 1537 (s), 1505 (m), 1477 (s), 1455 (s), 1441 (m), 1425 (m), 1372 (m), 1341 (s), 1194 (m), 1160 (s), 1095 (m), 1052 (w), 1011 (w), 927 (w), 802 (w), 769 (s), 652 (w), 576 (w), 481 (w).

3.3. X-Ray Crystal Structures

The crystal structures were determined at 150 K on a Stoe IPDS II diffractometer with graphite-monochromated Mo-Kα radiation. The structures were solved by direct methods and refined by full-matrix least-squares using the SHELXTL [34] program suite.

Crystallographic data (excluding structure factors) for the structures in this paper have been deposited with the Cambridge Crystallographic Data Centre as supplementary publication Nos. CCDC 1442826-1442827. Copies of the data can be obtained, free of charge, on application to CCDC, 12 Union Road, Cambridge CB2 1EZ, UK: https://summary.ccdc.cam.ac.uk/structure-summary-form.

3.4. Magnetic Measurements

The magnetic susceptibility measurements were obtained using a Quantum Design SQUID magnetometer MPMS-XL (LOT–Quantum Design, Darmstadt, Germany) in the temperature range

1.8–300 K. Measurements were performed on polycrystalline samples constrained in grease. Magnetization isotherms were collected at 2, 3, 5 K between 0 and 7 T. Alternating current (ac) susceptibility measurements were performed with an oscillating field of 3 Oe and ac frequencies ranging from 1 to 1500 Hz. The magnetic data were corrected for the sample holder and the diamagnetic contribution.

4. Conclusions

In summary, we have utilized 6-hydroxymethyl-pyridine-2-carbaldehyde based mono-hydrazone Schiff-base ligand to assemble two Dy_2 compounds bridged by alkoxide O with an eight-coordinate N_2O_6 hula-hoop like coordination environment. The dc magnetic measurements show that two complexes are antiferromagnetically coupled. Remarkably distinct dynamic magnetization was observed. Compound **1** show slow magnetic relaxation with anisotropic barriers of 35.36 K under 2 kOe dc field, while compound **2** shows slow magnetic relaxation with energy barriers of 38.46 K under zero dc field. These results provide important evidence that the dynamic behaviour of complexes can be modulated by careful tuning of the structural environments.

Supplementary Materials: Supplementary materials can be accessed at www.mdpi.com/2304-6740/4/1/2/s1.

Acknowledgments: We thank the Deutsche Forschungsgemeinschaft (DFG) for funding via the SFB TR88 "3MET" project.

Author Contributions: The preparation of the manuscript was made by all authors. Yan Peng: Syntheses, IR and Powder X-ray diffraction studies; Valeriu Mereacre: Magnetization studies; Christopher E. Anson: Single crystal X-ray diffraction study; Annie K. Powell: Careful follow-up and improvement of the manuscript, general idea and plan for the publication.

Conflicts of Interest: The authors declare no conflict of interest.

References

1. Sessoli, R.; Powell, A.K. Strategies towards single molecule magnets based on lanthanide ions. *Coord. Chem. Rev.* **2009**, *253*, 2328–2341. [CrossRef]
2. Osa, S.; Kido, T.; Matsumoto, N.; Re, N.; Pochaba, A.; Mrozinski, J. A Tetranuclear 3d-4f single molecule magnet: $[Cu^{II}LTb^{III}(hfac)_2]_2$. *J. Am. Chem. Soc.* **2003**, *126*, 420–421. [CrossRef] [PubMed]
3. Sorace, L.; Benelli, C.; Gatteschi, D. Lanthanides in molecular magnetism: Old tools in a new field. *Chem. Soc. Rev.* **2011**, *40*, 3092–3104. [CrossRef] [PubMed]
4. Rinehart, J.D.; Long, J.R. Exploiting single-ion anisotropy in the design of f-element single-molecule magnets. *Chem. Sci.* **2011**, *2*, 2078–2082. [CrossRef]
5. Guo, Y.-N.; Xu, G.-F.; Guo, Y.; Tang, J. Relaxation dynamics of dysprosium(III) single molecule magnets. *Dalton Trans.* **2011**, *40*, 9953–9963. [CrossRef] [PubMed]
6. Luzon, J.; Sessoli, R. Lanthanides in molecular magnetism: So fascinating, so challenging. *Dalton Trans.* **2012**, *41*, 13556–13567. [CrossRef] [PubMed]
7. Palii, A.V.; Tsukerblat, B.; Klokishner, S.; Dunbar, K.R.; Clemente-Juan, J.M.; Coronado, E. Beyond the spin model: Exchange coupling in molecular magnets with unquenched orbital angular momenta. *Chem. Soc. Rev.* **2011**, *40*, 3130. [CrossRef] [PubMed]
8. Palii, A.V.; Tsukerblat, B.S.; Clemente-Juan, J.M.; Coronado, E. Isotropic magnetic exchange between anisotropic Yb(III) ions. Study of $Cs_3Yb_2Cl_9$ and $Cs_3Yb_2Br_9$ crystals. *Inorg. Chem.* **2005**, *44*, 3984–3992. [CrossRef] [PubMed]
9. Kostakis, G.E.; Hewitt, I.J.; Ako, A.M.; Mereacre, V.; Powell, A.K. Magnetic coordination clusters and networks: Synthesis and topological description. *Philos. Trans. R. Soc. A* **2010**, *368*, 1509–1536. [CrossRef] [PubMed]
10. Long, J.; Habib, F.; Lin, P.-H.; Korobkov, I.; Enright, G.; Ungur, L.; Wernsdorfer, W.; Chibotaru, L.F.; Murugesu, M. Single-molecule magnet behavior for an antiferromagnetically superexchange-coupled dinuclear dysprosium(III) complex. *J. Am. Chem. Soc.* **2011**, *133*, 5319–5328. [CrossRef] [PubMed]

11. Guo, Y.-N.; Xu, G.-F.; Wernsdorfer, W.; Ungur, L.; Guo, Y.; Tang, J.K.; Zhang, H.-J.; Chibotaru, L.F.; Powell, A.K. Strong axiality and ising exchange interaction suppress zero-field tunneling of magnetization of an asymmetric Dy$_2$ single-molecule magnet. *J. Am. Chem. Soc.* **2011**, *133*, 11948–11951.

12. Lin, P.-H.; Sun, W.-B.; Yu, M.-F.; Li, G.-M.; Yan, P.-F.; Murugesu, M. An unsymmetrical coordination environment leading to two slow relaxation modes in a Dy$_2$ single-molecule magnet. *Chem. Commun.* **2011**, *47*, 10993–10995. [CrossRef] [PubMed]

13. Lin, P.-H.; Burchell, T.J.; Clérac, R.; Murugesu, M. Dinuclear dysprosium(III) single-molecule magnets with a large anisotropic barrier. *Angew. Chem. Int. Ed.* **2008**, *4*, 8848–8851. [CrossRef] [PubMed]

14. Zou, L.; Zhao, L.; Chen, P.; Guo, Y.-N.; Guo, Y.; Li, Y.-H.; Tang, J. Phenoxido and alkoxido-bridged dinuclear dysprosium complexes showing single-molecule magnet behaviour. *Dalton Trans.* **2012**, *41*, 2966–2971. [CrossRef] [PubMed]

15. Guo, Y.-N.; Chen, X.-H.; Xue, S.; Tang, J. Modulating magnetic dynamics of three Dy$_2$ complexes through Keto-Enol tautomerism of the *o*-vanillin picolinoylhydrazone ligand. *Inorg. Chem.* **2011**, *50*, 9705–9713. [CrossRef] [PubMed]

16. Guo, Y.-N.; Chen, X.-H.; Xue, S.; Tang, J. Molecular assembly and magnetic dynamics of two novel Dy$_6$ and Dy$_8$ aggregates. *Inorg. Chem.* **2012**, *51*, 4035–4042.

17. Runschke, C.; Meyer, G. [La$_2$I$_2$(OH)$_2$(Dibenzo-18-Krone-6)$_2$]I(I$_3$), ein kationischer, dimerer *in-cavity*-Komplex mit Iodid und Triiodid als Anionen. *Z. Anorg. Allg. Chem.* **1997**, *623*, 1493–1495. [CrossRef]

18. Ruiz-Martinez, A.; Casanova, D.; Alvarez, S. Polyhedral structures with an odd number of vertices: Nine-coordinate metal compounds. *Chem. Eur. J.* **2008**, *14*, 1291–1303. [CrossRef] [PubMed]

19. Kahn, M.L.; Sutter, J.-P.; Golhen, S.; Guionneau, P.; Ouahab, L.; Kahn, O.; Chasseau, D. Systematic investigations of the nature of the coupling between a Ln(III) ion (Ln = Ce(III) to Dy(III)) and its aminoxyl radical ligands. Structural and magnetic characteristics of a series of {Ln(organic radical)$_2$} compounds and the related {Ln(nitrone)$_2$} derivatives. *J. Am. Chem. Soc.* **2000**, *122*, 3413–3421.

20. Kahn, M.L.; Ballou, R.; Porcher, P.; Kahn, O.; Sutter, J.-P. Analytical determination of the {Ln-aminoxyl radical} exchange interaction taking into account both the ligand-field effect and the spin-orbit coupling of the lanthanide ion (Ln = Dy^{III} and Ho^{III}). *Chem. Eur. J.* **2002**, *8*, 525–531. [CrossRef]

21. Abbas, G.; Lan, Y.; Kostakis, G.E.; Wernsdorfer, W.; Anson, C.E.; Powell, A.K. Series of isostructural planar lanthanide complexes [Ln$^{III}_4$(μ_3-OH)$_2$(mdeaH)$_2$(piv)$_8$] with single molecule magnet behavior for the Dy$_4$ analogue. *Inorg. Chem.* **2010**, *49*, 8067–8072. [CrossRef] [PubMed]

22. Bi, Y.F.; Wang, X.T.; Liao, W.P.; Wang, X.W.; Deng, R.P.; Zhang, H.-J.; Gao, S. Thiacalix[4]arene-supported planar Ln$_4$ (Ln = Tb^{III}, Dy^{III}) clusters: Toward luminescent and magnetic bifunctional materials. *Inorg. Chem.* **2009**, *48*, 11743–11747. [CrossRef] [PubMed]

23. Yan, P.-F.; Lin, P.-H.; Habib, F.; Aharen, T.; Murugesu, M.; Deng, Z.-P.; Li, G.-M.; Sun, W.B. Planar tetranuclear Dy(III) single-molecule magnet and its Sm(III), Gd(III), and Tb(III) analogues encapsulated by Salen-type and β-diketonate ligands. *Inorg. Chem.* **2011**, *50*, 7059–7065. [CrossRef] [PubMed]

24. Tuna, F.; Smith, C.A.; Bodensteiner, M.; Ungur, L.; Chibotaru, L.F.; McInnes, E.J.L.; Winpenny, R.E.P.; Collison, D.; Layfield, R.A. A High anisotropy barrier in a sulfur-bridged organodysprosium single-molecule magnet. *Angew. Chem. Int. Ed.* **2012**, *51*, 6976–6980. [CrossRef] [PubMed]

25. Habib, F.; Brunet, G.; Vieru, V.; Korobkov, I.; Chibotaru, L.F.; Murugesu, M. Significant enhancement of energy barriers in dinuclear dysprosium single-molecule magnets through electron-withdrawing effects. *J. Am. Chem. Soc.* **2013**, *135*, 13242–13245. [CrossRef] [PubMed]

26. Zhang, P.; Guo, Y.-N.; Tang, J. Recent advances in dysprosium-based single molecule magnets: Structural overview and synthetic strategies. *Coord. Chem. Rev.* **2013**, *257*, 1728–1763. [CrossRef]

27. Woodruff, D.N.; Winpenny, R.E.P.; Layfield, R.A. Lanthanide single-molecule magnets. *Chem. Rev.* **2013**, *113*, 5110–5148. [CrossRef] [PubMed]

28. Habib, F.; Murugesu, M. Lessons learned from dinuclear lanthanide nano-magnets. *Chem. Soc. Rev.* **2013**, *42*, 3278–3288. [CrossRef] [PubMed]

29. Martínez-Lillo, J.; Tomsa, A.R.; Li, Y.; Chamoreau, L.M.; Cremades, E.; Ruiz, E.; Barra, A.L.; Proust, A.; Verdaguer, M.; Gouzerh, P. Synthesis, crystal structure and magnetism of new salicylamidoxime-based hexanuclear manganese(III) single-molecule magnets. *Dalton Trans.* **2012**, *41*, 13668–13681. [CrossRef] [PubMed]

30. Pinsky, M.; Avnir, D. Continuous symmetry measures, V: The classical polyhedra. *Inorg. Chem.* **1998**, *37*, 5575–5582. [CrossRef] [PubMed]

31. Casanova, D.; Cirera, J.; Llunell, M.; Alemany, P.; Avnir, D.; Alvarez, S. Minimal distortion paths in polyhedral rearrangements. *J. Am. Chem. Soc.* **2004**, *126*, 1755–1763. [CrossRef] [PubMed]

32. Cirera, J.; Ruiz, E.; Alvarez, S. Generalized interconversion coordinates. *Chem. Eur. J.* **2006**, *12*, 3162–3167. [CrossRef] [PubMed]

33. Bagai, R.; Wernsdorfer, W.; Abboud, K.A.; Christou, G. Exchange-biased dimers of single-molecule magnets in OFF and ON states. *J. Am. Chem. Soc.* **2007**, *129*, 12918–12919. [CrossRef] [PubMed]

34. Sheldrick, G.M. Crystal structure refinement with *SHELXL*. *Acta Cryst. C* **2015**, *C71*, 3–8. [CrossRef] [PubMed]

inorganics

MDPI

Article

A Structural and Spectroscopic Study of the First Uranyl Selenocyanate, [Et$_4$N]$_3$[UO$_2$(NCSe)$_5$]

Stefano Nuzzo, Michelle P. Browne, Brendan Twamley, Michael E. G. Lyons and Robert J. Baker *

School of Chemistry, Trinity College, University of Dublin, 2 Dublin, Ireland; nuzzos@tcd.ie (S.N.);
Brownm6@tcd.ie (M.P.B.); twamleyb@tcd.ie (B.T.); melyons@tcd.ie (M.E.G.L.)
* Correspondence: bakerrj@tcd.ie; Tel.: +353-1-896-3501; Fax: +353-1-671-2826

Academic Editors: Stephen Mansell and Steve Liddle
Received: 30 October 2015; Accepted: 4 February 2016; Published: 16 February 2016

Abstract: The first example of a uranyl selenocyanate compound is reported. The compound [Et$_4$N]$_3$[UO$_2$(NCSe)$_5$] has been synthesized and fully characterized by vibrational and multinuclear (^{1}H, ^{13}C{^{1}H} and ^{77}Se{^{1}H}) NMR spectroscopy. The photophysical properties have also been recorded and trends in a series of uranyl pseudohalides discussed. Spectroscopic evidence shows that the U–NCSe bonding is principally ionic. An electrochemical study revealed that the reduced uranyl(V) species is unstable to disproportionation and a ligand based oxidation is also observed. The structure of [Et$_4$N]$_4$[UO$_2$(NCSe)$_5$][NCSe] is also presented and Se···H–C hydrogen bonding and Se···Se chalcogen–chalcogen interactions are seen.

Keywords: uranyl; structural determination; photophysics

---

1. Introduction

The chemistry of uranium in its highest oxidation state has held scientists fascination for a long period of time. The uranyl moiety, [UO$_2$]$^{2+}$, is well studied in aqueous phases due, in part, to relevance in the nuclear waste treatment. Moreover, the photophysical properties of uranyl were first used in ancient roman times in colored glass [1], whilst comprehensive understanding of the bonding, and therefore photophysical properties, has come from both experiment and theory. An authoritative review by Denning summarizes these fundamental developments [2], and further reviews cover recent results [3–6]. The photophysical properties of the uranyl ion have been elucidated from these studies and the optical properties are due to a ligand-to-metal charge transfer (LMCT) transition involving promotion of an electron from a bonding –yl oxygen orbital (σ_u, σ_g, π_u and π_g) to a non-bonding 5f$_\delta$ and 5f$_\phi$ orbital on uranium. De-excitation of this $^3\Pi_u$ triplet excited state causes the characteristic green emission at *ca.* 500 nm. Visible on the absorption and emission bands are the vibronic progression arising from strong coupling of the ground state Raman active symmetric vibrational O=U=O (ν_1) mode with the $^3\Pi_u$ electronic triplet excited state. Time resolved studies allow sometimes complex speciation in water to be deconvoluted [7], whilst in non-aqueous media the positions of the emission maxima and lifetimes can be used as electronic and structural probes. For instance in the family of complexes *trans*-[UO$_2$X$_2$(O=PPh$_3$)$_2$] (X = Cl, Br, I) the photoluminescent properties do not vary [8], but for the compounds *trans*-[UO$_2$Cl$_2$L$_2$] (L = Ph$_3$P=NH, Ph$_3$P=O and Ph$_3$As=O) a red shift in the O$_{yl}$→U LMCT band is observed, in line with the increased donor strength of the ligand [9]. A further interesting photophysical property of certain uranyl compounds are thermochromic effects. Thus the compound [C$_4$mim]$_3$[UO$_2$(NCS)$_5$] (C$_4$mim = 1-butyl-3-methylimidazolium) is thermochromic in ionic liquids [10] but in organic solvents [Et$_4$N]$_3$[UO$_2$(NCS)$_5$] is not [11]. We have reported on the latter compound recently and now extend our study to the selenocyanate [NCSe]$^-$ derivatives which have not been reported. Indeed there is only one structurally characterized U–NCSe complex, *viz.* [Pr$_4$N]$_4$[U(NCSe)$_8$] [12]. In this work we have synthesized [Et$_4$N]$_3$[UO$_2$(NCSe)$_5$] and have characterized this by vibrational and multinuclear NMR spectroscopy and a photophysical investigation. X-ray diffraction of the compound $[Et_4N]_4[UO_2(NCSe)_5][NCSe]$ is also reported.

2. Results and Discussion

The synthesis of $[Et_4N]_3[UO_2(NCSe)_5]$, **1**, was conducted in a comparable way to that for the thiocyanate derivatives. Thus uranyl nitrate was treated with five equivalents of K[NCSe] followed by three equivalents of Et_4NCl in acetonitrile. A yellow precipitate was formed which was soluble in dichloromethane or acetone. We have noted that whilst this compound is air and moisture stable, it is somewhat light sensitive so reactions were conducted in the dark; the uranium(IV) compound $[^nPr_4N]_4[U(NCSe)_8]$ was also reported to be light sensitive [12]. Decomposition to red selenium powder was sometimes observed but the fate of the uranium was not determined. An alternative route to this compound was to treat a THF solution of $[UO_2Cl_2(THF)_3]$ sequentially with K[NCSe] and Et_4NCl. **1** was characterized by spectroscopic methods and single crystals were grown from slow evaporation of an acetonitrile solution. Unfortunately, crystals grown from different solvents always proved to be twinned so refinement to a satisfactory standard was not possible, however it did prove atom connectivity (Figure S1). During the course of one experiment, a few single crystals which had a different morphology were observed; these were separated by hand and the structure was solved to be $[Et_4N]_4[UO_2(NCSe)_5][NCSe]$, **2**.

The solid state structure of **2** is shown in Figure 1 and the packing shown in Figure 2. The structure of **2** contains disorder in two of the Et_4N^+ cations and the uncoordinated $[NCSe]^-$ anion which were modelled with restraints and constraints. The geometry around the uranyl in **2** are a typical pentagonal bipyramid with linear NCSe fragments and the N$\cdots$N intramolecular distances are similar to that seen in $[Et_4N]_3[UO_2(NCS)]_5$ (**2**: 2.89 Å; NCS: 2.87 Å) [13]. The U=O bond length is 1.771(2) Å and average U–N, N=C and C=Se bond lengths of 2.459 Å, 1.149 Å and 1.794 Å respectively can be compared to the uncoordinated $[N=C=Se]^-$ ion (N=C: 1.081(14) Å and C=Se: 1.846(7) Å) in **2**. Upon coordination to the uranyl ion the N=C bond lengthens slightly and the C=Se bond shortens slightly, suggesting a reorganization in the π-framework of the ligand; this effect has also been observed in uranyl thiocyanates experimentally and theoretically [11,14]. The average U–N bond in $[Et_4N]_3[UO_2(NCS)]_5$ is 2.443 Å [13], whilst in a suite of $[pyridinium][UO_2(NCS)_4(H_2O)]$ compounds the U–N bond lengths are 2.454(3) Å and 2.437(4) Å [15]. As has been previously described for $[Et_4N]_4[An(NCS)_8]$ (An = Th, U, Pu) [16], the lack of perturbation of the π-system in the $[NCS(e)]^-$ ligands suggests no π-overlap in the U–N bond.

The packing diagram (Figure 2) shows that the structure is a layer type where the cationic components sit between layers of uranyl ions. Hydrogen bonds between the U=O and H–C of the cations link these layers ($d_{C\cdots O}$ = 3.175–3.300Å), as now commonly observed [17]. There are also number of Se$\cdots$H–C short contacts. The most recent IUPAC definition of a hydrogen bond states that "*in most cases, the distance between H and Y are found to be less than the sum of their van der Waals radii*" [18]. According to this criterion, and using van der Walls radii taken from reference [19], H$\cdots$Se distances of less than 3.02 Å are classed as hydrogen bonds. These form a link between the layers via a C–H of an ethyl group and a Se atom in the coordinated and non-coordinated [NCSe] anion ($d_{C\cdots Se}$ = 3.687(16)–3.856(10); C–H$\cdots$Se = 142°–155°) [20]. Also present in the structure are close contacts between a coordinated selenium atom and the selenium of the non-coordinated $[NCSe]^-$ (3.427(1) Å) that are shorter than the van der Waals radii (3.64 Å) [19]. Chalcogen–chalcogen interactions have been studied both experimentally [21,22], and in the case of $[NCS]^-$ also for uranyl (S$\cdots$S = 3.536(2) Å) [15], and theoretically [23,24]. This may explain the difficulty in growing single crystals of **1** as these weak interactions may be important. Further studies are underway in our laboratory and will be reported on in due course.

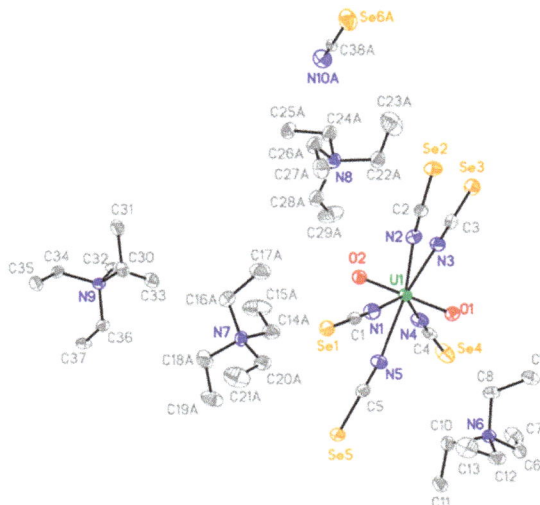

Figure 1. ORTEP plot of the structure of **2** refined with 70; 65; 55% occupancy for C14a–C21a; C22a–C29a; N10a–Se6a respectively. Thermal displacement shown at 50% occupancy and hydrogen atoms omitted for clarity. Selected bond lengths (Å): U(1)–O(1): 1.771(2); U(1)–N(1): 2.448(3); U(1)–N(2): 2.466(3); U(1)–N(3): 2.474(3); U(1)–N(4): 2.440(2); U(1)–N(5): 2.468(3); N(1)–C(1): 1.158(4); N(2)–C(2): 1.161(5); N(3)–C(3): 1.154(5); N(4)–C(4): 1.120(5); N(5)–C(5): 1.151(6); C(1)–Se(1): 1.791(3); C(2)–Se(2): 1.782(4); C(3)–Se(3): 1.798(4); C(4)–Se(4): 1.805(4); C(5)–Se(5): 1.794(4); N(10a)–C(38a): 1.081(14); C(38a)–Se(6a): 1.846(7).

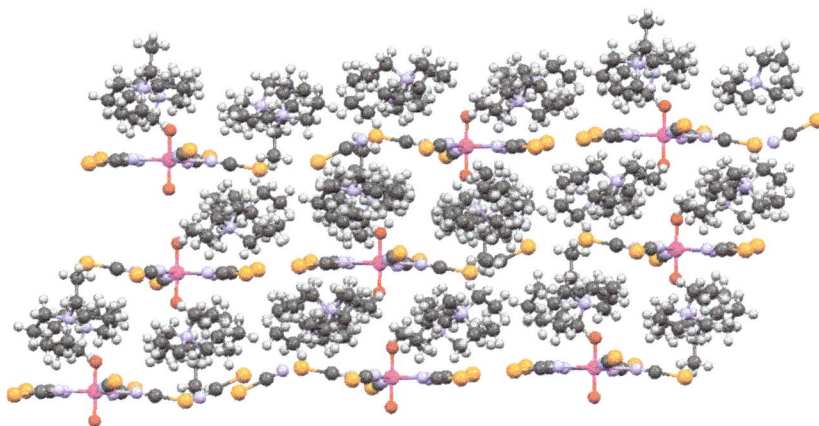

Figure 2. Packing diagram of **2** viewed down the *a* axis. Color code: U—pink; N—blue; C—Grey; S—yellow; O—red.

1 has been spectroscopically characterized, whilst for **2** there was not enough material. The uranyl group has characteristic vibrations in both the infrared and Raman spectra (Figures S2 and S3). For **1** these bands occur at 921 cm^{-1} (IR) and 845 cm^{-1} (R) comparable to the thiocyanate analogue 924 (IR) and 849 cm^{-1} (R) respectively. The N=C stretching frequency at 2056 cm^{-1} (IR) and 2051, 2060, 2091 cm^{-1} (R) are also similar to the NCS compound [2063 cm^{-1} (IR) and 2088, 2058, 2044 cm^{-1} (R)] [11], whilst the C=Se stretch of **1** is visible in the Raman spectrum at 635 and 672 cm^{-1}. ^{1}H NMR spectroscopy was uninformative (Figure S4). ^{13}C{^{1}H} NMR spectroscopy shows the resonance attributable to the selenocyanate at 117.4 ppm whilst a single peak is observed at −342.4 ppm in the ^{77}Se{^{1}H} NMR spectrum. For comparison, in our hands these peaks occur in K[NCSe] at 119.2 and −314.2 ppm respectively. Therefore, on the basis of the metric parameters from the X-ray structure, vibrational and NMR spectroscopic data we suggest that the bonding in these compounds are ionic with little perturbation of the [NCSe]$^-$ anionic fragment upon coordination. Our recent theoretical study of the [NCS]$^-$ compound suggested a predominantly ionic interaction [11].

The photophysics of this compound has also been investigated (Figure 3). The electronic absorption spectrum of **1** (Figure 3a) shows a broad featureless band at 320 nm (ε = 1,132 mol$^{-1}\cdot$cm^{-1}) assigned to transitions due to the [NCSe]$^-$ fragment and a weak vibronically coupled band at 460 nm (ε ~ 100 mol$^{-1}\cdot$cm^{-1}) due to the LMCT uranyl band. Excitation at 340 nm gives an emission spectrum typical for a uranyl moiety (Figure 3b). Pertinent properties are recorded in Table 1, along with a comparison for the uranyl thiocyanate and other pseudohalides. The average vibronic progression of the emission bands are coupled to the Raman active vibrational modes, which at 861 cm^{-1} is in close agreement with that measured in the Raman spectrum (849 cm^{-1}). The individual spacing (828, 868 and 888 cm^{-1}) reflect the transition of the vibronic parabola from harmonic to anharmonic. The luminescence lifetime of **1** was determined by the correlated single photon counting on the microsecond scale following excitation at 372 nm with a nanoLED (Table 1). The kinetic decay profile was fitted to a mono-exponential decay and the luminescence lifetime for **1** was measured to be 1.30 ± 0.02 μs. No significant change in lifetime was observed for the different pseudohalide systems given in Table 1. Given the ionicity of the U–N bond, ligand exchange processes may be faster than the lifetime of the uranyl excited state and so contributes to the shorter lifetime [25].

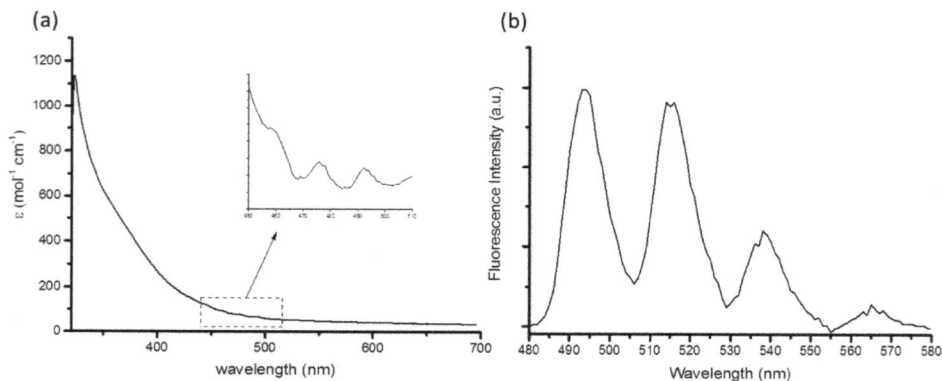

Figure 3. (a) UV-Vis absorption spectrum of **1** in acetone; (b) emission spectrum of **1** in acetone (λ_{ex} = 340 nm).

Table 1. Comparison of photophysical properties of selected uranyl halides and pseudohalides.

Compound (Solvent)	λ_{abs} U=O (nm)	λ_{em} (nm)	E_{0-0} (cm^{-1})	τ (μs)	χ^2	Ref.
1 (MeCN)	460	514	20,267	1.30	1.40	This work
[Et$_4$N]$_3$[UO$_2$(NCS)$_5$] (MeCN)	440	520	20,072	1.40	1.02	[11]
[UO$_2$Cl$_2$(OPPh$_3$)$_2$] (MeCN)	440	515	20,325	1.08	1.07	[8]
[UO$_2$Cl$_4$]$^{2-}$ (MeBu$_3$N[Tf$_2$N])		509	20,329	0.7		[26]

We have also briefly examined the electrochemistry of **1** (Figure 4). Cyclic voltammetry of a solution of **1** in acetonitrile containing 0.1 M [^{n}Bu$_4$N][BPh$_4$] shows an irreversible cathodic wave at $E_{p,c} = -0.95$ V (*vs.* Fc/Fc$^+$) ascribed to the unstable $[UO_2]^{2+}/[UO_2]^+$ redox couple, in line with known formal redox potentials of U(VI)/U(V) reduction. For comparison the uranyl [NCS]$^-$ analogue displayed the reduction at -1.45 V [11]. The putative 1 e$^-$ reduced uranyl(V) species [Et$_4$N]$_4$[UO$_2$(NCSe)$_5$] would be predicted to be quite unstable as it is now established that good π-donors and/or sterically bulky groups in the equatorial plane are required for stabilization of this unusual oxidation state [27,28]; although, there is evidence for the kinetic stabilization of the [UO$_2$]$^+$ ion in ionic liquids [29–31]. Any instability would manifest itself in an irreversible reduction, which is indeed what is observed. Also observed in this voltammogram is a broad, poorly defined irreversible oxidation at $E_{p,a} = +0.09$ V (*vs.* Fc/Fc$^+$) which is not observed at low scan rates, indicating the instability of this species. Given that the metal is in its highest oxidation state, it can be assigned as ligand based; we have observed similar behavior in the uranyl thiocyanate analogue ($E_{p,a} = +0.30$ V *vs.* Fc/Fc$^+$) and extends the family of uranyl coordinated to redox non-innocent ligands [11,32].

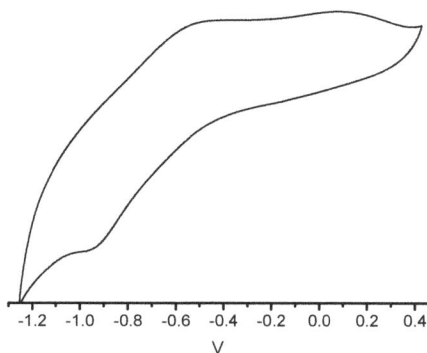

Figure 4. Cyclic Voltammogram of [1] *vs.* Fc/Fc$^+$ in MeCN at 293 K, with 0.1 M [Bu$_4$N][BPh$_4$] as a supporting electrolyte (scan rate = 0.1 V·s^{-1}).

3. Experimental Section

Caution! Although depleted uranium was used during the course of the experimental work, as well as the radiological hazards uranium is a toxic metal and care should be taken with all manipulations. Experiments were carried out using pre-set radiological safety precautions in accordance with the local rules of Trinity College Dublin.

^{1}H, ^{13}C{^{1}H} and ^{77}Se{^{1}H} NMR spectra were recorded on an AV400 spectrometer (Bruker, Karlsruhe, Germany) operating at 400.23, 155.54 and 76.33 MHz respectively, and were referenced to the residual ^{1}H resonances of the solvent used or external Me$_2$Se. IR spectra were recorded on a Spectrum One spectrometer (Perkin Elmer, Norwalk, CT, USA) with attenuated total reflectance (ATR) accessory. Raman spectra were obtained using 785-nm excitation on a 1000 micro-Raman system (Renishaw, Wotton-under-Edge, UK) in sealed capillaries. X-ray crystallography data were measured on an Apex diffractometer (Bruker). The structures were solved by direct methods and refined on F2 by full matrix least squares (SHELX97) using all unique data. CCDC 1424467 contains the supplementary crystallographic data for this paper. This data can be obtained free of charge from The Cambridge Crystallographic Data Centre via www.ccdc.cam.ac.uk/data_request/cif. UV-Vis measurements were made on a Lambda 1050 spectrophotometer (Perkin Elmer USA), using fused silica cells with a path length of 1 cm. Steady-state photoluminescence spectra were recorded on a Fluorolog-3 spectrofluorimeter (Horiba–Jobin–Yvon, Stanmore, UK). Luminescence lifetime data were recorded following 372 nm excitation, using time-correlated single-photon counting (a PCS900 plug-in PC card for fast photon counting). Lifetimes were obtained by tail fit on the data obtained, and the quality of fit was judged by minimization of reduced chi-squared and residuals squared.

Cyclic voltammetry measurements were conducted in a standard three-electrode cell using a high performance digital potentiostat (CH model 1760 D Bi-potentiostat system monitored using CH1760D electrochemical workstation beta software). All solutions were degassed for 15 min before commencing analysis. A platinum electrode with a diameter of 2 mm was employed as the working electrode, a platinum rod (together with internal referencing *vs.* Fc/Fc$^+$) was used as a reference electrode and a platinum wire electrode as counter electrode. The electrolyte was a solution of 0.1 M [^{n}Bu$_4$N][BPh$_4$] in CH$_3$CN. [UO$_2$Cl$_2$(THF)$_3$] was prepared via the literature procedure [33] whilst all other reagents and solvents were obtained from commercial sources.

Synthesis of 1

Method 1. To a solution of UO$_2$(NO$_3$)$_2$· 6H$_2$O (400 mg, 0.80 mmol) in acetonitrile (30 cm^3) were added KNCSe (576 mg, 4.0 mmol) and Et$_4$NCl (328 mg, 2.4 mmol). The solution mixture was stirred at room temperature for 60 min. The resulting orange solution was filtered and the solvent was reduced in volume. After 48 h at room temperature, the orange solution deposited orange-yellow crystals suitable for X-ray diffraction (408 mg, 0.34 mmol, yield = 43%).

Method 2. Under an atmosphere of high purity dry argon, to a solution of UO$_2$Cl$_2$THF$_3$ (50 mg, 0.090 mmol) in dry THF (20 cm^3) were added sequentially KNCSe (65 mg, 0.45 mmol) and Et$_4$NCl (45 mg, 0.27 mmol). The solution was stirred at room temperature for 60 min. The resulting orange solution was filtered and the solvent was reduced in volume. Placement at -20 °C overnight yielded an orange powder (33 mg, 0.036 mmol, yield = 40%).

IR (ν/cm^{-1}): 784 (C=Se), 921 (U=O), 2056 (C=N); Raman(ν/cm^{-1}): 635 and 672 (C=Se), 845 (U=O), 2051, 2060 and 2091 (C=N); δ_H (CD$_3$CN/ppm): 3.21 (q, 2H, $^3J_{H-H}$ = 7.28 Hz, CH$_2$), 1.25 (t, 3H, $^3J_{H-H}$ = 7.32 Hz, CH$_3$); δ_C (CD$_3$CN/ppm): 117.4 (N=C=Se), 29.9 (CH$_2$), 6.8 (CH$_3$); δ_{Se} (d-CH$_3$CN/ppm): -342.2 (N=C=Se).

4. Conclusions

To summarize, we have prepared and structurally characterized the first uranyl complexes of a selenocyanate ligand, which feature some unusual Se· · ·Se chalcogenide interactions and Se· · ·H–C hydrogen bonding. Vibrational and structural data suggest that the U–N bond is ionic and there is little perturbation of the [NCSe]$^-$ fragment compared to K[NCSe]. A photophysical investigation has shown that there is a small shift in the positions of the bands compared to the analogous [UO$_2$(NCS)$_5$]$^{3-}$ compound and the lifetime of the emission does not vary significantly with the nature of the pseudohalide. Finally, an electrochemical investigation revealed that the putative uranyl(V) compound is unstable with respect to disproportionation whilst there is a ligand based oxidation, similar to that observed in the [UO$_2$(NCS)$_5$]$^{3-}$ analogue.

Supplementary Materials: Supplementary materials can be accessed at www.mdpi.com/2304-6740/4/1/4/s1.

Acknowledgments: We thank Trinity College Dublin for funding this work.

Author Contributions: Steffano Nuzzo conducted the synthesis and data analysis, Brendan Twamley conducted the X-ray crystallography, Michelle P. Browne and Michael E. G. Lyons did the electrochemistry measurements, Robert J. Baker conceived the experiments and wrote the manuscript.

Conflicts of Interest: The authors declare no conflict of interest.

References

1.   Günther, R.T.; Manley, J.J. A mural glass mosaic from the Imperial Roman Villa near Naples. *Archaeologia* **1912**, *63*, 99–108. [CrossRef]

2.   Denning, R.G. Electronic structure and bonding in actinyl ions and their analogs. *J. Phys. Chem. A* **2007**, *111*, 4125–4143. [CrossRef] [PubMed]

3.   Liddle, S.T. The renaissance of non-aqueous uranium chemistry. *Angew. Chem. Int. Ed.* **2015**, *54*, 8604–8641. [CrossRef] [PubMed]

4. Jones, M.B.; Gaunt, A.J. Recent developments in synthesis and structural chemistry of nonaqueous actinide complexes. *Chem. Rev.* **2013**, *113*, 1137–1198. [CrossRef] [PubMed]

5. Baker, R.J. New reactivity of the uranyl ion. *Chem. Eur. J.* **2012**, *18*, 16258–16271. [CrossRef] [PubMed]

6. Natrajan, L.S. Developments in the photophysics and photochemistry of actinide ions and their coordination compounds. *Coord. Chem. Rev.* **2012**, *256*, 1583–1603. [CrossRef]

7. Drobot, B.; Steudtner, R.; Raff, J.; Geipel, G.; Brendler, V.; Tsushima, S. Combining luminescence spectroscopy, parallel factor analysis and quantum chemistry to reveal metal speciation—A case study of uranyl(VI) hydrolysis. *Chem. Sci.* **2015**, *6*, 964–972. [CrossRef]

8. Hashem, E.; McCabe, T.; Schulzke, C.; Baker, R.J. Synthesis, structure and photophysical properties of $[UO_2X_2(O=PPh_3)_2]$ (X = Cl, Br, I). *Dalton Trans.* **2014**, *43*, 1125–1131. [CrossRef] [PubMed]

9. Redmond, M.P.; Cornet, S.M.; Woodall, S.D.; Whittaker, D.; Collison, D.; Helliwell, M.; Natrajan, L.S. Probing the local coordination environment and nuclearity of uranyl(VI) complexes in non-aqueous media by emission spectroscopy. *Dalton Trans.* **2011**, *40*, 3914–3926. [CrossRef] [PubMed]

10. Aoyagi, N.; Shimojo, K.; Brooks, N.R.; Nagaishi, R.; Naganawa, H.; van Hecke, K.; van Meervelt, L.; Binnemans, K.; Kimura, T. Thermochromic properties of low-melting ionic uranyl isothiocyanate complexes. *Chem. Commun.* **2011**, *47*, 4490–4492. [CrossRef] [PubMed]

11. Hashem, E.; Platts, J.A.; Hartl, F.; Lorusso, G.; Evangelisti, M.; Schulzke, C.; Baker, R.J. Thiocyanate complexes of uranium in multiple oxidation states: A combined structural, magnetic, spectroscopic, spectroelectrochemical, and theoretical study. *Inorg. Chem.* **2014**, *53*, 8624–8637. [CrossRef] [PubMed]

12. Crawford, M.-J.; Karaghiosoff, K.; Mayer, P. The homoleptic $U(NCSe)_8^{4-}$ anion in $(Pr_4N)_4U(NCSe)_8 \cdot 2CFCl_3$ and $Th(NCSe)_4(OP(NMe_2)_3)_4 \cdot 0.5CH_3CN \cdot 0.5H_2O$: First structurally characterized actinide isoselenocyanates. *Z. Anorg. Allg. Chem.* **2010**, *636*, 1903–1906. [CrossRef]

13. Rowland, C.E.; Kanatzidis, M.G.; Soderholm, L. Tetraalkylammonium uranyl isothiocyanates. *Inorg. Chem.* **2012**, *51*, 11798–11804. [CrossRef] [PubMed]

14. Straka, M.; Patzschke, M.; Pyykkö, P. Why are hexavalent uranium cyanides rare while U–F and U–O bonds are common and short? *Theor. Chem. Acc.* **2003**, *109*, 332–340. [CrossRef]

15. Surbella, R.G.; Cahill, C.L. The exploration of supramolecular interactions stemming from the $[UO_2(NCS)_4(H_2O)]^{2-}$ tecton and substituted pyridinium cations. *CrystEngComm* **2014**, *16*, 2352–2364. [CrossRef]

16. Carter, T.J.; Wilson, R.E. Coordination chemistry of homoleptic actinide(IV)–thiocyanate complexes. *Chem. Eur. J.* **2015**, *21*, 15575–15582. [CrossRef] [PubMed]

17. Fortier, S.; Hayton, T.W. Oxo ligand functionalization in the uranyl ion (UO_2^{2+}). *Coord. Chem. Rev.* **2010**, *254*, 197–214. [CrossRef]

18. Arunan, E.; Desiraju, G.R.; Klein, R.A.; Sadlej, J.; Scheiner, S.; Alkorta, I.; Clary, D.C.; Crabtree, R.H.; Dannenberg, J.J.; Hobza, P.; *et al.* Definition of the hydrogen bond (IUPAC Recommendations 2011). *J. Pure Appl. Chem.* **2011**, *83*, 1637–1641. [CrossRef]

19. Alvarez, S. A cartography of the van der Waals territories. *Dalton Trans.* **2013**, *42*, 8617–8636. [CrossRef] [PubMed]

20. Michalczyk, R.; Schmidt, J.G.; Moody, E.; Li, Z.; Wu, R.; Dunlap, R.B.; Odom, J.D.; Silks, L.A., III. Unusual C–H $\cdots$ Se=C interactions in aldols of chiral *N*-acyl selones detected by gradient-selected ^{1}H–^{77}Se HMQC NMR spectroscopy and X-ray crystallography. *Angew. Chem. Int. Ed.* **2000**, *39*, 3067–3070. [CrossRef]

21. Uhl, W.; Wegener, P.; Layh, M.; Hepp, A.; Wuerthwein, E.-U. Chalcogen capture by an Al/P-based frustrated lewis pair: Formation of Al-E-P bridges and intermolecular tellurium–tellurium interactions. *Organometallics* **2015**, *34*, 2455–2462. [CrossRef]

22. Kobayashi, K.; Masu, H.; Shuto, A.; Yamaguchi, K. Control of face-to-face π–π stacked packing arrangement of anthracene rings via chalcogen–chalcogen interaction: 9,10-Bis(methylchalcogeno)anthracenes. *Chem. Mater.* **2005**, *17*, 6666–6673. [CrossRef]

23. Bleiholder, C.; Gleiter, R.; Werz, D.B.; Koeppel, H. Theoretical investigations on heteronuclear chalcogen–chalcogen interactions: On the nature of weak bonds between chalcogen centers. *Inorg. Chem.* **2007**, *46*, 2249–2260. [CrossRef] [PubMed]

24. Bleiholder, C.; Werz, D.B.; Koeppel, H.; Gleiter, R. Theoretical investigations on chalcogen–chalcogen interactions: What makes these nonbonded interactions bonding? *J. Am. Chem. Soc.* **2006**, *128*, 2666–2674. [CrossRef] [PubMed]

25. Fazekas, Z.; Yamamura, T.; Tomiyasu, H. Deactivation and luminescence lifetimes of excited uranyl ion and its fluoro complexes. *J. Alloys Compd.* **1998**, *271–273*, 756–759. [CrossRef]

26. Sornein, M.-O.; Cannes, C.; le Naour, C.; Lagarde, G.; Simoni, E.; Berthet, J.-C. Uranyl complexation by chloride ions. Formation of a tetrachlorouranium(VI) complex in room temperature ionic liquids [Bmim][Tf$_2$N] and [MeBu$_3$N][Tf$_2$N]. *Inorg. Chem.* **2006**, *45*, 10419–10421. [CrossRef] [PubMed]

27. Hardwick, H.C.; Royal, D.S.; Helliwell, M.; Pope, S.J.A.; Ashton, L.; Goodacre, R.; Sharrad, C.A. Structural, spectroscopic and redox properties of uranyl complexes with a maleonitrile containing ligand. *Dalton Trans.* **2011**, *40*, 5939–5952. [CrossRef] [PubMed]

28. Clark, D.L.; Conradson, S.D.; Donohoe, R.J.; Keogh, D.W.; Morris, D.E.; Palmer, P.D.; Rogers, R.D.; Tait, C.D. Chemical speciation of the uranyl ion under highly alkaline conditions. Synthesis, structures, and oxo ligand exchange dynamics. *Inorg. Chem.* **1999**, *38*, 1456–1466. [CrossRef]

29. Yaprak, D.; Spielberg, E.T.; Bäcker, T.; Richter, M.; Mallick, B.; Klein, A.; Mudring, A.-V. A roadmap to uranium ionic liquids: Anti-crystal engineering. *Chem. Eur. J.* **2014**, *20*, 6482–6493. [CrossRef] [PubMed]

30. Ogura, T.; Takao, K.; Sasaki, K.; Arai, T.; Ikeda, Y. Spectroelectrochemical identification of a pentavalent uranyl tetrachloro complex in room-temperature ionic liquid. *Inorg. Chem.* **2011**, *50*, 10525–10527. [CrossRef] [PubMed]

31. Ikeda, Y.; Hiroe, K.; Asanuma, N.; Shirai, A. Electrochemical studies on uranyl(VI) chloride complexes in ionic liquid, 1-butyl-3-methylimidazolium chloride. *J. Nucl. Sci. Technol.* **2009**, *46*, 158–162. [CrossRef]

32. Takao, K.; Tsushima, S.; Ogura, T.; Tsubomura, T.; Ikeda, Y. Experimental and theoretical approaches to redox innocence of ligands in uranyl complexes: What is formal oxidation state of uranium in reductant of uranyl(VI)? *Inorg. Chem.* **2014**, *53*, 5772–5780. [CrossRef] [PubMed]

33. Wilkerson, M.P.; Burns, C.J.; Paine, R.T.; Scott, B.L. Synthesis and crystal structure of UO$_2$Cl$_2$(THF)$_3$: A simple preparation of an anhydrous uranyl reagent. *Inorg. Chem.* **1999**, *38*, 4156–4158. [CrossRef]

![inorganics logo] *inorganics*

MDPI

Article

Optical Properties of Heavily Fluorinated Lanthanide Tris β-Diketonate Phosphine Oxide Adducts

Adam N. Swinburne [1], Madeleine H. Langford Paden [1], Tsz Ling Chan [1], Simon Randall [1], Fabrizio Ortu [1], Alan M. Kenwright [2] and Louise S. Natrajan [1,*]

[1] Centre for Radiochemistry Research, School of Chemistry, The University of Manchester, Oxford Road, Manchester M19 9PL, UK; answinburne@googlemail.com (A.N.S.); mhlp@madeleinelangford.co.uk (M.H.L.P.); tracyycart1202@gmail.com (T.L.C.); sjtrandall@gmail.com (S.R.); fabrizio.ortu@manchester.ac.uk (F.O.)

[2] Department of Chemistry, University of Durham, South Road, Durham DH1 3LE, UK; a.m.kenwright@durham.ac.uk

* Correspondence: louise.natrajan@manchester.ac.uk; Tel.: +44-161-275-1426

Academic Editors: Moris S. Eisen and Yi Luo
Received: 13 June 2016; Accepted: 9 August 2016; Published: 20 September 2016

Abstract: The construction of lanthanide(III) chelates that exhibit superior photophysical properties holds great importance in biological and materials science. One strategy to increase the luminescence properties of lanthanide(III) chelates is to hinder competitive non-radiative decay processes through perfluorination of the chelating ligands. Here, the synthesis of two families of heavily fluorinated lanthanide(III) β-diketonate complexes bearing monodentate perfluorinated tris phenyl phosphine oxide ligands have been prepared through a facile one pot reaction $[Ln(hfac)_3\{(Ar^F)_3PO\}(H_2O)]$ and $[Ln(F_7\text{-}acac)_3\{(Ar^F)_3PO\}_2]$ (where Ln = Sm^{3+}, Eu^{3+}, Tb^{3+}, Er^{3+} and Yb^{3+}). Single crystal X-ray diffraction analysis in combination with photophysical studies have been performed to investigate the factors responsible for the differences in the luminescence lifetimes and intrinsic quantum yields of the complexes. Replacement of both bound H_2O and C–H oscillators in the ligand backbone has a dramatic effect on the photophysical properties of the complexes, particularly for the near infra-red emitting ion Yb^{3+}, where a five fold increase in luminescence lifetime and quantum yield is observed. The complexes $[Sm(hfac)_3\{(Ar^F)_3PO\}(H_2O)]$ **(1)**, $[Yb(hfac)_3\{(Ar^F)_3PO\}(H_2O)]$ **(5)**, $[Sm(F_7\text{-}acac)_3\{(Ar^F)_3PO\}_2]$ **(6)** and $[Yb(F_7\text{-}acac)_3\{(Ar^F)_3PO\}_2]$ **(10)** exhibit unusually long luminescence lifetimes and attractive intrinsic quantum yields of emission in fluid solution (Φ_{Ln} = 3.4% **(1)**; 1.4% **(10)**) and in the solid state (Φ_{Ln} = 8.5% **(1)**; 2.0% **(5)**; 26% **(6)**; 11% **(10)**), which are amongst the largest values for this class of compounds to date.

Keywords: lanthanide; luminescence; perfluorinated β-diketonates; intrinsic quantum yield

1. Introduction

The unique optical properties of the trivalent lanthanides continue to garner appeal due to their numerous commercially exploitable applications in the biomedical and materials science fields [1,2]. Their inherent luminescent properties, which arise from parity forbidden intra f–f transitions, including long-lived emission (microsecond to millisecond range), high colour purity, insensitivity to dissolved molecular oxygen and resistance to photobleaching and blinking render these ions highly suitable as the emissive components in a number of growing technologies. These include, organic light emitting diodes [3], up and down-converting phosphors [4–6], luminescent sensors in biomedical diagnostics [7] and luminescent markers in cell microscopy [8–10]. More recently, the potential for near infra-red (nIR) emitting lanthanide chelates (Yb^{3+}, Nd^{3+}, Er^{3+}, Pr^{3+}) [11–16] to be exploited in mammalian cell imaging and in the telecommunications industry has been realized. Lanthanide ions that are emissive in the near infra-red region of the electromagnetic spectrum are particularly suited to biological imaging using confocal microscopy [17] and in optical amplifiers and waveguides [18–20] since their emission coincides with the more transparent wavelengths of biological tissue and silica. Above all however, the long-lived emission of the trivalent lanthanides is ideal for preventing spectral interference from scattered light and autofluorescence via time-gated luminescence detection [21] (pp. 6–8), which commonly occurs upon UV excitation in molecular complexes, particularly in biological media.

Lanthanide chelates that incorporate organic chromophores offer additional attractive properties for their use as optical materials, namely the use of low power excitation sources, a large Stokes' shift between excitation and emission and theoretically very high quantum yields of emission. This is principally because the parity forbidden f–f transitions can be overcome by efficient energy transfer from the triplet state of the chromophore of the chelated organic ligand (antenna) to the lanthanide excited emissive state upon UV–visible excitation; this is termed the "antenna effect" or "sensitised emission" [22]. In this regard, β-diketonates and their substituted derivatives [23] have become synonymous with the development of highly luminescent lanthanide complexes used in optoelectronic devices and in fluoroimmunoassays such as DELFIA (dissociation-enhanced lanthanide fluoroimmunoassay) [21,24–26] (p. 5085). These monoanionic ligands form very kinetically and thermodynamically stable 1:3 charge neutral metal:ligand chelates and the appended chromophores can very efficiently sensitise both the visible and nIR f–f based emission from lanthanide ions [27].

However, the majority of these complexes only take advantage of the green and red emission of Tb^{3+} and Eu^{3+} respectively due to the fact that the excited states of these ions are only marginally quenched by frequency matched vibrational harmonics of proximate O–H and N–H oscillators and the luminescence lifetimes are of millisecond order. By contrast, the relative lower energy of the excited states of the nIR emitting lanthanides means that the emission from these ions is considerably quenched by lower energy vibrational overtones of X–H bond vibrations (particularly C–H oscillators) present within the ligand architecture and/or by closely diffusing solvent molecules [12]. Since the magnitude of vibrational quenching of the emissive excited state of a lanthanide ion is dictated by the energy gap law [28,29], X–H quenching is much more pronounced in nIR emitting lanthanides compared to those that emit in the visible region. In general, vibrational quenching is considered negligible if the energy gap between the emissive state and the next lowest lying energy state is greater than or equal to the sixth harmonic of the fundamental vibrational mode. However, in the case of the nIR emissive lanthanides, the emissive states often lie within the first and third vibrational overtones of X–H oscillators [30]. In this regard, the fact that O–D oscillators whose fundamental modes vibrate at lower frequency (e.g., 3450 cm^{-1} in an OH bond compared to 2500 cm^{-1} in an OD bond), can be exploited to increase the luminescence lifetime and quantum yield of Ln^{3+} based emission. Horrocks [31,32] and others [11,33] have exploited this effect to develop an empirical equation to determine the number of O–H oscillators and therefore water or methanol molecules bound to a Ln^{3+} ion in aqueous or methanolic solution to great effect (Ln^{3+} = Eu^{3+}, Tb^{3+}, Yb^{3+} and in a series of isostructural lanthanide(III) complexes, Nd^{3+}). By extension, the luminescence lifetime of a given lanthanide complex can be increased by simply replacing the spectroscopic solvent to its deuterated counterpart. However, X–D vibrational quenching is often still operative in nIR emissive complexes; as an example, the third C–D overtone overlaps with the $^4I_{13/2}$ excited state of Er^{3+} (in comparison to the first vibrational harmonic of C–H bonds).

To further overcome the limitations of vibrational quenching and thereby significantly increase the luminescence quantum yield of nIR (and visible) emitting lanthanide complexes, synthetic approaches based on partial or full deuteration [27,34–37] of the C–H bonds in organic ligands can be accomplished. For example, Seitz and Platas-Iglesias have synthesized several series of selectively deuterated Lehn cryptand complexes of Pr^{3+}, Nd^{3+}, Er^{3+} and Yb^{3+} and evaluated the spectral overlap integrals of the excited states with aromatic C–H and C–D overtones to develop a comprehensive picture of lanthanide C–H quenching combinations for luminescence enhancement in nIR emitting lanthanide complexes [30]. An alternative strategy is partial or perfluorination of C–H bonds of the supporting ligand [38–41]. This has also been utilized very effectively to prepare partially fluorinated complexes mainly based on diamine adducts of tris β-diketonate hexafluoroacetylacetonate (hfac) and heptafluoroacetylacetanoate (F_7-acac) chelates that exhibit improved photophysical properties [23,42]. For example, the radiative lifetimes of the $^4I_{13/2} \to {}^4I_{13/2}$ transition at 1530 nm in crystalline samples of Er(hfac)$_3$(H$_2$O)$_2$ and Cs[Er(hfac)$_4$] have been determined to be ~9.5 and 23.4 ms respectively, reflecting the greater degree of fluorination in the latter [38].

Conversely, complexes of the lanthanides where all C–H bonds on all coordinated ligands have been fluorinated are rather rare [34–37]. Notable examples of sensitized nIR Ln^{3+} emission include the Er^{3+} complexes of tetrapentafluorophenylimidodiphosphinate (F-TPIP) [43,44], perfluorodiphenylphosphinic acid [41], perfluorinated nitrosopyrazolone [45] and perfluoro-benzophenone in Er^{3+} doped Zeolite L [46]. Remarkably, ligand sensitised nIR emission in the Er(F-TPIP)$_3$ complex resulted in extremely long-lived Er^{3+} nIR emission with a reported lifetime of 200 μs, while the luminescence lifetime following direct intra 4f excitation at 978 nm in the perfluorodiphenylphosphinic acid complex reported by Song et al. was measured as 0.336 ms. In this contribution, we describe the synthesis and enhanced optical properties of two families of fluorinated tris β-diketonate lanthanide complexes of the visible and nIR emitting lanthanides Sm^{3+}, Eu^{3+}, Tb^{3+}, Er^{3+} and Yb^{3+} where the eight coordinate coordination geometries are completed by the monodentate perfluorinated tris phenyl phosphine oxide (C$_6$F$_5$)$_3$PO ((Ar^F)$_3$PO) [41,45,47].

2. Results and Discussion

2.1. Synthesis of the Complexes

A series of lanthanide tris hexafluoroacetylacetonate (hfac) and heptafluoroacetylacetonate (F_7-acac) bis perfluorinated tris aryl phosphine oxide complexes were originally targeted by two synthetic routes as outlined in Scheme 1. Either treatment of prepared Ln(hfac)$_3$·2H$_2$O [28] with two equivalents of (Ar^F)$_3$PO (tris(pentafluorophenyl)phosphine oxide) in CH$_2$Cl$_2$ at room temperature sonicated for one hour or a one pot reaction of the corresponding lanthanide acetate, protonated β-diketonate and two equivalents of (Ar^F)$_3$PO [41] at 2×10^{-3} M concentrations heated to reflux temperature for one hour yielded the complexes [Ln(hfac)$_3${(Ar^F)$_3$PO}(H$_2$O)] and (Ar^F)$_3$PO (Ln = Sm^{3+}, Eu^{3+}, Tb^{3+}, Er^{3+} and Yb^{3+}, **1–5**) and [Ln(F$_7$-acac)$_3${(Ar^F)$_3$PO}$_2$] (Ln = Sm^{3+}, Eu^{3+}, Tb^{3+}, Er^{3+} and Yb^{3+}, **6–10**) in moderate yields (26%–59%) after recrystallization from CH$_2$Cl$_2$ solutions (as verified by single crystal X-ray diffraction analysis). Interestingly, conducting the reactions in dry CH$_2$Cl$_2$ under air sensitive conditions at higher concentrations (1×10^{-2} M) in order to promote bis phosphine oxide substitution with the labile coordinated water molecules in the parent lanthanide acetate or β-diketonate complexes, resulted solely in the isolation of the mono phosphine oxide substituted complexes for the hexafluorinated β-diketonate hfac [47], whereas only the bis phosphine oxide substituted derivatives were isolated with the heptafluoro β-diketonate F$_7$-acac using both standard and dry solvents in similar conditions (Scheme 1). This interesting divergence in reactivity can be attributed to the greater electron withdrawing effects of the F$_7$-acac ligand relative to the hfac β-diketonate which promotes a higher affinity for the second phosphine oxide to bind to the electropositive lanthanide(III) metal centres. In the case of the hfac reactions, adjusting the stoichiometry of the phosphine oxide accordingly, resulted in slightly improved product yields in most cases apart from complex **4** (44%–53%).

The ^{31}P and ^{19}F NMR (nuclear magnetic resonance) spectra of all the complexes exhibited paramagnetically shifted phosphorous and fluorine resonances; the ^{19}F resonances of the β-diketonate being significantly more shifted and broadened than the corresponding resonances in the coordinated (Ar^F)$_3$PO ligands (see supplementary material). Notably, multinuclear NMR analysis of the crude powders from the hfac reactions taken after removal of all volatiles before recrystallization suggested no presence of a minor lanthanide species containing perfluorinated phosphine oxide ligands that could be formulated as [Ln(hfac)$_3${(Ar^F)$_3$PO}$_2$] as previously reported for Er^{3+} (vide infra) [47].

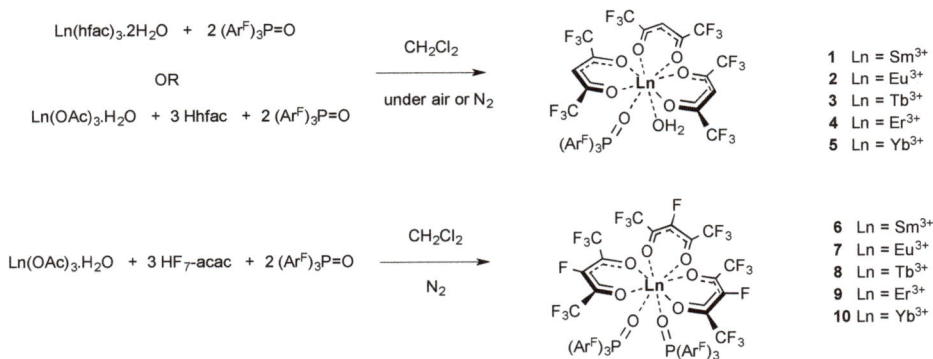

Scheme 1. Synthesis of the heavily fluorinated lanthanide(III) complexes [Ln(hfac)$_3${(Ar^F)$_3$PO}(H$_2$O)] (**1–5**) and [Ln(F$_7$-acac)$_3${(Ar^F)$_3$PO}$_2$] (**6–10**) described in this article.

2.2. Single Crystal X-ray Diffraction Analysis

Representative single crystals suitable for X-ray diffraction analysis were grown for the complexes [Ln(hfac)$_3${(Ar^F)$_3$PO}(H$_2$O)] (Ln = Tb^{3+}, **3** and Er^{3+}, **5**) and [Ln(F$_7$-acac)$_3${(Ar^F)$_3$PO}$_2$] (Ln = Sm^{3+}, **6**, Eu^{3+}, **7**, Er^{3+}, **9** and Yb^{3+}, **10**) by slow evaporation or slow cooling of saturated CH$_2$Cl$_2$ or CH$_2$Cl$_2$:pentane or hexane solutions to −18°C (Figure 1). A preliminary crystallographic data set was also collected for [Yb(hfac)$_3${(Ar^F)$_3$PO}(H$_2$O)], **5** (Table S1) and confirmed the same connectivity as the Tb^{3+} and Er^{3+} derivatives (complexes **3** and **4** respectively). Additionally, the single crystal X-ray structure of [Tm(F$_7$-acac)$_3${(Ar^F)$_3$PO}$_2$] was also determined and found to be isostructural to its congeners **6**, **7**, **9** and **10**.

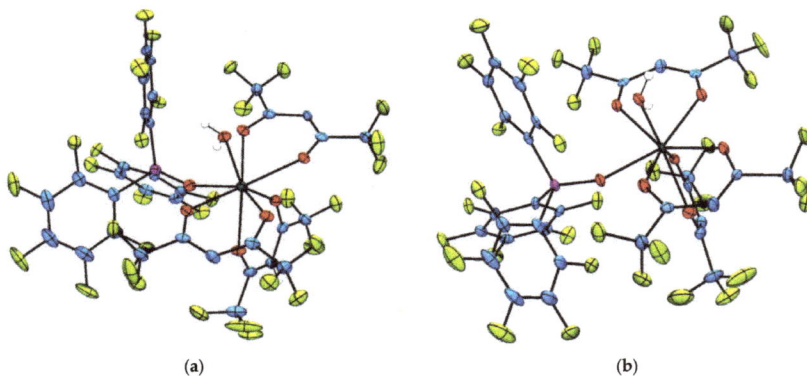

(a)   (b)

Figure 1. Representative solid state molecular structures of the [Ln(hfac)$_3${(Ar^F)$_3$PO}(H$_2$O)] complexes with thermal ellipsoids set at the 50% probability level and C–H atoms omitted for clarity. Atom colour key: C, blue, Ln^{3+}, dark green, O, red, H, white, P, pink, F, green. (**a**) Thermal ellipsoid plot of [Tb(hfac)$_3${(Ar^F)$_3$PO}(H$_2$O)], complex **3**; (**b**) Thermal ellipsoid plot of [Er(hfac)$_3${(Ar^F)$_3$PO}(H$_2$O)], complex **4**.

The molecular structures of the two hfac complexes [Ln(hfac)$_3${(Ar^F)$_3$PO}(H$_2$O)] (Ln = Tb^{3+}, **3** and Er^{3+}, **4**) are isostructural and the lanthanide(III) ions are octacoordinated by three bidentate hfac ligands, one monodentate perfluorinated triphenylphosphine oxide and one water molecule (Figure 1). The coordination geometries at the lanthanide centres are best described as distorted square antiprismatic with two hfac ligands occupying one square plane and the third hfac, the phosphine oxide and coordinated water molecule describing the second square plane. The average hfac–oxygen–Tb^{3+} and hfac–oxygen–Er^{3+}-bond distances are 2.383(3) and 2.357(5) Å respectively and lie in the range of bond distances previously reported [40,47]. Similarly, the $(Ar^F)_3PO–Tb^{3+}$ and $(Ar^F)_3PO–Er^{3+}$ bond distances of 2.314(3) and 2.284(5) Å compare well to those reported for the complex $[Er(hfac)_3\{(Ar^F)_3PO\}_2]$ (ave. = 2.323(4) Å) [34] and the $Tb^{3+}–OH_2$ and $Er^{3+}–OH_2$ bond distances are measured as 2.377(3) and 2.337(5) Å respectively, which are similar to the $Ln–OH_2$ bond distances commonly measured in trivalent lanthanide complexes of DOTA; range = 2.590 Å ($[CeDOTA]^-$) to 2.355 Å ($[YbDOTA]^-$), (DOTA = 1,4,7,10-tetraazacyclododecane-*N′,N″,N‴,N⁗*-tetraacetic acid) [48].

Single X-ray diffraction analysis of the fully perfluorinated lanthanide(III) complexes $[Ln(F_7-acac)_3\{(Ar^F)_3PO\}_2]$ (where Ln = Sm^{3+}, **6**, Eu^{3+}, **7**, Er^{3+}, **9**, and Yb^{3+}, **10**) confirmed the expected connectivity (Figure 2). Again, this series of complexes are isostructural and all crystallise with a single CH_2Cl_2 solvent molecule in the asymmetric unit cell. In each structure, the lanthanide(III) ion is eight coordinate and the coordination geometry best described as distorted square antiprismatic, where the perfluorinated phosphine oxide ligands lie in a mutually *cis* arrangement, but are located in different square planes, presumably to minimize steric interactions. Average F_7-acac oxygen–lanthanide(III) bond distances are 2.409(5) Å (**6**), 2.394(4) Å (**7**), 2.338(4) Å (**9**) and 2.317(4) Å (**10**) and follow the 8-coordinate lanthanide contraction as the series is progressed. Similarly, the average $(Ar^F)_3PO–Ln^{3+}$ distances become smaller traversing the 4f series (2.368(5) Å for **6** to 2.269(4) Å for **10**) and lie in the range of those reported previously for the complex $[Er(F_7-acac)_3\{(Ar^F)_3PO\}_2]$ which is a polymorph of **9** [47] and for the related complex $[Eu(F_7-acac)_3\{Ph_3PO\}_2]$ [49].

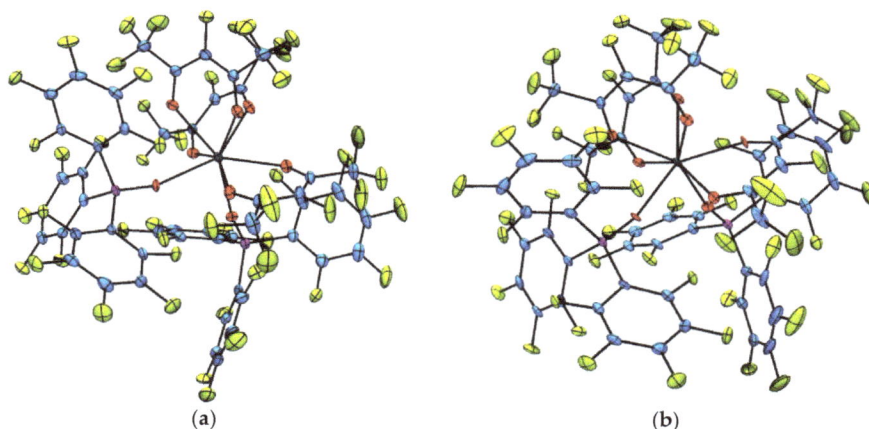

(a)   (b)

Figure 2. Representative solid state molecular structures of the $[Ln(F_7-acac)_3\{(Ar^F)_3PO\}_2]$ complexes with thermal ellipsoids set at the 50% probability level and lattice CH_2Cl_2 molecules omitted for clarity. Atom colour key: C, blue, Ln^{3+}, dark green, O, red, P, pink, F, green. (a) Thermal ellipsoid plot of $[Sm(F_7-acac)_3\{(Ar^F)_3PO\}_2]$, complex **6**; (b) Thermal ellipsoid plot of $[Yb(F_7-acac)_3\{(Ar^F)_3PO\}_2]$, complex **10**.

2.3. Photophysical Properties of the Complexes

With two families of complexes incorporating different degrees of fluorination in hand, we next resolved to investigate the effects of perfluorination on the optical properties of the complexes. The UV–visible absorption spectra of all the complexes **1–10** exhibit intense absorptions in the UV region of the electromagnetic spectrum with absorption maxima at ca. 230 and 300 nm (complexes **1–5**) attributable to the π–π* transitions of the perfluorinated-phenyl and β-diketonate chromophores respectively. In the absorption spectra of complexes **6–10**, an additional maximum centered at ca. 325 nm is also observed. Following UV excitation into all ligand absorption bands at 230, 280 or 355 nm, all the complexes exhibited lanthanide f-centered emission in the visible and near infra-red region at typical wavelengths for a given lanthanide(III) ion (Figure 3). In all cases, ligand sensitised emission was confirmed by recording the excitation spectrum at the emission maximum, which matched well to the absorption spectra indicating that both chromophores are involved in the sensitization process. Representative emission spectra for the $[Ln(F_7\text{-acac})_3\{(Ar^F)_3PO\}_2]$ family of complexes are illustrated in Figure 3. Interestingly, the relative intensities of the visible and near infra-red emission, in particular of the Er^{3+} and Yb^{3+} complexes (**9** and **10**), are much greater than those recorded for the Er^{3+} and Yb^{3+} nIR analogues of $[Ln(hfac)_3\{(Ar^F)_3PO\}(H_2O)]$ (**4** and **5**) reflecting the greater degree of vibrational quenching from the coordinated water molecule in the latter.

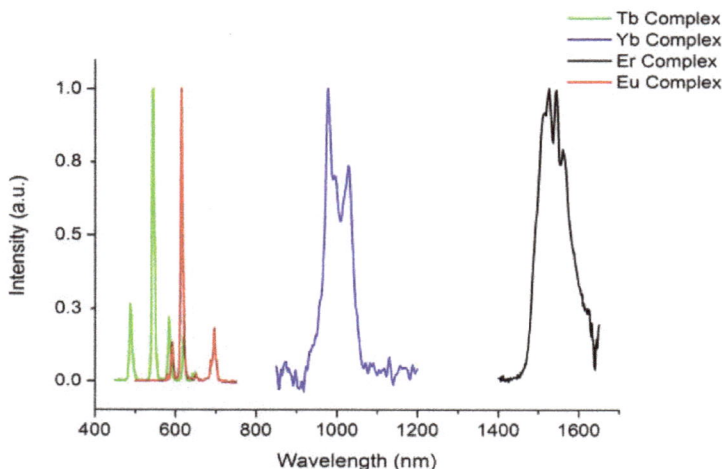

Figure 3. Corrected and normalized steady state emission spectra of $[Ln(F_7\text{-acac})_3\{(Ar^F)_3PO\}_2]$ complexes following 280 nm excitation in CH_2Cl_2 at 298 K (Ln^{3+} = Eu, **7**, Tb, **8**, Er, **9**, and Yb, **10**).

The luminescence lifetimes of the $[Ln(hfac)_3\{(Ar^F)_3PO\}(H_2O)]$ complexes recorded at the emission maxima are typical for both the visible and nIR emitting Ln^{3+} complexes, being in the microsecond range (Table 1). Interestingly, for the shorter lived Er^{3+} and Yb^{3+} complexes **4** and **5**, the kinetic traces following 337 nm excitation were best fitted to a biexponential decay, suggestive of two non-interconverting emissive species in solution on the timescale of the experiment. Given that the longer lived component of the kinetic trace of **5** (τ = 1.94 μs) can be compared to the related complex $Er(TPIP)_3$ (TPIP = tetraphenylimidodiphosphinate) [43], where τ_{CDCl3} = 5.0 μs, it seems reasonable to assume that the shorter lived component is due to the aqua species $[Ln(hfac)_3\{(Ar^F)_3PO\}(H_2O)]$, whereas the longer lived species is devoid of a coordinated solvent molecule. In the case of the longer lived visible emitting lanthanide ions in this series of hfac complexes, dynamic exchange of the labile water molecule is faster than the timescale of the experiment leading to the observation of an averaged solution lifetime [48]. It is worth noting here, that the luminescence lifetimes of all the complexes measured in solution are exceptionally sensitive to small amounts of water present in the solvent, therefore the solvent was thoroughly dried prior to use and reported values are reproducible in at least three independent measurements. Additionally, initial discrepancies in kinetic measurements led us to additionally record the photophysical properties of the complexes in the solid state (Table 1).

$$\Phi_{Ln} = \tau_{obs}/\tau_{rad} \tag{1}$$

Table 1. Photophysical properties of the complexes recorded in dry CH_2Cl_2 and in the solid state at 298 K upon excitation at 280 nm.

Complex	Complex	λ_{em} (nm)	τ_{CH2Cl2} (µs) [1]	Φ_{CH2Cl2} (%) [2]	τ_{solid} (µs) [1]	Φ_{solid} (%) [2]
[Sm(hfac)$_3${(Ar^F)$_3$PO}(H$_2$O)]	1	650	67.1	3.4	169	8.5
[Eu(hfac)$_3${(Ar^F)$_3$PO}(H$_2$O)]	2	617	5	5	116 (30%), 799 (70%)	53
[Tb(hfac)$_3${(Ar^F)$_3$PO}(H$_2$O)]	3	545	79.3	1.6	638	13
[Er(hfac)$_3${(Ar^F)$_3$PO}(H$_2$O)]	4	1550	1.94 (14%), 0.394 (86%) [4]	0.05 [3]	3.60	0.1
[Yb(hfac)$_3${(Ar^F)$_3$PO}(H$_2$O)]	5	980	3.31 (65%), 0.742 (35%) [4]	0.25 [3]	26.1	2.0
[Sm(F$_7$-acac)$_3${(Ar^F)$_3$PO}$_2$]	6	650	5	5	506	26
[Eu(F$_7$-acac)$_3${(Ar^F)$_3$PO}$_2$]	7	617	638	58	220 (14%), 741 (86%)	67 [3]
[Tb(F$_7$-acac)$_3${(Ar^F)$_3$PO}$_2$]	8	545	1520	30	421 (33%), 1400 (67%)	27 [3]
[Er(F$_7$-acac)$_3${(Ar^F)$_3$PO}$_2$]	9	1530	15.3 [6]	0.4	16.8	0.4
[Yb(F$_7$-acac)$_3${(Ar^F)$_3$PO}$_2$]	10	980	18.2 [6]	1.4	139	11

[1] Reported lifetimes are subject to an error of ±15%, indistinguishable data were obtained at 230 and 355 nm excitation; [2] The intrinsic quantum yield of Ln^{3+} centered emission was calculated using equation 1 using the following τ_{rad} (natural radiative lifetime) values: Sm^{3+} 1.98 ms ($^4G_{5/2}$) from reference [50]; Er^{3+} 4 ms ($^4I_{13/2}$) from reference [47]; Eu^{3+}, 1.11 ms ($^5D_0$) from reference [14]; Tb^{3+}, 5.1 ms ($^5D_4$) from reference [44]; and Yb^{3+} 1.3 ms ($^2F_{5/2}$) from reference [51]; [3] Value determined using the longest lifetime component of the emission; [4] Lifetime determined following 337 nm excitation using a ns pulsed N$_2$ laser; [5] Not determined due to very weak or lack of f-centered emission; [6] Complex unstable when excited with using a ns pulsed N$_2$ laser at 337 nm and 10 Hz.

Since we were unable to reliably determine the total quantum yield of emission for all the complexes, (due in part to the small f–f molar absorption extinction coefficients) we have instead estimated the intrinsic quantum yield of emission (Φ_{Ln}) based on calculated and published values of the Ln^{3+} radiative lifetime of a given emissive excited state in the absence of any non-radiative deactivation processes, τ_{rad} (Equation (1)) [2,37,40,47]. The quantum efficiency of the Ln^{3+} based emission can be determined by measuring the quantum yield based on emission following excitation into the f–f absorption bands. Alternatively, and in the absence of sufficiently intense emission upon direct excitation, the intrinsic quantum yield can be estimated from the ratio of the observed luminescence lifetime (τ_{obs}) with the radiative lifetime (τ_{rad}). However, since τ_{rad} values depend heavily on the coordination environment and refractive index of the medium of a given lanthanide ion, we have here chosen larger τ_{rad} values reported for structurally similar complexes as far as possible to give more reasonable estimations of Φ_{Ln} (Table 1) [14,15,39,50–54]. Nevertheless, caution must be exercised when interpreting these values since very small changes in the ligand structure, solvent and coordination geometry of the complex can lead to large discrepancies in τ_{rad} values. As a result, the values reported here therefore only serve as a guide. For the series of the $[Ln(hfac)_3\{(Ar^F)_3PO\}(H_2O)]$ complexes (1–5), generally the intrinsic quantum yields in fluid solution are typical for partially fluorinated β-diketonate complexes, but notably, the quantum yield values for the Sm^{3+} and Yb^{3+} derivatives 1 and 5 both in solution and in the solid state are relatively large (3.4%, 8.5%, 0.3% and 2% respectively). However, concentration quenching effects or local heating in the solid state samples cannot be ruled out and these values may in fact be considerably larger.

The effect of complete fluorination on the photophysical properties of the complexes is further evidenced by the significant increase in the luminescence lifetimes of the $[Ln(F_7-acac)_3\{(ArF)_3PO\}_2]$ series of complexes 6–10 (Table 1) of approximately one order of magnitude. This effect is particularly pronounced for the nIR emitting lanthanide ions Er^{3+} and Yb^{3+}, where in the case of the Er^{3+} derivative, 9, the lifetimes measured in solution and the solid state are very similar (15.3 and 16.8 μs respectively). This observation is in agreement with those of Monguzzi and co-workers who recorded identical lifetimes for this complex in $CDCl_3$ solution and in the solid state. For complex 9, the similar solution state and solid state lifetimes suggest that the Er^{3+} ions may be shielded to the same degree in their immediate coordination environment. Again, the detrimental effect of intermolecular quenching between neighbouring Er^{3+} ions and/or heating of the samples may result in an apparent lowering of the actual lifetime and quantum yield values. Here, the intrinsic quantum yield of 0.4% is considerably lower than the theoretical value where the radiative lifetime is equal to 4 ms [47], which suggests that residual CH_2Cl_2 solvent (as observed by X-ray crystallography) may play a role in lowering the overall quantum yield of emission. This observation is borne out to a certain degree by examination of the kinetic data for the Yb^{3+} complex 10, where the Yb^{3+} excited state is less susceptible to vibrational quenching by C–H oscillators [42]. This complex exhibits a marked enhancement (by a factor of >5) in both solution and solid state luminescence lifetimes and quantum yields when compared to the hexafluorinated aqua complex $[Yb(hfac)_3\{(Ar^F)_3PO\}(H_2O)]$, 5. Indeed, the fully perfluorinated derivative 10 exhibits a remarkably long luminescence lifetime of 139 μs in the solid phase (18 μs in CH_2Cl_2 solution) and an intrinsic quantum yield of Yb^{3+} based emission of 11%. This compares well to the Yb^{3+} bis-bipyridine *N*-oxide derivative of the Lehn cryptand described by Seitz that exhibits a room temperature solution luminescence lifetime of 26.1 μs in deuterated methanol [31]. In this system, perdeuteration of the ligand results in an extraordinarily long solution lifetime of 172 μs and the highest reported intrinsic quantum yield of Yb^{3+} emission in solution to date (26%). Together, these observations further highlight the fact that purposeful reduction in competitive vibrational quenching thereby substantially increasing the lifetime of the lanthanide based emission is key to achieving large nIR quantum yields of emission.

Interestingly, in the case of the Sm^{3+} complex 6, no f-centered emission was observed in solution at room temperature; the emission spectrum being dominated by residual ligand centered transitions. This clearly indicates that the F_7-acac ligand is a poor sensitizer for Sm^{3+} and that the C–F bond in the β-diketonate unit quenches the Sm^{3+} based emission considerably [44]. In the solid state however, complex **6** exhibits a typical Sm^{3+} based emission profile with a long luminescence lifetime of 506 µs and a large calculated intrinsic quantum yield of 26% [16,55].

3. Materials and Methods

3.1. General Details

All chemical reagents were obtained from Sigma-Aldrich chemical company (Dorset, England) apart from the lanthanide acetate hydrate salts ($Ln(OAc)_3 \cdot xH_2O$), which were obtained from Alfa Aesar (Heysham, UK) and 1,1,1,3,5,5,5-heptafluoroacetylacetone (F_7-acac) from Apollo Scientific (Manchester, UK) and were used as supplied or recrystallised before use. The compounds tris(pentafluorophenyl)phosphine oxide (Ar^F_3PO) [41] and lanthanide(III) tris(1,1,5,5,5-hexafluoro-2,4-pentanedionate) dihydrates ($Ln(hfac)_3 \cdot 2H_2O$, Ln = Sm^{3+}, Eu^{3+}, Tb^{3+}, Tm^{3+}, Er^{3+} and Yb^{3+}) [34] were prepared according to literature and modified literature procedures. Reagent grade anhydrous solvents were either obtained by passing through activated alumina columns (Innovative Technologies) or dried over potassium (hexane and pentane) or over CaH_2 (CH_2Cl_2) and were distilled and degassed prior to use. Deuterated solvents for multinuclear NMR studies were dried over 4 Å molecular sieves before use. All air sensitive experiments were performed using standard Schlenk line techniques.

NMR spectra were recorded on either a Bruker Avance 400 spectrometer (Bruker UK, Coventry, UK), operating frequency 400 MHz (^{1}H), 101 MHz (^{13}C), 162 MHz (^{31}P) and 376 MHz (^{19}F) variable temperature unit set at 295 K, or on a Varian Mecury-200 spectrometer (Varian Associates, Palo Alto, CA, USA) at 298 K (^{1}H at 200 MHz, ^{13}C at 50 MHz, ^{31}P at 81 MHz, ^{19}F at 188 MHz). Chemical shifts are reported in parts per million (ppm) relative to TMS (^{1}H), 85% H_3PO_4 (*ortho*-phosphoric acid) (^{31}P{^{1}H}) and $CFCl_3$ (^{19}F). All ^{1}H NMR spectra were internally referenced to residual proton resonances in $CDCl_3$ or d_6-acetone.

Mass spectra were obtained using either MALDI from CH_2Cl_2 solutions with a dithrinol matrix on a Shimadzu Axima confidence spectrometer (Shimazdu, Kratos site, Manchester, UK, for all complexes) or electrospray mass spectrometry, performed on a Micromass Platform II system (ligands and precursors). For all of the complexes **1–10**, no identifiable molecular ion peaks or fragmentation products were observable.

Elemental analyses on the compounds were performed by M. Jennings and colleagues in the microanalytical laboratory in the School of Chemistry at the University of Manchester; a Carlo ERBA Instruments CHNS–O EA1108 elemental analyzer (Carlo ERBA Instruments, Milan, Italy) was used for C, H and N analysis and a Fisons Horizon elemental analysis ICP-OED spectrometer (VG Elemental, Winsford, UK) for metals.

X-ray diffraction data for compounds **3**, **6**, **7**, and [Tm(F_7-acac)$_3${(Ar^F)$_3$PO}$_2$]) were collected using an Rigaku Oxford Diffraction Xcalibur, Sapphire2 diffractometer (Rigaku, Tokyo, Japan), utilising graphite-monochromated Mo Kα X-ray radiation (λ = 0.71073 Å); compounds 9 and 10, or on a Bruker X8 Prospector diffractometer (Bruker, Billerica, MA, USA) utilising graphite-monochromated Cu Kα X-ray radiation (λ = 1.54178 Å) and compound 4 on a Bruker AXS SMART diffractometer (Bruker, Billerica, MA, USA) using graphite-monochromated Mo Kα X-ray radiation (λ = 0.71073 Å). Data were corrected for Lorentz and polarisation factors and empirical absorption corrections applied [56,57]. Crystal data, data collection and structural refinement parameters are given in Table S1. The structures were solved by direct methods using the program SHELXT [58] or OLEX-2 [59]. Full matrix refinement on F^2 and all further calculations were performed using OLEX-2 or ShelXL [58]. The non-H atoms were refined anisotropically and hydrogen atoms were positioned in idealised sites and were allowed to ride on their parent C or N atoms.

Absorption spectra were recorded in dry CH_2Cl_2 on a T60U spectrometer (PG Instruments Ltd., Lutterworth, UK) using fused quartz cells with a path length of 1 cm or on a double-beam Cary Varian 500 scan UV–vis–nIR spectrophotometer over the range 300–1300 nm.

All solution luminescence measurements were recorded on compounds dissolved in dry CH_2Cl_2 solutions using screw or Teflon™ (Hellma UK Ltd., Southend on Sea, UK) capped fused quartz cuvettes with a 1 cm or 0.1 cm path length. All measurements were recorded within 30 min of sample preparation to avoid ingress of oxygen and moisture. Luminescence measurements of solid samples were recorded using finely divided powdered samples held in between two 10 cm^2 quartz plates. All steady state emission and excitation spectra were recorded on an Edinburgh Instrument FP920 Phosphorescence Lifetime Spectrometer (Edinburgh Instruments, Livingston, Scotland) equipped with a 450 watt steady state xenon lamp, a 5 watt microsecond pulsed xenon flashlamp, (with single 300 mm focal length excitation and emission monochromators in Czerny Turner configuration), a red sensitive photomultiplier in peltier (air cooled) housing (Hamamatsu R928P), and a liquid nitrogen cooled nIR photomultiplier (Hamamatsu, Hamamatsu City, Shizuoka Prefecture, Japan). Lifetime data were recorded following excitation with the microsecond flashlamp using time correlated single photon counting (PCS900 plug-in PC card for fast photon counting). Lifetimes were obtained by tail fit on the data obtained and quality of fit judged by minimization of reduced chi-squared and residuals squared. For the Er^{3+} and Yb^{3+} complexes **4** and **5**, the sample was excited using a pulsed nitrogen laser (337 nm) operating at 10 Hz. Light emitted at right angles to the excitation beam was focused onto the slits of a monochromator, which was used to select the appropriate wavelength. The growth and decay of the luminescence at selected wavelengths was detected using a germanium photodiode (Edinburgh Instruments, EI-P, Edinburgh Instruments, Livingstone, Scotland) and recorded using a digital oscilloscope (Tektronix TDS220, Tektronix Inc., Beaverton, OR, USA) before being transferred to a PC for analysis. Luminescence lifetimes were obtained by iterative reconvolution of the detector response (obtained by using a scatterer) with exponential components for growth and decay of the metal centred luminescence, using a spreadsheet running in Microsoft Excel. The details of this approach have been discussed elsewhere [60,61]. Unless otherwise stated, fitting to a double exponential decay yielded no improvement in fit as judged by minimisation of residual squared and reduced chi squared. Note that the Tm^{3+} complexes $[Tm(hfac)_3\{(Ar^F)_3PO\}(H_2O)]$ and $[Tm(F_7\text{-}acac)_3\{(Ar^F)_3PO\}_2]$ were found to be non-emissive in CH_2Cl_2 solutions at room temperature upon UV excitation (230–360 nm).

3.2. Synthetic Procedures

3.2.1. Preparation of Tris(pentafluorophenyl)phosphine Oxide, $(C_6F_5)_3PO$ (Ar^F_3PO)

According to a modification of a literature procedure [41], tris(pentafluorophenyl)phosphine (1.031 g, 1.9 mmol) was dissolved in chloroform (40 mL), cooled to 0 °C in an ice bath and stirred for 30 min. 3-Chloroperoxybenzoic acid (0.3013 g, 2.2 mmol) was dissolved in chloroform (10 mL) and then added dropwise to the reaction mixture over 10 min. The reaction mixture was again cooled to 0 °C in an ice bath and left to warm to room temperature and stirred for 48 h. After this time, saturated sodium hydrogen carbonate solution (50 mL) was added and the reaction mixture stirred for 30 min. The organic soluble products were extracted with chloroform (3 × 50 mL) and dried over $MgSO_4$. The organic layer was removed by rotary evaporation and the white powder dried in vacuo to give 0.89 g of tris(pentafluorophenyl)phosphine oxide (81% yield). All data were consistent with those documented in the literature. ES+ MS (MeCN) *m/z* 549 $[M + H]^+$ (44%), 571 $[M + Na]^+$ (100%), 1119 $[2M + Na]^+$ (62%), 1667 $(3M + Na)^+$ (67%).

NMR/$CDCl_3$ (400 MHz) δ_F: 131.30 (d, 2F, $^3J_{FF}$ = 22.4 Hz, *ortho*-CF), −141.32 (t, 1F, $^3J_{FF}$ = 18.6, 22.6 Hz, *para*-CF), −157.27 (tt, 2F, $^3J_{FF}$ = 22.6, 18.8 Hz, $^4J_{FF}$ = 7.5, 3.8 Hz, *meta*-CF); δ_P: −8.26 (s, Ar^F_3PO). UV–vis (CH_2Cl_2) λ = 230, 277 nm.

3.2.2. Preparation of $Ln(hfac)_3 \cdot 2H_2O$

According to a literature procedure reported for Nd^{3+} [34], The corresponding lanthanide acetate hydrate ($Ln(OAc)_3 \cdot xH_2O$) (15 mmol) was dissolved in deionised water (20 mL) with stirring in an ice bath. 1,1,1,5,5,5-hexafluoro-2,4-pentanedione (hfac) (5.0 g, 24 mmol) was dissolved in methanol and was added dropwise to the lanthanide acetate solution. The reaction mixture was stirred for 3 h in an ice bath and a further 65 h at room temperature. The solvent was then removed using a rotary evaporator and the resulting product was recrystallised from MeOH.

Yields: Tb(hfac)$_3$·2H$_2$O = 93% (pale green-blue powder), Sm(hfac)$_3$·2H$_2$O (pale yellow powder) = 93%, Er(hfac)$_3$·2H$_2$O (pink powder) = 86%, Yb(hfac)$_3$·2H$_2$O (cream powder) = 56%, Tm(hfac)$_3$·2H$_2$O (white powder) = 43%, Eu(hfac)$_3$·2H$_2$O (cream powder) = 69%. All spectroscopic data were consistent with the formulation of the compounds as Ln(hfac)$_3$·2H$_2$O as previously reported.

3.2.3. Preparation of [Ln(hfac)$_3${(Ar^F)$_3$PO}(H$_2$O)], Ln = Sm^{3+}, Eu^{3+}, Tb^{3+}, Er^{3+}, Yb^{3+} (**1–5**) from Ln(hfac)$_3$·2H$_2$O

In air, Ln(hfac)$_3$·2H$_2$O (0.31 mmol) and ^{F}Ph$_3$PO (0.31 mmol) were added to a round bottomed flask and dissolved in CH$_2$Cl$_2$ (150 mL). The reaction mixture was sonicated for 1 h. The reaction mixture was then filtered under gravity. All volatiles were removed by rotary evaporation and the resultant crude powders recrystallised from CH$_2$Cl$_2$ by slow evaporation and the crystalline products filtered under gravity, washed with CH$_2$Cl$_2$ and dried in vacuo. Once isolated, all the complexes are air stable for extended time periods (>4 years) are moderately soluble in acetone and sparingly soluble in CH$_2$Cl$_2$ and CHCl$_3$ precluding acquisition of NMR data in some cases.

[Sm(hfac)$_3${(Ar^F)$_3$PO}(H$_2$O)] (**1**): Isolated in 41% yield as pale yellow crystals (170 mg). NMR/CDCl$_3$ (400 MHz) δ_H: 7.51 (s, C*H*-hfac); δ_F: −77.79 (s, 18F, C*F*$_3$-hfac), −131.37 (s, 6F) *ortho*-C*F*), −140.37 (br, 3F, *para*-C*F*), −157.05 (m, 6F, *meta*-C*F*) δ_P {^{1}H}: −7.44 (s, (Ar^F)$_3$PO). UV–vis (CH$_2$Cl$_2$) λ = 234 nm ((Ar^F)$_3$PO), 306 nm ((Ar^F)$_3$PO and hfac). Anal. Calcd. For C$_{33}$H$_5$O$_8$F$_{33}$PSm: C 29.63, H 0.37, N 0.0, P 2.31, Sm 11.24. Found: C 29.52, H 0.0, N 0.0, P 2.06, Sm 11.63.

[Eu(hfac)$_3${(Ar^F)$_3$PO}(H$_2$O)] (**2**): Isolated in 38% yield as colourless crystals (158 mg). NMR/CDCl$_3$ (400 MHz) δ_H, δ_F, δ_P {^{1}H}: no signals observed. UV–vis (CH$_2$Cl$_2$) λ = 232 nm ((Ar^F)$_3$PO), 301 nm ((Ar^F)$_3$PO and hfac). Anal. Calcd. For C$_{33}$H$_5$O$_8$F$_{33}$PEu: C 29.59, H 0.38, N 0.0. Found: C 28.64, H 0.14, N 0.0.

[Tb(hfac)$_3${(Ar^F)$_3$PO}(H$_2$O)] (**3**): Isolated in 49% yield as pale green crystals (205 mg). NMR/CDCl$_3$ (400 MHz) δ_H: −0.19 (br, C*H*-hfac); δ_F: −45.54 (br, C*H*-hfac), −131.88 (br, *ortho*-C*F*), −141.34 (br, *para*-C*F*), −157.33 (br, *meta*–C*F*). δ_P {^{1}H}: −10.06 (br). UV–vis (CH$_2$Cl$_2$) λ = 231 nm ((Ar^F)$_3$PO), 303 nm ((Ar^F)$_3$PO and hfac). Anal. Calcd. For C$_{33}$H$_5$O$_8$F$_{33}$PTb: C 29.44, H 0.37, N 0.0, P 2.30, Tb 11.80. Found: C 29.32, H 0.0, N 0.0, P 1.99, Tb 10.80.

[Er(hfac)$_3${(Ar^F)$_3$PO}(H$_2$O)] (**4**): Isolated in 49% yield as pink crystals (206 mg). NMR/CDCl$_3$ (400 MHz) δ_H: no signal observed; δ_F: −94.62 (br, C*F*$_3$-hfac), −137.20 (br, *ortho*/*meta*/*para*-C*F*), −153.32 (br, 6F, *ortho*/*meta*/*para*-C*F*); δ_P {^{1}H}: no signal observed. UV–vis (CH$_2$Cl$_2$) λ = 233 nm ((Ar^F)$_3$PO), 302 nm ((Ar^F)$_3$PO and hfac). Anal. Calcd. For C$_{33}$H$_5$O$_8$F$_{33}$PEr: C 29.25, H 0.37, N 0.0, P 2.29, Er 12.38. Found: C 29.28, H 0.0, N 0.0, P 1.97, Er 12.10.

[Yb(hfac)$_3${(Ar^F)$_3$PO}(H$_2$O)] (**5**): Isolated in 59% yield as colourless crystals (249 mg). NMR/CDCl$_3$ (400 MHz) δ_H: 8.07 (br, C*H*-hfac); δ_F: −87.83 (s, 2F, *ortho*/*meta* C*F*), −123.75 (br, C*F*$_3$-hfac), −138.98 (s, 1F, *para*-C*F*), −155.68 (s, 2F, *ortho*/*meta* C*F*); δ_P {^{1}H}: no signal observed. UV–vis (CH$_2$Cl$_2$) λ = 232 ((Ar^F)$_3$PO), 302 nm ((Ar^F)$_3$PO and hfac). Anal. Calcd. For C$_{33}$H$_5$O$_8$F$_{33}$PYb: C 29.14, H 0.37, N 0.0, P 2.27. Found: C 28.82, H 0.0, N 0.0, P 1.73.

[Tm(hfac)$_3${(Ar^F)$_3$PO}(H$_2$O)]: Isolated in 26% yield as colourless crystals (109 mg). Anal. Calcd. For C$_{33}$H$_5$O$_8$F$_{33}$PTm: C 29.22, H 0.37, N 0.0, P 2.28, Tm 12.45. Found: C 29.18, H 0.0, N 0.0, P 2.05, Tm 12.36.

3.2.4. Preparation of [Ln(hfac)$_3${(Ar^F)$_3$PO}(H$_2$O)], Ln = Sm^{3+}, Eu^{3+}, Tb^{3+}, Er^{3+}, Yb^{3+} (**1–5**) from Ln(OAc)$_3 \cdot x$H$_2$O under N$_2$

Under N$_2$, an oven dried Schlenk was charged with 0.26 mmol of the appropriate lanthanide acetate hydrate (Ln(OAc)$_3 \cdot x$H$_2$O) and 0.26 mmol (142 mg) (Ar^F)$_3$PO and 25 mL of dry CH$_2$Cl$_2$ added by cannula or syringe. A solution of Hfac (0.78 mmol, 154 mg) in minimal CH$_2$Cl$_2$ (~2 mL) was then added dropwise and the reaction mixture heated at reflux temperature for 1 hour under a flow of N$_2$ and then cooled to room temperature. After this time, the solution was filtered and the filtrate layered with hexane or pentane and placed in the freezer at −18 °C. After several days, the crystalline products were isolated by filtration, washed with cold CH$_2$Cl$_2$ and dried in vacuo to yield the title product.

[Sm(hfac)$_3${(Ar^F)$_3$PO}(H$_2$O)] (**1**): Isolated in 53% yield (184 mg). All characterization data are consistent with the proposed structure.

[Tb(hfac)$_3${(Ar^F)$_3$PO}(H$_2$O)] (**3**): Isolated in 44% yield (155 mg). All characterization data are consistent with the proposed structure.

[Er(hfac)$_3${(Ar^F)$_3$PO}(H$_2$O)] (**4**): Isolated in 16% yield (55 mg). All characterization data are consistent with the proposed structure.

[Yb(hfac)$_3${(Ar^F)$_3$PO}(H$_2$O)] (**5**): Isolated in 48% yield (170 mg). All characterization data are consistent with the proposed structure.

3.2.5. Preparation of [Ln(F$_7$-acac)$_3${(Ar^F)$_3$PO}$_2$] Ln = Sm^{3+}, Eu^{3+}, Tb^{3+}, Er^{3+}, Yb^{3+} (**6–10**) from Ln(OAc)$_3 \cdot x$H$_2$O under N$_2$

Under N$_2$, an oven dried Schlenk was charged with 0.33 mmol of the appropriate lanthanide acetate hydrate (Ln(OAc)$_3 \cdot x$H$_2$O) and 0.66 mmol (142 mg) (Ar^F)$_3$PO. 25 mL of dry CH$_2$Cl$_2$ was then added by cannula or syringe. 1,1,1,3,5,5,5-heptfluoro-2,4-pentanedione (0.94 mmol) was subsequently rapidly added and the reaction mixture heated at reflux temperature for 1 hour under a flow of N$_2$, then cooled to room temperature. After this time, dry pentane (40 mL) was added and the Schlenk flask placed in the freezer at −18 °C. After one week, the crystalline products were isolated by filtration, washed with cold CH$_2$Cl$_2$ and dried under vacuum suction to yield the title compounds. The products are slightly hygroscopic when stored under ambient conditions for extended periods of time (>1 year). All complexes are soluble in acetone and sparingly soluble in CH$_2$Cl$_2$ and CHCl$_3$.

[Sm(F$_7$-acac)$_3${(Ar^F)$_3$PO}$_2$] (**6**): Isolated in 42% yield as a pale yellow crystalline solid (289 mg). NMR/d_6-acetone (200 MHz) δ_F: −74.67 (d, 18F, $^4J_{FF}$ = 18.3 Hz, CF$_3$-hfac), −134.15 (d, 12F, $^3J_{FF}$ = 10.9 Hz, $^4J_{FF}$ = 8.1 Hz, *ortho*-CF), −145.68 (m, 6F, *para*-CF), −161.28 (m, 12F, *meta*-CF), −192.30 (br, 3H, CF F$_7$-acac); δ_P {^{1}H}: −9.20 (s, (Ar^F)$_3$PO). UV–vis (CH$_2$Cl$_2$) λ = 230, 272 nm ((Ar^F)$_3$PO), 328, 356 (sh) nm (F$_7$-acac). Anal. Calcd. For C$_{51}$O$_8$F$_{51}$P$_2$Sm·2CH$_2$Cl$_2$: C 30.43, H 0.19, N 0.0, Sm 7.20. Found: C 30.25, H 0.0, N 0.0, Sm 6.85.

[Eu(F$_7$-acac)$_3${(Ar^F)$_3$PO}$_2$] (**7**): Isolated in 37% yield as a pale yellow crystalline solid (242 mg). NMR/d_6-acetone (200 MHz) δ_F: −78.14 (d, 18F, $^4J_{FF}$ = 17.4 Hz, CF$_3$ F$_7$-acac), −134.11 (dd, 12F, $^3J_{FF}$ = 22.0 Hz, $^4J_{FF}$ = 1.9 Hz, *ortho*-CF), −145.68 (m, 6F, *para*-CF), −161.29 (m, 12F, *meta*-CF), −170.01 (m, 3F, CF F$_7$-acac); δ_P {^{1}H}: −7.80 (s, (Ar^F)$_3$PO). UV–vis (CH$_2$Cl$_2$) λ = 232, 274 nm ((Ar^F)$_3$PO), 324, 358 (sh) nm (F$_7$-acac). For C$_{51}$O$_8$F$_{51}$P$_2$Eu·CH$_2$Cl$_2$: C 31.10, H 0.1, N 0.0, Eu 7.57. Found: C 30.73, H 0.0, N 0.0, Eu 8.57.

[Tb(F$_7$-acac)$_3${(Ar^F)$_3$PO}$_2$] (**8**): Isolated in 47% yield as a pale yellow crystalline solid (312 mg). NMR/d_6-acetone (200 MHz) δ_F: −58.84 (br, CF$_3$ F$_7$-acac), −134.20 (d, 6F, $^4J_{FF}$ = 20.7 Hz, *ortho*-CF), −145.65 (m, 3F, *para*-CF), −161.28 (m, 6F, *meta*-CF), −108.00 (br, CF F$_7$-acac); δ_P{^{1}H}: −11.61 (br, (Ar^F)$_3$PO). UV–vis (CH$_2$Cl$_2$) λ = 230, 276 nm ((Ar^F)$_3$PO), 330 nm (F$_7$-acac). Anal. Calcd. For C$_{51}$O$_8$F$_{51}$P$_2$Tb·CH$_2$Cl$_2$: C 30.99, H 0.1, N 0.0, Tb 7.89. Found: C 30.92, H 0.0, N 0.0, Tb 7.20.

[Er(F$_7$-acac)$_3${(Ar^F)$_3$PO}$_2$] (**9**): Isolated in 40% yield as a pale yellow crystalline solid (289 mg). NMR/d_6-acetone (200 MHz) δ_F: −87.28 (br, CF$_3$ F$_7$-acac), −133.90 (d, 6H, $^4J_{FF}$ = 21.8 Hz, *ortho*-CF), −145.47 (m, 3H, *para*-CF), −161.27 (m, 6H, *meta*-CF), −213.41 (br, CF F$_7$-acac); δ_P{^{1}H}: −10.63 (s, (Ar^F)$_3$PO). UV–vis (CH$_2$Cl$_2$) λ = 232, 274 nm ((Ar^F)$_3$PO), 322, 354 (sh) nm (F$_7$-acac). Anal. Calcd. For C$_{51}$O$_8$F$_{51}$P$_2$Er·3CH$_2$Cl$_2$: C 29.57, H 0.28, N 0.0, Er 7.63. Found: C 29.10, H 0.0, N 0.0, Er 7.22.

[Yb(F$_7$-acac)$_3${(Ar^F)$_3$PO}$_2$] (**10**): Isolated in 27% yield as a pale yellow crystalline solid (213 mg). NMR/d_6-acetone (200 MHz) δ_F: −87.37 (br, CF$_3$ F$_7$-acac), −133.99 (d, 6H, $^4J_{FF}$ = 20.0 Hz, *ortho*-CF), −145.66 (m, *para*-CF), −161.28 (m, 6H, *meta*-CF), −208.20 (br, CF F$_7$-acac); δ_P {^{1}H}: −8.93 (br, (Ar^F)$_3$PO). UV–vis (CH$_2$Cl$_2$) λ = 230, 274 nm ((Ar^F)$_3$PO), 322, 354 (sh) nm (F$_7$-acac). Anal. Calcd. For C$_{51}$O$_8$F$_{51}$P$_2$Yb·5CH$_2$Cl$_2$: C 28.39, H 0.43, N 0.0, Yb 7.30. Found: C 27.82, H 0.0, N 0.0, Yb 6.98.

[Tm(F$_7$-acac)$_3${(Ar^F)$_3$PO}$_2$]: Isolated in 48% yield as a white crystalline solid (321 mg). NMR/d_6-acetone (200 MHz) δ_F: −98.08 (br, CF$_3$ F$_7$-acac), −133.88 (s, 6H, *ortho*-CF), −145.47 (m, *para*-CF), −161.27 (m, 6H, *meta*-CF), −245.04 (br, CF F$_7$-acac); δ_P: −9.10 (br, (Ar^F)$_3$PO). Anal. Calcd. For C$_{51}$O$_8$F$_{51}$P$_2$Tm·CH$_2$Cl$_2$: C 30.84, H 0.0, N 0.0, Tm 8.34. Found: C 30.47, H 0.0, N 0.0, Tm 8.74.

3.3. X-ray Crystallographic Data for Compounds **3**, **4**, **6**, **7**, **9** *and* **10**

The X-ray crystallographic information files (cif) for all the single crystal X-ray structures reported herein have been deposited with the Cambridge Crystallographic Database (CCDC), numbers: 1472376–1472382. These data are available free of charge at www.ccdc.cam.ac.uk. For data collection and structural refinement details, see Table S1.

4. Conclusions

In summary, two families of emissive heavily fluorinated lanthanide(III) β-diketonate complexes bearing monodentate perfluorinated tris phenyl phosphine oxide ligands, [Ln(hfac)$_3${(Ar^F)$_3$PO}(H$_2$O)] and [Ln(F$_7$-acac)$_3${(Ar^F)$_3$PO}$_2$] (where Ln = Sm^{3+}, Eu^{3+}, Tb^{3+}, Er^{3+} and Yb^{3+}) have been prepared and characterized by NMR spectroscopy, single crystal X-ray diffraction and luminescence spectroscopy. In depth photophysical studies on the complexes in CH$_2$Cl$_2$ solution and in the solid state have shown that replacing both the bound inner sphere H$_2$O molecule and C–H oscillator in the β-diketonate ligand backbone in [Ln(hfac)$_3${(Ar^F)$_3$PO}(H$_2$O)] by a second (Ar^F)$_3$PO ligand and C–F bond respectively in [Ln(F$_7$-acac)$_3${(Ar^F)$_3$PO}$_2$] has a dramatic effect on the photophysical properties of the complexes. This effect is particularly notable for the near infra-red emitting ion Yb^{3+}, where a five fold increase in luminescence lifetime and quantum yield is observed in [Yb(F$_7$-acac)$_3${(Ar^F)$_3$PO}$_2$] (**10**) compared to [Ln(hfac)$_3${(Ar^F)$_3$PO}(H$_2$O)] (**5**). The presented data herein conclude that replacing all C–H oscillators in the immediate coordination environment of the Ln^{3+} ions with C–F bonds substantially reduces the degree of competitive vibrational quenching, particularly for the nIR emitting trivalent lanthanides in solution which in turn leads to more intense and longer lived f-centered emission. Here, this is particularly evident for the Sm^{3+} and Yb^{3+} complexes **1**, **5** and **10**, which exhibit both long luminescence lifetimes and relatively large intrinsic quantum yields of emission. Such an approach combined with intentional reduction of the radiative lifetime of the lanthanide based emission as demonstrated by Seitz [37] may find increasing use in the development of optical imaging probes and highly emissive materials for optoelectronics amongst other applications in the near future.

Supplementary Materials: The following are available online at www.mdpi.com/2304-6740/4/3/27/s1, Table S1: Single Crystal X-ray Data Collection and Structural Refinement for the Complexes and unit cell parameters for **5**; Figures S1–S7: ^{19}F and ^{31}P{^{1}H} NMR spectra for selected complexes (**1**, **3**, **4**, **5**, **8**, **9** and **10**).

Acknowledgments: We thank the EPSRC for funding a Career Acceleration Fellowship (Louise S. Natrajan), postdoctoral funding (Adam N. Swinburne) and a studentship (Simon Randall) (grant number EP/G004846/1). We also thank the Leverhulme Trust for additional postdoctoral funding (Adam N. Swinburne, Fabrizio Ortu) (RL-2012-072) and a research Leadership award (Louise S. Natrajan) and the University of Manchester for project student support. This work was also funded by the EPSRC (grant number EP/K039547/1). We are additionally grateful to Stephen Faulkner for the loan of his facilities for the nIR measurements of complexes **4** and **5**.

Author Contributions: Louise S. Natrajan conceived and designed the experiments; Adam N. Swinburne, Tsz Ling Chan, Madeleine H. Langford Paden and Louise S. Natrajan performed the experiments; Adam N. Swinburne, Louise S. Natrajan, Simon Randall, Fabrizio Ortu and Alan M. Kenwright analyzed the data; Alan M. Kenwright contributed analysis tools; Louise S. Natrajan wrote the manuscript.

Conflicts of Interest: The authors declare no conflict of interest.

Abbreviations

The following abbreviations are used in this manuscript:

NMR	nuclear magnetic resonance
nIR	near Infra-Red
UV/vis	Ultra-violet/visible
hfac	1,1,1,5,5,5-hexafluoro-2,4-pentanedione
F_7-acac	1,1,1,3,5,5,5-heptafluoroacetylacetone
$(Ar^F_3)PO$	tris(pentafluorophenyl)phosphine oxide
TPIP	tetraphenylimidodiphosphinate
F-TPIP	perfluorotetraphenylimidodiphosphinate
OAc	acetate
Ln	lanthanide
DOTA	1,4,7,10-tetraazacyclododecane- N',N'',N''',N''''-tetraacetic acid

References

1. Bünzli, J.-C.G.; Eliseeva, S.V. Intriguing aspects of lanthanide luminescence. *Chem. Sci.* **2013**, *4*, 1939–1949. [CrossRef]
2. Binnemans, K. Lanthanide-based luminescent hybrid materials. *Chem. Rev.* **2009**, *109*, 4283–4374. [CrossRef] [PubMed]
3. De Bettencourt-Dias, A. Lanthanide-based emitting materials in light-emitting diodes. *Dalton Trans.* **2007**, 2229–2241. [CrossRef] [PubMed]
4. Suyver, J.F.; Aebischer, A.; Biner, D.; Gerner, P.; Grimm, J.; Heer, S.; Krämer, K.E.; Reinhard, C.; Güdel, H.U. Novel materials doped with trivalent lanthanides and transition metal ions showing near-infrared to visible photon upconversion. *Opt. Mater.* **2005**, *27*, 1111–1130. [CrossRef]
5. Meijerink, A.; Wegh, R.; Vergeer, P.; Vlugt, T. Photon management with lanthanides. *Opt. Mater.* **2006**, *28*, 575–581. [CrossRef]
6. Harvey, P.; Oakland, C.; Driscoll, M.D.; Hay, S.; Natrajan, L.S. Ratiometric detection of enzyme turnover and flavin reduction using rare-earth upconverting phosphors. *Dalton Trans.* **2014**, *43*, 5265–5268. [CrossRef] [PubMed]
7. Bünzli, J.-C.G. Lanthanide luminescence for biomedical analyses and imaging. *Chem. Rev.* **2010**, *110*, 2729–2755. [CrossRef] [PubMed]
8. Montgomery, C.P.; Murray, B.S.; New, E.J.; Pal, R.; Parker, D. Cell-penetrating metal complex optical probes: Targeted and responsive systems based on lanthanide luminescence. *Acc. Chem. Res.* **2009**, *42*, 925–937. [CrossRef] [PubMed]
9. Beeby, A.; Clarkson, I.M.; Faulkner, S.; Botchway, S.; Parker, D.; Williams, J.A.G.; Parker, A.W. Luminescence imaging microscopy and lifetime mapping using kinetically stable lanthanide (III) complexes. *J. Photochem. Photobiol. B Biol.* **2000**, *57*, 83–89. [CrossRef]
10. Vuojola, J.; Soukka, T. Luminescent lanthanide reporters: New concepts for use in bioanalytical applications. *Methods Appl. Fluoresc.* **2014**, *2*, 1–28. [CrossRef]
11. Beeby, A.; Burton-Pye, B.P.; Faulkner, S.; Motson, G.R.; Jeffery, J.C.; McCleverty, J.A.; Ward, M.D. Synthesis and near-IR luminescence properties of neodymium(III) and ytterbium(III) complexes with poly(pyrazolyl)borate ligands. *J. Chem. Soc. Dalton Trans.* **2002**, 1923–1928. [CrossRef]
12. Faulkner, S.; Carrie, M.-C.; Pope, S.J.A.; Squire, J.; Beeby, A.; Sammes, P.G. Pyrene-sensitised near-IR luminescence from ytterbium and neodymium complexes. *Dalton Trans.* **2004**, 1405–1409. [CrossRef] [PubMed]

13. Shavaleev, N.M.; Scopelliti, R.; Gumy, F.; Bünzli, J.-C. Surprisingly bright near-infrared luminescence and short radiative lifetimes of ytterbium in hetero-binuclear Yb_Na chelates. *Inorg. Chem.* **2009**, *48*, 7937–7946. [CrossRef] [PubMed]

14. Moudam, O.; Rowan, B.C.; Alamiry, M.; Richardson, P.; Richards, B.S.; Jones, A.C.; Robertson, N. Europium complexes with high total photoluminescence quantum yields in solution and in PMMA. *Chem. Commun.* **2009**, 6649–6651. [CrossRef] [PubMed]

15. Magennis, S.W.; Ferguson, A.J.; Bryden, T.; Jones, T.S.; Beeby, A.; Samuel, I.D.W. Time-dependence of erbium(III) tris(8-hydroxyquinolate) near-infrared photoluminescence: Implications for organic light-emitting diode efficiency. *Synth. Metals* **2003**, *138*, 463–469. [CrossRef]

16. Lunstroot, K.; Nockemann, P.; Van Hecke, K.; Van Meervelt, L.; Görller-Walrand, C.; Binnemans, K.; Driesen, K. Visible and near-infrared emission by Samarium(III)-containing ionic liquid mixtures. *Inorg. Chem.* **2009**, *48*, 3018–3026. [CrossRef] [PubMed]

17. Liao, Z.; Tropiano, T.; Mantulnikovs, K.; Faulkner, S.; Vosch, T.; Sørensen, T.J. Spectrally resolved confocal microscopy using lanthanide centred near-IR emission. *Chem. Commun.* **2015**, *51*, 2372–2375. [CrossRef] [PubMed]

18. Digonnet, M. (Ed.) *Rare Earth Doped Fiber Lasers and Amplifiers*; Marcel Dekker: New York, NY, USA, 1993.

19. Desurvire, E. *Erbium Doped Fiber Amplifiers*; Wiley: New York, NY, USA, 1994.

20. Desurvire, E. The golden age of optical fiber amplifiers. *Phys. Today* **1994**, *47*, 20–27. [CrossRef]

21. Faulkner, S.; Burton-Pye, B.P.; Pope, S.J.A. Lanthanide complexes for luminescence imaging applications. *Appl. Spectrosc. Rev.* **2005**, *40*, 1–31. [CrossRef]

22. Faulkner, S.; Natrajan, L.S.; Perry, W.S.; Sykes, D. Sensitised luminescence in lanthanide containing arrays and d–f hybrids. *Dalton Trans.* **2009**, 3890–3899. [CrossRef] [PubMed]

23. Binnemans, K. Rare-earth β-diketonates. In *Handbook on the Physics and Chemistry of Rare Earths*; Gschneidner, K.A., Jr., Bünzli, J.-C.G., Pecharsky, V.K., Eds.; Elsevier: Amsterdam, The Netherlands, 2005; Volume 35, pp. 107–271.

24. Hemilla, A. *Applications of Fluorescence in Immunoassays*; Wiley Interscience: New York, NY, USA, 1991.

25. Yuan, J.; Wang, G. Lanthanide complex-based fluorescence label for time-resolved fluorescence bioassay. *J. Fluoresc.* **2005**, *15*, 559–568. [CrossRef] [PubMed]

26. Sy, M.; Nonat, A.; Hildebrandt, N.; Charbonnière, L.J. Lanthanide-based luminescence biolabelling. *Chem. Commun.* **2016**, *52*, 5080–5095. [CrossRef] [PubMed]

27. Browne, W.R.; Vos, J.G. The effect of deuteriation on the emission lifetime of inorganic compounds. *Coord. Chem. Rev.* **2001**, *219–221*, 761–787. [CrossRef]

28. Stein, G.; Wurzberg, E. Energy gap law in the solvent isotope effect on radiationless transitions of rare earth Ions. *J. Chem. Phys.* **1975**, *62*, 208–213. [CrossRef]

29. Heller, A. Formation of hot OH Bonds in the radiationless relaxations of excited rare earth ions in aqueous solutions. *J. Am. Chem. Soc.* **1966**, *88*, 2058–2059. [CrossRef]

30. Doffek, C.; Alzakhem, N.; Bischof, C.; Wahsner, J.; Güden-Silber, T.; Lugger, J.; Platas-Iglesias, C.; Seitz, M. Understanding the quenching effects of aromatic C–H– and C–D– oscillators in Near-IR Lanthanoid Luminescence. *J. Am. Chem. Soc.* **2012**, *134*, 16413–16423. [CrossRef] [PubMed]

31. Horrocks, W.D.; Sudnick, D.R. Lanthanide ion luminescence probes of the structure of biological macromolecules. *Acc. Chem. Res.* **1981**, *14*, 384–392. [CrossRef]

32. Supkowski, R.M.; Horrocks, W.D. On the determination of the number of water molecules, q, coordinated to europium(III) ions in solution from luminescence decay lifetimes. *Inorg. Chim. Acta* **2002**, *340*, 44–48. [CrossRef]

33. Beeby, A.; Clarkson, I.M.; Dickins, R.S.; Faulkner, S.; Parker, D.; Royle, L.; de Sousa, A.S.; Williams, J.A.G.; Woods, M. Non-radiative deactivation of the excited states of europium, terbium and ytterbium complexes by proximate energy-matched OH, NH and CH oscillators: An improved luminescence method for establishing solution hydration states. *J. Chem. Soc. Perkin Trans. 2* **1999**, 493–503. [CrossRef]

34. Yasuchika Hasegawa, H.; Kimura, Y.; Murakoshi, K.; Wada, Y.; Kim, J.-H.; Nakashima, N.; Yamanaka, T.; Yanagida, S. Enhanced emission of deuterated tris(hexafluoroacetylacetonato)neodymium(III) complex in solution by suppression of radiationless transition via vibrational excitation. *J. Phys. Chem.* **1996**, *100*, 10201–10205. [CrossRef]

35. Wahsner, J.; Seitz, M. Perdeuterated 2,2'-Bipyridine-6,6'-dicarboxylate: An extremely efficient sensitizer for thulium luminescence in solution. *Inorg. Chem.* **2013**, *52*, 13301–13303. [CrossRef] [PubMed]

36. Bischof, C.; Wahsner, J.; Scholten, J.; Troslen, S.; Seitz, M. Quantification of C–H quenching in near-IR luminescent ytterbium and neodymium cryptates. *J. Am. Chem. Soc.* **2010**, *132*, 14334–14335. [CrossRef] [PubMed]

37. Doffeck, C.; Seitz, M. The radiative lifetime in near-IR-luminescent ytterbium cryptates: The key to extremely high quantum yields. *Angew. Chem. Int. Ed.* **2015**, *54*, 9719–9721. [CrossRef] [PubMed]

38. Ye, H.-Q.; Peng, Y.; Li, Z.; Wang, C.-C.; Zheng, Y.-X.; Motevalli, M.; Wyatt, P.B.; Gillin, W.P.; Hernańdez, I. Effect of fluorination on the radiative properties of Er^{3+} organic complexes: An opto-structural correlation study. *J. Phys. Chem. C* **2013**, *117*, 23970–23975. [CrossRef]

39. Quochi, F.; Orrú, R.; Cordella, F.; Mura, A.; Bongiovanni, G.; Artizzu, F.; Deplano, P.; Mercuri, M.L.; Pilia, L.; Serpe, A. Near infrared light emission quenching in organolanthanide complexes. *J. Appl. Phys.* **2006**, *99*, 053520. [CrossRef]

40. Congiu, M.; Alamiry, M.; Moudam, O.; Ciorba, S.; Richardson, P.R.; Maron, L.; Jones, A.C.; Bryce, S.; Richards, B.S.; Robertson, N. Preparation and photophysical studies of [Ln(hfac)$_3$DPEPO], Ln = Eu, Tb, Yb, Nd, Gd; of total photoluminescence quantum yields. *Dalton Trans.* **2013**, *42*, 13537–13545. [CrossRef] [PubMed]

41. Song, L.; Hu, J.; Wang, J.; Liu, X.; Zhen, Z. Novel perfluorodiphenylphosphinic acid lanthanide (Er or Er–Yb) complex with high NIR photoluminescence quantum yield. *Photochem. Photobiol. Sci.* **2008**, *7*, 689–693. [CrossRef] [PubMed]

42. Tan, R.H.C.; Motevalli, M.; Abrahams, I.; Wyatt, P.B.; Gillin, W.P. IR luminescence of Er, Nd and Yb β-diketonates. *J. Phys. Chem. B* **2006**, *110*, 24476–24479. [CrossRef] [PubMed]

43. Mancino, G.; Ferguson, A.J.; Beeby, A.; Long, N.J.; Jones, T.S. Dramatic increases in the lifetime of the Er^{3+} ion in a molecular complex using a perfluorinated imidodiphosphinate sensitizing ligand. *J. Am. Chem. Soc.* **2005**, *127*, 524–525. [CrossRef] [PubMed]

44. Glover, P.B.; Bassett, A.P.; Nockemann, P.; Benson, M.; Kariuki, M.; Van Deun, R.; Pikramenou, Z. Fully fluorinated imidodiphosphinate shells for visible- and NIR-emitting lanthanides: Hitherto unexpected effects of sensitizer fluorination on lanthanide emission properties. *Chem. Eur. J.* **2007**, *13*, 6308–6320. [CrossRef] [PubMed]

45. Beverina, L.; Crippa, M.; Sassi, M.; Monguzzi, A.; Meinardi, F.; Tubino, R.; Pagani, G.A. Perfluorinated nitrosopyrazolone-based erbium chelates: A new efficient solution processable NIR emitter. *Chem. Commun.* **2009**, *34*, 5103–5105. [CrossRef] [PubMed]

46. Mech, A.; Monguzzi, A.; Meinardi, F.; Mezyk, J.; Macchi, G.; Tubino, R. Sensitized NIR Erbium(III) emission in confined geometries: A new strategy for light emitters in telecom applications. *J. Am. Chem. Soc.* **2010**, *132*, 4574–4576. [CrossRef] [PubMed]

47. Monguzzi, A.; Tubino, R.; Meinardi, F.; Biroli, A.O.; Maddalena Pizzotti, M.; Demartin, F.; Quochi, F.; Cordella, F.; Loi, M.A. Novel Er^{3+} perfluorinated complexes for broadband sensitized near infrared emission. *Chem. Mater.* **2009**, *21*, 128–135. [CrossRef]

48. Natrajan, L.S.; Khoabane, N.M.; Dadds, B.L.; Muryn, C.A.; Pritchard, R.G.; Heath, S.L.; Alan, M.; Kenwright, A.M.; Ilya Kuprov, I.; Faulkner, S. Probing the structure, conformation, and stereochemical Exchange in a family of lanthanide complexes derived from tetrapyridyl-appended cyclen. *Inorg. Chem.* **2010**, *49*, 7700–7710. [CrossRef] [PubMed]

49. Petrov, V.A.; Marshall, W.J.; Grushin, V.V. The first perfluoroacetylacetonate metal complexes: As unexpectedly robust as tricky to make. *Chem. Commun.* **2002**, 520–521. [CrossRef]

50. Chen, X.Y.; Jensen, M.P.; Liu, G.K. Analysis of energy level structure and excited-state dynamics in a Sm^{3+} complex with soft-donor ligands: Sm(Et$_2$Dtc)$_3$(bipy). *J. Phys. Chem. B* **2005**, *109*, 13991–13999. [CrossRef] [PubMed]

51. Aebischer, A.; Gumy, F.; Bünzli, J.-C.G. Intrinsic quantum yields and radiative lifetimes of lanthanide tris(dipicolinates). *Phys. Chem. Chem. Phys.* **2009**, *11*, 1346–1353. [CrossRef] [PubMed]

52. Werts, M.H.V.; Jukes, R.T.F.; Verhoeven, J.W. The emission spectrum and the radiative lifetime of Eu^{3+} in luminescent lanthanide complexes. *Phys. Chem. Chem. Phys.* **2002**, *4*, 1542–1548. [CrossRef]

53. He, H.; Sykes, A.G.; May, P.S.; He, G. Structure and photophysics of near-infrared emissive ytterbium(III) monoporphyrinate acetate complexes having neutral bidentate ligands. *Dalton Trans.* **2009**, *36*, 7454–7461. [CrossRef] [PubMed]

54. Duhamel-Henry, N.; Adam, J.L.; Jacquier, B.; Linarès, C. Photoluminescence of new fluorophosphate glasses containing a high concentration of terbium (III) ions. *Opt. Mater.* **1996**, *5*, 197–207. [CrossRef]

55. Natrajan, L.S.; Weinstein, J.A.; Wilson, C.; Arnold, P.L. Synthesis and luminescence studies of mono- and C3-symmetric, tris(ligand) complexes of Sm(III), Y(III) and Eu(III) with sulfur-bridged binaphtholate ligands. *Dalton Trans.* **2004**, 3748–3755. [CrossRef] [PubMed]

56. Sheldrick, G.M. *SADABS, Empirical Absorption Correction Program Based upon the Method of Blessing*; University of Göttingen: Göttingen, Germany, 1997.

57. Blessing, R.H. An empirical correction for absorption anisotropy. *Acta Crystallogr.* **1995**, *51*, 33–38. [CrossRef]

58. Sheldrick, G.M. Integrated space-group and crystal-structure determination. *Acta Crystallogr.* **2015**, *A71*, 3–8. [CrossRef] [PubMed]

59. Dolomanov, O.V.; Bourhis, L.J.; Gildea, R.J.; Howard, J.A.K.; Puschmann, H. *OLEX2*: A complete structure solution, refinement and analysis program. *J. Appl. Cryst.* **2009**, *42*, 339–341. [CrossRef]

60. Beeby, A.; Faulkner, S. Luminescence from neodymium(III) in solution. *Chem. Phys. Lett.* **1997**, *266*, 116–122. [CrossRef]

61. Beeby, A.; Faulkner, S.; Parker, D.; Williams, J.A.G. Sensitised luminescence from phenanthridine appended lanthanide complexes: Analysis of triplet mediated energy transfer processes in terbium, europium and neodymium complexes. *J. Chem. Soc. Perkin Trans. 2* **2001**, 1268–1273. [CrossRef]

MDPI AG
St. Alban-Anlage 66
4052 Basel, Switzerland
Tel. +41 61 683 77 34
Fax +41 61 302 89 18
http://www.mdpi.com

Inorganics Editorial Office
E-mail: inorganics@mdpi.com
http://www.mdpi.com/journal/inorganics

www.ingramcontent.com/pod-product-compliance
Lightning Source LLC
Chambersburg PA
CBHW051724210326
41597CB00032B/5602